Multi-Agent Systems

Multi-Agent Systems

Special Issue Editors

Vicent Botti
Andrea Omicini
Stefano Mariani
Vicente Julian

MDPI • Basel • Beijing • Wuhan • Barcelona • Belgrade

MDPI

Special Issue Editors

Vicent Botti
Universitat Politècnica de València
Spain

Andrea Omicini
Department of Computer Science and Engineering (DISI)
Alma Mater Studiorum—Università di Bologna
Italy

Stefano Mariani
Università degli Studi di Modena e Reggio Emilia
Italy

Vicente Julian
Universitat Politècnica de València
Spain

Editorial Office
MDPI
St. Alban-Anlage 66
4052 Basel, Switzerland

This is a reprint of articles from the Special Issue published online in the open access journal *Applied Sciences* (ISSN 2076-3417) from 2017 to 2019 (available at: https://www.mdpi.com/journal/applsci/special_issues/Multi-Agent_Systems)

For citation purposes, cite each article independently as indicated on the article page online and as indicated below:

LastName, A.A.; LastName, B.B.; LastName, C.C. Article Title. *Journal Name* **Year**, *Article Number*, Page Range.

ISBN 978-3-03897-924-1 (Pbk)
ISBN 978-3-03897-925-8 (PDF)

Contents

About the Special Issue Editors

Vicent Botti is a professor of computer science at the Universitat Politècnica de València (Spain). He has been working in the area of artificial intelligence and multi-agent systems for 30 years. His main research areas have included knowledge-based systems, autonomous agents, multi-agent systems, agreement technologies, agent-based social simulation, emotional agents, real-time artificial intelligence, real-time systems, and soft computing. He has published over 350 international refereed works. Of these, 65 are SCI-JCR Journal publications, 2 are research books (Springer), 200 are international conference publications, and 29 are chapters in international research books. He has taken part in 69 research projects, including 6 EU projects and 30 national projects, and he was the principal investigator (IP) in 30. In the context of these projects, he has been responsible for the development of seven large applications that have produced technologies that have been transferred to seven industrial companies and technology-based firms.

Andrea Omicini is a professor at DISI, the Department of Computer Science and Engineering of the Alma Mater Studiorum–Università di Bologna, Italy. He holds a PhD in computer and electronic engineering, and his main research interests include coordination models, multi-agent systems, intelligent systems, programming languages, autonomous systems, middleware, simulation, software engineering, pervasive systems, and self-organization. On those subjects, he has published over 300 articles, edited a number of international books, guest-edited several Special Issues of international journals, and held many invited talks and lectures at international conferences and schools.

Stefano Mariani is an assistant professor of computer science at the University of Modena and Reggio Emilia. He earned his PhD in computer science from the University of Bologna in 2016. He has been involved in EU FP7 Project SAPERE and currently is involved in EU H2020 Project CONNECARE. His research interests include coordination models and languages, agent-oriented technologies, pervasive computing, self-organization mechanisms, and socio-technical systems.

Vicente Julian is a professor of computer science at the Universitat Politècnica de València (UPV), where he has taught since 1996. Vicente Julian is member of the GTI-IA research group and Deputy Director of the Computer Science Department at the UPV. His research on artificial intelligence has included four international projects, two international excellence networks, 21 Spanish projects, and four technology transfer projects. He has published more than 70 works in journals with outstanding positions in the Journal Citation Reports or in conference proceedings that have a system of external peer review and dissemination of knowledge comparable to journals indexed in relevant positions. Moreover, he has presented more than 200 contributions in international conferences or workshops.

Preface to "Multi-Agent Systems"

After more than 20 years of academic research on multi-agent systems (MASs), agent-oriented models and technologies have been promoted as the most suitable candidates for the design and development of distributed and intelligent applications in complex and dynamic environments. In order to actually become "the next big thing", however, MASs need to complete their transition from a (mostly) academic product to the industry mainstream. To this end, a huge number of aspects and issues relating to MAS techniques and methods must be scrutinized and explored within the many relevant application scenarios where complex intelligent systems are required. This is one of the main motivations behind the Applied Sciences Special Issue "Multi-Agent Systems", from which this book stems.

Already planned to continue as a series in 2019, this Special Issue gathers original research articles reporting results on the steadily growing area of agent-oriented computing and multi-agent systems technologies. In particular, the 17 contributions have been categorized by the guest editors' as belonging to the following topics: agent-based modeling and simulation, situated multi-agent systems, socio-technical multi-agent systems, and semantic technologies applied to multi-agent systems. Papers in the first category emphasize that whenever the system under study is decentralized and open to the dynamic join and leave of participants, who must be modeled as autonomous and loosely coupled entities (that is, related by soft dependencies), agent abstraction is the preferred method. Those in the second category show the fundamental role of the environment that the MAS is modeling, or within which the MAS is executing, in affecting its functionalities: there are properties outside of agents' control that they should sense in order to proceed to decision making in an informed way. Papers in the third category unsurprisingly highlight how agent abstraction is a perfect fit for either modeling human behavior or effectively dealing with human users: goal-orientation, autonomy, structured communication capabilities, and even mental attitudes, if we consider BDI-like architectures, are the most cited components necessary to do so. Finally, the last category links agents to the notion of intelligence through the need for semantics to make them understand the concepts they are manipulating.

With respect to both their quality and range, the papers in this Special Issue already represent a meaningful sample of the most recent advancements in the field of agent-oriented models and technologies. The guest editors are thus confident that the readers of *Applied Sciences*, including academic researchers and industry practitioners, will be able to appreciate the growing role that MASs will play in the design and development of the next generation of complex intelligent systems.

Vicent Botti, Andrea Omicini, Stefano Mariani, Vicente Julian
Special Issue Editors

applied
sciences

MDPI

Article

Special Issue "Multi-Agent Systems": Editorial

Stefano Mariani [1,*,†] **and Andrea Omicini** [2,†]

1   Department of Sciences and Methods of Engineering, Università di Modena e Reggio Emilia,
    42122 Modena, Italy
2   Department of Computer Science and Engineering, Università di Bologna, 47521 Cesena, Italy;
    andrea.omicini@unibo.it
*   Correspondence: stefano.mariani@unimore.it; Tel.: +39-0522-52-2660
†   These authors contributed equally to this work.

Received: 1 March 2019; Accepted: 1 March 2019; Published: 6 March 2019

check for
updates

Abstract: Multi-agent systems (MAS) allow and promote the development of distributed and intelligent applications in complex and dynamic environments. Applications of this kind have a crucial role in our everyday life, as witnessed by the broad range of domains they are deployed to—such as manufacturing, management sciences, e-commerce, biotechnology, etc. Despite heterogeneity, those domains share common requirements such as autonomy, structured interaction, mobility, and openness—which are well suited for MAS. Therein, in fact, goal-oriented processes can enter and leave the system dynamically and interact with each other according to structured protocols. This special issue gathers 17 contributions spanning from agent-based modelling and simulation to applications of MAS in situated and socio-technical systems.

Keywords: multi-agent systems; agent-based modelling; agent-based simulation; agent-oriented technologies; coordination; Artificial Intelligence; computer science

1. Introduction

Social, political, and technological pressure towards intelligent systems able to help and support humans in any non-trivial process and activity is leading the way for new programming paradigms, providing suitable abstractions and mechanisms for modelling and designing complex software systems. More the twenty years of academic research on multi-agent systems (MAS) have promoted agent-oriented models and technologies as the most suitable candidates for the design and development of distributed and intelligent applications in complex and dynamic environments.

To actually become "the next big thing", however, MAS need to complete their transition from a (mostly) academic product to the industry mainstream. To this end, a huge number of aspects and issues relating to MAS techniques and methods have to be scrutinised and explored within the many relevant application scenarios where complex intelligent systems are required: this is one of the main motivations behind the special issue.

Before delving into the individual contributions gathered, a few general statistics and observations are useful to have an overview of the content and outreach of this special issue:

- 55 papers have been submitted for peer review, out of which 17 were finally published, resulting in an acceptance rate of ≈31%
- the median article processing time to publish, intended as the time passed from submission to online availability, is 40 days, with a standard deviation of ≈17 days—dates are publicly available on the special issue web page (https://www.mdpi.com/journal/applsci/special_issues/Multi-Agent_Systems)

- papers generated an average of 0.76 citations (1.71 standard deviation) and ≈687 downloads (≈326 standard deviation) per year—citations, downloads, and also views count are publicly available on each paper own web page, accessible starting from the special issue one
- published papers have been co-authored by authors coming from 13 countries, covering Europe, Asia, and South America. Among these, Spain is the most represented, having 5 papers with at least one local author

When considering that this is the first year of the special issue, and that Applied Sciences is relatively new to the field of research in agent-oriented models and technologies, we are very happy with both the number of submissions and their quality, as well as with the number and quality of the papers finally published. After their publication, papers are typically valued by the number of citations they get; obviously, they need a bit more time than less than a year to start generating mentions. Nevertheless, some of the papers published here have already started gaining attention.

Figure 1 shows the wordcloud generated from the full pdfs of the published papers.

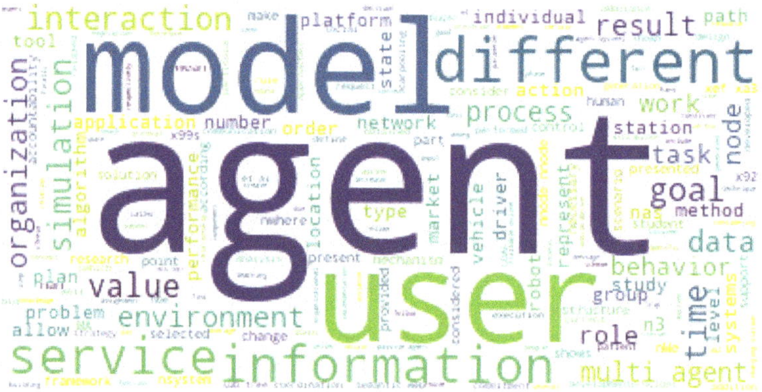

Figure 1. Wordcloud generated from the PDFs of each publication of the special issue (Python code available on request).

Unsurprisingly, the most mentioned word is "agent", followed by "model" and "user". The latter one may have a quite generic meaning, thus is difficult to interpretate, but the second one already highlights one of the relevant areas of application of MAS, that is, agent and multi-agent based modelling—which is one of the four main topics emerging from the analysis of the content of the published papers. Other highly mentioned words working as clues for relevant application areas are the following ones, presented along with our interpretation:

- "organisation", "role", which point to the *social dimension* of agenthood
- "simulation", which complements agent-based modelling with agent and multi-agent based simulation
- "environment", "time", which emphasise the *situated dimension* of MAS

Accordingly, in fact, the four main topics we identified by examining the content of the papers are:

agent-based modelling and simulation — as the disciplines of modelling systems by adopting the agent abstraction [1], possibly along the other two pillars of MAS—that is, environment and society—and of simulating systems as an ensemble of loosely-coupled, goal-oriented autonomous entities (indeed, agents), either competing or collaborating by exchanging messages.

situated systems — as the application of agent-oriented models and technologies to systems highly intertwined with their environment [2], be it virtual or physical, there including the space-time fabric, hence leveraging the situated nature of agents' actions, which are deeply dependant on the context of performance [3].

socio-technical systems — as the application of agent-oriented models and technologies to those kind of systems where humans play a fundamental role, where they are (also) a functional part of the system itself, bearing with them all the complexities of human behaviour, organisational structures, ethical issues, cognitive aspects, etc. [4].

semantic technologies — as the discipline of making computational devices able to interpret and understand the semantics of objects, concepts, processes, etc., usually in the context of the Semantic Web vision [5].

In the following sections we classify published papers according to the categories above, and summarise their main contributions.

2. Agent-Based Modelling and Simulation

In [6] the authors apply *genetic programming* techniques to an agent-based model of training, education, and entertainment applications with the aim of automatically generating *agent behaviour trees*. By acknowledging shortcomings of genetic programming application to evolve behaviour trees—such as search space explosion and intensive knowledge engineering efforts—they complement the approach with both static and dynamic constraints to ease exploration of the search space while still being fairly domain-independent. To demonstrate efficacy of the proposal, they experiment with behaviour generation for the Pac-man game, achieving better behaviour in fewer evolutionary rounds w.r.t. other state-of-art approaches. As a bonus, they also get more readable behaviour specifications, suitable to be later refined by domain experts.

In [7] the author applies agent-based simulation to achieve fair purchase prices in the context of perishable goods markets. They aim at reforming the current market model, in which sellers are penalised by the rapid perishing of their goods (i.e., fish and vegetables) which forces them to accept buyers offers sooner or later—otherwise the goods will be wasted. To overcome this issue, they propose a *double auction model* in which buyers are penalised each time they fail an offer, so as to promote fair bidding prices. By simulating traders in different market conditions they show that their approach achieves fair prices for the allocation of goods.

In [8] the authors developed an agent-based *simulation software tool* to investigate how *students' sociograms*—a representation of social relationships—evolve during time. It has been demonstrated, in fact, that different sociograms contribute in different ways to the academic performance of students, thus arranging classes and lectures so as to promote such sociograms would be beneficial. Nevertheless, knowing the sociogram beforehand—before the education period starts—is not possible. The authors' work overcomes this limitation by enabling educators to simulate student sociograms as an agent-based model built out of students' psychological profile. The authors corroborate the hypothesis that simulated sociograms are sufficiently close to real ones through statistical binomial testing.

In [9] the authors propose TELEKA, an *agent-based model and architecture* for *network traffic* analysis and optimisation. The model is intended to exploit *negotiation* techniques to advance network management practices towards the fifth level of IBM's degree of automaticity in network management [10], which fosters network policies and goals able to autonomously adapt to the contingencies arising during operation. The proposed architecture is vertically decomposed in three layers: the lower one is in charge of fine-grained monitoring and low-level control of individual devices, the middle one is devoted to classify and aggregate perceptions coming from the monitoring so as to deliver higher level information to the decision making module, that is, the upper level of the architecture, in charge of triggering the SEHA negotiation algorithm for congestion resolution and traffic optimisation based on information and alerts coming from the lower levels. Effectiveness of the model is evaluated through simulations in NetLogo (https://ccl.northwestern.edu/netlogo/), and an analysis of sensitivity to different topologies is included.

In [11] the author formally investigates the problem of reaching consensus in presence of either transmission or processing delays. In particular, they focus on *scaled consensus*, where consensus is not about reaching agreement on an absolute common value, but rather about achieving given relative proportions [12]. They prove that scaled consensus can be achieved regardless of transmission delays as long as the network contains a spanning tree, whereas the same does not hold in presence of processing delays, which can hinder convergence to consensus when too large. In case consensus is reached, the scaled consensus values are the same regardless of the delays being due to transmission or processing. Numeric simulations confirm these formal results.

Wrap up. Despite their huge heterogeneity, all the research works just described perfectly sum up the circumstances that call for agent-based modelling and simulation: whenever the system under study is *decentralised* and *open* to dynamic join and leave of participants, who must be modelled as *autonomous* and *loosely* coupled entities (that is, related by soft dependencies), the agent abstraction is the way to go.

3. Situated Systems

In [13] the authors approach the problem of *coordinated control* of a fleet of autonomous hovercrafts at both the individual and the collective level: for the former, they propose a controller based on a Radial Basis Function Neural Network for tolerating non-modelled terms while following a given path, for the latter they apply *multi-agent consensus* to achieve movement in a desired formation. Simulations of the approach confirm effectiveness in a few selected communication topologies, such as cascade-directed communication graph and parallel-directed communication graph.

In [14] the authors apply *multi-agent based modelling* to the domain of *urban planning*, in particular, for supporting decision making about the design and deployment of an electric charging stations infrastructure in a city. The proposed multi-agent system features several agents in charge of complementary functionalities, such as querying open data portals of local administrations to gather info about potential offer and demand for charging stations as well as average traffic conditions, crawling social networks to rank potential locations where to put charging stations, and execute optimisation algorithms to find the best spots among a set of candidates. A graphical interface is also available to set various parameters guiding the optimisation process, which is based on *genetic algorithms*. The authors validate their approach with a case study implemented in the city of Valencia, Spain.

Also [15] concerns the domain of urban planning: there, the authors propose a multi-agent system providing decision support, in the form of *demand prediction services*, in the context of a *bike sharing* application. The system analyses heterogeneous data such as availability of bikes at stations, trip information, and weather forecasts to build a demand prediction model and a dashboard for historical data visualisation. The prediction model is built by comparing regression techniques such as Random Forest and Gradient Boosting, evaluated by means of the Root Mean Square Logarithmic Error. The resulting system has been deployed to the city of Salamanca, Spain.

In [16] the authors exploit *agent-based modelling* and simulation techniques to investigate how different agent behaviours and interactions affect the *negotiation process* in a car-pooling scenario. The proposed multi-agent system adopts an organisational metaphor to arrange and analyse the social relationships and individual behaviour of agents; in particular, the *Capacity, Role, Interaction, and Organization* meta-model [17]. The proposed system works in three main phases: *exploration*, during which agents search the carpooling social network seeking for opportunities to carpool, either as drivers or passengers, and get matched depending on user profiles, trip data, time of constraints, etc.; *negotiation*, where agents being matched start a negotiation process to fine-tune the details of the joint voyage, such as the actual departure time, the path to be followed, and the driver; *carpooling*, where the actual trip takes place and contingencies should be handled (such as disbanding non-compliant agents). To validate the approach, the FEATHERS operative activity-based traffic demand model is used to generate synthetic data and agents in order to test different negotiation settings.

In [18] the authors propose a novel *navigational strategy* for moving robots able to avoid collision with multiple passive agents featuring a (at most) partially predictable behaviour. The main application scenario envisioned is safe autonomous driving in presence of pedestrians, and the approach proposed—featuring a detailed kinodynamic model for the robot—is compared with acceleration-velocity obstacles [19] and generalised velocity obstacles [20] to showcase its gain in performance.

Wrap up. Also in this category papers tend to vary a lot in the actual topic of their contribution, yet they all adhere to a common principle: there are properties of the *environment* that the MAS is modelling, or within which the MAS is executing, which affect its functionalities, thus must be properly modelled. Be it the unpredictable oscillations of the physical environment—as in the case of the former and latter papers—or the need to account for spatial constraints while simulating a urban infrastructure, there are properties *outside of agents' control* that they should sense in order to proceed to decision making in an informed way.

4. Socio-Technical Systems

In [21] the authors argue that multi-agent systems featuring *agreement technologies* for coordinating agents' interactions are a good fit for many socio-technical systems in the common realm of *smart cities*. The motivation for considering agreement technologies stems from the *open* nature of most applications therein, when users may join and leave the system anytime, and whose behaviour is at most *partially controllable* and predictable—in contrast with software-only systems. They substantiate their claim with several use cases including intersection management, emergency evacuation, and healthcare, each accompanied with experimental results coming from either extensive simulations or actual deployments.

In [22] the authors develop a *personal assistant* agent leveraging and complementing ambient intelligence systems to safely navigate people with *cognitive disabilities* in unfamiliar environments. The proposed system features location tracking, an orientation system, and speculative reasoning to enable monitoring of patients' locations by caregivers and relatives, increase autonomy of patients, and pro-actively detect potential mistakes leading to the patient getting lost. Also, the proposed solution exploits a learning module based on past trajectories mining to build and keep updated a patient profile of familiar and preferred routes. Finally, to ease usage and understanding while lowering the cognitive workload of users, the system is served through an *augmented reality* interface which overlays crucial information on the physical world by means of a mobile device (such as a smartphone or tablet).

In [23] the authors target socio-technical systems where *accountability* of actions is a must-have feature. They propose a framework specific for multi-agent systems—named ADOPT—where accountability is enabled by the notion of *social commitment* [24], and can be enforced by design through a specific interaction protocol leveraging a shared environment and the notion of *role*—played by agents in a MAS organisation. In particular, a two-stages protocol is adopted, and defined as a sequence of two FIPA Contract-Net protocols [25]: in the first one, the role played by an agent while joining a computational organisation is established, while in the second one, the goals to pursue are negotiated. The proposed framework is then implemented on top of the JaCaMo agent development platform [26], featuring BDI agents programmed in Jason [27] and environment engineering based on CArtAgO artefacts [28], and conceptually validated in a toy scenario about an ensemble of agents cooperating to build a house.

In [29] the authors deploy agents in an interactive museum exhibit scenario, with the task of evaluating *interaction levels* [30] (the quality of interaction, frequency, average time, etc.) and improving user experience. In particular, three distinct agent sub-systems are designed: one representing the users, hence visitors attending the exhibit, thus modelling their behaviour and interactions with the museum facilities, one representing the interactive exhibits, and the latter implementing the self-evaluation functionality—delivered through a fuzzy inference system. Results of deployment of the MAS in practice are also shown, regarding 500 users visiting the "El Trompo" interactive museum in Tijuana, Mexico.

In [31] the authors used agent-based modelling for the *3D simulation* of work environments, with the goal of investigating potential accessibility problems w.r.t. people with disabilities. In particular, a multi-agent system has been developed in JADE [32] and integrated with Unity3D game engine, leveraging its 3D modelling capabilities, to enable *interactive simulations* for "what-if" analyses of potential architectural barriers. The overall system is thus, essentially, a decision support tool for Human Resources management offices. Accordingly, evaluation of the system is done against Indra Sistemas S.A. offices in Salamanca, Spain.

Wrap up. The lesson learnt here is that, unsurprisingly, the agent abstraction is a perfect fit for either modelling human behaviour or effectively deal with human users: for the former, *goal-orientation, autonomy, structured communication* capabilities, even *mental attitudes* if we consider BDI-like architectures [33], are all facets that characterise humans and that agents are able to mimic (at least, to some extent); for the latter, the same agents' abilities are useful to give human users a more natural peer for their interactions with technology.

5. Semantic Technologies

In [34] the authors build a novel *agent development platform* explicitly focussed on enabling development of *linked data* [35] aware agents, and deployment of mobile devices such as smartphones. In reference to the former feature, it implies that agents are able to gather information from a linked data graph, and store it in their own belief base, where it can be used to trigger execution of plans. Indeed, agents adopt the *BDI architecture* for inner reasoning, and *FIPA compliant* speech acts for communicating with others—FIPA standards are also adopted for the agent management functionality, thus the proposed platform features typical services such as white and yellow pages for, respectively, agents and services discovery. The proposed development platform has been evaluated in a toy scenario concerning an auction for exchanging products in an electronic market.

In [36] the authors propose a *model-driven development* platform and methodology for MAS, rooted in the domain-specific language *SEA_ML*, featuring BDI agents and *automatic code generation*, and specifically targeted at developing *Semantic Web* enabled agents. The platform is able to generate actual source code for JADE [32], JADEX [37], and JACK [38], thanks to a pipeline of model-to-model plus model-to-code transformations, as well as OWL-S [38] and WSMO [39] documents. SEA_ML meta-model and methodology articulates along 8 different aspects called "viewpoints": Internal for agents' inner behaviour, Interaction for their communication, MAS for organisational properties, Role for access control, Environment for handling of resources, Plan for plan actions and tasks definition, Ontology for knowledge specification, Agent–SWS Interaction for definition of entities and relations for handling Semantic Web Services.

Wrap up. This category is less represented and possibly describes a narrow area of application for agent-oriented techniques. Nevertheless, there is an important aspect that links semantic technologies with agents: intelligence. One acceptance of the term implies the ability to understand the *meaning* of concepts, not solely how to manipulate them. That is, one of the many possible interpretation of the term "intelligence" requires agents to understand what they are doing, what information they are communicating, etc., not only how to do so without any clue on the semantics behind actions and data. Under this perspective, semantic technologies are a great complement to agents' innate rational capabilities.

6. Conclusions

For both their quality and range, the papers in this special issue already represent a meaningful sample of the most recent advancements in the field of agent-oriented models and technologies. In fact, it is surprising to witness how such a limited portion of MAS research already highlights the most relevant usage of agent-based models and technologies, as well as their most appreciated characteristics. For instance, modelling and simulation straightforwardly substantiate our opening claim that agent-oriented abstractions are widely recognised as the most rich and useful for conceiving and designing complex systems, while situatedness directly puts MAS in relation with the dynamic and unpredictable nature of the physical environment.

We are then confident that the readers of Applied Intelligence will be able to understand the growing role that MAS are going to play in the design and development of the next generation of complex intelligent systems. Yet, the large number of submissions to this first instalment of the MAS special issue have made it clear that there is a huge space that could be covered by another special issue of the same sort. This is why the special issue has been converted to a yearly series, of which the new call is already available at the publisher website: https://www.mdpi.com/journal/applsci/special_issues/Multi-Agent_Systems_2019.

Author Contributions: Conceptualization, S.M. and A.O.; methodology, S.M. and A.O.; investigation, S.M.; writing—original draft preparation, S.M.; writing—review and editing, A.O.; supervision, A.O.

Acknowledgments: The guest editors would like to thank the Applied Sciences Editorial Office, in particular the reference contact Daria Shi, for the extreme efficiency and attention devoted to the handling of papers, from submission to publication, through the peer review process. We would also like to thank the many reviewers participating in the selection process (3 to 4 on average) for their valuable constructive criticism, often appreciated by the authors themselves. Last but not least, our gratitude goes to the authors who submitted their papers, and to the many readers who already generated citations and downloads.

Conflicts of Interest: The authors declare no conflict of interest.

References

1. Macal, C.M.; North, M.J. Tutorial on agent-based modeling and simulation. In Proceedings of the 37th Winter Simulation Conference, Orlando, FL, USA, 4–7 December 2005; p. 14, doi:10.1109/WSC.2005.1574234. [CrossRef]
2. Weyns, D.; Holvoet, T. A formal model for situated multi-agent systems. *Fundam. Inf.* **2004**, *63*, 125–158.
3. Suchman, L.A. *Plans and Situated Actions: The Problem of Human-Machine Communication*; Cambridge University Press: New York, NY, USA, 1987.
4. Whitworth, B. Socio-technical systems. *Encycl. Hum. Comput. Interact.* **2006**, 533–541.
5. Berners-Lee, T.; Hendler, J.; Lassila, O. The semantic web. *Sci. Am.* **2001**, *284*, 34–43. [CrossRef]
6. Zhang, Q.; Yao, J.; Yin, Q.; Zha, Y. Learning behavior trees for autonomous agents with hybrid constraints evolution. *Appl. Sci.* **2018**, *8*, 1077, doi:10.3390/app8071077. [CrossRef]
7. Miyashita, K. Incremental design of perishable goods markets through multi-agent simulations. *Appl. Sci.* **2017**, *7*, 1300, doi:10.3390/app7121300. [CrossRef]
8. García-Magariño, I.; Lombas, A.S.; Plaza, I.; Medrano, C. ABS-SOCI: An agent-based simulator of student sociograms. *Appl. Sci.* **2017**, *7*, 1126, doi:10.3390/app7111126. [CrossRef]
9. Raya-Díaz, K.; Gaxiola-Pacheco, C.; Castañón-Puga, M.; Palafox, L.E.; Castro, J.R.; Flores, D.L. Agent-based model for automaticity management of traffic flows across the network. *Appl. Sci.* **2017**, *7*, 928, doi:10.3390/app7090928. [CrossRef]
10. IBM. *An Architectural Blueprint for Autonomic Computing*; Technical Report; IBM: Armonk, NY, USA, 2005.
11. Shang, Y. On the delayed scaled consensus problems. *Appl. Sci.* **2017**, *7*, 713, doi:10.3390/app7070713. [CrossRef]
12. Roy, S. Scaled consensus. *Automatica* **2015**, *51*, 259–262, doi:10.1016/j.automatica.2014.10.073. [CrossRef]
13. Duan, K.; Fong, S.; Zhuang, Y.; Song, W. Artificial neural networks in coordinated control of multiple hovercrafts with unmodeled terms. *Appl. Sci.* **2018**, *8*, 862, doi:10.3390/app8060862. [CrossRef]
14. Jordán, J.; Palanca, J.; del Val, E.; Julian, V.; Botti, V. A multi-agent system for the dynamic emplacement of electric vehicle charging stations. *Appl. Sci.* **2018**, *8*, 313, doi:10.3390/app8020313. [CrossRef]
15. Lozano, Á.; De Paz, J.F.; Villarrubia González, G.; Iglesia, D.H.D.L.; Bajo, J. Multi-agent system for demand prediction and trip visualization in bike sharing systems. *Appl.Sci.* **2018**, *8*, 67, doi:10.3390/app8010067. [CrossRef]

16. Hussain, I.; Khan, M.A.; Baqueri, S.F.A.; Shah, S.A.R.; Bashir, M.K.; Khan, M.M.; Khan, I.A. An organizational-based model and agent-based simulation for co-traveling at an aggregate level. *Appl. Sci.* **2017**, *7*, 1221, doi:10.3390/app7121221. [CrossRef]

17. Cossentino, M.; Gaud, N.; Hilaire, V.; Galland, S.; Koukam, A. ASPECS: An agent-oriented software process for engineering complex systems. *Auton. Agents Multi-Agent Syst.* **2010**, *20*, 260–304, doi:10.1007/s10458-009-9099-4. [CrossRef]

18. Zuhaib, K.M.; Khan, A.M.; Iqbal, J.; Ali, M.A.; Usman, M.; Ali, A.; Yaqub, S.; Lee, J.Y.; Han, C. Collision avoidance from multiple passive agents with partially predictable behavior. *Appl. Sci.* **2017**, *7*, 903, doi:10.3390/app7090903. [CrossRef]

19. Van den Berg, J.; Snape, J.; Guy, S.J.; Manocha, D. Reciprocal collision avoidance with acceleration-velocity obstacles. In Proceedings of the 2011 IEEE International Conference on Robotics and Automation, Shanghai, China, 9–13 May 2011; pp. 3475–3482, doi:10.1109/ICRA.2011.5980408. [CrossRef]

20. Wilkie, D.; van den Berg, J.; Manocha, D. Generalized velocity obstacles. In Proceedings of the 2009 IEEE/RSJ International Conference on Intelligent Robots and Systems, St. Louis, MO, USA, 10–15 October 2009; pp. 5573–5578, doi:10.1109/IROS.2009.5354175. [CrossRef]

21. Billhardt, H.; Fernández, A.; Lujak, M.; Ossowski, S. Agreement technologies for coordination in smart cities. *Appl. Sci.* **2018**, *8*, 816, doi:10.3390/app8050816. [CrossRef]

22. Ramos, J.; Oliveira, T.; Satoh, K.; Neves, J.; Novais, P. Cognitive assistants—An analysis and future trends based on speculative default reasoning. *Appl. Sci.* **2018**, *8*, 742, doi:10.3390/app8050742. [CrossRef]

23. Baldoni, M.; Baroglio, C.; May, K.M.; Micalizio, R.; Tedeschi, S. Computational accountability in MAS organizations with ADOPT. *Appl. Sci.* **2018**, *8*, 489, doi:10.3390/app8040489. [CrossRef]

24. Castelfranchi, C. *Commitments: From Individual Intentions to Groups and Organizations*; ICMAS: Maryville, TN, USA, 1995; Volume 95, pp. 41–48.

25. Foundation for Intelligent Physical Agents. *FIPA Contract Net Interaction Protocol Specification*; Foundation for Intelligent Physical Agents: Geneva, Switzerland, 2002.

26. Boissier, O.; Bordini, R.H.; Hübner, J.F.; Ricci, A.; Santi, A. Multi-agent oriented programming with JaCaMo. *Sci. Comput. Programm.* **2013**, *78*, 747–761, doi:10.1016/j.scico.2011.10.004. [CrossRef]

27. Bordini, R.H.; Hübner, J.F.; Wooldridge, M.J. *Programming Multi-Agent Systems in AgentSpeak Using Jason*; Wiley: Hoboken, NJ, USA, 2007.

28. Ricci, A.; Piunti, M.; Viroli, M.; Omicini, A. Environment programming in CArtAgO. In *Multi-Agent Programming: Languages, Tools and Applications*; El Fallah Seghrouchni, A., Dix, J., Dastani, M., Bordini, R.H., Eds.; Springer: Boston, MA, USA, 2009; pp. 259–288, doi:10.1007/978-0-387-89299-3_8.

29. Rosales, R.; Castañón-Puga, M.; Lara-Rosano, F.; Flores-Parra, J.M.; Evans, R.; Osuna-Millan, N.; Gaxiola-Pacheco, C. Modelling the interaction levels in HCI using an intelligent hybrid system with interactive agents: A case study of an interactive museum exhibition module in Mexico. *Appl. Sci.* **2018**, *8*, 446, doi:10.3390/app8030446. [CrossRef]

30. Gayesky, D.; Williams, D. Interactive video in higher education. In *Video in Higher Education*; Kogan Page: London, UK, 1984.

31. Barriuso, A.L.; De la Prieta, F.; Villarrubia González, G.; De La Iglesia, D.H.; Lozano, Á. MOVICLOUD: Agent-based 3D platform for the labor integration of disabled people. *Appl. Sci.* **2018**, *8*, 337, doi:10.3390/app8030337. [CrossRef]

32. Bellifemine, F.L.; Poggi, A.; Rimassa, G. JADE—A FIPA-compliant agent framework. In *Proccedings of the 4th International Conference and Exhibition on the Practical Application of Intelligent Agents and Multi-Agent Technology (PAAM-99)*; The Practical Application Company Ltd.: London, UK, 1999; pp. 97–108.

33. Rao, A.S.; Georgeff, M.P. Modeling rational agents within a BDI-architecture. In *Proceedings of the Second International Conference on Principles of Knowledge Representation and Reasoning*; Morgan Kaufmann Publishers Inc.: San Francisco, CA, USA, 1991; pp. 473–484.

34. Boztepe, İ.S.; Erdur, R.C. Linked data aware agent development framework for mobile devices. *Appl. Sci.* **2018**, *8*, 1831, doi:10.3390/app8101831. [CrossRef]

35. Berners-Lee, T. Personal View on Linked Data for Semantic Web: Architectural Design Issues. Available online: https://www.w3.org/DesignIssues/LinkedData.html (accessed on 5 March 2019).

36. Challenger, M.; Tezel, B.T.; Alaca, O.F.; Tekinerdogan, B.; Kardas, G. Development of semantic web-enabled BDI multi-agent systems using SEA_ML: An electronic bartering case study. *Appl. Sci.* **2018**, *8*, 688, doi:10.3390/app8050688. [CrossRef]

37. Pokahr, A.; Braubach, L.; Lamersdorf, W. Jadex: A BDI reasoning engine. In *Multi-Agent Programming: Languages, Platforms and Applications*; Bordini, R.H., Dastani, M., Dix, J., El Fallah Seghrouchni, A., Eds.; Springer: Boston, MA, USA, 2005; pp. 149–174, doi:10.1007/0-387-26350-0_6.

38. Howden, N.; Ronnquist, R.; Hodgson, A.; Lucas, A. Jack intelligent agents- summary of an agent infrastructure. In Proceedings of the 5th International Conference on Autonomous Agents, Montreal, QC, Canada, 28 May–1 June 2001.

39. Roman, D.; Keller, U.; Lausen, H.; de Bruijn, J.; Lara, R.; Stollberg, M.; Polleres, A.; Feier, C.; Bussler, C.; Fensel, D. Web service modeling ontology. *Appl. Ontol.* **2005**, *1*, 77–106.

applied sciences

MDPI

Editorial
Multi-Agent Systems

Vicente Julian *,† and Vicente Botti *,†

Departamento de Sistemas Informáticos y Computación, Universitat Politecnica de Valencia,
Camino de Vera s-n, 46980 Valencia, Spain
* Correspondence: vinglada@dsic.upv.es (V.J.); vbotti@dsic.upv.es (V.B.)
† These authors contributed equally to this work.

Received: 29 March 2019; Accepted: 29 March 2019; Published: 3 April 2019

check for
updates

Abstract: With the current advance of technology, agent-based applications are becoming a standard in a great variety of domains such as e-commerce, logistics, supply chain management, telecommunications, healthcare, and manufacturing. Another reason for the widespread interest in multi-agent systems is that these systems are seen as a technology and a tool that helps in the analysis and development of new models and theories in large-scale distributed systems or in human-centered systems. This last aspect is currently of great interest due to the need for democratization in the use of technology that allows people without technical preparation to interact with the devices in a simple and coherent way. In this Special Issue, different interesting approaches that advance this research discipline have been selected and presented.

Keywords: multi-agent systems; agent methodologies; agent-based simulation; ambient intelligence; smart cities

1. Introduction

The concept of n intelligent agent is a concept that is born from the area of artificial intelligence; in fact, a commonly-accepted definition relates the discipline of artificial intelligence with the analysis and design of autonomous entities capable of exhibiting intelligent behavior. From that perspective, it is assumed that an intelligent agent must be able to perceive its environment, reason about how to achieve its objectives, act towards achieving them through the application of some principle of rationality, and interact with other intelligent agents, being artificial or human [1].

Multi-agent systems are a particular case of a distributed system, and its particularity lies in the fact that the components of the system are autonomous and selfish, seeking to satisfy their own objectives. In addition, these systems also stand out for being open systems without a centralized design [2]. One main reason for the great interest and attention that multi-agent systems have received is that they are seen as an enabling technology for complex applications that require distributed and parallel processing of data and operate autonomously in complex and dynamic domains.

Research in the discipline of multi-agent systems (MAS) is based on the results of distributed computing asking new questions about how agents must interact with each other in order to coordinate their activities and solve complex problems. Most current research focuses on designing appropriate coordination mechanisms for managing coalitions or teams of agents. The programming of intelligent agents poses complex challenges to engineers because in addition to the complexity of designing concurrent and distributed systems, there is the added complexity that the components must have an architecture that includes aspects such as *reactivity*, *proactivity*, and *sociability*. These properties are not easy to program when the environment is dynamic and complex. In order to achieve real agent programming, a multitude of proposals have been made, by many researchers, for agent architectures, communication languages, and decision-making and coordination mechanisms. In the latter case, the need arises that agents have to be able to reach *agreements* in order to be able to operate in a multi-agent system. Here lies an important aspect of the programming complexity of intelligent agents.

This Special Issue attempts to advance the paradigm of multi-agent systems by proposing new works in different areas of interest. In this way, works are proposed in the areas of agent-oriented software engineering, multi-agent learning, agent based simulation, and agent applications in highly topical domains such as smart cities and ambient intelligence. The following sections detail the selected contributions in each of these areas.

2. MAS and Methodologies

Within the framework of artificial intelligence, multi-agent systems have been characterized by offering a possible solution to the development of complex problems with distributed characteristics. When approaching the development of multi-agent systems, there is undoubtedly a significant increase in complexity, as well as the need for adapting existing techniques, or in some situations, developing new techniques and tools. In recent years, different works have appeared trying to propose new processes and techniques for the development of multi-agent systems [3].

The construction of MAS integrates technologies from different areas of knowledge: software engineering techniques to structure the development process; AI techniques to provide systems with the capacity to deal with unexpected situations and to make decisions, and concurrent programming to address task coordination executed on different machines under different scheduling policies. Due to this mix of technologies, the development of MASis complicated. In this sense, during the last few years, there have been different development platforms and tools that provide partial solutions for the modeling and design of agent-based systems [4].

There is still much work to be done in this area. Thus, in this Special Issue, three works related to agent-oriented software engineering have been included. The first work [5] shows how accountability plays a central role in MAS engineering. Accountability is a well-known key resource inside human organizations, and the idea of this proposal is to propose the design of agent systems where accountability is a property that is guaranteed by design. The authors proposed an interaction protocol, called ADOPT, that allows the realization of accountable MAS organizations. The proposed protocol has been implements using JaCaMo [6], allowing one to demonstrate how to develop agents and the organization to which they belong, being mutually accountable.

The second work [7] proposed a new development methodology for the development of MAS working in semantic web environments. The proposed methodology is based on a domain-specific modeling language, called Semantic Web-Enabled Agent Modeling Language (SEA_ML) [8]. The study was demonstrated using a case study implemented using the well-known JACKplatform (http://aosgrp.com/products/jack/). The proposed example consists of a set of agents that exchange services or goods of owners according to their preferences without using any currency. Finally, the third work [9] introduced an agent development framework for mobile devices. The proposed framework allows users to build intelligent agents with the typical agent-oriented attributes of social ability, reactivity, proactivity, and autonomy. Actually, the main contribution is the linked data support of the framework. Linked data support corresponds to the ability to supply the agent's beliefs from the linked data environment and to use those beliefs during the planning process. According to the authors, these kinds of agent development frameworks, specifically addressed for mobile devices, will be of great importance in domains such as cyber physical systems and the Internet of Things in the near future.

3. MAS and Learning

Learning in MAS is a paradigm of great importance because a system capable of learning and changing its way of acting dynamically provides great potential to face many problems for which we do not know the behavior of other agents in the environment. This adds more levels of difficulty in tasks of consensus and coordination between agents, as they may be learning at all times and changing their behavior.

Multi-Agent Learning (MAL) [10] allows us to design certain guidelines from which an agent will be able to exploit the dynamics of its environment and use them to adapt to it. In a multi-agent environment, learning is both more important and more difficult, since the selection of actions has to be done in the presence of other agents, who do not necessarily have to follow the rules of the environment and can make non-deterministic decisions. These agents in turn will adapt their actions to those previously carried out by the other agents. The problems seen in the MAL have a strong relationship with the theory of games, in which an agent selects actions to maximize his/her advantage over the rest.

In the area of MAS learning, this Special Issue contributes with two papers. The first one [11] dealt with the problem of coordinated control of multiple hovercrafts. To address this problem, the authors proposed the design of coordinated control algorithms for multiple agents. For a single vehicle, they proposed to use Radial Basis Function Neural Networks (RBFNNs) to improve the robustness of the controller. For multiple vehicles, they considered the use of a directed topology, but considering that communication among vehicles is continuous.

The second work [12] proposed an approach to learn behavior models as behavior trees for autonomous agents. The main goal of the proposal is to facilitate behavior modeling for autonomous agents in simulation and computer games. The experiments, carried out on the Pac-Man game environment, showed the effectiveness of the proposal, although it is necessary to broaden the applications of the proposed approach in more complex scenarios and configurations.

4. MAS in Ambient Intelligence

Ambient Intelligence (AmI) was built to respond to the technological call to monitor and act in the homes of people with disabilities. Its aim was to create a nest of sensor systems that together could provide more information than alone, thus transforming data into knowledge [13–15]. This is achieved by incorporating digital environments that are sensitive to people's needs, can respond to their requirements, anticipate behaviors, and adjust the response accordingly. In the last few years, several projects have been developed to attend to the needs of the AmI, being very recently under the focus of the industry, a result of the advances of the sensor systems and their decrease of cost, as well as the introduction into mobile devices. Furthermore, the advances of the Internet of Things (IoT) have introduced a new hypothesis of communicating with the home appliances and, above all, introducing them in the home network where they can be remotely controlled [16–18].

Over the last few years, MAS have been employed as a tool for the development of many AmI frameworks. As examples, we can highlight the iGendaframework [19,20], which has as its main goal providing intelligent event management, consisting of a platform that receives events from other users and tries to schedule them according to its importance, having the ability to create, move, and delete events. Moreover, aiming for the implementation of the active aging concept, it schedules ludic activities in the users' free time, adjusted to their medical conditions, user shared events being the last implemented feature. Another example is ALZ-MAS [21], which consists of an Ambient Intelligence framework based on multi-agent technology aimed at enhancing the assistance and health care for Alzheimer patients.

In this Special Issue, three new contributions in this area have been presented. The first one [22] presented a tool, the goal of which is to assist in creating a work environment that is adapted to the needs of people with disabilities. The tool measures the degree of accessibility in the place of work and identifies the architectural barriers of the environment by considering the activities carried out by workers. A case study was conducted to assess the performance of the system, analyzing the accessibility of the different jobs in a real company. Although the tool was initially conceived of for the detection of accessibility problems in office environments, it can be considered a valid tool for the simulation of any agent production process representing human beings in the field of office environments.

The second contribution [23] explored the representation of user interaction levels using an intelligent hybrid system approach with agents. The authors considered the use of an intelligent hybrid approach to provide a decision-making system to an agent that self-evaluates interactions in interactive modules in a museum exhibition. The main goal of this work was to provide a solution to the problem of overcrowding in museums, making museums smart spaces with multi-user adaptive interaction exhibitions. As a case study, the authors built software agents that represented a high-level abstraction of a gallery, specifically an interactive exhibition module in a real museum in Mexico.

Finally, the third work [24] presented CogHelper, which is an orientation system for people with cognitive disabilities. The main idea is that using this system, people with cognitive disabilities may have a more active life, reducing the worry of getting lost both indoors and outdoors. CogHelper will guide users taking into consideration his preferences. To achieve this behavior, the proposed system applies a speculative computation module [25], which needs to be loaded by the traveling path before its execution. For the moment, the system is just a prototype, and authors will test the entire system in real case scenarios in the near future.

5. MAS and Simulation

Agent-based simulation is an approach to modeling systems, which focuses on the simulation of complex technical systems that are distributed and involve complex interaction between humans and machines [26]. It can be seen as a type of computational model that allows the simulation of actions and interactions of autonomous individuals within an environment and allows determining what effects they produce in the system as a whole. It combines elements of game theory, complex systems, emergency, computational sociology, multi-agent systems, and evolutionary programming. The models simulate the simultaneous operations of multiple entities (agents) in an attempt to recreate and predict the actions of complex phenomena. It is an emergency process from the most elementary level (micro) to the highest level (macro).

In this way, agent-based simulation can be seen as a powerful research method that allows dealing in a simple way with the complexity, the emergency, and the non-linearity typical of many social, political, and economic phenomena [27] through mechanisms that allude to the actions of agents and the structure of the interaction between them.

In this Special Issue, the work proposed in [28] took into account the problem of the simulation of Double Action (DA) markets [29]: both buyers and sellers submit their bids, and an auctioneer determines resource allocation and prices on the basis of their bids. Recent works have not considered the fluctuating nature of perishable goods markets, where supply and demand change dynamically and unpredictably. To solve the problem, the authors have developed an online DA market, in which multiple buyers and sellers dynamically tender their bids for trading commodities before their due dates. The experimental results using multi-agent simulation showed that the proposed DA mechanism was effective at promoting the truthful behavior of traders for realizing the fair distribution of large utilities between sellers and buyers. With this work, the authors hope to contribute to promoting successful deployment of electronic markets for fisheries and improve the welfare of people in the area by attracting more traders online.

Moreover, the work presented in [22], mentioned in the previous section, can also be seen as a simulation platform for a 3D environment, which is capable of modeling and enabling simulations in office environments.

6. MAS in Smart Cities

The concept of the smart city [30] arises from the need to find a solution to rapid population growth and the risks this entails for a city, economic risks such as unemployment or physical risks such as over-pollution. To solve these problems, different technologies, among many others, have been applied in order to find solutions in this area. Multi-agent systems together with the Internet of Things [18] are traditionally the most employed.

Typically, these concepts come together, designing intercommunicated networks that responds to the needs of citizens both individually and as a whole and also monitoring through sensors the levels of pollution, traffic, noise, etc. A smart city is a great intercommunicated organism that, together with an intelligent government, seeks to improve the quality of life of its citizens [31].

Therefore, a smart city would be full of sensors constantly collecting information about actions that happen in the city, humidity sensors, temperature sensors, noise, pollution, etc. All these sensors are part of a data collection system that will be responsible for processing information quickly and intelligently. It is for this point where the use of multi-agent systems make sense. The decentralized control of an MAS offers the possibility of managing all available information in a distributed way and also coordinating possible actions effectively over the city. Moreover, decision making processes, apart from being coordinated, can execute in parallel actions at different points of the city, without a strong centralized control, which gives greater flexibility and adaptation to the whole system.

This Special Issue contributes with three works in the area of smart cities. The first work [32] proposed a multi-agent system that provides visualization and prediction tools for bike sharing systems. The proposed MAS includes an agent that performs data collection and cleaning processes, and it is also capable of creating demand forecasting models for each bicycle station. The authors included a case study, which validated the proposed system, by implementing it in a public bicycle sharing system in Salamanca, called SalenBici. In the proposed solution, the information collected was employed by the agents who performed demand forecasting. Moreover, different regression algorithms have also been employed in the process of bike demand prediction. Additionally, a statistical analysis has been performed in order to show the differences in their performance and to determine the relevance of results. The second approach presented in this Special Issue regarding smart cities is the work done in [33], where a multi-agent system was proposed, in order to facilitate the analysis of different possible placement configurations for electric vehicles charging stations in a city. The MAS proposed in this paper integrates the information extracted from heterogeneous data sources as a starting point to specify the areas where future charging stations could potentially be placed. To do this, the proposed MAS integrates an optimization algorithm, which is in charge of the locating process. Finally, the third contribution, presented in [34], analyzed the use of agreement technologies [35], which envisions the next-generation of open distributed systems, where interactions between software components are based on the concept of agreement. In this sense, the authors increased the coordination among entities using agreement technologies in the domain of smart cities as a way to enable the development of novel applications. Concretely, they proposed the use of these techniques in a specific domain such as the coordination in emergency medical services, which includes many tasks that require flexible, on-demand negotiation, initiation, coordination, information exchange, and supervision among different involved entities. All these aspects can be solved through the use of, as previously mentioned, agreement technologies.

7. Conclusions

As it has been possible to observe throughout the analysis carried out on the accepted articles, research on MAS continues to provide technological solutions in a wide variety of domains. MAS researchers develop new advances that allow the development of more powerful, flexible, and adapted systems that allow predicting a fruitful future. This Special Issue of Applied Sciences gives us a precise view of the area covering different hot topics.

The high number of submissions and the quality of the selected works gives us an idea of the potential of the multi-agent systems area and their excellent health after more than two decades of research. In this way, the main goal of this Special Issue is considered to be more than reached, which has allowed us to extend it to new editions to continue disseminating high-quality works in this area.

Appl. Sci. **2019**, 9, 1402

Funding: This research received no external funding.

Acknowledgments: The Guest Editors would like to thank all the authors that have participated in this Special Issue and also the reference contact in MDPI, Daria Shi, for all the work dedicated to the success of this Special Issue.

Conflicts of Interest: The authors declare no conflict of interest.

References

1. Russell, S.; Norvig, P. *Artificial Intelligence: A Modern Approach*, 3rd ed.; Prentice Hall Press: Upper Saddle River, NJ, USA, 2009.
2. Wooldridge, M. *An Introduction to MultiAgent Systems*, 2nd ed.; Wiley Publishing: Hoboken, NJ, USA, 2009.
3. Shehory, O.; Sturm, A. *Agent-Oriented Software Engineering: Reflections on Architectures, Methodologies, Languages, and Frameworks*; Springer: Berlin/Heidelberg, Germany, 2014.
4. Kravari, K.; Bassiliades, N. A Survey of Agent Platforms. *J. Artif. Soc. Soc. Simul.* **2015**, *18*, 11. [CrossRef]
5. Baldoni, M.; Baroglio, C.; May, K.; Micalizio, R.; Tedeschi, S. Computational accountability in MAS organizations with ADOPT. *Appl. Sci.* **2018**, *8*, 489. [CrossRef]
6. Boissier, O.; Bordini, R.H.; Hübner, J.F.; Ricci, A.; Santi, A. Multi-agent Oriented Programming with JaCaMo. *Sci. Comput. Program.* **2013**, *78*, 747–761. [CrossRef]
7. Challenger, M.; Tezel, B.; Alaca, O.; Tekinerdogan, B.; Kardas, G. Development of semantic web-enabled BDI multi-agent systems using SEA_ML: an electronic bartering case study. *Appl. Sci.* **2018**, *8*, 688. [CrossRef]
8. Challenger, M.; Demirkol, S.; Getir, S.; Mernik, M.; Kardas, G.; Kosar, T. On the Use of a Domain-specific Modeling Language in the Development of Multiagent Systems. *Eng. Appl. Artif. Intell.* **2014**, *28*, 111–141. [CrossRef]
9. Boztepe, I.; Erdur, R. Linked Data Aware Agent Development Framework for Mobile Devices. *Appl. Sci.* **2018**, *8*, 1831. [CrossRef]
10. Shoham, Y.; Powers, R.; Grenager, T. If Multi-agent Learning is the Answer, What is the Question? *Artif. Intell.* **2007**, *171*, 365–377. [CrossRef]
11. Duan, K.; Fong, S.; Zhuang, Y.; Song, W. Artificial Neural Networks in Coordinated Control of Multiple Hovercrafts with Unmodeled Terms. *Appl. Sci.* **2018**, *8*, 862. [CrossRef]
12. Zhang, Q.; Yao, J.; Yin, Q.; Zha, Y. Learning Behavior Trees for Autonomous Agents with Hybrid Constraints Evolution. *Appl. Sci.* **2018**, *8*, 1077. [CrossRef]
13. Aarts, E.H.; Encarnação, J.L. *True Visions: The Emergence of Ambient Intelligence*; Springer Science & Business Media: Berlin/Heidelberg, Germany, 2006.
14. Aarts, E.; Harwig, E.; Schuurmans, M. Ambient Intelligence. In *The Invisible Future*; Denning, J.; Ed.; McGraw-Hill, Inc.: New York, NY, USA, 2001.
15. Cook, D.J.; Augusto, J.C.; Jakkula, V.R. Ambient intelligence: Technologies, applications, and opportunities. *Pervasive Mob. Comput.* **2009**, *5*, 277–298. [CrossRef]
16. Kranz, M.; Holleis, P.; Schmidt, A. Embedded interaction: Interacting with the internet of things. *IEEE Internet Comput.* **2010**, *14*, 46–53. [CrossRef]
17. Gershenfeld, N.; Krikorian, R.; Cohen, D. The internet of things. *Sci. Am.* **2004**, *291*, 76–81. [CrossRef] [PubMed]
18. Atzori, L.; Iera, A.; Morabito, G. The Internet of Things: A Survey. *Comput. Netw.* **2010**, *54*, 2787–2805. [CrossRef]
19. Costa, Â.; Novais, P.; Corchado, J.M.; Neves, J. Increased performance and better patient attendance in an hospital with the use of smart agendas. *Log. J. IGPL* **2011**, *20*, 689–698. [CrossRef]
20. Costa, Â.; Novais, P. An intelligent multi-agent memory assistant. In *Handbook of Digital Homecare*; Springer: Berlin/Heidelberg, Germany, 2011; pp. 192–221.
21. Tapia, D.I.; Corchado, J.M. An Ambient Intelligence Based Multi-Agent System for Alzheimer Health Care. *IJACI* **2009**, *1*, 15–26. [CrossRef]
22. Barriuso, A.; De la Prieta, F.; Villarrubia González, G.; De La Iglesia, D.; Lozano, Á. MOVICLOUD: Agent-Based 3D Platform for the Labor Integration of Disabled People. *Appl. Sci.* **2018**, *8*, 337. [CrossRef]
23. Rosales, R.; Castañón-Puga, M.; Lara-Rosano, F.; Flores-Parra, J.; Evans, R.; Osuna-Millan, N.; Gaxiola-Pacheco, C. Modelling the Interaction Levels in HCI Using an Intelligent Hybrid System with Interactive Agents: A Case Study of an Interactive Museum Exhibition Module in Mexico. *Appl. Sci.* **2018**, *8*, 446. [CrossRef]

24. Ramos, J.; Oliveira, T.; Satoh, K.; Neves, J.; Novais, P. Cognitive Assistants—An Analysis and Future Trends Based on Speculative Default Reasoning. *Appl. Sci.* **2018**, *8*, 742. [CrossRef]

25. Satoh, K. Speculative Computation and Abduction for an Autonomous Agent. *IEICE Trans. Inf. Syst.* **2005**, *E88-D*, 2031–2038. [CrossRef]

26. Davidsson, P. Multi Agent Based Simulation: Beyond Social Simulation. In *Proceedings of the Second International Workshop on Multi-Agent-Based Simulation-Revised and Additional Papers*; Springer: London, UK, 2001; pp. 97–107.

27. Uhrmacher, A.M.; Weyns, D. *Multi-Agent Systems: Simulation and Applications*, 1st ed.; CRC Press, Inc.: Boca Raton, FL, USA, 2009.

28. Miyashita, K. Incremental Design of Perishable Goods Markets through Multi-Agent Simulations. *Appl. Sci.* **2017**, *7*, 1300. [CrossRef]

29. Friedman, D. *The Double Auction Market: Institutions, Theories, and Evidence*; Routledge: Abingdon-on-Thames, UK, 2018.

30. Albino, V.; Berardi, U.; Dangelico, R.M. Smart Cities: Definitions, Dimensions, Performance, and Initiatives. *J. Urban Technol.* **2015**, *22*, 3–21. [CrossRef]

31. Roscia, M.; Longo, M.; Lazaroiu, G.C. Smart City by multi-agent systems. In Proceedings of the 2013 International Conference on Renewable Energy Research and Applications (ICRERA), Madrid, Spain, 20–23 October 2013; pp. 371–376. [CrossRef]

32. Lozano, Á.; De Paz, J.; Villarrubia González, G.; Iglesia, D.; Bajo, J. Multi-agent system for demand prediction and trip visualization in bike sharing systems. *Appl. Sci.* **2018**, *8*, 67. [CrossRef]

33. Jordán, J.; Palanca, J.; del Val, E.; Julian, V.; Botti, V. A Multi-Agent System for the Dynamic Emplacement of Electric Vehicle Charging Stations. *Appl. Sci.* **2018**, *8*, 313. [CrossRef]

34. Billhardt, H.; Fernández, A.; Lujak, M.; Ossowski, S. Agreement Technologies for Coordination in Smart Cities. *Appl. Sci.* **2018**, *8*, 816. [CrossRef]

35. Ossowski, S. *Agreement Technologies*; Springer: Dordrecht, The Netherlands, 2012.

*applied
sciences*

MDPI

Article

Learning Behavior Trees for Autonomous Agents with Hybrid Constraints Evolution

Qi Zhang, Jian Yao, Quanjun Yin * and Yabing Zha

College of Systems Engineering, National University of Defense Technology, Changsha 410073, Hunan, China; zhangqiy123@nudt.edu.cn (Q.Z.); markovyao@163.com (J.Y.); zhayabing@nudt.edu.cn (Y.Z.)
* Correspondence: yinquanjun@nudt.edu.cn; Tel.: +86-0731-8450-6327

Received: 7 May 2018; Accepted: 28 June 2018; Published: 3 July 2018

check for
updates

Featured Application: The proposed approach can learn transparent behavior models represented as Behavior Trees, which could be used to alleviate the heaven endeavor of manual agent programming in game and simulation.

Abstract: In modern training, entertainment and education applications, behavior trees (BTs) have already become a fantastic alternative to finite state machines (FSMs) in modeling and controlling autonomous agents. However, it is expensive and inefficient to create BTs for various task scenarios manually. Thus, the genetic programming (GP) approach has been devised to evolve BTs automatically but only received limited success. The standard GP approaches to evolve BTs fail to scale up and to provide good solutions, while GP approaches with domain-specific constraints can accelerate learning but need significant knowledge engineering effort. In this paper, we propose a modified approach, named evolving BTs with hybrid constraints (EBT-HC), to improve the evolution of BTs for autonomous agents. We first propose a novel idea of dynamic constraint based on frequent sub-trees mining, which can accelerate evolution by protecting preponderant behavior sub-trees from undesired crossover. Then we introduce the existing 'static' structural constraint into our dynamic constraint to form the evolving BTs with hybrid constraints. The static structure can constrain expected BT form to reduce the size of the search space, thus the hybrid constraints would lead more efficient learning and find better solutions without the loss of the domain-independence. Preliminary experiments, carried out on the Pac-Man game environment, show that the hybrid EBT-HC outperforms other approaches in facilitating the BT design by achieving better behavior performance within fewer generations. Moreover, the generated behavior models by EBT-HC are human readable and easy to be fine-tuned by domain experts.

Keywords: Behavior Trees (BTs); Genetic Programming (GP); autonomous agents; behavior modeling; tree mining

---

1. Introduction

Modern training, entertainment and education applications make extensive use of autonomously controlled virtual agents or physical robots [1]. In these applications, the agents must display complex intelligent behaviors to carry out given tasks. Until recently, those behaviors have always been developed using manually designed scripts, finite state machines (FSMs) or behavior trees (BTs) etc. However, these ways may not only impose intensive work on human designers when facing multiple types of agents, missions or scenarios, but also result in rigid and predictable agent behaviors [2,3]. An alternative way is using machine learning (ML) techniques to generate agent behaviors automatically [1,4,5]. Through providing sample traces or evaluation criterion of experts' desired behavior, an agent can learn behavior model from expert demonstration or trial-and-error experience respectively. Nevertheless, pure ML approaches, like neuron network (NN) or reinforcement learning (RL), usually generate behavior models as black box systems, which are difficult for domain experts to understand, validate and modify [6,7].

Having the advantages of modularity, reactiveness and scalability compared to FSMs, BTs have become a dominant approach to encode embodied agent behavior in computer games, simulation and robotics [8,9]. A BT can be regarded as a hierarchical goal-oriented reactive planner, which can represent not only a static task plan, but also a complex task policy through conditional checks of various situations. Moreover, due to the hierarchical and modular tree structure, BTs are compatible with genetic programming (GP) to perform sub-tree crossover and mutation, which can yield an optimized BT [6]. Furthermore, the well organized behavior model in BT formalism is accessible and easy to fine-tune for domain experts [10,11].

To balance between automatic generation and model accessibility, recently, several researchers are focusing on learning transparent behavior models as BTs, particularly using genetic programming [11–14]. GP is an evolutionary optimization approach to search an optimal program for a given problem through learning from experience repeatedly [15,16]. The evolving BT is a series of specific approaches which apply GP for agent behavior modeling in certain tasks. The learned model is represented and acted upon in the form of a BT, which is usually evaluated according to a fitness function defined by domain expert based on mission/task. While those approaches have achieved positive results, there are still some open problems [12,14]. For the standard evolving BTs approach (EBT), the global random crossover and mutation would result in dramatically growing trees with many nonsensical branches, which makes it fail to scale up and to provide good solutions. To efficiently generate a good BT solution, some approaches apply a set of domain-specific constraints to reduce the size of the search space, which may limit the application of evolving BTs approaches [13,17].

In this paper, we propose a modified approach, named evolving BTs with hybrid constraints (EBT-HC), to learn behavior models as BTs for autonomous agents. Firstly, a novel idea of dynamic constraint based on frequent sub-trees mining is presented to accelerate learning. It first identifies frequent sub-trees of the superior individuals with higher fitness, then adjusts node crossover probability to protect such preponderant sub-trees from undesired crossover. However, for the large global random search and increasing risk trapped in unwanted local optima, the evolving BTs with only dynamic constraint (EBT-DC) may find unstable solution. Secondly, we extend our dynamic constraint with the existing 'static' structural constraint (EBT-SC) [14] to form the evolving BTs with hybrid constraints. The static constraint can set structural guideline for expected BTs to limit the size of search space, thus the hybrid constraints would lead more efficient learning and find better solutions without the loss of the domain-independence. The experiments, carried out on the Pac-Man game environment, show that in most cases the dynamic constraint is effective to help both EBT-DC and EBT-HC to accelerate evolution and find comparable final solutions than the EBT and EBT-SC. However, the solutions found by EBT-DC become more unstable as the diversity of population decreases. The hybrid EBT-HC can outperform other approaches by yielding more stable solutions with higher fitness in fewer generations. Additionally, the resulting BTs and frequent sub-trees found by EBT-HC are comprehensible and easy to analyze and refine.

The remainder of this paper is organized as follows. Section 2 introduces background and related works of agent behavior modeling and evolving BTs. Section 3 describes the proposed evolving BTs with hybrid constraints approach. Section 4 tests the proposed approach in the Pac-Man AI game. Finally, Section 5 draws conclusions and suggests directions for future research.

2. Background and Related Works

In this section, we recall behavior trees and genetic programming as our necessary research background, and review some related works of agent behavior modeling and evolving behavior trees.

2.1. Behavior Trees

A behavior tree can be regarded as a hierarchical plan representation and decision-making tool to encode autonomous agent behavior. It is an intuitive alternative to FSM with modularity and scalability advantages. Thus, experts can decompose a complex task into simple and reusable low level task modules and build them independently. Nowadays BTs have been adopted dominantly to model the behavior of non-player characters (NPC) (a.k.a. computer generated forces (CGF) in simulation) in game industry, and also applied widespread on robotics [10].

A BT is usually defined as a directed rooted tree $BT =< V, E, \tau >$, where V is the set of all the tree nodes, E is the set of edges to connect tree nodes, $\tau \in V$ is the root node. For each connected node, we define *parent* as the outgoing node and *child* as the incoming node. The *root* has no parents and only one child, and the *leaf* has no child. A single leaf node represents a *primitive behavior*. In addition, a node between root and leaves represents a *composed behavior* combining of primitive behaviors and other composed behaviors, which corresponds to a *behavior hierarchy*. The execution of a BT proceeds as follows. Periodically, the *root* node sends a signal called *tick* to its children. This tick is then passed down to leaf nodes according to the propagation rule of each node type. Once a leaf node receives a tick, a corresponding behavior is executed. The node returns to its parent status *Running* if its execution has not finished yet, *Success* if it has achieved its goal, or *Failure* otherwise.

In this paper, we adopt the BTs building approach recommended in [18], whose components include *Condition*, *Selector*, *Sequence* and *Action* nodes.

Condition: The condition node checks whenever a condition is satisfied or not, returning *success* or *failure* accordingly. The condition node never returns *running*.

Selector: a selector node propagates the tick signal to its children sequentially. If any child returns *Success* or *Running*, the *Selector* stops the propagation and returns the received state. If all children return *Failure*, the Selector returns *Failure*.

Sequence: a sequence node also propagates the tick signal to its children sequentially. However, if any child returns *Failure* or *Running*, the *Sequence* stops the propagation and returns the received state. Only if all children return *Success*, the Sequence returns *Success*.

Action: The action node performs a primitive behavior, returning *success* if the action is completed and *failure* if the action cannot be completed. Otherwise it returns *running*.

Figure 1 shows the graphical representation of all types of nodes used in this paper.

condition   action   selector   sequence

Figure 1. The graphical representation of a behavior tree nodes used.

2.2. Genetic Programming

Genetic programming is a specialization of genetic algorithms which performs a stochastic search to solve a particular task inspired by Darwin's theories of evolution [15,16]. In GP, each individual within the evolving population represents a computer program, which typically is a tree structure such as a behavior tree. The evolving BTs approach applies GP to optimize a population of randomly-generated BTs for agent behavior modeling. Each BT represents a possible behavior model to control autonomous agent evaluated according to a fitness function defined by domain expert based on task. The learning goal is to find a BT controller which can maximize the fitness in the task. For a BT controller, possible states related to decision-making are encoded as condition nodes, available primitive actions in the task are encoded as action nodes, decision-making logic is controlled by BT composite nodes (such as selector, sequence or parallel), a behavior policy is a tree individual with ordered composition of control nodes and leaf nodes.

For BT populations, individuals are evolved using genetic operations of reproduction, crossover, and mutation. In each iteration of the algorithm, some fitter individuals are selected for reproduction directly. Some individuals take crossover operation where a random sub-tree from one individual is swapped with a random sub-tree from another individual and produce two new trees for the next generation. A mutation operator that randomly produces small changes to individuals is also used in order to increase diversity within the population. This process continues until the GP finds a BT that satisfies the goal (e.g., minimize the fitness function and satisfy all constraints).

Often, the crossover operation can bring on undesirable effect of rapidly increasing tree sizes for final generated BT. This phenomenon of generating a BT of larger size than necessary can be termed as bloat. Several approaches for dealing with bloat have been developed [16]. These approaches essentially have a fitness cost based on the size of the tree, thus increasing the tendency for more compact tree to be selected for reproduction.

2.3. Agent Behavior Modeling and Evolving Behavior Trees

In computer games and simulation, a variety of agent programming techniques have been employed to represent and embed agent behaviors, especially for decision-making process. Those techniques usually encode agent behaviors in well-defined structures/models based on domain expertise and customizable constrains, such as FSMs, hierarchical finite state machines (HFSMs), rule-based systems and BTs etc. [2,3]. Among which, BTs have come to the forefront recently for their modularity, scalability, reusability and accessibility [8–10]. However, most of the developments based on those scripting approaches rely on domain expertise and suffer from time-consuming, expensive and difficult endeavor of programming complexity [2,4,19].

On the other side of the spectrum, various approaches are emerged in machine learning community to generate adaptive agent behavior automatically [5,20–22]. This field has been studied basically from two perspectives [1,2]: learning from observation (LfO) (a.k.a, learning from demonstration, programming from demonstration) and learning from experience. The former allows agent to extract the behavior model of the target agent by observing the behavior trace of another agent (e.g., using NN and case-based learning) [4,20]. For example, Fernlund et al. [20] adopted LfO to build agents capable of driving a simulated automobile in a city environment. Ontañón et al. [23] use learning from demonstration for realtime strategy games in the context of case-based planning. While the later leads virtual agent to learn and optimize its behavior by interacting with environment repeatedly (e.g., reinforcement learning and evolutionary algorithms) [5,22]. The performance of the agent is measured according to how well the task is performed as expert's evaluation criterions, which may sometimes find creative solutions that are not found by humans termed as computational creativity [4]. For instance, Aihe and Gonzalez [24], propose Reinforcement learning (RL) to compensate for situations where the domain expert has limited knowledge of the subject being modeled. Teng et al. [22] use a self-organizing neural network to learn incrementally through real-time interactions with the environment, which can improve air combat maneuvering strategies of CGFs. Please note that most of those machine learning methods generate behavior model as a black box system [6,7]. As a result, domain expert could not produce a clear explanation of the relationship between behaviors and models, which is hard to analyze and validate.

To remedy the disadvantages, in both behavior learning perspectives, there are some attempts to generate behavior models represented as BTs from observation [7,25] or experience [11,12,14,26,27] automatically. In this paper, we are focusing on generate BTs through experiential learning, especially evolving BTs. Please note that comparing with other policy representation approaches (decision tree etc.), BT is a more flexible representation which allows explicitly a course of actions as a sub-policy for certain situation. Therefore, for evolving BTs, the scalability is still an open problem stemming from the random large space search [12,14]. In [12], the author points out it is too flexible for evolving BTs without structural guidelines, which would result in most trees that are quite inefficient and impossible to read. So the author constrains the crossover with fixed 'behavior block' sub-trees, which yield comparable reactive behavior. In [14], the authors investigate the effect of 'standard BT design' constraint on evolving BTs approach, which is domain-independent and efficient. They also point out most existing evolving BT approaches adopt different manual constraints to design what the BTs represent and the nature of the tree's constraints. Some of those approaches can speed up learning efficiently but need a lot of knowledge engineering works, which may limit the application of evolving BT approaches. For instance, Scheper et al. [17] apply genetic algorithm to generate improved BTs for a real-world robotic, the initial creation of the trees are not random but human design. In [13], the whole task in game DEFCON is decomposed into a series of sub-tasks and the learning task is just to evolve simple parameter for each sub-task respectively.

Even though the works mentioned above cover most aspects of the behavior modeling with evolving BTs, we intend to use a model-free dynamic constraint to accelerate evolution. We base our work on the standard evolving BTs approach. The main concern is around how to apply model-free constraint or heuristic to speed up BT evolving.

3. Methodology

In this section, we give details about our proposed approach in mainly tow folds. Firstly, we show an overview of the proposed framework, including its main components and basic workflow. Secondly, we elaborate the proposed dynamic constraint and how we extend it with the existing static structural constraint.

3.1. The Proposed Evolving Behavior Trees Framework

Our proposed approach, evolving BTs with hybrid constraints (EBT-HC), is outlined in Figure 2. As the figure shows, two new components, 'Static Structural Constraint' and 'Dynamic Constraint', are added and interacted with the standard evolving BTs process. For the static structural constraint, it set some tree rules to constrain expected BT structure in population initialization and crossover, which can avoid many meaningless and inefficient tree configurations in evolution. For the dynamic constraint, it first applies frequent sub-trees mining for a few higher ranked individuals in each generation, then adjusts nodes crossover probabilities based on the extracted frequent sub-trees, which can protect preponderant structures against undesirable crossover.

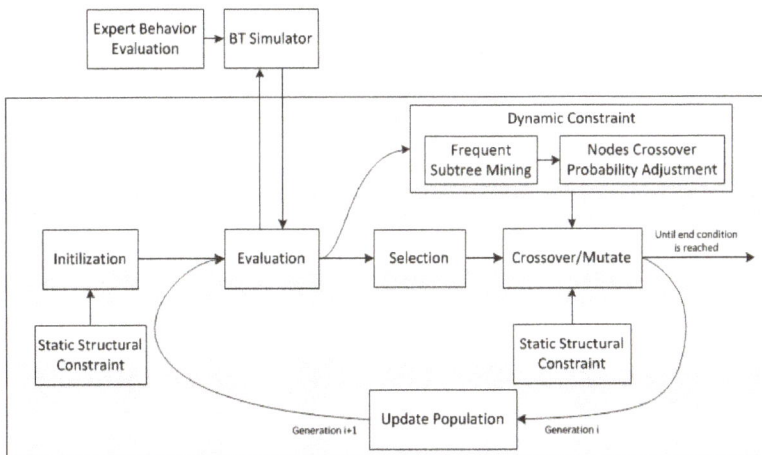

Figure 2. The proposed evolving behavior tree framework, behavior tree (BT).

21

In detail, the workflow of the proposed EBT-HC for agent behavior modeling can be described as follows:

At first, the GP system creates initial BT population individuals under static structural constraint. Unlike fully random combination of leaf nodes and control nodes in standard evolving BTs, the initial BTs are generated constrained by predefined BT syntax rules which will be elaborated in Section 3.3.

Secondly, the GP system evaluates each individual in the population respectively, which needs to run the BT simulator and calculate fitness according to the simulation results and behavior evaluation function. The BT simulator simulates the task execution with the agent controlled by the evaluated BT individual, the behavior evaluation function depicts desired behavior effect quantitatively, which will serve as the fitness measure base that determines the appropriateness of the individuals being evolved.

Thirdly and foremost, some superior individuals with higher fitness are selected to perform crossover and mutate operations to reproduce offsprings. Here we adopt tournament select, sub-tree crossover and single point mutation. Please note that in the sub-tree crossover, the select of crossover node should be constrained by both static structural constraint and dynamic constraint.

Before crossover, we execute a FREQT similar tree mining algorithm for the population and find frequent sub-trees as preponderant structure needing to protect. For each tree individual, according to whether a node belongs to a frequent sub-tree found, we classify nodes in this tree (except the root node) to two sets, protected nodes and unprotected nodes. Then we adjust selected crossover probability of each node accordingly. In brief, we increase the selected probability of unprotected nodes and reduce the selected probability of protected nodes to avoid undesired crossover.

After genetic operation, we update the population to next generation to continue evolution until the end condition is reached.

3.2. Dynamic Constrain Based on Frequent Sub-Tree Mining

In genetic programming, the learner selects preponderant individuals of the current population to reproduce offspring through select operator (e.g., tournament, wheel roulette). From another perspective, the evolution to find an optimal BT is also the process of preponderant structures combination, where a preponderant structure is usually a self-contained behavior sub-tree to deal with certain local situation correctly. So regarding the population individuals as dataset, in each generation, we can mine frequent sub-tree structures of higher fitness individuals. After that we adjust nodes crossover probability to protect such sub-trees against destroyed for faster experiential learning. We call such soft way as dynamic constraint based on frequent sub-tree mining. The intuition behind dynamic constraint is that a frequent sub-tree in superior individuals has a bigger chance to be required by most individuals with higher fitness, even as a sub-tree of the optimal target BT. Thus, we should give more chance to protect such preponderant sub-trees for inherited to next generation. Through preference of crossover nodes based on frequent sub-trees found, in the next generation, there will be more individuals containing those frequent sub-trees, which would lead more precise search around problem space based on those frequent sub-trees, and increase the chance to find a better solution.

In detail, there are two steps to apply dynamic constraint in evolution, frequent sub-tree mining and nodes crossover probability adjustment.

3.2.1. Frequent Sub-Tree Mining

In this section, an adaptation of FREQT [28] is used to mine frequent sub-tree structures in population. FREQT is a classic pattern mining algorithm to discover frequent tree patterns from a collection of labeled ordered trees (LOT). It adopts *rightmost expansion* technique to construct candidate frequent patterns incrementally. At the same time, frequencies of the candidates are computed by maintaining only the occurrences of the rightmost leaf efficiently. It has been demonstrated that FREQT can scale almost linearly in the total size of maximal tree patterns slightly depending on the size of the longest pattern [28,29].

A labeled ordered tree usually represents a semi-structured data structure such as XML. According to the structure and semantics, a behavior tree is a typical labeled ordered tree. Thus the formalism of BT can be expanded from definition of LOT as $BT_{LOT} =< V, E, \tau, L, \preceq >$, where $BT =< V, E, \tau >$ is the basic structure of a BT, $\tau \in V$ is the root node. The mapping $L : V \rightarrow \iota$ is the labeling function, ι includes the labels of root node, control nodes and leaf nodes (condition nodes and action nodes) of a BT. The binary relation $\preceq \subseteq V \times V$ represents a sibling relation for two nodes in a BT. For two nodes μ and v of the same parent, iff $\mu \preceq v$ then μ is an elder brother of v. The execution of BT is following order of depth first from left to right, $\preceq$ represents execution orders of two nodes. Thus, we can construct indexes for all the nodes as depth first in an LOT, which can be consistent with records in GP.

Let $T_D = \{T_1, T_2, ..., T_n\}$ be the dataset of tree mining, which includes a small fraction of individuals with higher fitness in current population. T_p is a candidate frequent pattern, which is usually a sub-tree in tree mining. $\delta_T(T_p)$ is the frequency of T_p in a tree T, $d_T(T_p)$ depicts whether T_p exists in T. There is $d_T(T_p) = 1$ if $\delta_T(T_p) > 0$, else $d_T(T_p) = 0$. $\sigma(T_p) = \sum_{T \in T_D} d_T(T_p)$ represents the number of trees where frequent pattern sub-tree T_p exists. $n_t(T_p)$ depicts terminal node size of frequent pattern tree T_p in tree T.

To adapt the notions from FREQT to BTs mining in GP system, we modify the rules to judge whether a sub-tree is frequent. According to BT syntax and its design pattern, a tree T_p can be regarded as a frequent pattern iff it satisfies all the following proposed rules.

1. $\sigma(T_p)/|T_D| > \sigma_{min}$ and $N_{T_{pmin}} < |T_p| < N_{T_{pmax}}$, where σ_{min} depicts the minimal support of a frequent pattern, $N_{T_{pmin}}$ depicts the minimal node size of a frequent pattern and $N_{T_{pmax}}$ depicts the maximal node size of a frequent pattern.
2. All terminal nodes in a frequent pattern $|T_p|$ must be leaf nodes (condition nodes or action nodes).
3. $n_t(T_p) > N_{T_{ptmin}}$, where $N_{T_{ptmin}}$ depicts the minimal terminal node size of a frequent pattern.

Rule 1 is the basic requirement of FREQT data mining algorithms. Rule 2 and rule 3 represent proposed form requirements of expected patterns in behavior modeling with BT. As a decision making tool, the core of a BT is rooted in the logic relation among its leaf nodes. Thus, in rule 2, we believe if a terminal node is a control node, it is meaningless for its located branch. For rule 3, if a frequent pattern has too few terminal nodes (for example only one terminal node), it shows trivial effect on the whole tree construction.

3.2.2. Nodes Crossover Probability Adjustment

After finding the frequent sub-trees collection, the crossover probability of each node is adjusted according to its relation to discovered frequent sub-trees, which can protect those preponderant structures to be inherited to the next generation more likely.

Formally, let T_{D_i} depict the set of the selected superior individuals of BTs population at generation i, T is a chromosome tree selected for crossover in T_{D_i}, $V(T)$ is the set of all the tree nodes in T except the root node τ. Let T_{Df} depict the set of the mined frequent sub-trees in T_{D_i}, T_p is a frequent sub-tree in T_{Df}, $V_p(T)$ is the set of all the tree nodes in T_p, where we define $V_p^r(T)$ as the root node of T_p, $V_p^{in}(T) = V_p(T) \setminus V_p^r(T)$ as the set of nodes in T_p.

Provided we find N distinct frequent sub-trees T_p^k in T, $k = 1, 2, ..., N$, $T_p^k \in T_{Df}$. Then for the tree T, we define the root node set $V^r(T) = \bigcup_{k=1}^N V_{p,k}^r(T)$, the inside node set $V^{in}(T) = \bigcup_{k=1}^N V_{p,k}^{in}(T)$, and the other node set $V^{neu}(T) = V(T) \setminus (V^r(T) \cup V^{in}(T))$. To protect the frequent sub-trees unbroken more likely in crossover, we can classify $V(T)$ to two sets, protected nodes set $V_{pro}(T)$ and unprotected nodes set $V_{unpro}(T)$. That is, $V_{pro}(T) = V^{in}(T)$, which stores nodes needing to be protected in T, and $V_{unpro}(T) = V^r(T) \cup V^{neu}(T)$, which stores nodes to be unprotected in T.

Obviously, to protect preponderant sub-trees inherited to the next generation, we should decrease the select probability of nodes in $V_{pro}(T)$ and increase the select probability of nodes in $V_{unpro}(T)$ as crossover point. We consider the fact that standard sub-tree crossover operation produce two child trees, as illustrated in Figure 3.

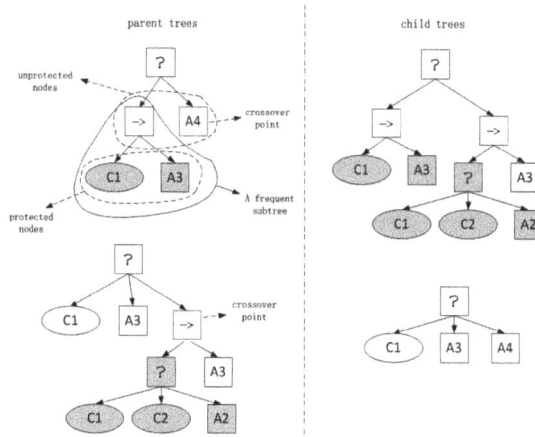

Figure 3. The proposed crossover operation with frequent sub-trees.

As we can see in the leftup parent tree, except the root node, its nodes are classified to unprotected nodes and protected nodes enclosed with two dashed curves respectively. The mined frequent sub-tree is enclosed with a non-dashed curve including all the protected nodes and the root node of the frequent sub-tree. Let us denote the select probability as a crossover point at a node v by $p_{cross}(v)$. The GP system picks up two individuals (e.g., by a tournament selection) from the population, and performs a crossover operation at a node v with the probability $p_{cross}(v)$, which has been modified and normalized as follows:

$$p_{cross}(v) = \frac{\gamma}{|V_{pro}(T)| + |V_{unpro}(T)|} \quad v \in V_{pro}(T) \tag{1}$$

$$p_{cross}(v) = \frac{1}{|V_{pro}(T)| + |V_{unpro}(T)|} + \frac{\frac{1-\gamma}{|V_{pro}(T)|+|V_{unpro}(T)|} * |V_{pro}(T)|}{|V_{unpro}(T)|} \quad v \in V_{unpro}(T) \tag{2}$$

where γ depicts the discount factor, which control the select probability preference for nodes in the protected nodes set.

In Figure 3, the light protected nodes have more chance to be selected in crossover, node 'A4' with red square in the figure. Then two sub-trees including preponderant structures will be combined in the right up child tree inherited to next generation. Besides, we can see in Equations (1) and (2), if we cannot find any frequent sub-trees, there is no effect on the standard evolving process. With generation increasing, the crossover probability adjustment would have bigger effect on exploiting frequent preponderant sub-trees.

3.3. Evolving BTs with Hybrid Constraints

Although the idea behind dynamic constraint based on frequent sub-tree mining is intuitive to accelerate evolution, we found it cannot achieve expected performance in some real applications. For standard evolving BT approach, the global random crossover and mutation result in dramatically growing trees with many nonsensical branches. Therefore, it is hard for the standard evolving BT approach to escape from the local minimum, and some frequent patterns found may be inefficient with inactive nodes never to be executed. In this section, we extend our dynamic constraint with the existing static structural constraint [14]. The static constraint sets structural guideline for generated BTs in initiation and crossover, the dynamic constraint adjusts nodes crossover probability to protect preponderant structure based on constrained configuration space, which can lead more efficient learning.

The static structural constraint is referred from paper [14], which enforce following tree rules as 'standard behavior tree design':

- Selector node may only be placed at depth levels that are even.
- Sequence node may only be placed at depth levels that are odd.
- All terminal child nodes of a node must be adjacent, and those child nodes must be one or more condition nodes followed by on or more action nodes. If there is only one terminal child node, it must be an action node.

Figure 4 is an example generated BT using above static structural constraint. The generated initial BT individuals are efficient and well understood. To ensure the static structural constraint conformed in evolution, the adjacent terminal child nodes of a node will be regarded as a sequential block to swap together.

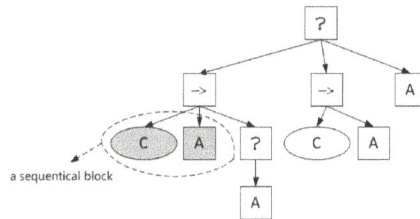

Figure 4. An example behavior tree designed using static structural constraint.

To combine dynamic constraint with static structural constraint in evolution, the following two points should be taken into account for Section 3.2:

1. The available selected units are changed in evolving BTs with hybrid constraint.

In evolving BTs based on dynamic constraint, we sort all nodes in a tree to either protected nodes set or unprotected nodes set. While in evolving BTs with hybrid constraints, the candidate nodes to be sorted are subset of all the tree nodes. On one hand, in each parent tree, we regard the adjacent terminal child nodes as a sequential block to crossover as Figure 4. So the size of candidate nodes to be sorted is the sum of control nodes and blocks. Under static constraint, the adjacent terminal child nodes are regarded as a sequential behavior block and the crossover is constrained only between nodes/blocks with the same type, so the possible behavior blocks will be unchanged in crossover, which will reduce the population diversity and limit the search for possible solution. Thus, we should set a high mutation probability to maintain the diversity of generated behavior blocks. On the other hand, for the crossover node in the first parent tree, candidate nodes can be control nodes or blocks, while for the crossover nodes in the second selected parent tree, the crossover node must be the same type as the selected node in the first tree to keep the static structural constraint unbroken, here types include sequence, selector or terminal block.

2. Nodes crossover probability is adjusted based on step 1.

After modifying the candidate nodes in step 1, the crossover probability of nodes should be adjusted accordingly in Section 3.2.2. It should be noted that iff all nodes in a sequential block are in a frequent sub-tree found, the block can be protected.

4. Experimental Section

In this section, a series of experiments are carried out in the Pacman AI open-source benchmark to test the performance of our approach in agent behavior modeling. The experiments are run single threaded on an Intel Core i7, 3.40 GHz CPU using the Windows 7 64-bit operating system. Four evolving BTs approaches with different constraints and a handcrafted BT are compared, the training and final test performance are monitored over time, along with other statistical measurements. The main goal of our experiments is to demonstrate whether our proposed dynamic constraint can help the original approaches to accelerate behavior trees generation and reach comparable behavior performance. Another goal is to ascertain whether we can get useful behavior sub-trees and well-designed final behavior model as handcrafted BTs.

4.1. Simulation Environment and Agents

Our experiments are tested in the 'Ms. Pac-Man vs Ghosts' game competition environment [30], which provides available AI API for the original arcade game Ms. Pac-Man. As Figure 5 shows, this game consists of 5 agents, a single Ms. Pac-Man and 4 Ghost agents. In the game, the player, controlling Pac-Man, must navigate a maze-like level to collect pills and avoid enemy ghosts or else lose a life. After collecting large 'power' pills, Pac-Man can consume the ghosts and score additional points in a limited period of time. When all the pills in the level are collected the player moves on to the next level, but if three lives are lost the game is over. The only actions available to the player are movement in a 2-dimensional space along the cardinal directions (up, down, left and right), which makes the action space very small. However, the behavior of an AI agent for this game can be quite complex, making it a suitable candidate for the experiments. The scoring method for the game is as follows: eating a normal pill earns Pac-Man 10 points, eating a power pill earns Pac-Man 50 points, and eating ghosts earn 200 points for the first ghost but doubling each time up to 1600 points for the fourth ghost.

Figure 5. The benchmark 'Ms. Pac-Man vs. Ghosts' used in the experiments.

To test our evolving BTs approach for agent behavior modeling, we integrate behavior trees and genetic programming into the 'Ms Pac-Man vs Ghost' API to model Pac-Man behavior. The ghosts are controlled by the basic script provided in the competition, in which ghosts can communicate to share their perception and choose action with a little randomness. The design of behavior trees for Pan-Man agent are modeled on [18], with the components used including sequence, selector, condition and action nodes. So the function set in GP contains 'sequence' and 'selector', and the terminal set contains several game-related conditions and actions. At each time step, the game environment requests a single move (up, down, left, right, no move) from the AI agent, which is returned by executing the behavior tree. The actions and conditions are defined as [14], which can be implemented easily by API provided:

- Conditions

  isInedibleGhostCloseVLow/Low/Med/High/VHigh/Long, six condition nodes which return 'true' if there is a ghost in the 'Inedible' state within a certain fixed distance range, as well as targeting that ghost.

  isEdibleGhostCloseVLow, Low/Med/High/VHigh/Long, six condition nodes which return 'true' if there is a ghost in the 'Edible' state within a certain fixed distance range, as well as targeting that ghost.

  isTargetGhostEdibleTimeLow/Med/High, three condition nodes which return 'true' if a previous condition node has targeted a ghost, which is edible and whose remaining time in the 'edible' state is within a certain fixed range.

  isGhostScoreHigh/VHigh/Max, three condition nodes which return 'true' if the current point value for eating a ghost is 400/800/1600.

- Actions

  moveToEatAnyPill: an action node which set Pac-Man's direction to the nearest pill or power pill, returning 'true' if any such pill exists in the level or 'false' otherwise.

  moveToEatNormalPill: an action node which set Pac-Man's direction to the nearest normal pill, returning 'true' if any such pill exists in the level or 'false' otherwise.

  moveToEatPowerPill: an action node which set Pac-Man's direction to the nearest power pill, returning 'true' if any such pill exists in the level or 'false' otherwise.

  moveAwayFromGhost: an action node which set Pac-Man's direction away from the nearest ghost that was targeted in the last condition node executed, returning 'true' if a ghost has been targeted or 'false' otherwise.

  moveTowardsGhost: an action node which set Pac-Man's direction towards the ghost that was targeted in the last condition node executed, returning 'true' if a ghost has been targeted or 'false' otherwise.

- Fitness Function

  The fitness function is the sum of averaged game score and a parsimony pressure value as formula $f_p(x) = f(x) - cl(x)$ [16]. Where x is the evaluated BT, $f_p(x)$ is the fitness value, $f(x)$ is the averaged game score for a few game runs, c is a constant value known as the parsimony coefficient, $l(x)$ is the node size of x. The simple parsimony pressure can adjust the original fitness based on the size of BT, which will increase the tendency for more compact tree to be selected for reproduction.

4.2. Experimental Setup

In the experiments, four evolving BTs approaches with different constraints are implemented to make comparison. Those are standard evolving BTs, evolving BTs with static constraint, evolving BTs with dynamic constraint, and evolving BTs with hybrid constraints, which are denoted simply as EBT, EBT-SC, EBT-DC and EBT-HC respectively. A handcrafted BT denoted as Hand is also created manually in order to provide a baseline comparison, which is provided by the competition [30]. The handcrafted BT follows some simple sequential rules: initially checking if any inedible ghosts were too close and moving away from them before moving to chase nearby edible ghosts. If there are no ghosts within range, Pac-Man would travel to the closest pill.

The parameter settings for four evolving BTs approaches are listed as Table 1. Please note that all four approaches use crossover operator to produce two child trees from two parent trees. The main differences are the crossover node select and mutation probability as follows. For the approach EBT, each node, except the root node, has equal chance to be selected as a crossover point. For the approach EBT-SC, the adjacent terminal nodes are regarded as a sequential block, all the control nodes and blocks has equal chance to be selected and swapped. The second crossover node must be the same type as the first selected one. For the approach EBT-DC, each node is selected according to adjusted probability. For the approach EBT-HC, the crossover is similar to the approach EBT-SC, but node select probability is adjusted based on frequent sub-trees found. For the approaches EBT-DC and EBT-HC, the minimal support σ_{min} for frequent sub-trees are set as 0.3, the minimal node size $N_{T_{pmin}}$ of a frequent sub-tree is set as 3, the minimal terminal node size $N_{T_{ptmin}}$ is set as 2, the maximal terminal node size $N_{T_{ptmax}}$ is set as 15. The discount factor is set as 0.9.

To validate the robustness of the proposed approach, a few GA parameters are selected to be variable for the same game scenario. Specifically, we vary three important GA parameters (crossover probability, new chromosomes, and mutation probability) and report 9 results of different combinations for the four evolving approaches. Please note that the sum of crossover probability and reproduction probability is always equal to 1. The number of full variable parameters combination can be very big, thus we adopt following combination strategy. First we set a group of common GA parameters as basis, with crossover proportion 0.9, new chromosomes 0.3, and mutation probability 0.01. Because under static constraint, the adjacent terminal child nodes are regarded as a sequential behavior block and the crossover is constrained only between nodes/blocks with the same type, so the population diversity is reduced greatly. Thus, we set a high mutation probability of 0.1 as basic value for the EBT-SC and EBT-HC to increase the diversity of generated behavior blocks. For example, when new chromosomes and mutation probability are fixed as 0.3 and 0.01/0.1 (EBT, EBT-DC/EBT-SC, EBT-HC correspondingly), the crossover probability is set as different values of 0.6, 0.7, 0.8 and 0.9 respectively. Similarly, the new chromosomes is set as different values of 0.1, 0.2 and 0.3 respectively, and the mutation probability is set as different values of 0.01 and 0.1 respectively. So we get 9(4 + 3 + 2) experimental results for all the evolving approaches.

Table 1. Parameter settings for different tested approaches, evolving BTs with only dynamic constraint (EBT-DC), evolving BTs with hybrid constraints (EBT-HC).

Approach	Parameter	Value
fixed to all approaches	population size	100
	generations	100
	initial min depth	2
	initial max depth	3
	selection tournament size	5%
	parsimony coefficient	0.7
variable to all approaches	new chromosomes	10/20/30%
	crossover probability	0.6/0.7/0.8/0.9
	reproduction probability	0.4/0.3/0.2/0.1
	mutation probability	0.01/0.1
EBT-DC/EBT-HC	superior individuals	50%
	the minimal support σ_{min}	0.3
	the minimal node size $N_{T_{pmin}}$	3
	the maximal node size $N_{T_{pmin}}$	15
	the minimal terminal node size $N_{T_{ptmin}}$	2
	the discount factor γ	0.9

For each evolving approach, agents are trained for 100 generations with corresponding configuration and the resulting BT with highest averaged fitness is then played 1000 game runs. Please note that in each generation, each individual is evaluated for 100 game runs to get an expected score as fitness, which is used to reduce the effect of game randomness. All above evolving processes are averaged across 10 trials.

4.3. Results and Analysis

During the learning process, we record all fitness values of individuals and frequent sub-trees found for each generation. After finishing learning, the final test results for generated best individual, the frequent sub-trees found and the final generated BTs are also recorded as results to evaluate the generated behavior models.

Figures 6–8 show the learning curves of mean best fitness for the tested approaches across 10 trials. Table 2 and Figures 9–11 show the performance of the best individual averaged for 1000 simulation tests across 10 trials. Table 2 shows average results of mean and standard deviation, and the Figures 9–11 are more intuitionistic box-plots reflecting results distribution.

As the dynamic constraint is proposed to accelerate learning directly, we first check the learning speed of different approaches under different parameters. Figure 6 shows the learning curves of mean best fitness with variable crossover probability 0.6, 0.7, 0.8 and 0.9 respectively, Figure 7 shows the learning curves of mean best fitness with variable new chromosomes 0.1, 0.2 and 0.3 respectively, and Figure 8 shows the learning curves of mean best fitness with variable mutation probability 0.01 and 0.1 respectively.

In all the learning curves, the approaches EBT-HC and EBT-SC are obviously faster than the approaches EBT and EBT-DC and get higher best mean fitness in the end of evolution. That is because the static constraint can provide well-designed possible tree structure based on common design pattern, which would reduce search space effectively and find a good solution easier. However, the static structure can only support limited use of control nodes, selector and sequence.

Figure 6 shows the learning curves of four evolving approaches under the values of crossover probability 0.6, 0.7, 0.8 and 0.9 respectively. We can see that, in all 4 subfigures, the approach EBT-HC is faster than the EBT-SC and achieves comparable best mean fitness in the end of evolution. When the crossover probability is 0.6, the fitness of EBT-DC climbs obviously faster than the EBT within the first 20 generations, but becomes slower after that. In the end of evolution, the EBT-DC gets a lower mean best value fitness than EBT. It indicates that the EBT-DC converges prematurely to a local minimal value. When the crossover probability grows to 0.7, the EBT-DC performs slower than EBT in most generations, but converges to a similar final mean best fitness with EBT. When the crossover probabilities are 0.8 and 0.9 respectively, the EBT-DC begins to show better performance on average than EBT at generations of 20 and 10 respectively, and finally achieve a higher mean best fitness at generation 100. The results show that the dynamic constraint is robust to help EBT-SC to accelerate learning, while in partial values 0.8, 0.9 of crossover probability, it can help EBT to accelerate learning.

Figure 7 show the learning curves under the values of new chromosome proportion 0.1, 0.2 and 0.3 respectively. We can see that in all 3 subfigures, the EBT-DC is obviously faster than EBT to achieve higher fitness within limited generations. When the new chromosomes is 0.1, the EBT-HC shows similar performance with EBT-SC in term of learning speed and final mean best fitness. As the new chromosomes grow to 0.2 and 0.3, the EBT-HC learns faster at early stage of generation 10 and middle stage of generation 60, and finaly achieve a slight higher final best fitness than EBT-SC. The results show that the dynamic constraint can accelerate learning of EBT and get a better final best fitness, while for EBT-SC, it can help to accelerate learning in new chromosomes 0.2 and 0.3.

Table 2. Mean and standard deviation of the best individual of four evolving approaches under different parameters settings. The mean and standard deviation of the baseline handcrafted BT are 5351.3 and 47.6 (with 95% confidence interval), evolving BTs approach (EBT), existing 'static' structural constraint (EBT-SC).

		EBT	EBT-DC	EBT-SC	EBT-HC
Crossover probability	0.6	6011.8 ± 320.2	5734.6 ± 347.3	7344.8 ± 428.4	7324.5 ± 221.8
	0.7	5768.0 ± 341.6	5930.9 ± 572.2	7174.5 ± 454.0	7371.0 ± 224.0
	0.8	5781.7 ± 267.9	6274.8 ± 728.5	7169.6 ± 353.1	7419.5 ± 218.5
	0.9	5978.8 ± 361.1	6168.4 ± 609.4	7173.5 ± 402.2	7406.5 ± 258.2
new chromosomes	0.1	5886.4 ± 282.9	5962.3 ± 311.3	7353.6 ± 274.9	7350.0 ± 221.4
	0.2	5471.7 ± 769.9	6105.8 ± 592.1	7197.0 ± 369.3	7300.5 ± 344.9
	0.3	5978.8 ± 361.1	6168.4 ± 609.4	7173.5 ± 402.2	7406.5 ± 258.2
Mutation probability	0.01	5978.8 ± 361.1	6168.4 ± 609.4	6720.7 ± 606.6	6737.2 ± 262.6
	0.1	6456.9 ± 523.0	6087.7 ± 473.0	7173.5 ± 402.2	7406.5 ± 258.2

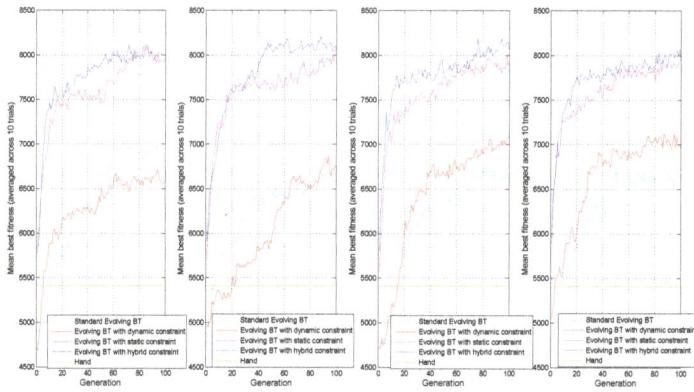

Figure 6. Learning curves of different approaches with variable crossover probability 0.6, 0.7, 0.8 and 0.9.

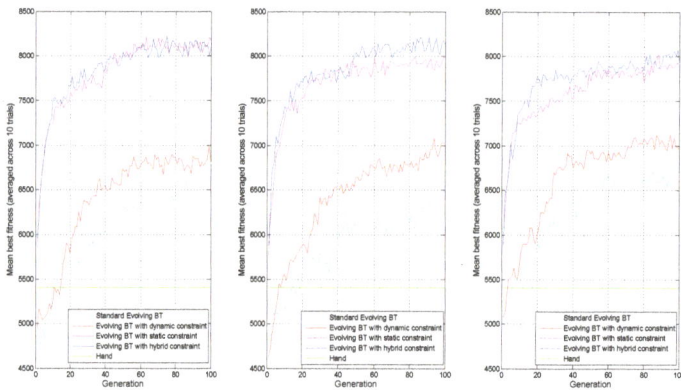

Figure 7. Learning curves of different approaches with variable new chromosomes 0.1, 0.2 and 0.3.

In Figure 8 we can see that, when the mutation probability is 0.01, the EBT-DC learns faster than EBT and converges to a higher mean best fitness in the end of evolution. For the EBT-HC, it climbs faster than EBT-SC within generation 10 and gets a comparable mean best fitness. When the mutation probability is 0.1, the EBT-DC is faster than EBT at early stage but converges to a lower value than EBT in the end of evolution. The EBT-HC learns faster than EBT-SC at early stage and both two converge to a similar mean best fitness in the end. It should be noted that, the final mean best fitness of EBT-SC and EBT-HC under mutation probability 0.01 are obviously lower than those under mutation probability of 0.1. In both the approaches EBT-SC and EBT-HC, the adjacent terminal child nodes are regarded as a sequential behavior block and the crossover is carried out only between nodes/blocks with the same type, thus the possible behavior blocks will remain unchanged and the population diversity declines to a great extent. The results indicate that under both mutation probabilities, the dynamic constraint can help EBT and EBT-SC to accelerate learning, but the approaches EBT-SC and EBT-HC need a bigger mutation probability to maintain higher diversity in generated blocks.

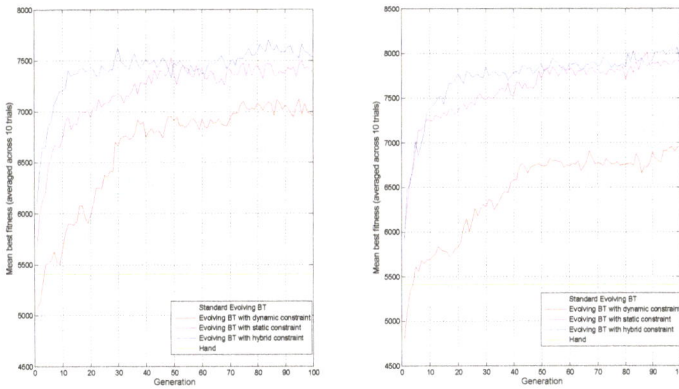

Figure 8. Learning curves of different approaches with variable mutation probability 0.01 and 0.1.

On the whole, in most tested GA parameters, the dynamic constraint can help standard evolving BT and evolving approach with static constraint to accelerate learning speed and achieve better individuals with higher final fitness.

For agent behavior modeling, another important concern is whether we can get an expected behavior model satisfying the evaluation criterions of domain expert. In evolving BTs for behavior modeling specifically, the goal is to produce the best BT individual with higher fitness than other existing approaches in simulation tests. Thus, we will check the average fitness of the best individual generated by different approaches. Table 2 shows all the average fitness results of mean and standard deviation under different GA parameters. Figure 9 shows the results distribution of mean best fitness with variable crossover probability 0.6, 0.7, 0.8 and 0.9 respectively, Figure 10 shows the results distribution of mean best fitness with variable new chromosomes 0.1, 0.2 and 0.3 respectively, and Figure 11 shows the results distribution of mean best fitness with variable mutation probability 0.01 and 0.1 respectively..

From Table 2 we can see that, when crossover probabilities are 0.7, 0.8 and 0.9, the EBT-HC can achieve bigger means of 7371.0, 7419.5 and 7406.5 than the EBT-SC with 7174.5, 7169.6 and 7173.5 respectively, and lower standard deviation of 224.0, 218 and 258.2 than the EBT-SC with 454.0, 353.1 and 402.2 respectively. In the boxplot Figure 9 we also can see that, the whole distribution of all the individuals in EBT-HC is above the distribution in EBT-SC. All three lower adjacent values in EBT-HC are higher than the lower whiskers in EBT-SC, which indicates the lowest individual in EBT-HC is better than more than 25% of individuals in EBT-SC. When the crossover probability is 0.6, the EBT-HC gets slightly smaller values of mean and standard deviation than EBT-SC. In terms of EBT-DC and EBT, we can see when the crossover probability is 0.6, the EBT-DC gets a lower 5734.6 than EBT with 6011.8. When the value grows to 0.7, the EBT-DC achieves similar mean with EBT, and when the value is 0.8 and 0.9, the EBT-DC achieves higher mean than EBT. In most cases, the standard deviation of EBT-DC is bigger than EBT. Those results indicate that the dynamic constraint can help EBT-DC and EBT-HC achieve better final solutions than EBT and EBT-SC respectively under bigger crossover probability like 0.8 and 0.9. The EBT-DC can achieve comparable final solution under crossover probability 0.8 and 0.9, but result in unstable solution. While for EBT-HC, it achieves higher and more stable final solution than EBT-SC.

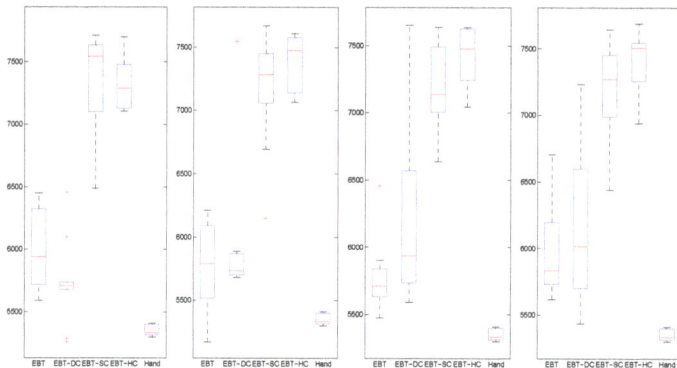

Figure 9. Boxplot results for the best generated BTs with variable crossover probability 0.6, 0.7, 0.8 and 0.9.

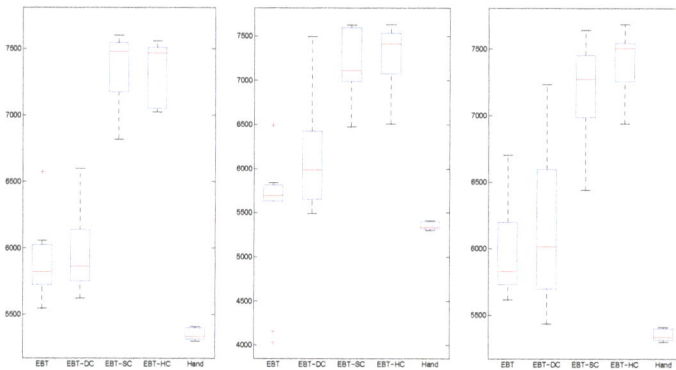

Figure 10. Boxplot results for the best generated BTs with variable new chromosomes 0.1, 0.2 and 0.3.

In Table 2 and Figure 10, when the new chromosomes is 0.1, the EBT-HC and EBT-DC get similar results with EBT-SC and EBT respectively. When the new chromosomess are 0.2 and 0.3, the EBT-HC and EBT-DC get higher mean and median than EBT-SC and EBT respectively. The EBT-DC gets bigger standard deviation than EBT, while EBT-HC gets smaller standard deviation than EBT-SC. That may be because when new chromosome proportion is lower as 0.1, the population diversity is declined, which may increase the risk of dynamic constraint trapped local minimum. Thus, we should use slight bigger new chromosome proportion like 0.2 and 0.3.

In Table 2 and Figure 11, when the mutation probability is 0.01, the EBT-DC gets higher mean and median but larger distribution. Comparing with EBT-SC, EBT-HC gets higher median, similar mean and smaller standard deviation. When the mutation probability is 0.1, EBT-DC gets obviously lower mean and median than EBT, EBT-HC gets higher median and mean than EBT-SC. It should be noted that the values of EBT-SC and EBT-DC under mutation probability 0.01 are obviously lower than those values under mutation probability 0.1. That is because the static constraint on crossover will decline the diversity of swapped behavior blocks, we should set a big mutation probability for the EBT-SC and EBT-HC.

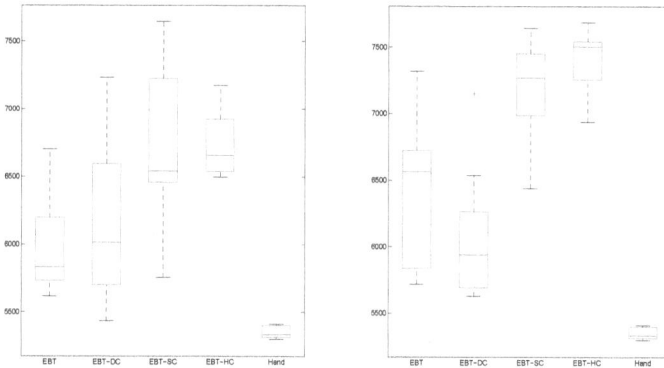

Figure 11. Boxplot results for the best generated BTs with variable mutation probability 0.01 and 0.1.

Those results show that the EBT-HC can achieve higher and more stable solution than EBT-SC in most tested parameter values. While for EBT-DC, the dynamic constraint can help it to achieve higher mean but bigger standard deviation when the population diversity is declined.

From above statistical experimental results of learning curves and final best individuals, we can draw conclusions that the EBT-DC with dynamic constraint can get faster learning speed and comparable final solution than EBT. However the results are more unstable. The EBT-HC can learn faster and achieve higher and more stable final solution than EBT-SC. Considering the sensitivity for variable parameters, the EBT-DC is a bit sensitive to the learning parameters, especially the crossover probability, the EBT-HC is more robust than the EBT-DC under tested GA parameters. Both EBT-HC and EBT-DC need a bigger crossover probability, a slight bigger new chromosome and a bit bigger mutation probability to increase the population diversity, which is important for approaches with dynamic constraint, especially EBT-DC.

The static structural constraint can effectively restrict the search space to find better solution faster. At the same time, it is easier to escape from the local optimal under existing genetic operators. When applying the dynamic constraint in the evolving approach with static constraint, it is easier to find valuable and tidy frequent sub-trees to accelerate learning for the hybrid approach, which will accelerate optimal tree structure composition. While in standard evolving BT approach, the search space is very big and many solutions found are inefficient with inactive nodes never executed. When the changes of GA parameters reduce the population diversity, such as a small crossover probability of 0.6, it is harder for the standard evolving BT approach to escape from the local minimum than the approach with static constraint. When applying dynamic constraint in standard evolving approach, the preference for crossover may increase the chance trapped in local minimum, which leads to the result more unstable.

In the experiment, we also record the final generated BTs and the mined frequent sub-trees to check the intuitive products of behavior modeling. Figure 12 shows a BT generated by the approach EBT-HC. We can see that it is easy to understand the logic behind the controller. It divides the decision-making of full state space into some specific situations to deal with. For example, when the distance to the nearest inedible ghost is low, the agent chooses to evade from the ghost. When the distance to the nearest inedible ghost is very high, the agent chooses to move to eat the nearest power pill. When all the above conditions are not meted, the action 'moveToEatAnyPill' is executed with the lowest priority to execute. Comparing with the handcrafted BT, it seems to be plausible. An interesting phenomenon is that the generated BT does not check the condition 'isEdibleGhostClose' and chooses action 'moveTowardsGhost' as the handcrafted BT does, it is out of our expectation but actually reaches a higher fitness than the handcrafted BT. It may be the expected result of large number of game runs considering the randomness. This phenomenon acting in a different way as human can be regarded as computational creativity that are not found by humans [5].

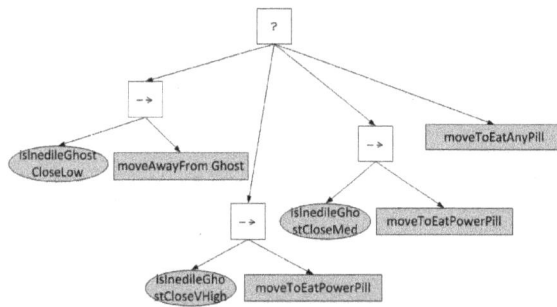

Figure 12. A BT generated by the proposed EBT-HC approach.

Figure 13 shows the sub-trees evolution of the proposed evolving BT approach with hybrid constraints over generations. Because of the limitation of space, we just show the distinct sub-trees found at generation 1, 10, 50 and 100. We can see that at generation 1, only a simple sub-tree is mined which leads agent to move to a pill, which can be formed in most initial individuals with higher ranked fitness. Please note that if a frequent sub-tree is a child of another frequent sub-tree, we only record the one with bigger size. So in fact the action nodes 'moveToAnyPill' and 'moveToNormalPill' are both frequent. For the next generation, the higher ranked individuals with the frequent sub-tree will protect the structure with a little preference, which encourage individuals to search better solution around this structure. At generation 10, we can see that the preponderant sub-tree of generation 1 is no longer frequent. More valuable frequent sub-trees are found which represent the reactive decision-making or action priority for some local situations. For example, when the distance to the nearest inedible ghost is very high, the agent will choose to seek to eat power pill which can provide the agent attack capacity. At generation 50, we can see that some bigger sub-trees become frequent based on the frequent sub-tree found at generation 10, some sub-trees are still frequent those in generation 10. At the same time, some more appropriate action priority is found to replace original structure, for example the node 'moveToAnyPill' is set as the node with lowest priority, which seems to be reasonable. At generation 100, some sub-trees at generation 10 are still frequent, which indicate that they are really necessary building blocks of optimal solution. While some new frequent sub-trees are found such as 'moveAwayFromGhost', some frequent sub-trees in generation 50 are changed. Comparing with the final structure in Figure 12, we can see that most frequent sub-trees mined at generation 100 are the sub-trees of the final best model. The evolution of sub-trees reflects transparently that how the final full model be composed of the frequent sub-tree found generation by generation, which can justify our approach further. Additionally, those sub-trees can be used to facilitate the BT design by domain expert.

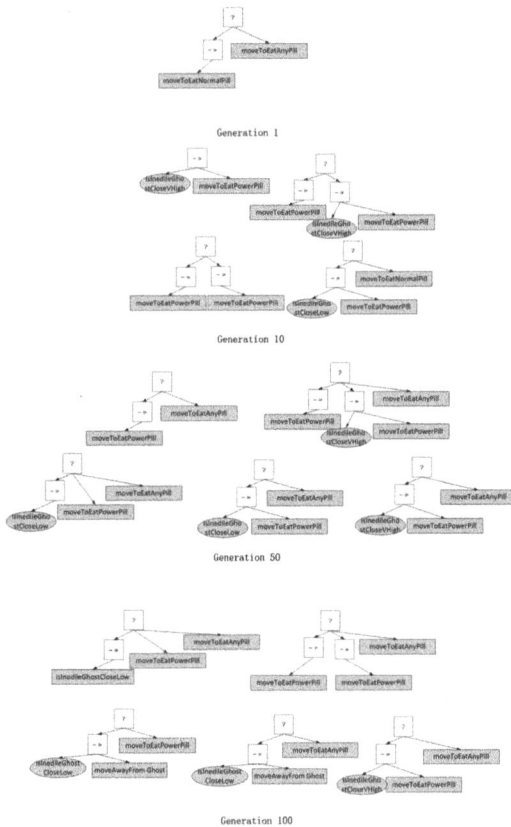

Figure 13. An evolution of frequent sub-trees found over generation by EBT-HC.

5. Conclusions and Future Works

This paper proposed a modified evolving behavior trees approach, named evolving BTs with hybrid constraints (EBT-HC), to facilitate behavior modeling for autonomous agents in simulation and computer games. Our main contribution is a novel idea of dynamic constraint to improve the evolution of Behavior Trees, which discovers the frequent preponderant sub-trees and adjusts nodes crossover probability to accelerate preponderant structure combination. To improve the evolving BT with only dynamic constraint further, we proposed the evolving BTs approach with hybrid constraints by combing the existing static structural constraint with our dynamic constraint. The hybrid EBT-HC can further accelerate behavior learning and find better solutions without the loss of the domain-independence. Preliminary experiments on 'Ms Pac-Man vs Ghosts' benchmark showed that the proposed EBT-HC approach can produce plausible behaviors than other approaches. The stable final best individual with higher fitness satisfies the goal of generating better behavior policy based on evaluation criteria provided by domain expert. The fast and stable learning curve showed the advantage of hybrid constraints to speed up convergence. From the perspective of tree design and implementation, the generated BTs are human readable and easy to analyze and fine-tune, which can be a promising initial step for transparent behavior modeling automatically.

There are some avenues of research to improve this study. Firstly, the proposed approach should be validated in more complex task scenarios and configurations for behavior modeling automatically. In current work, the Pac-Man game is configured as a simple and typical simulation environment to validate proposed approach. However, the interaction between the learning technique and the agent environment is nontrivial. The environmental model, behavior evaluation function, perception, and action sets are critical for behavior performance. Thus, more complex scenarios, such as bigger state-space representation, partial observation or multiple agents in real-time strategy game [31], should be considered to provide rich agent learning environment to validate the proposed approach. On the other hand, it is important to research automatic learning for appropriate parameters setting in GP systems. It is necessary to broaden the applications of proposed approach in more scenarios and configurations.

Another interesting research topic is learning behavior trees from observation. For behavior modeling through experiential learning, it is measured by how well the task is performed based on the evaluation criteria provided by experts. However, the optimal behavior may be inappropriate or unnatural. Thus, there are a few works of learning BTs from observation emerged but still an open problem [7,25]. We believe GP is a promising method and the frequent sub-tree mining can be a potential tool to facilitate behavior block building and accelerate learning. The possible issue behind the method is the similarity metric to evaluate the generated behavior. In [32], the authors investigate a gamalyzer-based play trace metric to measure the difference between two play traces in computer games. Based on above techniques, we can develop a model-free framework to generate BT by learning from training examples.

Author Contributions: Q.Z. and J.Y. conceived and designed the paper structure and the experiments; Q.Z. performed the experiments; Q.Y. and Y.Z. contributed with materials and analysis tools.

Funding: This work was partially supported by the National Science Foundation (NSF) project 61473300, CHINA.

Acknowledgments: This work was partially supported by the National Science Foundation (NSF) project 61473300, CHINA. The authors would like to thank the helpful discussions and suggestions with Xiangyu Wei, Kai Xu and Weilong Yang.

Abbreviations

The following abbreviations are used in this manuscript:

BTs   Behavior Trees
FSMs   Finite State Machines
NPC   Non-player Characters
CGF   Computer-Generated Force
GP   Genetic Programming
RL   Reinforcement Learning
LfO   Learning from Observation
NN   Neural Network
NSF   National Science Foundation

References

1. Tirnauca, C.; Montana, J.L.; Ontanon, S.; Gonzalez, A.J.; Pardo, L.M. Behavioral Modeling Based on Probabilistic Finite Automata: An Empirical Study. *Sensors* **2016**, *16*, 958. [CrossRef] [PubMed]
2. Toubman, A.; Poppinga, G.; Roessingh, J.J.; Hou, M.; Luotsinen, L.; Lovlid, R.A.; Meyer, C.; Rijken, R.; Turcanik, M. Modeling CGF Behavior with Machine Learning Techniques: Requirements and Future Directions. In Proceedings of the 2015 Interservice/Industry Training, Simulation, and Education Conference, Orlando, FL, USA, 30 November–4 December 2015; pp. 2637–2647.
3. Diller, D.E.; Ferguson, W.; Leung, A.M.; Benyo, B.; Foley, D. Behavior modeling in commercial games. In Proceedings of the 2004 Conference on Behavior Representation in Modeling and Simulation (BRIMS), Arlington, VA, USA, 17–20 May 2004; pp. 17–20.

4. Kamrani, F.; Luotsinen, L.J.; Løvlid, R.A. Learning objective agent behavior using a data-driven modeling approach. In Proceedings of the IEEE International Conference on Systems, Man, and Cybernetics, Budapest, Hungary, 9–12 October 2017; pp. 002175–002181.
5. Luotsinen, L.J.; Kamrani, F.; Hammar, P.; Jändel, M.; Løvlid, R.A. Evolved creative intelligence for computer generated forces. In Proceedings of the IEEE International Conference on Systems, Man, and Cybernetics, Budapest, Hungary, 9–12 October 2017; pp. 003063–003070.
6. Yao, J.; Huang, Q.; Wang, W. Adaptive human behavior modeling for air combat simulation. In Proceedings of the 2015 IEEE/ACM 19th International Symposium on Distributed Simulation and Real Time Applications (DS-RT), Chengdu, China, 14–16 October 2015; pp. 100–103.
7. Sagredo-Olivenza, I.; Gómez-Martín, P.P.; Gómez-Martín, M.A.; González-Calero, P.A. Trained Behavior Trees: Programming by Demonstration to Support AI Game Designers. *IEEE Trans. Games* **2017**. [CrossRef]
8. Sekhavat, Y.A. Behaivor Trees for Computer Games. *Int. J. Artif. Intell. Tools* **2017**, *1*, 1–27.
9. Rabin, S. *AI Game Programming Wisdom 4*; Nelson Education: Scarborough, ON, Canada, 2014; Volume 4.
10. Colledanchise, M.; Ögren, P. Behavior Trees in Robotics and AI: An Introduction. *arXiv* **2017**, arXiv:1709.00084.
11. Nicolau, M.; Perezliebana, D.; Oneill, M.; Brabazon, A. Evolutionary Behavior Tree Approaches for Navigating Platform Games. *IEEE Trans. Comput. Intell. AI Games* **2017**, *9*, 227–238. [CrossRef]
12. Perez, D.; Nicolau, M.; O'Neill, M.; Brabazon, A. Evolving behaviour trees for the mario AI competition using grammatical evolution. In Proceedings of the European Conference on the Applications of Evolutionary Computation, Torino, Italy, 27–29 April 2011; pp. 123–132.
13. Lim, C.U.; Baumgarten, R.; Colton, S. Evolving behaviour trees for the commercial game DEFCON. In Proceedings of the European Conference on the Applications of Evolutionary Computation, Torino, Italy, 7–9 April 2010; pp. 100–110.
14. Mcclarron, P.; Ollington, R.; Lewis, I. Effect of Constraints on Evolving Behavior Trees for Game AI. In Proceedings of the International Conference on Computer Games Multimedia & Allied Technologies, Los Angeles, CA, USA, 15–18 November 2016.
15. Press, J.R.K.M. Genetic programming II: Automatic discovery of reusable programs. *Comput. Math. Appl.* **1995**, *29*, 115.
16. Poli, R.; Langdon, W.B.; Mcphee, N.F. *A Field Guide to Genetic Programming*; lulu.com: Morrisville, NC, USA, 2008; pp. 229–230.
17. Scheper, K.Y.W.; Tijmons, S.; Croon, G.C.H.E.D. Behavior Trees for Evolutionary Robotics. *Artif. Life* **2016**, *22*, 23–48. [CrossRef] [PubMed]
18. Champandard, A.J. Behaivor Trees for Next-gen Game AI. Available online: http://aigamedev.com/insider/presentations/behavior-trees/ (accessed on 12 December 2007).
19. Zhang, Q.; Yin, Q.; Xu, K. Towards an Integrated Learning Framework for Behavior Modeling of Adaptive CGFs. In Proceedings of the IEEE 9th International Symposium on Computational Intelligence and Design (ISCID), Hangzhou, China, 10–11 December 2016; Volume 2, pp. 7–12.
20. Fernlund, H.K.G.; Gonzalez, A.J.; Georgiopoulos, M.; Demara, R.F. Learning tactical human behavior through observation of human performance. *IEEE Trans. Syst. Man Cybern. Part B Cybern.* **2006**, *36*, 128. [CrossRef]
21. Stein, G.; Gonzalez, A.J. *Building High-Performing Human-Like Tactical Agents Through Observation and Experience*; IEEE Press: Piscataway, NJ, USA, 2011; p. 792.
22. Teng, T.H.; Tan, A.H.; Tan, Y.S.; Yeo, A. Self-organizing neural networks for learning air combat maneuvers. In Proceedings of the IEEE International Joint Conference on Neural Networks (IJCNN), Brisbane, Australia, 10–15 June 2012; pp. 1–8.
23. Ontañón, S.; Mishra, K.; Sugandh, N.; Ram, A. Case-Based Planning and Execution for Real-Time Strategy Games. In Proceedings of the International Conference on Case-Based Reasoning: Case-Based Reasoning Research and Development, Northern Ireland, UK, 13–16 August 2007; pp. 164–178.
24. Aihe, D.O.; Gonzalez, A.J. Correcting flawed expert knowledge through reinforcement learning. *Expert Syst. Appl.* **2015**, *42*, 6457–6471. [CrossRef]
25. Robertson, G.; Watson, I. Building behavior trees from observations in real-time strategy games. In Proceedings of the International Symposium on Innovations in Intelligent Systems and Applications, Madrid, Spain, 2–4 September 2015; pp. 1–7.

26. Dey, R.; Child, C. Ql-bt: Enhancing behaviour tree design and implementation with q-learning. In Proceedings of the IEEE Conference on Computational Intelligence in Games (CIG), Niagara Falls, ON, Canada, 11–13 August 2013; pp. 1–8.

27. Zhang, Q.; Sun, L.; Jiao, P.; Yin, Q. Combining Behavior Trees with MAXQ Learning to Facilitate CGFs Behavior Modeling. In Proceedings of the 4th International Conference on IEEE Systems and Informatics (ICSAI), Hangzhou, China, 11–13 November 2017.

28. Asai, T.; Abe, K.; Kawasoe, S.; Arimura, H.; Sakamoto, H.; Arikawa, S. Efficient Substructure Discovery from Large Semi-structured Data. In Proceedings of the Siam International Conference on Data Mining, Arlington, Arlington, VA, USA, 11–13 April 2002.

29. Chi, Y.; Muntz, R.R.; Nijssen, S.; Kok, J.N. Frequent Subtree Mining—An Overview. *Fundam. Inf.* **2005**, *66*, 161–198.

30. Williams, P.R.; Perezliebana, D.; Lucas, S.M. Ms. Pac-Man Versus Ghost Team CIG 2016 Competition. In Proceedings of the IEEE Conference on Computational Intelligence and Games (CIG), Santorini, Greece, 20–23 September 2016.

31. Christensen, H.J.; Hoff, J.W. Evolving Behaviour Trees: Automatic Generation of AI Opponents for Real-Time Strategy Games. Master's Thesis, Norwegian University of Science and Technology, Trondheim, Norway, 2016.

32. Osborn, J.C.; Samuel, B.; Mccoy, J.A.; Mateas, M. Evaluating play trace (Dis)similarity metrics. In Proceedings of the AIIDE, Raleigh, NC, USA, 3–7 October 2014.

applied sciences

MDPI

Article

Incremental Design of Perishable Goods Markets through Multi-Agent Simulations

Kazuo Miyashita

National Institute of Advanced Industrial Science and Technology, 1-1-1 Umezono, Tsukuba, Ibaraki 305-8568, Japan; k.miyashita@aist.go.jp; Tel.: +81-29-861-5963

Received: 28 November 2017; Accepted: 11 December 2017; Published: 14 December 2017

Abstract: In current markets of perishable goods such as fish and vegetables, sellers are typically in a weak bargaining position, since perishable products cannot be stored for long without losing their value. To avoid the risk of spoiling products, sellers have few alternatives other than selling their goods at the prices offered by buyers in the markets. The market mechanism needs to be reformed in order to resolve unfairness between sellers and buyers. Double auction markets, which collect bids from both sides of the trades and match them, allow sellers to participate proactively in the price-making process. However, in perishable goods markets, sellers have an incentive to discount their bid gradually for fear of spoiling unsold goods. Buyers can take advantage of sellers' discounted bids to increase their profit by strategic bidding. To solve the problem, we incrementally improve an online double auction mechanism for perishable goods markets, which promotes buyers' truthful bidding by penalizing their failed bids without harming their individual rationality. We evaluate traders' behavior under several market conditions using multi-agent simulations and show that the developed mechanism achieves fair resource allocation among traders.

Keywords: online double auction; mechanism design; perishable goods; multi-agent simulation

1. Introduction

In the research of multi-agent simulations, several types of auction mechanisms have been investigated extensively to solve large-scale distributed resource allocation problems [1] and several applications have been proposed in different kinds of markets [2–4].

In the auctions, resources are generally supposed to have clear capacity limitations, and in some cases they also have explicit temporal limitations on their value [5]. In other words, the resources are modeled to be *perishable* in some problem settings. This study discusses the problem of trading perishable goods, such as fish and vegetables, in a market.

Several enterprises produce perishable goods or services and make profits by allocating them to dynamic demand before they deteriorate. In the services industries, *revenue management* techniques [6] have been investigated. Their objective is to maximize the revenues of a single seller because revenue management is typically practiced by the seller for its own profit. Therefore, it is difficult to apply those techniques to the markets, in which perishable products of multiple sellers must be traded in a coordinated manner to maximize social utility.

Agricultural and marine products are usually traded at *spot markets* (The opposite of spot markets are *forward markets*, which trade goods before production.) that deal in already-produced goods because their quantity and quality are highly uncertain in advance of production. Hence, their production costs are *fixed* in the markets. Their *salvage value* is zero because they perish when they remain unsold in the markets. Therefore, their production costs are *sunk* and sellers' marginal costs are zero in the traditional one-shot markets for perishable goods. This justifies extensive use of one-sided auctions, in which only buyers submit bids, for deciding allocations and prices of perishable goods in fresh markets [7]. However, in such markets, sellers cannot straightforwardly influence price making to obtain fair profits [8].

In *double action* (DA) markets [9], both buyers and sellers submit their bids and an auctioneer determines resource allocation and prices on the basis of their bids. Recent studies [10–12] have applied multi-attribute double auction mechanisms to perishable goods supply chain problems using mixed integer linear programming. Although their approach is powerful in dealing with idiosyncratic properties of perishable goods, such as lead time and a shelf life, they have not considered the fluctuating nature of perishable goods markets, where supply and demand change dynamically and unpredictably.

To solve the problem, we develop an *online DA* market, in which multiple buyers and sellers dynamically tender their bids for trading commodities before their due dates. Our online DA market is developed as an instance of a *call market*, which collects bids continuously and clears the market periodically using predetermined rules. Since bids in the call market have multiple matching chances, the perishable goods in the call market can hold certain salvage values until their time limits. Therefore, sellers can participate in the price-making process of the online DA market, in which their reservation prices are equal to remaining values of the goods. However, since values of the perishable goods decrease progressively, sellers have an incentive to discount their price for fear of spoiling unsold goods. Taking advantage of such an incentive of the sellers, buyers can increase their surplus by bidding a lower price. Nevertheless, buyers need to bid a moderate price for securing the goods before time limits for procurement.

Related Research

Until recently there has been little research undertaken on the online DA. As a preliminary study for an online DA, an efficient and truthful mechanism for a static DA with temporally constrained bids was developed using weighted bipartite matching in graph theory [13]. In addition, for online DAs, some studies have addressed several important aspects of the mechanism, such as design of matching algorithms with good worst-case performance within the competitive analysis framework [14], construction of a general framework that facilitates a truthful dynamic DA by extending static DA rules [15], and an application to electric vehicle charging problems [16]. Although these research results are innovative and significant, we cannot directly apply their mechanisms to our online DA problem because their models incorporate the assumption that trade failures never cause a loss to traders, which is not true in perishable goods markets.

It is demonstrated that no efficient and Bayesian–Nash incentive compatible exchange mechanism can be simultaneously budget balanced and individually rational [17]. For a static DA mechanism with temporal constraints, it is shown that the mechanism is dominant-strategy incentive compatible (or strategy-proof) if and only if the following three conditions are met: (1) its allocation policy is *monotonic*, where a buyer (seller) agent that loses with a bid (ask) v, arrival a and departure d also loses all bids (asks) with $v' \leq v$ ($v' \geq v$), $a' \geq a$ and $d' \leq d$ (When we must distinguish between claims made by buyers and sellers, we refer to the *bid* from a buyer and the *ask* from a seller.); (2) every winning trader pays (or is paid) her *critical value* at which the trader first wins in some period; and (3) the payment is zero for losing traders [13]. In the perishable goods market, the third condition cannot be satisfied because losing sellers have to give up the value of their unsold goods. Even when buyers bid truthfully, the sellers have a strong incentive to discount their valuation when their departure time is approaching and their goods remain unsold. With the knowledge of the sellers' incentive to discount their goods, buyers have an incentive to take advantage of sellers' discounts by underbidding. A mechanism that fails to induce truthful behavior in its participants cannot be efficient, because it does not have the information necessary to make welfare-maximizing decisions. In addition, for online DA markets, it is impossible to achieve efficiency because of the imperfect foresight about upcoming bids [18]. Therefore, neither perfect efficiency nor incentive compatibility can be achieved in the online DA market for perishable goods. In order to develop stable and socially profitable markets for perishable goods, we need to investigate a mechanism that imposes (weak) budget balance and individual rationality while promoting reasonable efficiency and moderate incentive compatibility.

In perishable goods markets, sellers have an incentive to discount their price for fear of spoiling unsold goods. Taking advantage of such an incentive of the sellers, buyers can increase their surplus by bidding a lower price. In our previous research [19], we designed a new online DA mechanism called the criticality-based allocation policy. In order to reduce trade failures, the mechanism prioritizes bids with closer time limits over bids with farther time limits that might produce more social surplus. The proposed mechanism achieves fair resource allocation in the computer simulations, assuming that agents report their temporal constraints truthfully. However, in the proposed mechanism, traders have an incentive to bid late in order to improve allocation opportunities by increasing the criticality of their bids. This causes unpredictable behavior of traders and leads to deterioration of market efficiency.

In this study, extending our previous research, we incrementally develop a heuristic online DA mechanism for perishable goods with a standard greedy price-based allocation policy and achieve fair resource allocation by penalizing buyers' untruthful bids while maintaining their individual rationality. In the penalized online DA mechanism, buyers have an incentive to report a truthful value because they must pay penalties for their failed bids. At the same time, buyers' individual rationality is not harmed because buyers are asked to pay their penalty only on the occasion of a successful matching, in which the sum of a market-clearing price and the penalty does not exceed the valuation of the buyers' bids.

The rest of this paper is organized as follows: Section 2 introduces our market model. Section 3 explains the settings of multi-agent simulations to be used in the following sections. Section 4 presents a naive mechanism design in our online DA market and experimentally analyzes buyers' behavior trading perishable goods. Section 5 improves the market mechanisms and evaluates their performance based on market equilibria in an incremental fashion. Finally, Section 6 concludes.

2. Market Model

In this section, we build a model of online DA markets for trading perishable goods, discuss strategic bidding by traders, and define their utility in the markets.

2.1. Notations

In our online DA market, we consider discrete time rounds, $T = \{1, 2, \dots\}$, indexed by t. For simplicity, we assume the market is for a single commodity. The market has two types of agents, either sellers (S) or buyers (B), who arrive and depart dynamically over time. In each round, the agents trade multiple units of the commodity. The market is cleared at the end of every round to find new allocations.

Each agent i has private information, called *type*, $\theta_i = (v_i, q_i, a_i, d_i)$, where v_i is agent i's valuation of a single unit of the good, q_i is the quantity of the goods that agent i wants to trade, a_i is the arrival time, and d_i denotes the departure time. The gap between the arrival time and departure time defines the agent's trading period $[a_i, d_i]$, indexed by p, during which the agent can modify the valuation of its unmatched bid. Moreover, agents can repeatedly participate in the market over several trading periods.

We model our market as a wholesale spot market. In the market, seller i submits a bid of her goods at arrival time a_i. At departure time d_i, the salvage value of the goods evaporates because of its perishability unless it is traded successfully. Seller i must bring her goods to the market before her arrival. Therefore, seller i incurs a production cost in her trading period and considers it as valuation v_i of the goods at arrival time a_i. Because of advance production and perishability, sellers face the distinct risk of failing to recoup the production cost in the trade. Buyers procure the goods to resell them in their retail markets. Arrival time a_j is the first time when buyer j values the item. For buyer j, valuation v_j represents her assumed budget to procure the goods. In addition to trade surplus, buyer j gains profits by retailing the goods if she succeeds in procuring them before her departure time d_j, which is the due time for a retail opportunity.

Let $\hat{\theta}^t$ denote the set of all the agents' types reported in round t; a complete reported type profile is denoted as $\hat{\theta} = (\hat{\theta}^1, \hat{\theta}^2, \ldots, \hat{\theta}^t, \ldots)$; and $\hat{\theta}^{\leq t}$ denotes the reported type profile restricted to the agents with a reported arrival of no later than round t. In each trading period p, agent i deals with a new trade and has a specific type $\theta_i^p = (v_i^p, q_i^p, a_i^p, d_i^p)$. Report $\hat{\theta}_i^t = (\hat{v}_i^t, \hat{q}_i^t, \hat{a}_i^p, \hat{d}_i^p)$ is a bid made by agent i in round t within trading period p (i.e., $t \in [a_i^p, d_i^p]$). To be noted is that successful trades in previous rounds of period p reduce the current quantity of goods to $q_i^t \leq q_i^p$.

In the market, a seller's ask and a buyer's bid can be matched when they satisfy the following condition.

Definition 1 (Matching condition). *Seller i's ask $\hat{\theta}_i^t = (\hat{v}_i^t, \hat{q}_i^t, \hat{a}_i^p, \hat{d}_i^p)$ and buyer j's bid $\hat{\theta}_j^t = (\hat{v}_j^t, \hat{q}_j^t, \hat{a}_j^p, \hat{d}_j^p)$ are matchable when*

$$(\hat{v}_i^t \leq \hat{v}_j^t) \wedge ([\hat{a}_i^p, \hat{d}_i^p] \cap [\hat{a}_j^p, \hat{d}_j^p] \neq \varnothing) \wedge (\hat{d}_i^p > \hat{d}_j^p) \wedge (\hat{q}_i^t > 0) \wedge (\hat{q}_j^t > 0). \tag{1}$$

The third term in the matching condition (1) (i.e., $\hat{d}_i^p > \hat{d}_j^p$) requires that seller's goods must not perish before the buyer's due date.

Among matchable bids and asks, the market decides the final allocations based on its mechanism M, which is composed of an *allocation policy* π and a *pricing policy* x. The allocation policy π is defined as $\{\pi^t\}^{t \in T}$, where $\pi_{i,j}^t(\hat{\theta}^{\leq t}) \in \mathbb{I}_{\geq 0}$ represents the quantity traded by agents i and j in round t, given reports $\hat{\theta}^{\leq t}$. The pricing policy x is defined as $\{x^t\}^{t \in T}$, $x^t = (s^t, b^t)$, where $s_{i,j}^t(\hat{\theta}^{\leq t}) \in \mathbb{R}_{\geq 0}$ represents a price seller i receives as a result of the trade in round t, given reports $\hat{\theta}^{\leq t}$ with buyer j who pays a price $b_{i,j}^t(\hat{\theta}^{\leq t}) \in \mathbb{R}_{>0}$.

2.2. Agent's Utility

Standard DA markets assume that agents have simple quasi-linear utility: $\sum_{j \in B}(s_{i,j} - v_i)\pi_{i,j}$ for seller i and $\sum_{i \in S}(v_j - b_{i,j})\pi_{i,j}$ for buyer j. However, to represent the characteristics of a wholesale market for perishable goods, we define the idiosyncratic utility for sellers and buyers.

For seller i, when $\pi_{i,j}^t$ units of goods are sold to buyer j at price $s_{i,j}^t$ in round t within period p, seller i obtains income $s_{i,j}^t$. Because the production cost of the seller is v_i^p, the seller's surplus is $(s_{i,j}^t - v_i^p)\pi_{i,j}^t$. If a unit of goods perishes in round t without being traded, seller i loses valuation v_i^p. Therefore, seller i's utility is defined as follows.

Definition 2 (Seller i's utility at time round t).

$$U_i(\hat{\theta}^{\leq t}) = \sum_{\{p | a_i^p \leq t\}} \sum_{t' \in [a_i^p, d_i^p]} \sum_{j \in B \wedge \pi_{i,j}^{t'} > 0} s_{i,j}^{t'}(\hat{\theta}^{\leq t})\pi_{i,j}^{t'}(\hat{\theta}^{\leq t}) - \sum_{\{p | d_i^p \leq t\}} q_i^p v_i^p. \tag{2}$$

The second term in Equation (2) represents that sellers' production cost is sunk. The sunk cost term can be removed from the utility function without making any influence on seller's behaviors, but we intentionally put this term in th utility function to explicitly represent sellers' difficulty of making profits in the market since the value of her goods is completely lost because of perishability whether they are sold or not.

When buyer j has a budget v_j^p for purchasing one unit of goods and succeeds at procuring $\pi_{i,j}^t$ units of the goods at price $b_{i,j}^t$, she obtains surplus $(v_j^p - b_{i,j}^t)\pi_{i,j}^t$. When $(1 + \gamma_j^p)v_j^p$ represents a retail price of the goods, buyer j earns additional profit $\gamma_j^p v_j^p \pi_{i,j}^t$ by retailing the procured goods. We set 1.0 as the value of γ_j^p for the empirical evaluation in Section 3.

Thus, buyer j's utility is defined as follows.

Definition 3 (Buyer j's utility at time round t).

$$U_j(\hat{\theta}^{\leq t}) = \sum_{\{p | a_j^p \leq t\}} \sum_{t' \in [a_j^p, d_j^p]} \sum_{i \in S \wedge \pi_{i,j}^{t'} > 0} \left(v_j^p - b_{i,j}^{t'}(\hat{\theta}^{\leq t}) + \gamma_j^p v_j^p \right) \pi_{i,j}^{t'}(\hat{\theta}^{\leq t}). \tag{3}$$

Agents are modeled as risk neutral and utility maximizing. Equation (2) shows that the seller's bidding strategy is intricate because she can enhance her utility by either raising the market price with higher valuation bidding or preventing the goods from perishing with lower valuation bidding. Equation (3) reveals that the buyer also has difficulty finding an optimal bidding strategy because she can improve her utility by either bringing down the market price with lower valuation bidding or increasing successful trades and retail opportunities with higher valuation bidding.

2.3. Strategic Bidding by Agents

Agents are self-interested and their types are private information. At the beginning of a trading period, agent i submits a bid by making a claim about its type $\hat{\theta}_i = (\hat{v}_i, \hat{q}_i, \hat{a}_i, \hat{d}_i) \neq \theta_i$ to the auctioneer. An agent's self-interest is exhibited in its willingness to misrepresent its type when this will improve the outcome of the auction in its favor. However, misrepresenting its type is not always beneficial or feasible for the agent.

Reporting an earlier arrival time is infeasible for both sellers and buyers because the arrival time is the earliest timing at which they decide to sell or buy the goods in the market. Reporting a later arrival time or an earlier departure time can only reduce the chance of successful trades for the agents. For example, even though buyers like to take advantage of sellers' time discounting behavior in perishable goods markets, buyers' belated arrival is not beneficial for them because they might possibly miss seller's low-priced bids that could be matched if buyers bid earlier. Buyers can get better quotes by submitting a low-valued bid as early as possible. In other words, the buyers do not lose their chance of buying at a discounted price by biding low prices without misreporting their arrival time.

For a seller, it is impossible to report a later departure time $\hat{d}_i > d_i$ since the goods to be sold in the market perish by time d_i. For a buyer, misreporting a later departure time $\hat{d}_j > d_j$ may delay retailing the procured goods.

As for quantity, it is impossible for a seller to report a larger quantity $\hat{q}_i > q_i$ because the sold goods must be delivered immediately after trade in a spot market. Moreover, it is unreasonable for a buyer to report a larger quantity $\hat{q}_j > q_j$ because excess orders may produce dead stocks for her.

In addition, although a seller can misreport a smaller quantity $\hat{q}_i < q_i$ with the intention of raising the market price, in that case, she needs to throw out some of her goods already produced for sale. If a buyer misrepresents a smaller quantity $\hat{q}_j < q_j$ to lower the market price, she loses a chance of retailing more goods. Although these misreports might create larger profits for sellers and buyers, finding the optimal quantity for increasing their profits is not a straightforward task. Therefore, in this study, we assume that the agents do not misrepresent a quantity value in their type.

On the other hand, we assume that an agent has incentives to misreport its valuation for increasing its profit because this is the most instinctive way for the agent to influence trades in a market. In a spot market for perishable goods, a seller may report a lower valuation $\hat{v}_i < v_i$ when she desperately wants to sell the goods before they perish. In addition, a buyer may report a lower valuation $\hat{v}_j < v_j$ to increase her profit when she can take advantage of the sellers' discounted bids.

Consequently, in this study, we assume that agent i may report its limit price strategically (i.e., $\hat{v}_i^t$ may not be equal to v_i) but reports truthful values of quantity q_i, arrival time a_i, and departure time d_i in any round t.

3. Multi-agent Simulation

Our online DA market is considerably more complex than traditional DA markets, for which theoretical analysis is intractable. Therefore, we need to evaluate the market empirically using multi-agent simulations.

3.1. Simulation Settings

In the simulation, 15 sellers and 15 buyers, each of which has one unit of demand or supply, dynamically participate in the market every day, which makes total quantities of demand and supply balanced in the market. We use seven types of markets with distinctive demand–supply curves. In the n-th market, buyer j has a unit of demand, whose value is $v_j = 60 - 2(j - 2n)$, and seller i has a unit of supply, whose value is $v_i = 60 + 2(i - 2n)$. Figure 1a–c represent markets with a different risk of trade failures. It is noteworthy that all the markets have an equal competitive equilibrium price as 60 despite their diversified risk of trade failures.

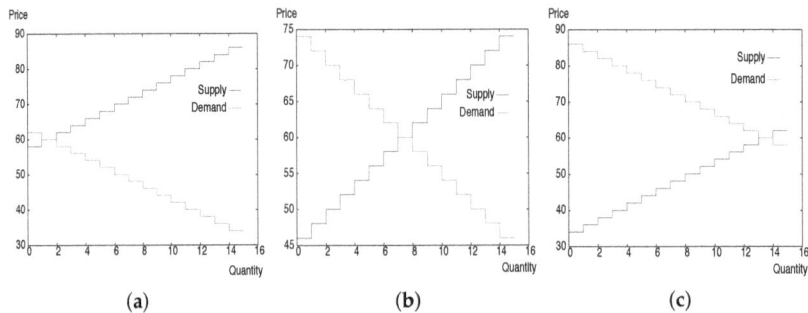

Figure 1. Market conditions. (**a**) High-risk market ($n = 1$); (**b**) Medium-risk market ($n = 4$); (**c**) Low-risk market ($n = 7$).

Each simulation runs for 28 days, in which the market is cleared every hour (i.e., duration of a time round is 1 h). Agents submit their bid to the market at the random timing every day. The submitted bid expires 48 h after their submission (i.e., duration of a trading period is 48 time rounds for every bid). Therefore, every agent has two valid bids in the market unless successfully matched bids are removed from the market.

In the time interval of a trading period, agents can freely modify the reported valuation in their unmatched bid. Fifty randomized trials are executed to simulate the diversified temporal patterns of agents' biddings in seven types of markets.

3.2. Agent's Bidding Strategies

As discussed in Section 4, it is impossible to achieve incentive compatibility in the online DA market for perishable goods, in which sellers suffer a loss when their goods remain unsold. In such markets, sellers adopt markdown pricing strategies to discount their asks progressively. Responding to the sellers' discounting behavior, buyers try to find the lowest matchable valuation to bid, considering market conditions and their true valuation. However, in order to achieve stable and fair trades, markets should give buyers a plausible incentive to bid truthfully. Therefore, we plan to design a market, in which "for buyers, always bidding untruthfully makes less profits than always bidding truthfully," even when sellers bid time discounting values. We call this property *quasi incentive compatibility* on the buyer's side. To behave optimally in such a market, buyers cannot stick to untruthful bidding but should switch to truthful bidding based on a market situation. However, we assume that in a complex and unpredictable online market, dynamically adapting the bidding strategy correctly is beyond average buyers' capability. Consequently, if a market satisfies quasi incentive compatibility for buyers, the buyers are supposed to maintain truthful bidding in the hope of being better off. In addition, we expect that sellers can make proper profits thanks to buyers' truthful bidding.

Experimental analyses of complex markets generally require many cycles of simulations using a large population of heterogeneous agents with sophisticated strategies [20]. However, in this study, we do not consider any fancy learning-based strategy, in which agent's bidding behavior is dynamically determined based on a statistically updated market model, because of the following reasons:

- Since agents' population fluctuates randomly in online markets, accurate status of a market is statistically unpredictable in the simulation.
- Since a market condition and agent's true valuation are fixed in each simulation run, agents can presume that a market is an approximately static environment in the simulation.

As for buyers, we assume that buyer j constantly bids with a certain deviation from her true valuation v_j^p as follows:

1. **Constant difference strategy (COND α):**

    Buyer j always reports her valuation as

    $$\hat{v}_j^t = v_j^p - \alpha.$$

    The value of α is constant for buyer j and $0 \leq \alpha \leq v_j^p$.

2. **Constant rate strategy (CONR β):**

    Buyer j always reports her valuation as

    $$\hat{v}_j^t = \beta v_j^p.$$

    The value of β is constant for buyer j and $0 \leq \beta \leq 1.0$.

As for sellers, we restrict our attention only to strategies that do not bid higher than their true valuation because sellers have a considerable risk of trade failures under our market conditions in which total quantities of demand and supply are balanced. Hence, in the experiments, we use the following strategies for sellers:

1. **Modest strategy (MODE):**

    Seller i always reports her valuation as
    $$\hat{v}_i^t = 0.$$

2. **Truth-telling strategy (TRUE):**

    Seller i always reports her valuation truthfully as

    $$\hat{v}_i^t = v_i^p.$$

3. **Step discount strategy (STEP):**

    Seller i bids as
    $$\hat{v}_i^t = \begin{cases} v_i^p & \text{when } t \in [a_i^p, (a_i^p + d_i^p)/2], \\ 0 & \text{otherwise.} \end{cases}$$

4. **Monotonous discount strategy (MONO):**

Seller i bids as

$$\hat{v}_i^t = \begin{cases} v_i^p & \text{when } t \in [a_i^p, (a_i^p + d_i^p)/2], \\ v_i^p \frac{2.0(d_i^p - t)}{d_i^p - a_i^p} & \text{otherwise.} \end{cases}$$

MODE strategy is developed to simulate one-sided auction markets in which only buyers submit their bids. STEP strategy is a stepwise discount rule, in which a seller reports zero valuation after the midpoint of trading periods. MONO strategy is a typical instance of markdown pricing that simulates sellers' progressive time-discounting behavior in perishable goods markets.

4. Primary Market Design

In the perishable goods markets described in Section 2, agents have to manipulate their valuation carefully in order to increase their utility. Our goal is to design a market mechanism that secures desirable outcomes for both individual agents and the entire market without strategic bidding by the agents.

The market mechanism has traditionally been developed by designers who formulate a problem mathematically and characterize desirable mechanisms analytically in that framework. However, the mechanism design of complex markets, such as our perishable goods markets, is analytically intractable. Recently, computational methodologies for mechanism design, such as automated mechanism design [21], have been advocated. They solve a constrained optimization problem to ensure desirable properties of the mechanism, such as strategy-proofness, for every possible input of agents. We cannot apply these methods to solve our problem because we need to find a *satisficing* but not optimal mechanism for our market, in which neither perfect incentive compatibility nor complete efficiency can be achieved. Instead, we take a heuristic approach known as an incremental mechanism design [22], which starts with a naive mechanism and incrementally improves it over a sequence of iterative evaluations.

4.1. Naive DA Mechanism

As the naive DA mechanism, we adopt standard allocation and pricing policies, which are commonly used in many studies on DA, in order to delineate idiosyncratic problems in traders' behavior in perishable goods markets.

A common goal of the allocation policy in DA markets is to compute trades that maximize social surplus, which is the sum of the difference between bid and ask prices for all matched bids, with the assumption that traders bid truthfully. Thus, the standard allocation policy arranges the asks in ascending order of the seller's price and the bids in descending order of the buyer's price, and matches the asks and bids in sequence. We refer to this allocation rule as a *greedy allocation policy*.

The greedy allocation policy is monotonic and is efficient in durable goods markets, in which agents do not lose utility when they fail to trade. However, sellers of perishable goods lose the value of perished goods when they fail to sell them during the trading period. Consequently, in addition to increasing social surpluses from trades, increasing the number of successful trades is important in the perishable goods markets for maximizing social utility. In our previous research [19], we advocated a criticality-based allocation policy that gives higher priorities to the expiring bids. Since the criticality-based allocation policy is not monotonic, traders have an incentive to misreport their valuation and temporal information.

The pricing policy is important to secure truthfulness and increase market efficiency. Nevertheless, since obtaining truthfulness in online DA markets for perishable goods is impossible, we impose both budget balance and individual rationality, and we promote reasonable efficiency and adequate truthfulness. We adopt the *k-DA pricing policy* [23], in which a clearing price for both sellers and buyers is determined as $(1 - k)\hat{v}_j^t + k\hat{v}_i^t$, thereby making the naive DA market budget balanced (i.e., $s_{i,j}^t = b_{i,j}^t$). We set the value of k as 0.5 for experiments in this paper.

4.2. Analyzing Buyers' Behaviors

Our goal in the research is to empirically design a perishable goods market that satisfies quasi incentive compatibility for buyers. As an initial step toward the goal, we evaluate the frailty of the naive DA mechanism against buyer's strategic behavior by analyzing buyers' utility in the seven types of markets when they have asymmetric strategy profiles. In the experiments, sellers adopt MONO strategy and buyers adopt COND α strategy. In the n-th market of the experiments, seller i's valuation is $v_i = 60 + 2(i - 2n)$ and all the buyers have a common valuation of $v_j = 60 - 2(1 - 2n)$. It should be noted that, in the following experiments in which buyers have asymmetric strategy profiles, buyer j does not have an individual true valuation $v_j = 60 - 2(j - 2n)$ as explained in Section 3.1. With an incentive to increase the trade surplus, buyer j is assumed to misreport her valuation as smaller than the true valuation by an idiosyncratic value of $\alpha = 2(j - 1)$.

4.2.1. Simulating Static Markets with Naive DA Mechanism

Before investigating online markets, we analyze buyers' utility in static markets using the naive mechanism composed of a greedy allocation policy and a k-DA pricing policy. In the static markets, every agent k enters and leaves the market simultaneously as $a_k^p = 24(p - 1), d_k^p = 24(p + 1)$.

Figure 2a–c show buyers' surplus, resale profit and total utility, respectively. The x-axes of the graphs represent 15 buyers in the markets. A buyer with a larger number bids a lower value than its truthful valuation (i.e., $\hat{v}_j^t = v_j - 2(j - 1)$). The graphs in each figure show the seven results obtained in distinctive market conditions.

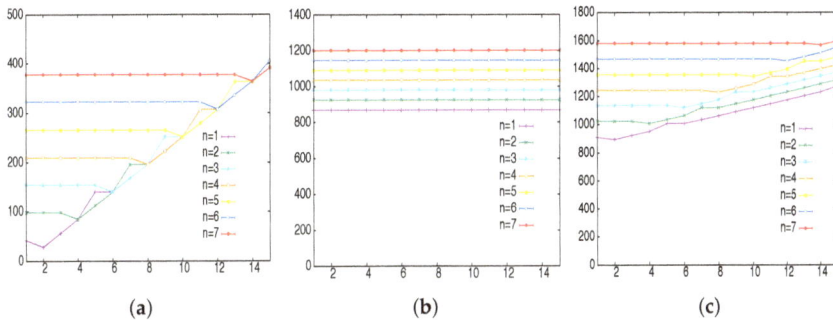

Figure 2. Simulation results in static markets with naive DA mechanism. (**a**) Buyers' surplus; (**b**) Buyers' resale profit; (**c**) Buyers' utility.

In the simulations, all the bids can be matched within the trading periods, because (1) sellers discount their asks monotonously to zero; and (2) quantities of supply and demand are always balanced in the markets. Thus, buyers' resale profits are constant, as shown in Figure 2b. Furthermore, buyers' surpluses are increased by buyers' strategic bidding, as shown in Figure 2a, because low-valued bids match discounted asks and yield large surpluses from their true valuation. Therefore, as shown in Figure 2c, we find that in static markets with a naive DA mechanism, buyers can obtain larger utility by misreporting their value than by bidding truthfully, especially in markets with high risk of trade failures (i.e., markets with a small n value).

4.2.2. Simulating Online Markets with Naive DA Mechanism

As a next step, we investigate buyers' utility in seven types of online markets with the naive DA mechanism.

Figure 3a–c show buyers' surplus, resale profit, and total utility, respectively. Vertical lines in the graphs show standard deviations of the obtained results.

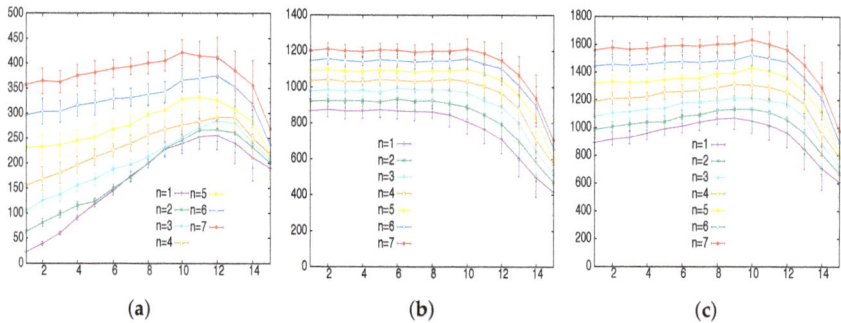

Figure 3. Simulation results in online markets with naive DA mechanism. (**a**) Buyers' surplus; (**b**) Buyers' resale profit; (**c**) Buyers' utility.

In online markets, where demand and supply are not always balanced due to dynamically changing agents' population, low-valued buyers' bids have a high possibility of not being able to find sellers' asks that satisfy the matching condition (1), which is not the case with static markets when sellers adopt a MONO bidding strategy. Therefore, increases of buyers' surplus by misreporting their value saturate at certain points, as shown in Figure 3a, and buyers' misreporting deteriorates resale profits, as shown in Figure 3b. Adding the results of buyers' surplus and resale profit, Figure 3c shows that modestly untruthful buyers still have higher utility than the truth-telling buyers in any type of online markets.

From the experimental results in this section, we find that the naive DA mechanism cannot satisfy quasi incentive compatibility for buyers in perishable goods markets.

5. Incremental Improvement of Market Mechanism

We incrementally improve the naive DA mechanism and investigate the agents' equilibrium behavior in the market using multi-agent simulations.

5.1. Imposing a Penalty on Buyers' Trade Failures

In perishable goods markets, sellers have an incentive to discount their bids to reduce unsold goods. Buyers tend to underbid in the markets and wait for sellers' prices to fall, as shown in the simulation results in Section 4. To encourage buyers' high valuation, we impose a penalty on buyers for their matching failures. We advocate a *penalized payment policy* $\{x^t\}^{t \in T}, x^t = (s^t, b^t, p^t)$, where $p_j^t(\hat{\theta}^{\leq t}) \in \mathbb{R}_{>0}$ represents a penalty imposed on buyer j as a result of her trade failures until round t, given reports $\hat{\theta}^{\leq t}$. Although imposing the penalty on buyers seems unrealistic in practical market situations, it can be considered as a market entry fee for the buyers, which is refunded when they successfully trade in the market.

The penalty of buyer j is an average price gap between a buyer's bid and sellers' asks in the past matching failures, which is updated every time the market is cleared, as shown in Algorithm 1.

Algorithm 1 Update buyer's penalty

1: **procedure** UPDATEPENALTY(j, *ave*, *cyc*)
2:   ▷ Buyer j's penalty information is stored in *ave* and *cyc*.
3:     $sum \leftarrow 0$
4:     $num \leftarrow 0$
5:     **for all** $i \in S \wedge ([a_i{}^p, d_i^p] \cap [a_j^p, d_j^p] \neq \varnothing) \wedge (d_i^p > d_j^p)$ **do**
6:         $gap = \hat{v}_i^t - \hat{v}_j^t$
7:         **if** $gap > 0$ **then**
8:             $sum \leftarrow sum + gap$
9:             $num \leftarrow num + 1$
10:        **end if**
11:    **end for**
12:    **if** $sum > 0$ **then**
13:       $ave \leftarrow ave + sum/num$
14:       $cyc \leftarrow cyc + 1$
15:    **end if**
16:    $p_j^t \leftarrow ave/cyc$ ▷ Updated buyer j's penalty.
17: **end procedure**

The penalized pricing policy requires a buyer to pay the sum of a clearing price and the penalty when the buyer's bid is executed. If the required amount exceeds the buyer's bid price, the buyer refuses the payment to cancel the execution and proceed to another bidding. Therefore, the penalized pricing policy does not cause buyers' individual rationality to deteriorate and the matching condition of the market changes as follows.

Definition 4 (Matching condition with buyer's penalty). *Seller i's ask $\hat{\theta}_i^t = (\hat{v}_i^t, q_i^t, a_i^p, d_i^p)$ and buyer j's bid $\hat{\theta}_j^t = (\hat{v}_j^t, q_j^t, a_j^p, d_j^p)$ are matchable when*

$$(\hat{v}_i^t \leq \hat{v}_j^t) \wedge ([a_i{}^p, d_i^p] \cap [a_j^p, d_j^p] \neq \varnothing) \wedge (d_i^p > d_j^p) \wedge (q_i^t > 0) \wedge (q_j^t > 0) \wedge \left(b_{i,j}^t(\hat{\theta}^{\leq t}) + p_j^t(\hat{\theta}^{\leq t}) \leq \hat{v}_j^t \right). \quad (4)$$

The last term in the matching condition (4) guarantees the buyer's individual rationality.

   When the penalized pricing policy is adopted, the buyer's utility is reduced by the penalty for trade failures as follows.

Definition 5 (Buyer j's penalized utility in time round t).

$$U_j(\hat{\theta}^{\leq t}) = \sum_{\{p|a_j^p \leq t\}} \sum_{t' \in [a_j^p, d_j^p]} \sum_{i \in S \wedge \pi_{i,j}^{t'} > 0} \left(v_j^p - b_{i,j}^{t'}(\hat{\theta}^{\leq t}) + \gamma_j^p v_j^p - p_j^{t'}(\hat{\theta}^{\leq t}) \right) \pi_{i,j}^{t'}(\hat{\theta}^{\leq t}). \quad (5)$$

Hence, with the penalized pricing policy, buyers have an incentive to report a high valuation for avoiding matching failures.

   We evaluate a penalized DA mechanism that replaces the simple k-DA pricing policy in the naive DA mechanism with the penalized pricing policy. We execute the same simulations described in Section 4.2.2 to investigate the effects of imposing a penalty on buyers when their bids fail to match sellers' asks.

   Figure 4a,b,d represent buyers' surplus, resale profit and total utility, respectively, in the same way as the previous experiments. Figure 4c shows the penalty imposed on each buyer by the mechanism.

   Comparing Figures 3b and 4b, we find that, after imposing penalties on buyers, misreporting decreases buyers' resale profits drastically, because the matching condition (4) increases matching failures and reduces resale opportunities for penalized buyers. In addition, Figure 4a,c show that the increase of buyers' surplus by misreporting is canceled out by the penalty imposed on the buyers.

Therefore, by comparing Figures 3c and 4d, it is shown that the proposed penalized DA mechanism largely succeeds in eliminating buyers' utility gain by misreporting their values in the markets.

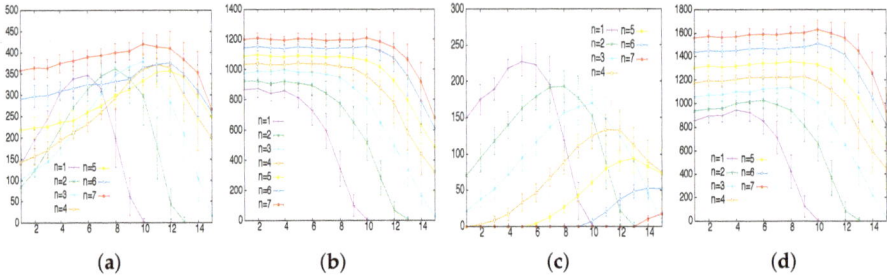

Figure 4. Simulation results in online markets with penalized DA mechanism. (**a**) Buyers' surplus; (**b**) Buyers' resale profit; (**c**) Buyers' penalty; (**d**) Buyers' utility.

5.2. Analyzing Agents' Equilibrium Behavior

The experimental results of the penalized DA mechanism demonstrate that truth-telling buyers are better off than untruthful buyers in the heterogeneous population of buyers. For the next step, we need to understand the interactions between buyers and sellers with various bidding strategies to ensure quasi incentive compatibility of buyers along with other favorable properties, such as efficiency and stability, in our online markets of perishable goods.

The Nash equilibrium is an appropriate solution concept for understanding and characterizing the strategic behavior of self-interested agents. However, computing the exact Nash equilibria is intractable for a dynamic market with non-deterministic aspects [24]. We analyze the market through simulations across the restricted strategy space to evaluate its quasi incentive compatibility for buyers.

In the experiments, buyers adopt CONR β strategies, in which buyer j bids $\hat{v}_j^t = \beta v_j^p$, with five different values of β (i.e., 0.2, 0.4, 0.6, 0.8, and 1.0), and sellers use four types of bidding strategies (i.e., MODE, TRUE, MONO, and STEP) explained in Section 3.2. In the n-th market of the experiments, seller i's valuation is $v_i = 60 + 2(i - 2n)$ and buyer j's valuation is $v_j = 60 - 2(j - 2n)$ as explained in Section 3.1. Because agents on each side are faced with the same utility maximization problem, they follow the same strategy in the equilibrium analysis. Therefore, we consider symmetric strategy profiles for each side of agents.

Table 1 shows the payoff matrix between sellers and buyers in the online market with low risk of trade failures shown in Figure 1c. Buyers have more chances of increasing their surpluses by underbidding in the lower-risk markets.

Each cell in the table represents the result of interactions between the sellers and buyers with the corresponding strategy. The cell is separated into the following two parts:

- The upper part of the cell shows the following three types of information: (1) the average clearing price; (2) the average matching rate; and (3) the total utility that is the sum of traders' utility and penalties paid by buyers to the auctioneer.
- The lower part of the cell represents utility of trader agents. The bottom left corner shows the averaged utility of seller agents, and the top right corner reveals that of buyer agents. Standard deviation of the average utility is shown inside parentheses.

In the table, numbers in boldface represent utilities of the agent's best response to the other agent's bidding strategy, and cells in gray show Nash equilibria.

It is noteworthy that the results in the first row of the table, where sellers adopt MODE strategy, are the same as those by the naive DA mechanism because there are no matching failures and thus, no penalty for buyers when sellers bid zero valuation.

Table 1. Payoff matrix in the low-risk market. MODE: modest strategy; TRUE: truth-telling strategy; STEP: step discount strategy; MONO: monotonous discount strategy.

Seller\Buyer	CONR 0.2	CONR 0.4	CONR 0.6	CONR 0.8	CONR 1.0
MODE	7 96% 18,972 **27,738** (280) **−8766** (101)	14 95% 18,763 26,182 (285) **−7419** (107)	20 94% 18,739 24,826 (282) −6087 (109)	27 94% 18,739 23,516 (270) −4777 (112)	33 94% 18,739 22,218 (257) −3479 (119)
TRUE	— 0% −10,077 0 (0) −10,077 (52)	33 1% −9682 302 (148) −9999 (69)	43 33% 1,008 7915 (425) −7106 (179)	51 73% 13,211 15,238 (232) −2168 (185)	57 93% 18,473 **17,307** (170) 1143 (161)
STEP	— 0% −10,077 0 (0) −10,077 (52)	17 28% −443 7810 (460) −9095 (108)	28 93% 18,432 **21,297** (398) −4625 (127)	45 95% 18,828 19,401 (271) −1086 (139)	57 94% 18,726 17,495 (204) 1201 (166)
MONO	— 0% 10,077 0 (0) −10,077 (52)	20 22% −2275 6238 (473) −9156 (81)	36 79% 14,749 17,550 (473) **−4081** (124)	50 91% 17,955 **18,089** (283) **−568** (165)	57 94% 18,684 17,416 (188) **1240** (172)

Table 1 shows that there are two Nash equilibria strategy profiles: (MODE, CONR 0.2), and (MONO, CONR 0.8). Among them, adopting the CONR 0.2 strategy is risky for buyers because it produces zero utility for them if sellers avoid using the MODE strategy, which produces a large loss for sellers even when buyers bid high values. Therefore, (MONO, CONR 0.8) is a more feasible strategy profile to be executed in the low-risk market with the penalized DA mechanism. From the above equilibrium analysis, it is found that buyers obtain more profits by strategically misreporting their value 20% less than the true valuation and the penalized DA mechanism fails to achieve quasi incentive compatibility for buyers.

5.3. Adjusting Buyers' Penalty Based on Market Condition

Since buyers are more likely to underbid for increasing their surpluses in the lower-risk markets, as shown in Table 1, larger penalties must be imposed on buyers' trade failures in the lower-risk markets in order to prevent buyers' strategic bidding. For this purpose, we modify the buyer's penalty on matching failures as follows:

$$\tilde{p}_j^t = p_j^t \times 0.5/(1.0 - MatchingRate()) \tag{6}$$

In Equation (6), *MatchingRate()* is a function that calculates a current successful matching rate of bids and asks in the market. If the matching rate is above 50%, the market is considered to have a low risk of trade failures and the adjusted buyer's penalty $\tilde{p}_j^t$ has a larger value than the original penalty p_j^t. We call the DA mechanism with the modified penalty an *adaptively penalized DA mechanism*.

Table 2 shows the payoff matrix between sellers and buyers in the low-risk market shown in Figure 1c with the adaptively penalized DA mechanism. The table shows that there are three Nash equilibria strategy profiles: (MODE, CONR 0.2), (STEP, CONR 1.0), and (MONO, CONR 1.0). To be noted is that we consider (MONO, CONR 1.0) is a Nash equilibrium because its seller's utility (i.e., 1051) is not significantly smaller than the seller's utility in (STEP, CONR 1.0) (i.e., 1058) based on its standard deviation (i.e., 147). Among them, adopting CONR 0.2 strategy is risky for buyers for the same reason explained in Section 5.2. Therefore, (STEP, CONR 1.0) and (MONO, CONR 1.0) are more feasible strategy profiles to be executed in the low-risk market with the adaptively penalized DA mechanism. From the above equilibrium analysis, it is found out that buyers maximize their profits by truthfully reporting their true valuation and the adaptively penalized DA mechanism achieves quasi incentive compatibility for buyers.

The adaptively penalized DA mechanism also succeeds in achieving quasi incentive compatibility for buyers in riskier market conditions, such as Figure 1a,b, as shown in Tables 3 and 4.

As explained, the results in the first row of the tables are the same as those of the naive DA mechanism. (MODE, CONR 0.2) is a dominant strategy equilibrium when sellers are not allowed to bid in the naive DA market, which is the case with the conventional perishable goods markets. Therefore, by comparing the results of (MONO, CONR 1.0) strategy profile with those of (MODE, CONR 0.2) strategy profile, we can clarify the effects of applying the adaptively penalized DA mechanism to perishable goods markets. From the abovementioned results, we find that the adaptively penalized DA mechanism increases sellers' utility while it depresses buyers' utility and decreases total utility especially in higher-risk markets. This means the adaptively penalized DA mechanism makes markets fairer, but it also makes markets less efficient in high risk conditions due to low matching rates. However, since the inefficiency of markets largely comes from sellers' loss of perished goods (i.e., 7984 in the high-risk market and 3797 in the medium-risk market), sellers can improve their profits and the market's efficiency by controlling their supplies based on the estimated market condition. In the adaptively penalized DA market, sellers can make a reasonable estimation of the market conditions from their matching rates, which is not possible with the conventional markets in which almost all of the supplies are sold at equally low prices regardless of the market conditions (i.e., 4 in the high-risk market, 5 in the medium-risk market, and 7 in the low-risk market).

Table 2. Payoff matrix in the low-risk market.

Seller\Buyer	CONR 0.2	CONR 0.4	CONR 0.6	CONR 0.8	CONR 1.0
MODE	7 96% 18,972 **27,738** (280) **−8766** (101)	14 95% 18,763 26,182 (285) −7419 (107)	20 94% 18,739 24,826 (282) −6087 (109)	27 94% 18,739 23,516 (270) −4777 (112)	33 94% 18,739 22,218 (257) −3479 (119)
TRUE	— 0% −10,077 0 (0) −10,077 (52)	33 1% −9682 309 (151) −9999 (69)	43 35% 1777 8497 (358) −6910 (160)	52 70% 13,344 14,646 (243) −2461 (180)	58 90% 17,779 **16,874** (200) 873 (175)
STEP	— 0% −10,077 0 (0) −10,077 (52)	16 42% 4153 11,624 (397) −8652 (108)	31 69% 12,137 16,121 (350) −5564 (135)	46 84% 16,208 **17,172** (301) −1839 (141)	57 94% 18,611 **17,374** (207) **1058** (155)
MONO	— 0% −10,077 0 (0) −10,077 (52)	20 39% 3182 10,607 (381) −8416 (75)	37 65% 10,989 14,835 (338) −5075 (101)	49 81% 15,311 16,373 (302) **−1663** (145)	57 92% 18,235 **17,090** (176) **1051** (147)

Table 3. Payoff matrix in the high-risk market.

Seller\Buyer	CONR 0.2	CONR 0.4	CONR 0.6	CONR 0.8	CONR 1.0
MODE	4 95% 4249 **18,530** (176) −14,280 (153)	9 95% 4192 17,634 (184) −13,442 (149)	13 94% 4157 16,741 (178) −12,584 (154)	17 94% 4157 15,881 (171) −11,724 (155)	21 94% 4157 15,075 (166) −10,917 (154)
TRUE	— 0% −15,117 0 (0) −15,117 (77)	— 0% −15,117 0 (0) −15,117 (77)	— 0% −15,117 0 (0) −15,117 (77)	— 0% −15,117 0 (0) −15,117 (77)	60 3% −14,259 417 (113) −14,695 (133)
STEP	— 0% −15,117 0 (0) −15,117 (77)	— 0% −15,117 0 (0) −15,117 (77)	19 6% −13,466 1152 (164) −14,855 (111)	23 33% −7060 5171 (146) −13,553 (141)	29 51% −3261 **6938** (155) −12,057 (151)
MONO	— 0% −15,117 0 (0) −15,117 (77)	— 0% −15,117 0 (0) −15,117 (77)	19 8% −13,095 1426 (208) −14,802 (83)	29 32% −7348 4691 (204) −13,194 (132)	40 49% −3798 **5569** (174) −11,028 (128)

Table 4. Payoff matrix in the mid-risk market.

Seller\Buyer	CONR 0.2	CONR 0.4	CONR 0.6	CONR 0.8	CONR 1.0
MODE	5 95% 11,572 **23,093** (230) **−11,521** (130)	11 94% 11,453 21,888 (241) **−10,435** (128)	17 94% 11,448 20,780 (230) **−9331** (131)	22 94% 11,448 19,696 (221) −8247 (131)	27 94% 11,448 18,646 (211) −7198 (134)
TRUE	— 0% -12,597 0 (0) −12,597 (65)	— 0% -12,597 0 (0) −12,597 (65)	— 0% −12,597 0 (0) −12,597 (65)	52 20% −6592 3,665 (251) −10,373 (168)	60 50% 1541 **7684** (168) −6319 (179)
STEP	— 0% -12,597 0 (0) −12,597 (65)	15 5% −11,175 1,134 (223) −12,449 (90)	20 42% −496 8,932 (215) −10,839 (114)	32 64% 4815 **11,314** (235) −8394 (153)	47 76% 7597 **11,217** (194) −5095 (164)
MONO	— 0% -12,597 0 (0) −12,597 (65)	15 5% −10,936 1327 (255) −12,420 (68)	26 40% −1106 8069 (274) −10,387 (105)	41 60% 3915 9910 (272) −7493 (125)	53 72% 6720 **10,192** (191) **−4560** (139)

In perishable goods markets, it is intuitively reasonable for sellers to offer discounted asks when there is little time left for matching with buyers' bids. The above-described equilibrium analysis results show that only the MONO strategy is a best response for sellers to buyers' truthful bidding in all the three types of markets. We investigate the buyers' best response to sellers' MONO strategy in seven types of markets.

The y-axes in Figure 5a–c show buyers' utility, sellers' utility, and total utility, respectively. The x-axis represents seven market conditions with a larger number corresponding to a smaller risk of trade failures as explained in Section 3.1. Graphs in each figure show the results obtained in the trades with buyers adopting the CONR strategy with five different β values (i.e., 0.2, 0.4, 0.6, 0.8 and 1.0).

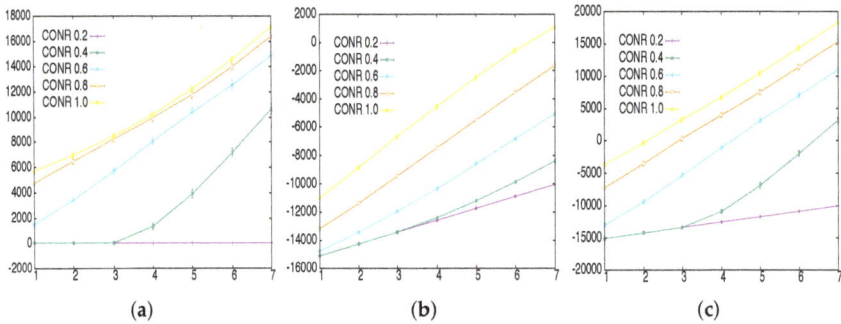

Figure 5. Agents' utility under seven market conditions. (a) Buyers' utility; (b) Sellers' utility; (c) Total utility.

Figure 5a shows that buyers can obtain the largest utility by bidding truthfully (i.e., $\beta = 1.0$) in all market conditions. Figure 5b reveals that sellers can also achieve the largest utility when buyers bid truthfully in every market condition. As a result, as Figure 5c shows, trades produce the largest social profit when buyers report their truthful valuation in all conditions of the market. Therefore, the experiments demonstrate that the adaptively penalized DA mechanism holds quasi incentive compatibility for buyers and succeeds in improving the profits for sellers in perishable goods markets.

6. Conclusions

We incrementally developed an online DA mechanism for the market of perishable goods to realize fair resource allocation among traders by considering the losses due to trade failures. We explained that sellers have a high risk of losing money by trade failures because their goods are perishable. To reduce trade failures in the perishable goods market, our DA mechanism encourages buyers' truthful bidding by penalizing buyers' failed bids without spoiling their individual rationality. The experimental results using multi-agent simulation showed that our DA mechanism was effective in promoting truthful behavior of traders for realizing the fair distribution of large utilities between sellers and buyers.

The experimental results in this study are very limited for any comprehensive conclusion on the design of online DA for perishable goods. For example, we assumed that buyers accept the penalty imposed by the mechanism as long as buyers' payment does not exceed their bid (thus, not violating their individual rationality). However, their tendency to accept the penalty might depend on possibility of their successful trades and potential profits from their resales. The influence of traders' sophisticated behavior in the market needs to be investigated further in future research. We need to investigate other types of bidding practices and test them under various experimental settings. In addition, human behavior in the market must be examined carefully to evaluate the designed mechanism [25].

In developing countries, there is strong demand for improving efficiency in agricultural markets [26] and this is also true for rural societies in developed countries, including Japan.

As a real-life application of our online DA mechanism, we have been operating a substantiative e-marketplace for trading oysters cultivated in the north-eastern coastal area of Japan (Miyagi prefecture) for five years. Only local traders used to participate in the markets run by the fisheries cooperatives because the highly perishable nature of fishery products prevents their trade from being open to wider participants. This has led to low incomes and the collapse of local fishery industries, which also suffered severe damage from a massive earthquake and tsunami in 2011. Although we do not have sufficient data to show the effectiveness of our online DA mechanism, we hope to contribute to promoting successful deployment of electronic markets for fisheries and improve the welfare of people in the area by attracting more traders online.

Acknowledgments: This work was supported by Japan Society for the Promotion of Science Grant-in-Aid for Scientific Research (C) Grant Number 00358128.

Conflicts of Interest: The author declares no conflict of interest.

References

1.   Shoham, Y.; Leyton-Brown, K. *Multiagent Systems: Algorithmic, Game-Theoretic and Logical Foundations*; Cambridge University Press: New York, NY, USA, 2009.
2.   Su, P.; Park, B. Auction-based highway reservation system an agent-based simulation study. *Transp. Res. Part C* **2015**, *60*, 211–226.
3.   García-Magariño, I.; Lacuesta, R. Agent-based simulation of real-estate transactions. *J. Comput. Sci.* **2017**, *21*, 60–76.
4.   Esmaeili Aliabadi, D.; Kaya, M.; Sahin, G. Competition, risk and learning in electricity markets: An agent-based simulation study. *Appl. Energy* **2017**, *195*, 1000–1011.
5.   Parsons, S.; Rodriguez-Aguilar, J.A.; Klein, M. Auctions and bidding. *ACM Comput. Surv.* **2011**, *43*, 1–59.
6.   Talluri, K.T.; van Ryzin, G.J. *The Theory and Practice of Revenue Management*; Kluwer Academic Publishers: Norwell, MA, USA, 2004.
7.   Kambil, A.; Van Heck, E. Reengineering the Dutch Flower Auctions: A Framework for Analyzing Exchange Organizations. *Inf. Syst. Res.* **1998**, *9*, 1–19.
8.   Bastian, C.; Menkhaus, D.; O'Neill, P.; Phillips, O. Supply and Demand Risks in Laboratory Forward and Spot Markets: Implications for Agriculture. *J. Agric. Appl. Econ.* **2000**, *32*, 159–173.
9.   Friedman, D.; Rust, J. *The Double Auction: Market Institutions, Theories, and Evidence*; Perseus Publishing: Cambridge, UK, 1993.
10. Viswanadham, N.; Chidananda, S.; Narahari, H.; Dayama, P. Mandi electronic exchange: Orchestrating Indian agricultural markets for maximizing social welfare. In Proceedings of the IEEE International Conference on Automation Science and Engineering, Seoul, Korea, 20–24 August 2012; pp. 992–997.
11. Devi, S.P.; Narahari, Y.; Viswanadham, N.; Kiran, S.V.; Manivannan, S. E-mandi implementation based on gale-shapely algorithm for perishable goods supply chain. In Proceedings of the IEEE International Conference on Automation Science and Engineering, Gothenburg, Sweden, 24–28 August 2015; pp. 1421–1426.
12. Cheng, M.; Xu, S.X.; Huang, G.Q. Truthful multi-unit multi-attribute double auctions for perishable supply chain trading. *Transp. Res. Part E* **2016**, *93*, 21–37.
13. Zhao, D.; Zhang, D.; Perrussel, L. Mechanism Design for Double Auctions with Temporal Constraints. In Proceedings of the IJCAI, Catalonia, Spain, 16–22 July 2011; pp. 472–477.
14. Blum, A.; Sandholm, T.; Zinkevich, M. Online algorithms for market clearing. *J. ACM* **2006**, *53*, 845–879.
15. Bredin, J.; Parkes, D.C.; Duong, Q. Chain: A dynamic double auction framework for matching patient agents. *J. Artif. Intell. Res.* **2007**, *30*, 133–179.
16. Gerding, E.; Stein, S.; Robu, V.; Zhao, D.; Jennings, N. Two-sided online markets for electric vehicle charging. In Proceedings of the AAMAS-2013, St. Paul, MN, USA, 6–10 May 2013; pp. 989–996.
17. Myerson, R.; Satterthwaite, M. Efficient Mechanisms for Bilateral Trading. *J. Econ. Theory* **1983**, *29*, 265–281.
18. Parkes, D.C. Online Mechanisms. In *Algorithmic Game Theory*; Nisan, N., Roughgarden, T., Tardos, E., Vazirani, V., Eds; Cambridge University Press: Cambridge, UK, 2007; Volume 53, pp. 411–439.

19. Miyashita, K. Online double auction mechanism for perishable goods. *Electron. Commer. Res. Appl.* **2014**, *13*, 355–367.

20. Jordan, P.; Schvartzman, L.; Wellman, M. Strategy exploration in empirical games. In Proceedings of the 9th International Conference on Autonomous Agents and Multiagent Systems, Toronto, ON, Canada, 10–14 May 2010; pp. 1131–1138.

21. Sandholm, T.W. Automated Mechanism Design: A New Application Area for Search Algorithms. In Proceedings of the International Conference on Principles and Practice of Constraint Programming (CP '03), Kinsale, Ireland, 29 September–3 October 2003; pp. 17–24.

22. Conitzer, V.; Sandholm, T. Incremental mechanism design. In Proceedings of the IJCAI, Hyderabad, India, 6–12 January 2007; pp. 1251–1256.

23. Satterthwaite, M.A.; Williams, S.R. Bilateral Trade with the Sealed Bid k-Double Auction: Existence and Efficiency. *J. Econ. Theory* **1989**, *48*, 107–133.

24. Walsh, W.; Das, R.; Tesauro, G.; Kephart, J. Analyzing complex strategic interactions in multi-agent systems. AAAI-02 Workshop on Game-Theoretic and Decision-Theoretic Agents, Edmonton, AB, Canada, 28 July–1 August 2002; pp. 109–118.

25. Krogmeier, J.; Menkhaus, D.; Phillips, O.; Schmitz, J. An experimental economics approach to analyzing price discovery in forward and spot markets. *J. Agric. Appl. Econ.* **1997**, *2*, 327–336.

26. Ssekibuule, R.; Quinn, J.A.; Leyton-Brown, K. A mobile market for agricultural trade in Uganda. In Proceedings of the 4th Annual Symposium on Computing for Development—ACM DEV-4 '13, Cape Town, South Africa, 6–7 December 2013; pp. 1–10.

applied sciences

MDPI

Article

ABS-SOCI: An Agent-Based Simulator of Student Sociograms

Iván García-Magariño [1,2,*], Andrés S. Lombas [3], Inmaculada Plaza [2,4] and Carlos Medrano [2,4]

1. Department of Computer Science and Engineering of Systems, University of Zaragoza, 44003 Teruel, Spain
2. Instituto de Investigación Sanitaria Aragón, University of Zaragoza, 50009 Zaragoza, Spain; inmap@unizar.es (I.P.); ctmedra@unizar.es (C.M.)
3. Department of Psychology and Sociology, University of Zaragoza, 44003 Teruel, Spain; slombas@unizar.es
4. Department of Electronics Engineering and Communications, University of Zaragoza, 44003 Teruel, Spain
* Correspondence: ivangmg@unizar.es; Tel.: +34-978645348

Received: 22 September 2017; Accepted: 27 October 2017; Published: 1 November 2017

Abstract: Sociograms can represent the social relations between students. Some kinds of sociograms are more suitable than others for achieving a high academic performance of students. However, for now, at the beginning of an educative period, it is not possible to know for sure how the sociogram of a group of students will be or evolve during a semester or an academic year. In this context, the current approach presents an Agent-Based Simulator (ABS) that predicts the sociogram of a group of students taking into consideration their psychological profiles, by evolving an initial sociogram through time. This simulator is referred to as ABS-SOCI (ABS for SOCIograms). For instance, this can be useful for organizing class groups for some subjects of engineering grades, anticipating additional learning assistance or testing some teaching strategies. As experimentation, ABS-SOCI has been executed 100 times for each one of four real scenarios. The results show that ABS-SOCI produces sociograms similar to the real ones considering certain sociometrics. This similarity has been corroborated by statistical binomial tests that check whether there are significant differences between the simulations and the real cases. This experimentation also includes cross-validation and an analysis of sensitivity. ABS-SOCI is free and open-source to (1) ensure the reproducibility of the experiments; (2) to allow practitioners to run simulations; and (3) to allow developers to adapt the simulator for different environments.

Keywords: agent-based simulation; agent-based social simulation; multi-agent system; agent-oriented software engineering; sociogram

1. Introduction

Instructors should take several aspects into account in order to obtain a high quality level of education customized for a particular group of students. One of the most relevant ones is the proper adaption of the lectures to the academic level of students. Another factor is to consider the social relations among the students, as these can condition their academic performance. For instance, isolated students may have a lower academic performance. Instructors can mitigate the isolation of certain students by making teamwork activities assuring that the potential isolated students are well integrated in their teams [1]. For achieving this purpose, it is essential that teachers have the appropriate tools for estimating the relations among students. This is the reason why a simulator of student sociograms can be helpful for teachers in order to personalize the education regarding the social statuses of students.

The current work presents an agent-based simulator (ABS) called ABS-SOCI for predicting sociograms of students classified by their activity roles.

ABS-SOCI assists teachers in ameliorating the isolation of students. It can also allow teachers to anticipate potential cases of bullied students. Notice that bullying has been considered as one of the most worrying problems in education [2], which, in the worst cases, can imply dramatic outcomes such as suicides.

ABS-SOCI has some advantages over the application of general-purpose ABS tools such as NetLogo [3] and Repast [4]. For instance, ABS-SOCI can directly use a customized representation of sociograms, so that ABS researchers do not need to deal with the implementation of their graphical representations. The graphical sociogram representation of ABS-SOCI includes different kinds of arrows (representing respectively selection and rejection relations) with different intensities (represented with different widths), and the individuals are represented in a circled layout as commonly done in sociograms [5]. In addition, ABS-SOCI simulates the evolution of social relations from a matrix of affinities between student activity roles, considering aspects such as the duration of the relations, their trends, and the common reciprocity. The implementation of these aspects are not straightforward in general-purpose ABS tools, to the best of the authors' knowledge.

The current research focuses on validating that the simulator forecasts sociograms are similar to the real ones. For this purpose, we have assessed the current approach by comparisons in (1) three different real scenarios taken from the literature and (2) another scenario from the experience of one of the current authors. ABS-SOCI has provided the base for the simulation of teaching strategies [6] and the clustering of students from their features for later simulating their social relations [7].

The next section introduces the related work. Section 3 introduces ABS-SOCI describing its theoretical bases and its underlying behavior. Section 4 describes the experiments conducted with ABS-SOCI, and Section 5 mentions the conclusions and future lines of research.

2. Background

Sociograms graphically represent the social relations among individuals. Sociograms can analyze several kinds of relations, such as selections and rejections. Selection relations can be obtained by asking for example "Who would you choose as a workmate?", while rejection relations can be obtained with questions such as "Who would you avoid to work with?" Sociograms [8] have proven to be especially useful for analyzing group dynamics, performing comparisons between groups, and studying the repercussion of the moderator (or instructor) technique.

Several works associated the analysis of social networks with the group performance. For instance, Grund [9] presented a detailed analysis of the different network structures and their relation with the team performance. The results of a league of soccer teams were used as experimental basis to measure the performance, and this was compared with the known measures of network structures such as network intensity (passes per minute) and network centralization. Furthermore, the work of Lin and Lai [10] not only related the network with the performance, but also increased the performance based on the network. In particular, this work presented a tool-supported technique for improving academic performance of students being aware of social networks through the analysis of sociograms. In online formative assessments, their system recommended students with low results to strengthen their social connections and academically collaborate with their closest peers. Their study showed that their system improved academic performance. However, none of these works allowed practitioners to simulate the evolution of these sociograms through time in a realistic way so that some possible sociogram structures could be forecasted given certain students with certain profiles.

There are different ways of analyzing sociograms. In particular, sociometrics are useful for performing this analysis. For instance, Sweet and Zheng [11] proposed a metric for measuring subgroup insularity or segregation, in order to quantify the separation between subgroups. In addition, Barrasa and Gil [12] proposed a software application that measured a wide range of sociometrics in sociograms. They proposed sociometrics that measured (a) different factors of the social status of each individual as well as (b) group properties such as cohesion and dissociation. The current approach used some of these metrics for measuring the simulation outcomes.

Several works have used sociograms in education. For example, Roberts [5] extracted sociograms from a group of students to analyze the influence of the teacher when encouraging student interaction.

She performed their analysis in the nursing education, and she analyzed the activity roles of student. She provided some hints about how to encourage student interaction from her experience. In addition, Dombrovskis et al. [13] used the sociograms for analyzing the interpersonal relations of young students in sport lessons. Their goal was to promote cooperation skills among children in order to evoke positive attitude towards learning and avoiding negative emotions in isolated ones. The current work shares some of the goals with these works, and its innovation is the proposal of a simulator tool for predicting sociograms of students beforehand.

Moreover, some works have studied the influence of grouping in education. For example, Yu et al. [14] performed a study in a large city in China, and showed how the low academic-level students increased their performance when they were included in heterogeneous academic-level groups. Furthermore, Chen et al. [15] applied a Genetic Algorithm (GA) to group partners based on the analysis of a social network, considering sociometrics for assessing the social statuses of students. They applied the GA to optimize the grouping satisfying certain constraints related to the diversity of grades, social status and genders. They compared the results of the GA with the solutions from Branch and Bound (B&B), which obtain optimal solutions. They showed that the GA worked properly as it frequently obtained optimal solutions or nearly optimal. These works provided alternative solutions for one of the applications of the sociogram simulations, which is the grouping of students based on their profiles. However, they did not provide solutions for other applications of sociogram simulations, such as the estimation about the need of teamwork activities.

There are some ABSs and agent-based models (ABMs) for simulating sociograms. For instance, Arentze et al. [16] presented an ABM for generating social networks. Their approach was based on the formation of friendships considering aspects such as geographic proximity and similarity between individuals. They used the random-utility-maximizing (RUM) framework, and they found that the simulations regularly obtained similar results to the real ones in the Swiss context. In addition, Hassan et al. [17] presented an ABM for simulating friendship relations by means of fuzzy logic. They modeled the mental states of people and simulated their changes of mind. They also simulated the relationships in social networks considering the similarity between individuals among other aspects. Furthermore, Dobson et al. [18] presented an interactive ABS with virtual reality for training teams. This simulator provided sociograms that represented the relations among agents. They simulated emergency situations related respectively with railways and nuclear power generators. Nevertheless, none of these works focused on the simulation of sociograms from the student roles about their activity in lectures and learning exercises.

Although most simulations of sociograms are performed with ABSs, there are also very few works that apply different approaches for this purpose. For example, McHardy and Vershinina [19] defined a simulation model inspired by micro-biology for simulating sociograms that represented networks of learners in industry. They simulated the transfer of knowledge based on the principles of autocatalysis, in order to explain the collective generation of promising ideas. Nevertheless, the current approach adheres to the major trend of using ABSs for simulating sociograms.

3. ABS-SOCI

ABS-SOCI is useful for estimating social structures from the student activity roles. This simulator can assist teachers in knowing which groups of students are more probable to have inconvenient social structures. In these cases, teachers can design collaborative learning activities for (a) mitigating the isolation of some students and (b) facilitating the social structures that imply a higher academic performance.

ABS-SOCI is able to simulate sociograms of student groups that consider the two relation kinds of selections and rejections, in which each relation can have a different intensity. Some inputs are the numbers of students associated with each activity role (quiet, participant, students that drive discussions to tangent topics, students that frequently make jokes, obstructive, and occasionally participant). Other main inputs are the number of lectures (represented as iterations) and the initial sociogram if known at the beginning of the course. ABS-SOCI simulates the evolution of the sociogram showing a visual animation. The main output of ABS-SOCI is the final state of the sociogram.

Researchers can calibrate the internal parameters of the simulator to obtain different kinds of simulation evolutions. They can change the internal parameters about (a) the coherence of relations in terms of reciprocity; (b) the frequency in which relations can change their trends; (c) the speed for evolving the intensity of selection and rejection relations; (d) the time from which the simulator starts reducing the alterations in order to achieve stability. Researchers can also calibrate the affinities and antipathies between every pair of student activity roles.

These internal parameters can be trained given certain known input conditions and certain real resulting sociograms following ATABS (a technique for Automatically Training ABSs) [20].

In a research data repository, our public dataset of ABS-SOCI [21] freely provides the executable file of this tool, its open source, its HTML documentations, and an example of initial sociogram. This section describes the different inner facets of this ABS. It also explains the main decisions and their advantages over other possible alternatives.

3.1. Applied Theoretical Bases

The interpersonal relationships depend on the student profiles according the previous findings in Ref. [22]. Hence, the current system simulates these relation establishments based on student profiles. In particular, the current work uses a classification of students inspired by the work of Roberts [5]. This classification has been selected as that work analyzed the corresponding student profiles in relation with the social relations between students. In addition, it was also useful to match the student profiles detected in other works about sociograms.

There are different grades of relationships, and, consequently, ABS-SOCI represents these grades. Specifically, the variations of the grades of relationships usually follow some trends as demonstrated by Leinhardt [23], although these trends vary sometimes. Hence, ABS-SOCI incorporates the concept of relation trends, considering positive, neutral or negative ones. The grades of relationships normally evolve following these trends with smooth transitions. The system simulates some eventual changes of trends.

The changes of relation trends depend on some aspects such as the durations of relationships. In particular, long-term relationships change less frequently their trends than the short-term relationships [24]. In the current simulator, the decisions of changing relation trends are simulated in a nondeterministic way considering the relation duration among other aspects.

The conformation of some relationships between the different profiles depends on the group size according to Ref. [5]. The presented system simulates this fact by using different probability values for simulating the social decisions regarding the group size.

The relationships are independently represented in the two directions between each pair of individuals. However, it is well known that a high proportion of relationships are reciprocal [25]. More specifically, Davis [26] proved that a person referred to as X is more likely to choose a person denoted as Y when Y chooses X. In order to properly simulate this fact, when an individual changes their relation trend towards another individual, the former corresponding agent communicates it to the latter. In this manner, the receiver agent decides whether to follow back this relation, it being possible to indirectly adjust the proportion of relations that are reciprocal.

3.2. User Interface of the Tool

Figure 1 shows the main frame of the user interface (UI) of ABS-SOCI. In this frame, users can introduce the input parameters of simulations. Each simulation can start either from (a) a neutral sociogram (i.e., with only neutral relationships) or from (b) an existing sociogram with specific selection, neutral and rejection relations. In the former case, users introduce the number of students of each role. In the latter case, users determine this sociogram with a text file (see an example in Figure 2).

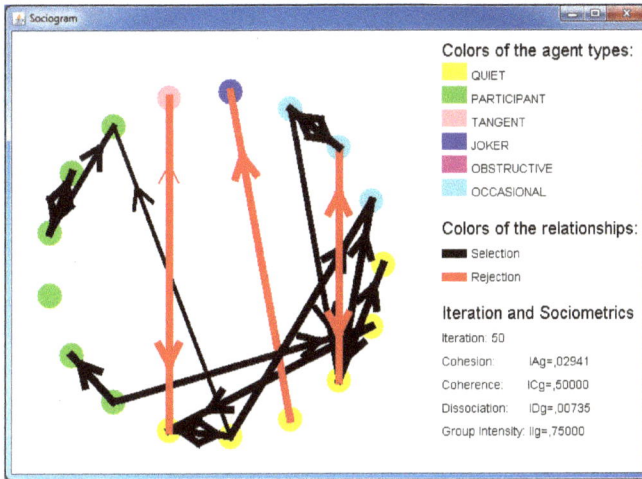

Figure 1. Frame for showing the graphical evolution of sociograms.

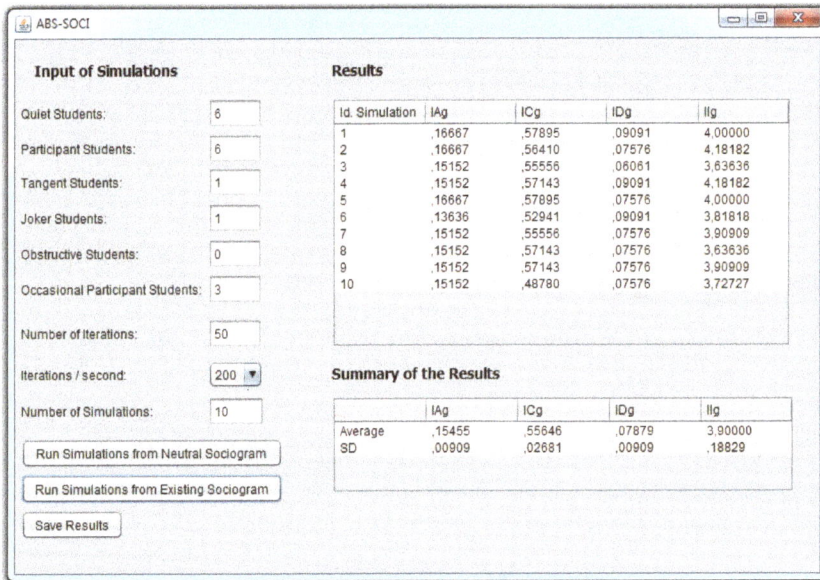

Figure 2. Main frame of the user interface (UI) of the Agent-Based Simulator of Student Sociograms (ABS-SOCI).

The number of iterations represents the number of teaching hours. Users can also select the animation speed (i.e., number of iterations per second), which does not have any repercussion in the actual simulation. Users can either observe all the steps and analyze these with a slow simulation, or go through the whole simulation quickly to obtain the final results. Users can repeat simulations with the same input parameters, by indicating the corresponding number of simulations.

Once a user has initiated the simulations, a new frame pops up as the one in Figure 3. For each simulation, this frame sequentially presents the evolution of the sociogram showing the results of all the iterations from the first one to the last one. The sociogram is presented by situating all the individuals in a circle, and connecting these with arrows when there is either a relationship of selection or rejection. This representation is common in the works about sociograms. The relationships have directions, and these directions are indicated with the corresponding arrowheads. The relationships have grades of intensity, and these are graphically represented with the thicknesses of the corresponding arrows.

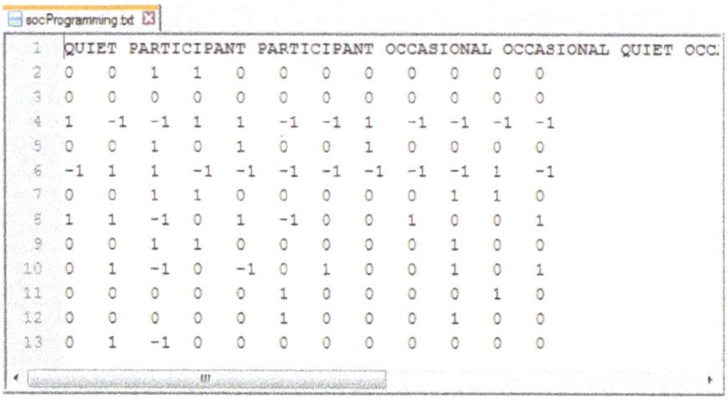

```
socProgramming.txt
1   QUIET PARTICIPANT PARTICIPANT OCCASIONAL OCCASIONAL QUIET OCC.
2   0    0   1   1   0   0   0   0   0   0   0   0
3   0    0   0   0   0   0   0   0   0   0   0   0
4   1   -1  -1   1   1  -1  -1   1  -1  -1  -1  -1
5   0    0   1   0   1   0   0   1   0   0   0   0
6  -1    1   1  -1  -1  -1  -1  -1  -1  -1   1  -1
7   0    0   1   1   0   0   0   0   1   1   0
8   1    1  -1   0   1  -1   0   0   1   0   0   1
9   0    0   1   1   0   0   0   0   1   0   0
10  0    1  -1   0  -1   0   1   0   0   1   0   1
11  0    0   0   0   0   1   0   0   0   0   1   0
12  0    0   0   0   0   1   0   0   0   1   0   0
13  0    1  -1   0   0   0   0   0   0   0   0   0
```

Figure 3. Initial sociogram represented with a text file.

In the right side, there is a legend for presenting some color notations. Each individual is represented as a small filled circle with the color that represents its student agent types. The selection and rejection relationships are respectively specified with black and red. The bottom right side of the frame indicates the number of the iteration and the measurement values of certain sociometrics.

The sociometric results of all the simulations are presented on the right side of the main frame with their averages and standard deviations (SD). The results can be saved into a text file by pressing the corresponding button.

The results of ABS-SOCI are measured with some of the sociometrics introduced by Barrasa and Gil [12] that are related to features of the group. The Index of Association of a group (IAg) measures the cohesion of the group. The Index of Dissociation of a group (IDg) measures the dislike between members of the group. The Index of Coherence of a group (ICg) measures the number of selections that are reciprocally divided by all the actual selections (either reciprocal or not). The Index of Intensity of a group (IIg) measures the grade of the emission of selections and rejections by the members of the group.

3.3. Architecture

The architecture of ABS-SOCI is presented using the Ingenias language [27], since this simulator has been modeled and developed as a multi-agent system (MAS). Figure 4 shows the main symbols of the Ingenias notation, so that readers can understand the next figures.

Agent   Role   Goal   Task   Frame Fact

Figure 4. Main symbols of the Ingenias notation.

The simulator was developed with the Process for Developing Efficient Agent-based Simulators (PEABS) [28] that uses the Ingenias language. Figure 5 defines the following agent roles of ABS-SOCI:

- *Student role*: This role impersonates the social behaviors of students. Its main goals are (1) to make friends; (2) to consider the affinities between the different types of students; (3) to follow the trends of their social relations in most cases; (4) to notice the duration of the relations for making it less probable that long-term relations change; (5) to maintain coherence in the ratio of relations that are reciprocal. The student activity roles have been obtained by analyzing the education literature [5] and selecting the following set of common student types based on the way of participating in lectures:

- – *Quiet Student agent*: This student is usually quiet and does not participate in group discussions during learning activities.
- – *Participant Student agent*: This student usually participates in class.
- – *Tangent Student agent*: This student usually drives the class discussions to tangent topics that are not relevant.
- – *Joker Student agent*: It usually makes jokes in class discussions.
- – *Obstructive Student agent*: It frequently mentions the disadvantages of the learning activities.
- – *Occasional Student agent*: This student occasionally participates in class.

- • *Sociogram Manager role*: This person is responsible for properly updating the sociometric status of the agents playing the student role. This role graphically presents the sociometric statuses with sociograms. It can save the relevant information into a text file.

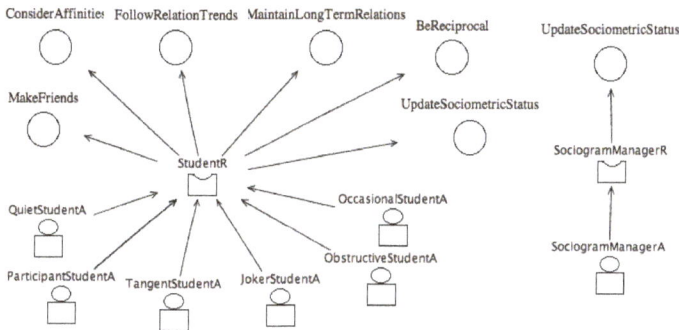

Figure 5. Definition of the roles and agents of ABS-SOCI.

3.4. Internal Functioning of the ABS

The presentation of the internal functioning of ABS-SOCI has been divided into (a) the introduction of the student agents and their data structures in Section 3.4.1; (b) the presentation of the sociogram representation in Section 3.4.2; and (c) the description of the evolution rules that are applied during the simulation in Section 3.4.3.

3.4.1. Student Agents

Some student types are more suitable with certain student types than others. For instance, participant students usually get on well with other participant students, and joker students (who like to make jokes in classes) usually get on with tangent students (who tend to derive the class to an irrelevant subject). However, other student types are frequently unsuitable with some student types. For example, participant students usually reject obstructive students.

Moreover, the size of the group influences the probability that an individual selects another as a friend. In general, for larger groups, the probability that two specific individuals are friends is usually less, as they have higher numbers of people they can choose as friends. However, some specific roles as jokers may be the opposite as two jokers may have more probabilities to be friends when they share more students in the group that listen to their jokes. For this reason, this simulator defines different size levels determined with specific intervals, and some matrices and constants are separately defined for each size level.

For representing all these scenarios, ABS-SOCI has the following vector and matrices:

- *"Size levels" vector*: It defines the interval for each size level.
- *"Suitable" matrix*: For each pair of student types *t1* and *t2* in a particular size level, it contains the probability that a student of *t1* type establishes a positive relation trend towards a student of *t2* type.
- *"Unsuitable" matrix*: For each pair of student types *t1* and *t2* in a specific size level, it contains the probability that a student of *t1* type establishes a negative relation trend towards a student of *t2* type.

The two matrices referred as Suitable and Unsuitable have the restriction that, for each pair of types *t1* and *t2* in a particular size level, the sum of the corresponding values in the two matrices must be less or equal to one. The probability of setting a neutral trend is implicitly obtained as the subtraction of Suitable and Unsuitable matrices values from one for each pair of type and size level.

3.4.2. Sociogram Representation

A sociogram is implemented similarly to a weighted and directed graph with an adjacency matrix. The weight between each pair of individuals *i1* and *i2* is a real number in the $[-1, 1]$ interval where -1 is the maximum level that *i1* can reject *i2* as a friend, and $+1$ is the maximum level that *i1* can select *i2* as a friend. As an alternative, we considered to have two different values for selection and rejection in two different adjacency matrices. However, this representation had the disadvantage that it would allow an individual *i1* to select *i2* as a friend and at the same time to reject *i2* as a friend, which is not consistent. For this reason, we decided to use the first representation with a unique value within the $[-1, 1]$ interval.

We used the thresholds of 0.5 for counting selections and -0.5 for rejections, when applying the sociometrics of Barrasa and Gil [12]. The other values were considered as neutral.

At the beginning of each simulation, the adjacency matrix can be either imported from a file or set to neutral values, depending on whether some individuals knew each before the simulation.

3.4.3. Simulation Evolution Rules

In each iteration, ABS-SOCI considers the following aspects to update the relationship of selection/rejection between each pair of students *i1* and *i2*:

- *Affinity of the type of i1 with the type of i2 in the particular size level.* This is represented within the student agent implementation with the Suitable and Unsuitable matrices.
- *The trend of the relationship.* For instance, if the value of a relationship was increasing, it is more likely that it continues increasing, and similarly for decreasing trends.
- *The duration of the relationship.* The long-term relationships between students are less probable to change their trend than short-term relationships.
- *The possibility of being a reciprocal relationship.* Most relationships are reciprocal. In other words, when an individual *i1* selects *i2* as a friend, *i2* usually selects back *i1* as a friend.

The simulation of sociogram evolutions use the following additional data structures:

- *"Trend" matrix*: This matrix is defined for determining the trends of the relationships for every pair of individuals. The values of this matrix are restricted to the $[-1, 1]$ interval. For instance, matrix $+1$ indicates that the value of the adjacency matrix of the corresponding pair of individuals is increasing, while -1 indicates the opposite. Zero indicates that the value of the adjacency matrix remains with the same value.
- *"Change" matrix*: This matrix determines the probabilities of changing trends between student types. In particular, each value of this matrix represents the probability that two individuals of the corresponding student types change their trend in an iteration.

The relationships between individuals are updated in each iteration in two phases: (1) updating the trends; (2) updating the relationships according to the trends.

The algorithm for updating the trends is represented with the dataflow diagram of Figure 6. This diagram uses "less than" comparisons. This diagram mainly receives input from the current student agent and another one. It also uses the matrices and constants previously defined. In this dataflow, the previous trend of the relationship is considered. If any of the first two conditional bifurcations goes to the negative branch, then the trend is not changed, and consequently the relationship continues with the same trend (either decreasing, neutral or increasing). More concretely, the first bifurcation considers the probability of changing the trend of the corresponding agent types.

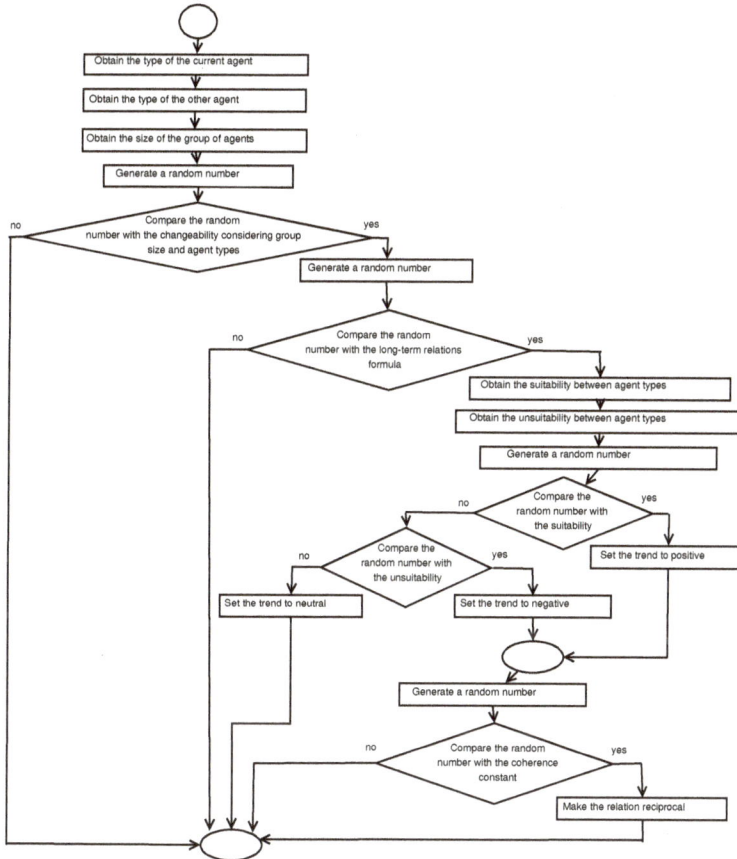

Figure 6. Dataflow diagram for updating the trend.

The second bifurcation is included to reduce the probability that long-term relationships change in comparison to short-term relationships. For this purpose, ABS-SOCI uses the relative position of the iteration in the simulation, represented with the 'iteration" variable that goes from one to the total number of iterations. The probability of changing the relation trend is determined with Equation (1). The probability is lower in later iterations, since it uses an inverse function of the iteration variable:

$$p = min(1, CST\_TREND * \frac{1}{iteration}).$$ (1)

The *CST_TREND* constant is applied to properly weight the effect of distinguishing different relation durations. This constant determines from which iteration the probability starts decreasing. The min function makes the outputted probability be one when the iteration position is lower than this constant. For example, assuming a common value of this constant like ten, for the first ten iterations, the probability of changing the trend according to relation duration will be the same (i.e., one). From this iteration forward, the change probability decreases.

Several alternatives were compared before choosing Equation (1) for considering the duration of relations. For instance, an alternative was to decrease the probability of changing trends in a linear way such as the one indicated in Equation (2):

$$p = max(0, 1 - iteration * CST_1).$$ (2)

CST_1 represents an alternative constant that modulates the impact of simulation durations on the simulations. However, its values should be much lower than one, and can be obtained by dividing one by the number of iterations that is considered a long relation. This equation was discarded because the probability of change would become absolutely zero in a certain number of iterations. This does not represent the reality and did not satisfy our goals.

The goal was to reduce the probability, and not to make the change impossible with a probability of zero.

Going back to the dataflow of Figure 6, in the cases that go through the previous bifurcations in the positive branches, the trend is reestablished considering probabilities of the suitability and unsuitability matrices following the formulas for nondeterministic decisions of TABSAOND (the Technique for developing ABS Apps and Online tools with Nondeterministic Decisions) [29].

If the trend is established to a non-zero value, the trend can become a reciprocal relationship (either positive or negative). In these cases, the $CST\_COHERENCE$ constant is used as the probability for simulating the establishment of reciprocal relations.

It is worth mentioning that both constants CST_TREND and $CST\_COHERENCE$ are actually vectors, since these have different values for each level of group size.

Once the phase about trend selection has finished, the relation between the two corresponding individuals is updated according to the trend following the dataflow of Figure 7. The variation of the relationship value is calculated as a portion of the distance to the approaching limit, and this portion is represented with the CST_INC value. CST_INC is set to 0.15 since this value allows the relationship values to surpass the selection and rejection thresholds from the initial zero value in a few iterations (i.e., five iterations), or go from a selection to a rejection or vice versa in a non-excessive number of iterations (i.e., seven to nine iterations). These numbers of iterations allow smooth transitions and can take place in simulations of common subject durations (i.e., from 20 to 90 class hours).

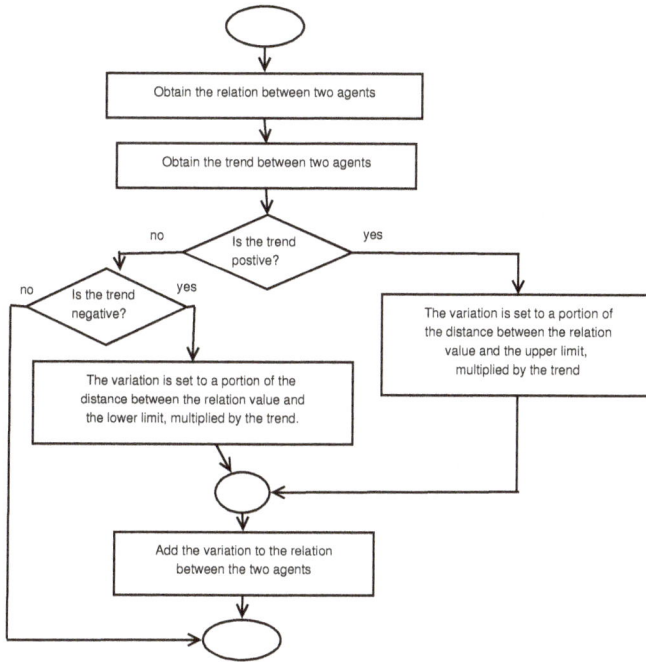

Figure 7. Dataflow diagram for updating the relationships according to their trends.

4. Experimentation

The current approach has been evaluated with several scenarios related to groups of students. These scenarios were extracted from both (a) external works reported in [5,12,30] and (b) an experiment of one of the current authors in a Programming Course in the University of Zaragoza. The simulations have been tested starting from both (1) initial sociograms with only neutral relations (in the scenarios from the external works); and (2) an initial existing sociogram where people already knew each other and there were different kinds of relations (in the Programming Course scenario). The current approach has been evaluated in both online education [30] and face-to-face education (in the other scenarios).

The goal of the scenarios extracted from the literature was to assess the prediction capability of the ABS-SOCI tool. In these scenarios, ABS-SOCI could have been used for estimating potential students in risk of isolation and the possible group cohesion in order to reinforce the courses with collaborative activities. In the experiment of the Programming Course, ABS-SOCI was actually used to estimate the cohesion of the group and the social structures represented as sociograms. The teacher added some collaborative exercises in his teaching schedule in order to reinforce collaboration among students and ameliorate the potential isolation cases.

Firstly, ABS-SOCI has been calibrated with two real sociogram cases, which are the ones presented in the work of Barrasa and Gil [12] and the work of Roberts [5]. These sociograms are different in terms of sample size, and some constants and matrix values have been adjusted to obtain similar results. This calibration of constants and matrix values has been performed following the White Box Calibration method proposed by Fehler et al. [31]. According to this calibration method, the current approach determines their main influence of these internal values in the simulation results. In particular, this approach considers the direct influence of the Suitable matrix on (1) the selection relations between individuals of certain types and on (2) the cohesion sociometric in general. It also considers the opposite influence of the Unsuitable matrix in these two aspects. It takes into account the influence of the Change matrix and the CST_TREND value in the evolution of sociometrics for a specific number of iterations. Finally, it adjusts the *CST_COHERENCE* value considering its repercussion in the coherence sociometric. This calibration phase can also be considered as a training phase.

Table 1 determines these internal constant values depending on the sample size after this calibration phase. As an example, Figure 8 visually shows the values of Suitable and Unsuitable matrices for the large groups (i.e., size ≥9) after the calibration phase. All of these values were the results of the calibration phase, and were used as the setting of the simulator for running the next simulations.

Table 1. Values of constants after the calibration process

	Size ≤8	Size ≥9
CST_TREND	10.0	10.0
CST_COHERENCE	0.50	0.33

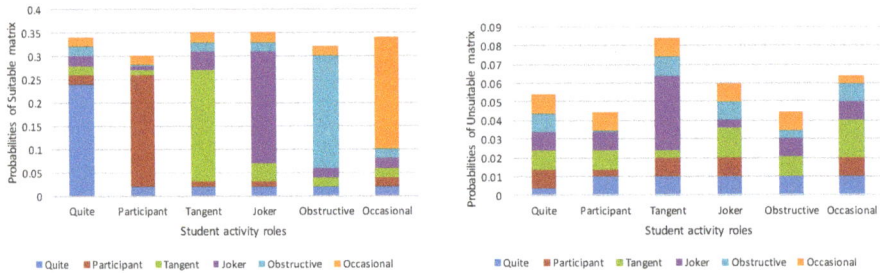

Figure 8. Charts about Suitable (in the left) and Unsuitable (in the right) matrices for large groups (i.e., size ≥9)

After the calibration, ABS-SOCI was run and validated with the new scenario introduced by Dawson [30], besides the aforementioned works. There were 100 executions for each case for obtaining a representative set of results.

Table 2 presents the input parameters of the scenarios of these experiments and the remaining ones. The number of iterations was 50 representing 50 h of teaching activities in all of the scenarios except one. The number of iterations from the first three scenarios correspond to the common number of teaching hours of these subjects in the corresponding academic level and countries reported by the original sources. In particular, the subjects belonged to either higher education or university education, from either Spain or United Kingdom.

Table 2. Input parameters of the simulator for the different scenarios.

	Barrasa and Gil [12]	Roberts [5]	Dawson [30]	Programming Course
Initial Sociogram	Neutral	Neutral	Neutral	Existing
Total Number of Students	6	15	24	12
Quiet Students	1	6	0	3
Participant Students	1	6	24	3
Tangent Students	1	1	0	0
Joker Students	1	1	0	1
Obstructive Students	1	0	0	0
Occasional Students	1	3	0	5
Number of Iterations	50	50	50	20

For each simulation, the IAg and ICg sociometric indexes were estimated. The other metrics (IDg and IIg) are not presented as the analyzed works did not include information about rejections, which are necessary for calculating these metrics. To determine if the estimations of IAg and ICg sociometrics provided by ABS-SOCI were statistically different from those obtained by the published works, a binomial test was carried out with a nominal significance level of 0.05 because these sociometrics represent proportions. Reported alpha levels were adjusted using the Bonferroni procedure to control for family-wise error rate, considering all simulations within the same index a family of tests.

In order to obtain an effect size for simulations, the difference of arcsine transformation of proportions was calculated, as it is described by Cohen [32]. This effect size was interpreted according to the mentioned work. That is, values of 0.2, 0.5 and 0.8 were interpreted as small, medium and large, respectively.

Since a non-significant effect could be attributed to a lack of statistical power, a sensitivity analysis was performed using G*Power 3 [33]. This analysis calculated the minimum effect size to which the test was able to detect with a power $(1-\beta)$ of 0.80, a significance level (α) of 0.05, and the sample size (n) of the used scenarios. The software offers a Hedges' g effect size, which was also interpreted according to the guidelines offered by Cohen [32]. Specifically, its interpretation is similar to the effect size based on the arcsine transformation of difference of proportions.

To summarize the results obtained on the analysis, the current work calculated the percentages of non-significant simulations. It also calculated the percentages of simulations with small, medium and large effect sizes. Table 3 presents all these percentages, along with the true value, the mean estimated value, standard deviation of estimated values, and Hedges' g obtained in the sensitivity analysis.

Table 3. Real and simulated values, percentages of non-significant simulations and of simulations with small, medium and large effect size, and Hedges' g obtained in sensitivity analysis for the Index of Association of a group (IAg) and the Index of Coherence of a group (ICg) in calibration and validation studies. Note: n = sample size of calibration and validation studies.

Index	n	Real Value	Mean of Simulated Values (SD)	Non-Significant Simulations Based on Binomial Test (%)	Hedges' g Obtained in Sensitivity Analysis	Effect Size of Simulation		
						Small (%)	Medium (%)	Large (%)
			First experimental phase *Calibration with Barrasa and Gil [12]*					
IAg	6	0.2667	0.2353 (0.1099)	100	0.73	57	33	10
ICg	6	0.6667	0.6346 (0.1872)	100	0.33	44	38	18
			Calibration with Roberts [5]					
IAg	17	0.0368	0.0196 (0.0101)	100	0.39	82	18	0
ICg	17	0.5556	0.4840 (0.2168)	87	0.43	35	38	27
			Validation with Dawson [30]					
IAg	24	0.0688	0.0372 (0.0114)	100	0.33	84	16	0
ICg	24	0.6441	0.5016 (0.1055)	93	0.35	29	61	10
			Second experimental phase *Calibration with Barrasa and Gil [12]*					
IAg	6	0.2667	0.2273 (0.1003)	100	0.73	58	28	14
ICg	6	0.6667	0.6009 (0.1818)	96	0.33	65	28	7
			Calibration with Dawson [30]					
IAg	24	0.0688	0.0705 (0.0157)	100	0.33	99	1	0
ICg	24	0.6441	0.5882 (0.0725)	99	0.35	77	22	1
			Validation with Roberts [5]					
IAg	17	0.0368	0.0276 (0.0126)	90	0.39	90	10	0
ICg	17	0.5556	0.5585 (0.1691)	99	0.43	49	33	18
			Third experimental phase (with the same calibration as in the previous phase) *Validation with the Programming Course*					
IAg	12	0.1667	0.1461 (0.0118)	100	0.41	99	1	0
ICg	12	0.6111	0.5539 (0.0324)	99	0.32	96	4	0

As an illustrative example of the temporal evolution of the sociometric status, Figure 9 contains the temporal evolution of IAg and ICg sociometrics of one of the executions for simulating the scenario of Barrasa and Gil. Figure 10 shows the temporal evolution of the sociometrics for a simulation in the scenario of Roberts. Figure 11 introduces the temporal evolution in the validation for the case of Dawson. As one can observe, the sociometrics relatively stabilized around certain values after some iterations.

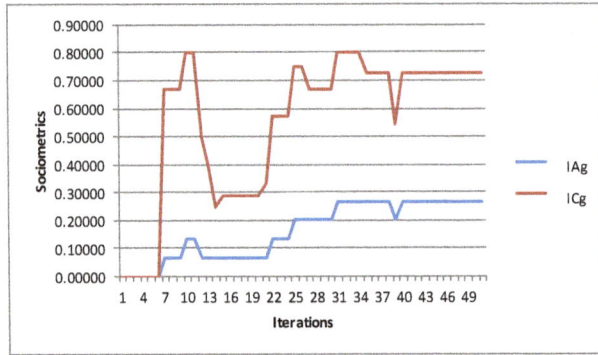

Figure 9. Temporal evolution of the Index of Association of a group (IAg) and the Index of Coherence of a group (ICg) sociometrics in a simulation execution for the scenario of Barrasa and Gil [12].

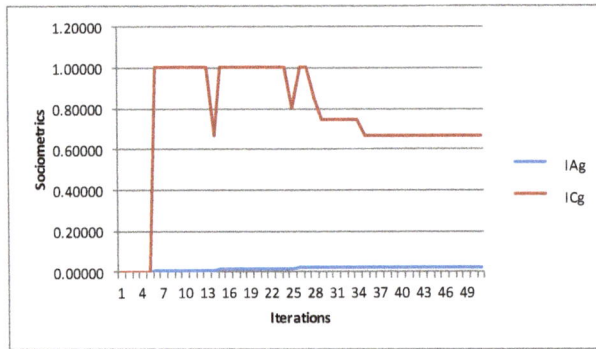

Figure 10. Temporal evolution of IAg and ICg sociometrics in a simulation execution for the scenario of Roberts [5].

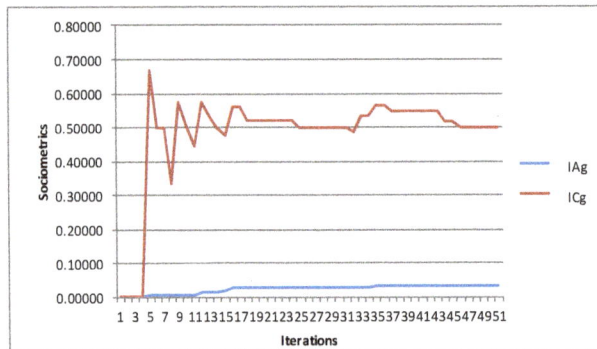

Figure 11. Temporal evolution of IAg and ICg sociometrics in a simulation execution for the scenario of Dawson [30].

In order to obtain a second validation, calibration and validation cases were interchanged. In this second phase of experimentation, ABS-SOCI was calibrated with the case of Barrasa and Gil and the case of Dawson, changing its constant values and matrices. Then, ABS-SOCI was executed and compared with the case of Roberts, which was not used in the calibration of this second experimental phase. The case of Roberts was selected for this validation instead of Barrasa and Gil because the results were more representative due to its larger number of students. In addition, the current work discarded the possibility of performing a validation with the scenario of Barrasa and Gil due to its small sample size, which yields a low sensitivity as shown in Table 3. Figure 12 shows the temporal evolution of sociometrics in one of the simulation executions in the validation of Roberts' scenario.

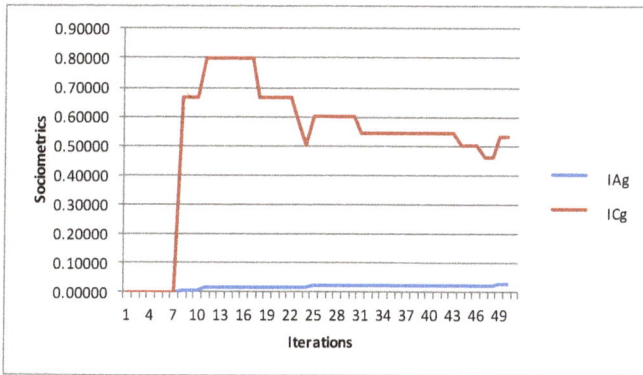

Figure 12. Temporal evolution of sociometrics in the scenario of Roberts [5] in the second validation experimentation, in which this scenario was not used in the training process.

ABS-SOCI was further assessed in a third experimental phase. The main difference of this experimentation from the previous ones is that the simulator departed from an existing sociogram that is not neutral. In particular, the initial sociogram had a cohesion of $IAg = 0.1364$ and a coherence of $ICg = 0.5454$. The initial sociogram was extracted from a group of students of a Programming Course in the University of Zaragoza. This experimentation used the same calibration as in the previous experimental phase. The simulated outcomes were compared with the real outcomes after 20 class hours (represented with 20 simulation iterations).

Results of experimentation showed that the difference between the real value and the mean estimated values for all simulations ranged from 0.001 to 0.14. In addition to that, percentage of non-significant simulations, on both calibration and validation phases, were high (ranging from 87% to 100%). The sensitivity of analysis on IAg was low in the scenario of Barrasa and Gil, since the minimum effect size able to detect was large, probably due to the small sample size of this work. However, the sensitivity analysis was adequate in (1) the remaining indexes and scenarios of the calibration phases and (2) in the validation phases. This is so because the Hedges' gs obtained on sensitivity analyses were small or medium. In addition, estimations obtained by ABS-SOCI appear to be relatively efficient, since, in both calibration and validation phases, a high percentage (ranging from 73% to 100%) of simulations had small or medium effect sizes. Thus, although it cannot be discarded, the possibility that the high percentage of non-significant analyses on the work of Barrasa and Gil was due to lack of statistical power, the high percentage of simulations with small or medium effect sizes provided evidence of a good performance. The performance was also good on the validation phases. This is supported by a high percentage of non-significant simulations, which cannot be attributed to a lack of statistical power, and a high percentage of simulations with small or medium effect sizes. The good performance found on validation phases was especially relevant because it discarded the possibility of over-fitting by performing simulations on a different sample in which the calibration was conducted. Furthermore, the experiments showed that the simulator obtained a high performance departing from both (a) neutral sociograms representing groups of students that did not know each other, and (b) an existing sociogram with different kinds of relations representing a group of students who already knew each other.

5. Conclusions

ABS-SOCI is a novel ABS that simulates the temporal evolution of sociograms of students starting from either a neutral state or an existing sociogram. The user has to determine the number of students in each category (quiet, participant, etc.) or to load the initial sociogram. ABS-SOCI considers the different affinities between types of students, the trends of the relationships among each pair of individuals in each specific time frame, and the duration of the relationships distinguishing between short-term and long-term ones. It also takes into account different levels of group sizes, and the coherence in relation to reciprocal relationships. The behaviors of the individual agents impersonating students rely on certain algorithms, constants and matrices. The emergent behavior has been tested with four real scenarios taken from either the literature or the experience of the authors. There were three experimental phases, in each of which the system was calibrated with some cases and was validated with a different one. In all of the experimental phases, the simulations of ABS-SOCI produced sociograms similar to the real ones according to the binomial tests. The sensitivity of the validations of these phases was adequate according to the Hedges' g and the guidelines of Cohen.

As future work, it is planned to improve the current version of ABS-SOCI by evaluating an enlarged dataset of sociograms. This dataset will include new sociograms from higher education in the grades of (a) computer science engineering; (b) electronics and automation engineering; and (c) psychology. In the same group of students, the sociograms will be extracted in different stages of the academic course, in order to assess whether the simulated evolution is similar to the actual one considering more states than only the final one.

Moreover, the simulator can be extended to consider more specific features of students such as (1) their gender; (2) whether they need a job to pay their studies; (3) their missing knowledge; (4) their religious or sexual orientation; and (5) other psychological profiles. The simulator will sequentially incorporate these features, experiencing and analyzing the repercussion of each one independently. In this way, the future research will determine the utility of incorporating each feature in the current approach.

ABS-SOCI is also planned to be extended considering groups of students with a high ratio of international students. This simulator extension will represent information such as the mother-tongue language of each student and the language of the lessons. It will also distinguish between national and international students.

Acknowledgments: This work acknowledges the research project "Desarrollo Colaborativo de Soluciones AAL" with reference TIN2014-57028-R funded by the Spanish Ministry of Economy and Competitiveness. This work has been supported by the program "Estancias de movilidad en el extranjero José Castillejo para jóvenes doctores" funded by the Spanish Ministry of Education, Culture and Sport with reference CAS17/00005. We also acknowledge support from "Universidad de Zaragoza", "Fundación Bancaria Ibercaja" and "Fundación CAI" in the "Programa Ibercaja-CAI de Estancias de Investigación" with reference IT24/16. It has also been supported by "Organismo Autónomo Programas Educativos Europeos" with reference 2013-1-CZ1-GRU06-14277. We also aknowledge support from project "Sensores vestibles y tecnología móvil como apoyo en la formación y práctica de mindfulness: prototipo previo aplicado a bienestar" funded by University of Zaragoza with grant number UZ2017-TEC-02. Furthermore, we acknowledge the "Fondo Social Europeo" and the "Departamento de Tecnología y Universidad del Gobierno de Aragón" for their joint support with grant number Ref-T81.

Author Contributions: All of the authors conceived and designed the experiments; all the authors proposed ideas for developing the simulator; Iván García-Magariño programmed the simulator; all the authors contributed in the analysis and interpretation of the data; and all the authors collaboratively wrote the paper coordinated by Iván García-Magariño. In addition, Inmaculada Plaza especially contributed in the definition of students' behaviors. Andrés S. Lombas specially contributed in the statistical analysis and the inclusion of the theory and metrics about sociograms. Carlos Medrano specially contributed in the design of the experiments in several calibration-validation phases. Iván García-Magariño specially contributed in the agent-based modeling of the system and its implementation.

Conflicts of Interest: The authors declare no conflict of interest.

References

1. Santos, S.J.D.; Hardman, C.M.; Barros, S.S.H.; Santos, C.D.F.B.F.; Barros, M.V.G.D. Association between physical activity, participation in Physical Education classes, and social isolation in adolescents. *J. Pediatr.* **2015**, *91*, 543–550.
2. Nikolaou, D. Do anti-bullying policies deter in-school bullying victimization? *Int. Rev. Law Econ.* **2017**, *50*, 1–6.
3. Thiele, J.C.; Grimm, V. NetLogo meets R: Linking agent-based models with a toolbox for their analysis. *Environ. Model. Softw.* **2010**, *25*, 972–974.
4. Cicirelli, F.; Furfaro, A.; Giordano, A.; Nigro, L. HLA_ACTOR_REPAST: An approach to distributing RePast models for high-performance simulations. *Simul. Model. Pract. Theory* **2011**, *19*, 283–300.
5. Roberts, S. Using practitioner research to investigate the role of the teacher in encouraging student interaction within group work. *Nurse Educ. Today* **2008**, *28*, 85–92.
6. García-Magariño, I.; Plaza, I. FTS-SOCI: An agent-based framework for simulating teaching strategies with evolutions of sociograms. *Simul. Model. Pract. Theory* **2015**, *57*, 161–178.
7. García-Magariño, I.; Medrano, C.; Lombas, A.S.; Barrasa, A. A hybrid approach with agent-based simulation and clustering for sociograms. *Inf. Sci.* **2016**, *345*, 81–95.
8. Drahota, A.; Dewey, A. The sociogram: A useful tool in the analysis of focus groups. *Nurs. Res.* **2008**, *57*, 293–297.
9. Grund, T.U. Network structure and team performance: The case of English Premier League soccer teams. *Soc. Netw.* **2012**, *34*, 682–690.
10. Lin, J.W.; Lai, Y.C. Online formative assessments with social network awareness. *Comput. Educ.* **2013**, *66*, 40–53.
11. Sweet, T.M.; Zheng, Q. A mixed membership model-based measure for subgroup integration in social networks. *Soc. Netw.* **2017**, *48*, 169–180.
12. Barrasa, A.; Gil, F. A software application for the calculus and representation of sociometric indexes and values (In Spanish Un programa informático para el cálculo y la representación de índices y valores sociométricos). *Psicothema* **2004**, *16*, 329–335.
13. Dombrovskis, V.; Guseva, S.; Capulis, S. Cooperation and Learning Effectiveness of First Graders during Sports Lessons. *Procedia Soc. Behav. Sci.* **2014**, *112*, 124–132.
14. Yu, Z.; Dongsheng, C.; Wen, W. The heterogeneous effects of ability grouping on national college entrance exam performance—Evidence from a large city in China. *Int. J. Educ. Dev.* **2014**, *39*, 80–91.
15. Chen, R.C.; Chen, S.Y.; Fan, J.Y.; Chen, Y.T. Grouping Partners for Cooperative Learning Using Genetic Algorithm and Social Network Analysis. *Procedia Eng.* **2012**, *29*, 3888–3893.
16. Arentze, T.A.; Kowald, M.; Axhausen, K.W. An agent-based random-utility-maximization model to generate social networks with transitivity in geographic space. *Soc. Netw.* **2013**, *35*, 451–459.
17. Hassan, S.; Salgado, M.; Pavón, J. Friendship dynamics: Modelling social relationships through a fuzzy agent-based simulation. *Discret. Dyn. Nat. Soc.* **2011**, *2*, 118–124.
18. Dobson, M.; Pengelly, M.; Sime, J.A.; Albaladejo, S.; Garcia, E.; Gonzales, F.; Maseda, J. Situated learning with co-operative agent simulations in team training. *Comput. Hum. Behav.* **2001**, *17*, 547–573.
19. McHardy, P.; Vershinina, N.A. The role of autocatalysis in learner's networks. *Int. J. Manag. Educ.* **2014**, *12*, 271–282.
20. García-Magariño, I.; Palacios-Navarro, G. ATABS: A technique for automatically training agent-based simulators. *Simul. Model. Pract. Theory* **2016**, *66*, 174–192.
21. García-Magariño, I.; Lombas, A.S.; Plaza, I.; Medrano, C. *Source Code of the Agent-Based Simulator of Student Sociograms Called ABS-SOCI*, version 1; Mendeley Data; Mendeley: London, UK, 2017. Available online: http://dx.doi.org/10.17632/ffy73gfkzc.1 (accessed on 20 September 2017).
22. DeGenaro, J.J. Condensing the Student Profile. *Acad. Ther.* **1988**, *23*, 293–296.

23. Leinhardt, S. Developmental change in the sentiment structure of children's groups. *Am. Sociol. Rev.* **1972**, *37*, 202–212.

24. Goldstone, S.; Boardman, W.K.; Lhamon, W.T.; Fason, F.L.; Jernigan, C. Sociometric status and apparent duration. *J. Soc. Psychol.* **1963**, *61*, 303–310.

25. Engel, A.; Coll, C.; Bustos, A. Distributed Teaching Presence and communicative patterns in asynchronous learning: Name versus reply networks. *Comput. Educ.* **2013**, *60*, 184–196.

26. Davis, J.A. Sociometric triads as multi-variate systems. *J. Math. Soc.* **1977**, *5*, 41–59.

27. Gomez-Sanz, J.J.; Fuentes, R.; Pavón, J.; García-Magariño, I. INGENIAS development kit: A visual multi-agent system development environment. In Proceedings of the 7th International Joint Conference on Autonomous Agents and Multiagent Systems: Demo Papers, Estoril, Portugal, 12–16 May 2008; International Foundation for Autonomous Agents and Multiagent Systems: Richland, DC, USA, 2008; pp. 1675–1676.

28. García-Magariño, I.; Gómez-Rodríguez, A.; González-Moreno, J.C.; Palacios-Navarro, G. PEABS: A process for developing efficient agent-based simulators. *Eng. Appl. Artif. Intell.* **2015**, *46*, 104–112.

29. García-Magariño, I.; Palacios-Navarro, G.; Lacuesta, R. TABSAOND: A technique for developing agent-based simulation apps and online tools with nondeterministic decisions. *Simul. Model. Pract. Theory* **2017**, *77*, 84–107.

30. Dawson, S. 'Seeing' the learning community: An exploration of the development of a resource for monitoring online student networking. *Br. J. Educ. Technol.* **2010**, *41*, 736–752.

31. Fehler, M.; Klügl, F.; Puppe, F. Techniques for analysis and calibration of multi-agent simulations. In *Engineering Societies in the Agents World V*; Lecture Notes Computer Science; Springer: Berlin, Germany, 2005; Volume 3451, pp. 305–321.

32. Cohen, J. *Statistical Power Analysis for the Behavioral Sciences*, 2nd ed.; Lawrence Erlbaum Associates: Hillsdale, NJ, USA, 1988.

33. Faul, F.; Erdfelder, E.; Lang, A.G.; Buchner, A. G* Power 3: A flexible statistical power analysis program for the social, behavioral, and biomedical sciences. *Behav. Res. Methods* **2007**, *39*, 175–191.

applied
sciences

MDPI

Article

Agent-Based Model for Automaticity Management of Traffic Flows across the Network

Karina Raya-Díaz [1,*], **Carelia Gaxiola-Pacheco** [1], **Manuel Castañón-Puga** [1], **Luis E. Palafox** [1],
Juan R. Castro [1] **and Dora-Luz Flores** [2]

[1]   Facultad de Ciencias Químicas e Ingeniería, Universidad Autónoma de Baja California,
      Calzada Universidad 14418, Tijuana 22390, Mexico; cgaxiola@uabc.edu.mx (C.G.-P.);
      puga@uabc.edu.mx (M.C.-P.); lepalafox@uabc.edu.mx (L.E.-P.); jrcastror@uabc.edu.mx (J.R.C.)
[2]   Facultad de Ingeniería, Arquitectura y Diseño, Universidad Autónoma de Baja California,
      Carretera Transpeninsular Ensenada, Tijuana 3917, Colonia Playitas, Ensenada, Baja California, Mexico;
      dflores@uabc.edu.mx
*   Correspondence: karina.raya@uabc.edu.mx; Tel.: +52-664-979-7500 (ext. 54303)

Received: 1 July 2017; Accepted: 4 September 2017; Published: 9 September 2017

Abstract: This paper presents an agent-based model that performs the management of traffic flows in a network with the purpose of observing in a simulation of distinctive congestion scenarios how the automation of the monitoring task improves the network performance. The model implements a decision-making algorithm to determine the path that the data flows will follow to reach their destination, according to the results of the negotiation between the agents. In addition, we explain how the behavior of the network is affected by its topology. The aim of this paper is to propose an agent-based model that simplifies the management of the traffic flows in a communications network towards the automaticity of the system.

Keywords: agent-based modeling; complex network; multi-agent system; network management

1. Introduction

The communication networks are complex systems that support a wide variety of applications, many of them in real time; this has led to a change in the network management scheme. Network administrators must deploy multiple tools to obtain the visibility and control necessary to operate today's complex networks, some of these tools are: software agents, active networks and policy languages [1].

The integration of intelligent agents into the management system facilitates the evaluation of the parameters that determine the network performance. An agent has the capacity to detect changes in the environment, and respond in accordance, in order to isolate or anticipate failures making the system proactive; thus, the direct intervention of the network administrator becomes unnecessary in the situations that the agents can handle.

Agent-based modeling is a powerful tool for complex system simulation. A computer network can be described as a complex system because it is dynamic, has nonlinear interactions, and its components exhibit unpredictable reactions that results in behavior patterns [2,3]. In the proposed model, when a micro behavior, as a congestion, emerges in a network node, the agent needs to make a decision according to its preferences.

The topology of a network contributes valuable information to analyze its structure. Network managing considering the complex networks approach allows the identification and classification of the nodes using the metrics of centrality and clustering degree, among others.

Autonomic network management is a complex task and consists in the automaticity of the system to protect, configure, optimize and heal itself without human intervention. In networks, the property of automaticity allows self-management, given a set of policies from the administrator, which depicts the agent's high-level goals [4]. The proposed model applies a distributed management to avoid attacks over the central nodes in the network.

This paper is structured as follows: in Section 2, autonomic networking management is defined, in Section 3, network metrics based on graph theory principles are presented. Next, in Section 4, we introduce the architecture of the agent-based model and the algorithm proposed is explained. In Section 5, the simulation experiment and the results obtained are presented. Finally, concluding remarks are given in Section 6.

2. Autonomic Networking Management

Today, people are using networks to communicate, and networks have become bigger and more complex. The network management is a complicated task that goes beyond technical support and device configuration. To simplify the network management, the automation of the management tasks are critical. The International Business Machines Corporation (IBM) has proposed five levels according to the degree of automaticity in the network management [4,5].

- Level 1: Refers to the continuous monitoring of the elements of the system by means of software tools handled by people who provide manual support in the event of a failure.
- Level 2: Consists of a set of monitoring applications that intelligently collect information to reduce the workload of the network administrator, and is known as the management level.
- Level 3: This level can recognize certain behavior patterns of the network in addition to suggesting actions to be taken by the support staff; this level is also called predictive.
- Level 4: Uses level 3 tools but attempts to minimize human intervention by helping with Service Level Agreements (SLAs); this level is known as adaptive.
- Level 5: Business policies and objectives dynamically manage the system at this level. Its disadvantage is that the automaticity is very closed only based on certain rules, and this level is recognized as autonomous.

The proposed agent-based model is a step forward to reach the fifth level of automaticity, by the integration of intelligent agents that determine the path that will follow the data flows when a congested node situation emerges. The rules used for these agents rely on the preferences of each network node.

3. Measures of Centrality in Networks

A network is a set of nodes and links [6]. Modeling a network helps to determine the structural issues relevant at a given time. The topology of the network can be illustrated applying an adjacency matrix or adjacency list. The decision of which one has to be used depends on the analysis type.

An adjacency matrix is a mathematical representation of a network, which allows the analysis of the clustering relation between the nodes. When search algorithms are used to enumerate components, the disadvantage of this matrix is the inefficient use of memory, but this is not a problem when a particular element is accessed [7,8]. Different techniques for network interpretation and visualization are used nowadays; one of these techniques is based on clustering algorithms, in which the use of an adjacency matrix is more convenient.

One of the measures that is applied to adjacency matrix is the centrality, which captures the hierarchy of the network nodes [9]. The awareness of the node relevance is crucial to manage any network because this will lead to the determination of the main nodes. The nodes are classified by their betweenness centrality. Eigenvalue centrality is an extension of degree centrality [8], and the topology of the network has an effect on both measures. The use of a node classification based on the centrality allows the network manager to realize the importance of the nodes. A definition of eigenvalue centrality of a node i is shown in Equation (1):

$$x'_i = \sum_j A_{ij} x_j, \tag{1}$$

where x'_i is defined by the sum of the $i's$ the centrality of i neighbors, and A_{ij} is an element of the adjacency matrix. The interpretation of eigenvalues tells if a node is growing or shrinking according to the amount of neighbors. In the model, this measure is used to identify the hierarchy of each node. Links allow communication to reach a remote node. Here, the betweenness centrality of a node has a relevant role. Clustering coefficient C_i describes how extended a node is, Equation (2) shows how to calculate it, where Equation (3) is the degree of the node $k(i)$ and $N(i)$ is the amount of neighbors of the node i [2,10]:

$$C_i = \frac{k_i(k_i - 1)}{2}, \tag{2}$$

$$k(i) = |N(i)|. \tag{3}$$

In a heterogeneous network, the nodes with higher clustering degree are called hubs. In [11], it is affirmed that high degree nodes play a very significant role because the packets are routed through these hubs, and, in a social network, the people highly connected distribute information better than isolated people [12].

Figure 1 shows a representation of hub nodes in a network where the size of the nodes is proportional to its degree, defined as $k(i) = |N(i)|$, where $N(i)$ is the amount of the neighbors of a node i [10], using this metric, a node hierarchy is obtained. The proposed agent-based model detects clusters of nodes when the flow preferences are assigned.

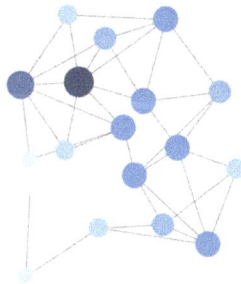

Figure 1. Example of network with representation of nodes by degree.

4. Architecture of the Agent-Based Model

An agent-based model is useful to analyze an approach for decision-making under conditions of deep uncertainty because it has the ability to connect heterogeneous micro and macro behaviors [13]. In a network model, the micro behaviors represent the interactions of different types of flows (data, video and voice), each one gives information for understanding the network behavior. The flow analysis is critical to the network availability and performance [14]. Because of this, the proposed agent-based model incorporates into the nodes a set of collaborative agents to collect the information requested by the decision-making level.

4.1. Multi-Agent Architecture for Management of Network Flows

The architecture of the proposed multi-agent model is shown in Figure 2. The Multi-Agent Architecture for Management of Network Flows, TELEKA, is organized by hierarchy using three levels: Network, Control and Negotiation.

- Network: In this level, all nodes are set up with the type of flows that will transmit and their destination. In addition, the agents of each node are initializing with its hierarchy obtained by Equation (1), and these preferences are assigned by the network administrator depending on their policies, and their utilities are set to zero. The agents collect information related to congestion status by sensing the current state of the nodes. When a congestion is recognized by exceeding a threshold, an *ECG* flag changed their value to one, and this will trigger the activation of the negotiation level.
- Control: Here, the agents detect and classify the flows that arrive at each node. After this, the agents provide the data to the Negotiation level, which holds the decision-making module.
- Negotiation: The algorithm SEHA (Social Election with Hidden Authorities) is triggered in this level when a congestion situation emerges. As a result, a set of actions to be performed by the lower-level agents is selected, achieving with this the optimization of the network status.

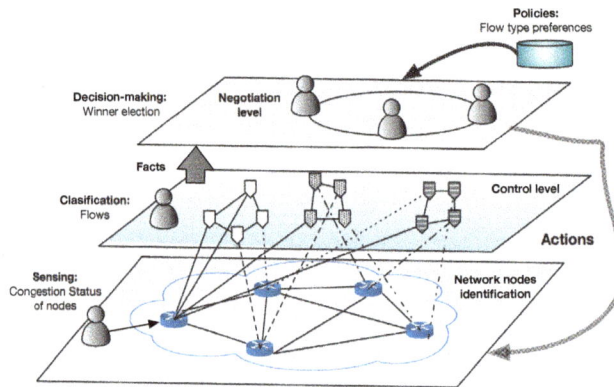

Figure 2. Multi-Agent Architecture for Management of Network Flows.

4.2. SEHA Social Election with Hidden Authorities Algorithm

According to the social choice theory, the group decision-making process in a multi-agent system should consider that each agent will have their preferences and will vote to choose a winner [15]. The proposed model considers a set of agents A = $\{Ag_1, ..., Ag_n\}$ in a network environment composed by n number of nodes and k links with different costs c.

The tuple that represents an agent is $Ag_i = < \Omega, \beta, h >$, where Ω is a set of preferences' relations $\Omega = \{\omega_1 \succ \omega_2 \succ \omega_3\}$, they are arranged according to the flow type to transmit and its priority β. Both parameters Ω and β are configured by the network administrator during the initialization process. The hierarchy of a node h is calculated by the centrality of each node among the network using the Equation (1).

The proposed algorithm is activated when a congestion emerges in a node, as is described in Algorithm 1. The SEHA algorithm returns a winner ω flow type (data, video or voice) after a voting process between the neighbor's nodes. The winner flow will have the higher priority in the queue of the congested node. The next step is transmitting the flow through one of its neighboring nodes using the shortest path with minimum cost c to its destination, the agent of the selected node obtains an utility u of +1. The rest of the non-preferred flows in the queue of the congested node picks randomly a neighboring node to continue their paths, but the utility u of the agent is set to -1 because they are not transmitting the preferred flows.

Algorithm 1 SEHA Social Election with Hidden Authorities Algorithm

---

> **function** FINDWINNER(Ω, β, h)
> > **set** *choices* [ω with max β] of *my_neighbors*
> > **set** *choices* modes of *choices*
> > **if** choices length > 1
> > > **set** prefered_choice [ω with max β] of *my_neighbors* with max h
> >
> > **else**
> > > **set** *prefered_choice choices*
> >
> > **return** *prefered_choice*
>
> **end function**

---

5. Simulated Scenarios and Results

The proposed model analyses the behavior of a network when a congestion situation emerges. In particular, this model aims to enable the implementation of SEHA (Social Election with Hidden Authorities Algorithm) algorithm to automatically manage the traffic flows across the network. The model validates the TELEKA architecture, which was illustrated in Figure 2.

Figure 3 presents the user interface in Netlogo of the model, and also shows the nodes' parameters as number, degree, centrality, queues threshold of congestion, and topology of the network. The environment where the agents coexist is a network in which the interconnection of the nodes and links are generated by scale-free distribution [16–18].

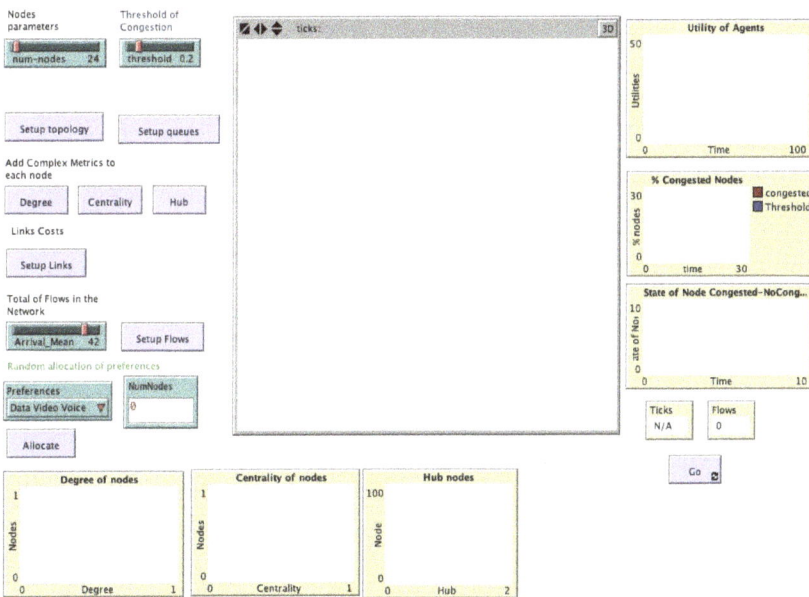

Figure 3. Netlogo interface of the model.

The experiment has the initialization values listed in Table 1, the hierarchy h, which corresponds to the values of centrality of each node, are calculated using the Equation (1), and the values of the preferences ω of each node are listed. Link costs are listed in Table A1 in Appendix A.

Table 1. Initialization parameters in the network.

Node	h	Preferences			Node	h	Preferences		
		ω_1	ω_2	ω_3			ω_1	ω_2	ω_3
0	0.34	Voice	Video	Data	12	0.15	Data	Voice	Video
1	1.00	Voice	Video	Data	13	0.079	Video	Data	Voice
2	0.604	Data	Voice	Video	14	0.069	Voice	Data	Video
3	0.533	Data	Video	Voice	15	0.272	Video	Data	Voice
4	0.181	Data	Voice	Video	16	0.34	Data	Video	Voice
5	0.232	Data	Voice	Video	17	0.104	Data	Video	Voice
6	0.34	Data	Voice	Video	18	0.34	Video	Voice	Data
7	0.442	Voice	Video	Data	19	0.092	Voice	Data	Video
8	0.272	Video	Voice	Data	20	0.104	Video	Data	Voice
9	0.204	Data	Video	Voice	21	0.035	Video	Data	Voice
10	0.15	Voice	Data	Video	22	0.035	Voice	Video	Data
11	0.181	Voice	Data	Video	23	0.092	Data	Video	Voice

In Figure 4, the color of each node represents its preferences (see Table A2 in Appendix A). The sizes of the nodes are proportional to their degree k. Finally, the dots inside of each node represents the flows waiting to be transmitted to their destination nodes.

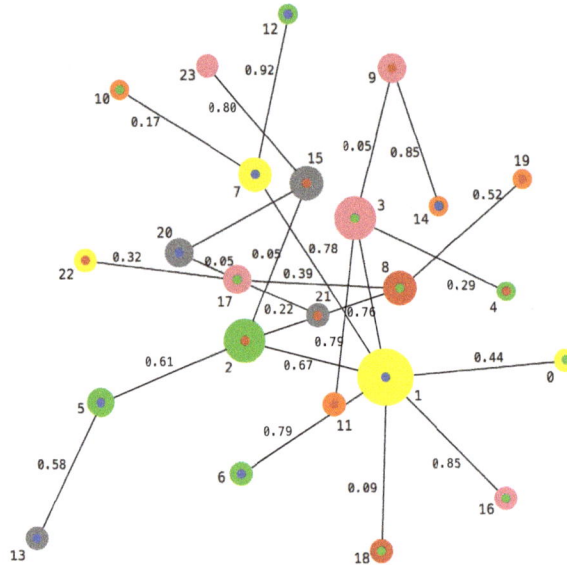

Figure 4. Topology with scale free distribution.

In Netlogo, the next process to activate after the setup of the world is called *Go*, and this is illustrated in Figure 5. The flow diagram represents the actions that are executed within a tick.

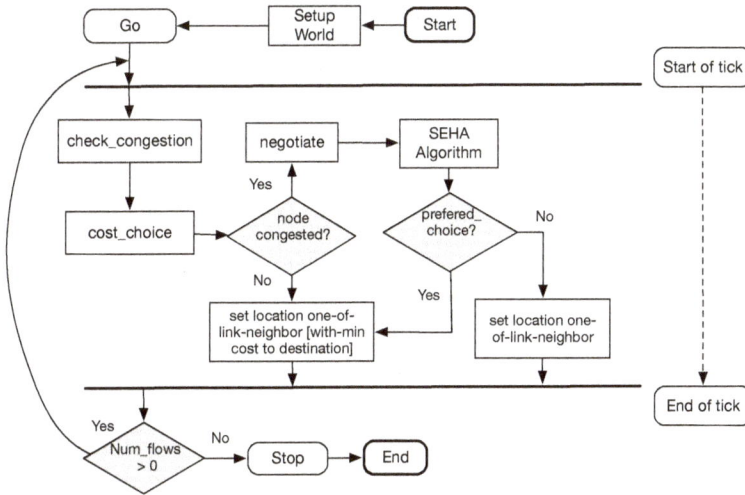

Figure 5. Flowchart of Go process.

5.1. Analysis of Results

Four different scenarios were simulated to analyze the results of the Netlogo model. Two topologies were used: the first topology considers a scale-free distribution and the second one considers a grade distribution. With each topology, the Algorithm 1 was applied, creating the first two scenarios, and it was not applied creating the two remaining scenarios.

5.2. Simulation Using the Topology with Scale Free Distribution

Figure 6 illustrates the histograms obtained from the measures of centrality and the classification of all nodes as hubs or not hubs. The histogram of degree allows the identification of the nodes with the higher degrees as those which are important to consider when a congestion starts.

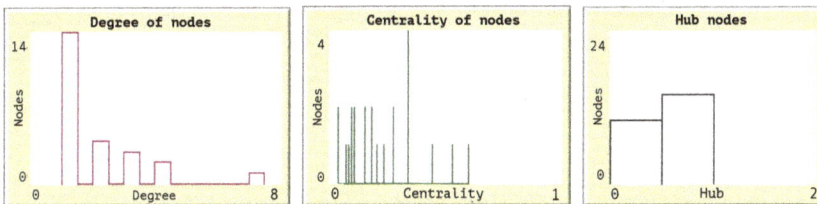

Figure 6. Histogram of complex metrics obtained from the model.

The network during the simulation maintains a maximum congestion limit of 30%. The congestion threshold of each node queue is set to 20%. This means that a value above this will change the status of the node to congested. Figure 7 shows the congestion status of the network when the SEHA Algorithm 1 was applied and Figure 8 shows the congestion status of the network without using a SEHA Algorithm 1.

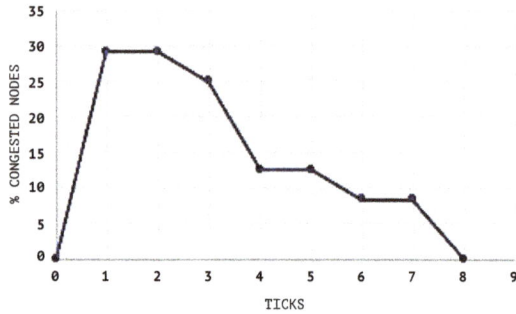

Figure 7. Congestion status of the network over the simulation applying the SEHA (Social Election with Hidden Authorities Algorithm).

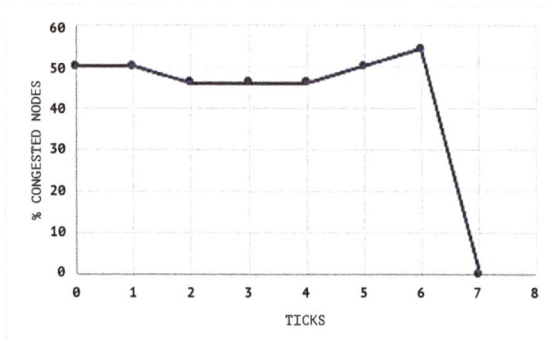

Figure 8. Congestion status of the network in simulation using Random distribution.

In Figures 9 and 10, the congestion averages by node are shown, and it can be noticed that the congested nodes amount are higher than those in the scenarios where SEHA Algorithm 1 was applied.

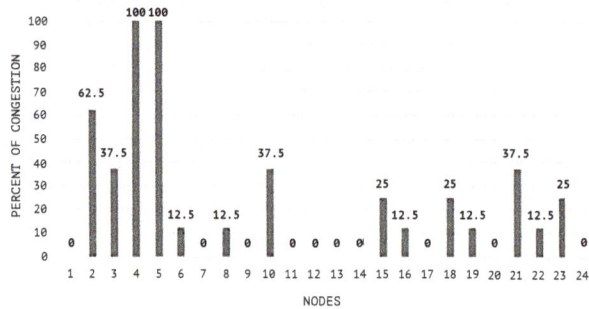

Figure 9. Congestion average by node during the simulation using the SEHA (Social Election with Hidden Authorities Algorithm).

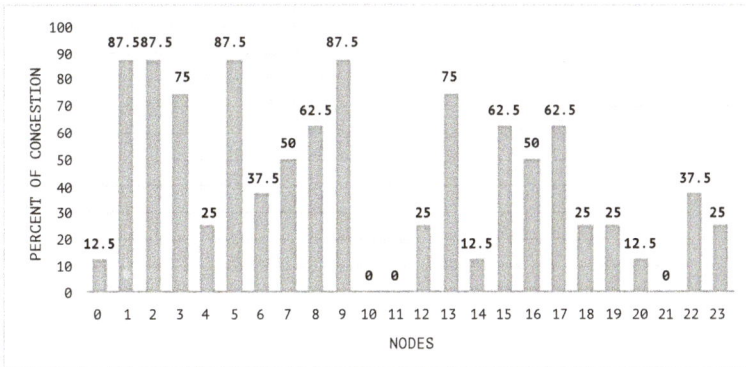

Figure 10. Congestion average by node during the simulation with Random distribution.

After each agent negotiates, they obtain a utility reward, and Figure 11 shows the utility of all agents during the simulation time using the SEHA algorithm. During the simulation with a random flow distribution, the agents receive no utility. Both models complete the distribution of 42 flows through the network to its destination node at the eighth tick.

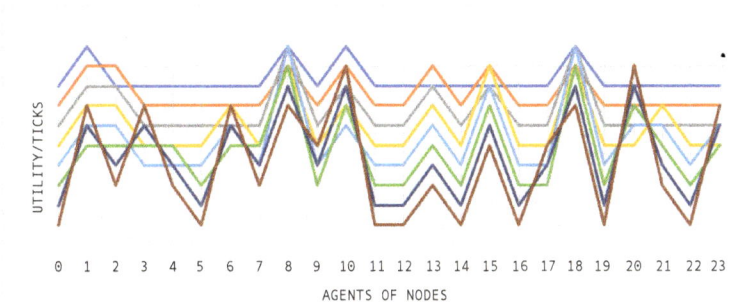

Figure 11. Utilities of each agent when the SEHA (Social Election with Hidden Authorities Algorithm) is applied.

The values obtained to generate this graph are shown in Table A3 of Appendix A.

5.3. Simulation Using Topology with Degree Distribution

The following simulation was configured with different topology, using the degree of six as a parameter to generate it. Figure 12 shows the view of the network in the simulator.

Figure 13 shows the complex metrics of degree, centrality and a classification of nodes in hubs or no hubs obtained by the model when the simulation was initialized.

The results obtained by the second simulation are illustrated in Figures 14 and 15. Compared with the results shown in Figure 14, the higher congested status of the network was less than 26% using the SEHA Algorithm 1.

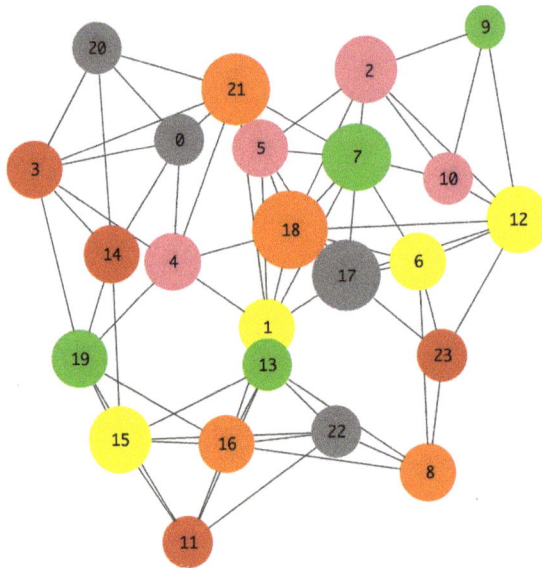

Figure 12. Topology with degree distribution each color represents the type of preferences.

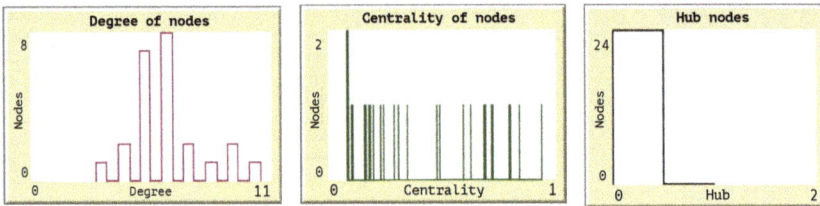

Figure 13. Histograms of the complex metrics obtained.

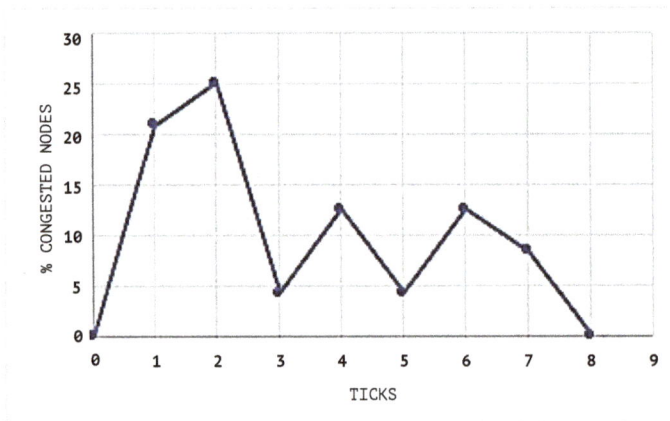

Figure 14. Average of congestion status of the network in the simulation applying the SEHA (Social Election with Hidden Authorities Algorithm).

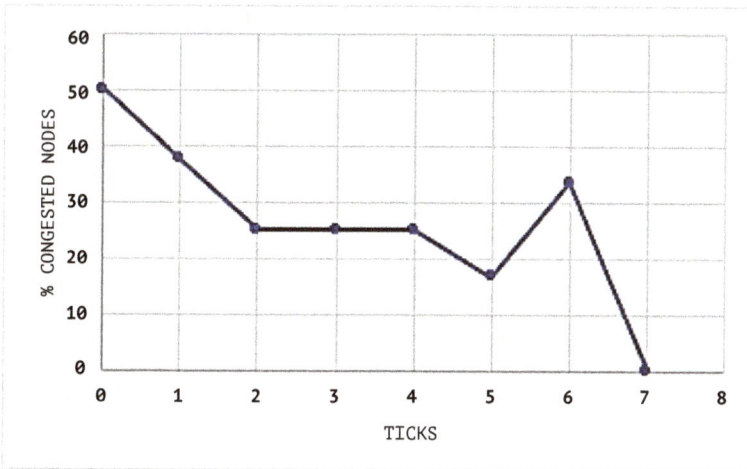

Figure 15. Average of congestion in the network during simulation using random distribution.

Detailed information of the congestion average by node during both simulations is shown in Figures 16 and 17.

According to the illustrations in Figures 16 and 17, the impact of congestion on the nodes of the network is lower using the SEHA (Social Election with Hidden Authorities Algorithm) than when a random distribution is used.

Finally, in Figure 18, the utilities of agents during the simulation are shown. The utilities obtained by the agents represent the most visited nodes when a congestion emerges using the SEHA Algorithm. Random distribution does not obtain utilities because there is no negotiation between agents to determine the new hop of the flow. For better description of the result values, see Table A4 in Appendix A.

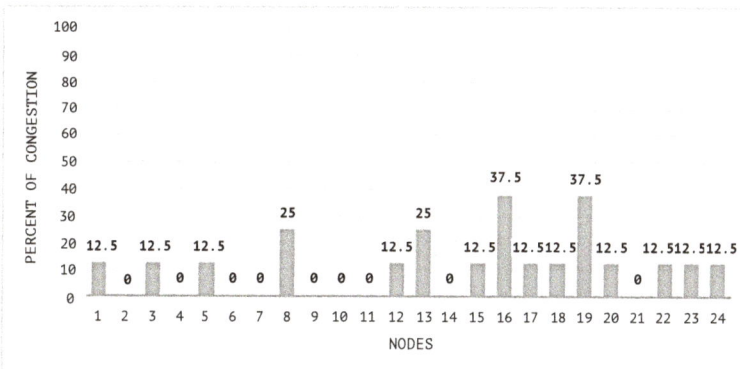

Figure 16. Congestion average by node during the simulation using the SEHA (Social Election with Hidden Authorities Algorithm).

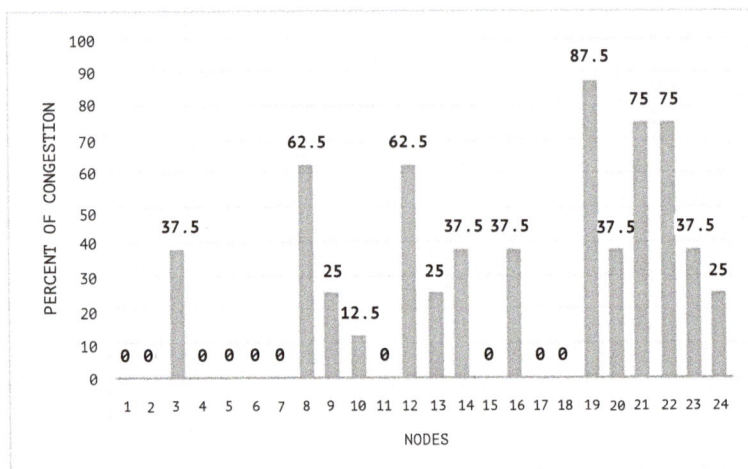

Figure 17. Congestion average by node during the simulation with Random distribution.

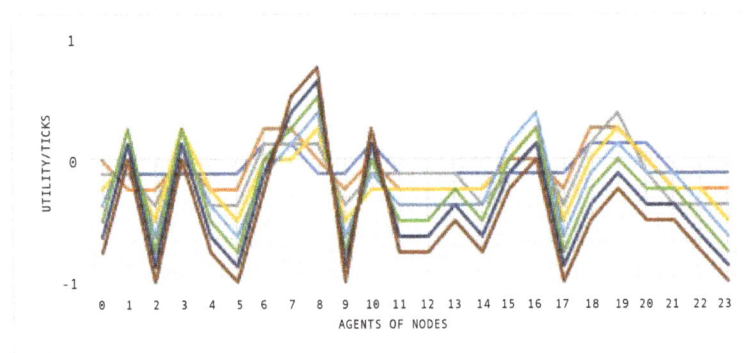

Figure 18. Utilities of each agent when applying the SEHA (Social Election with Hidden Authorities Algorithm).

5.4. Comparison of Results

In Table 2, the outcome of 600 simulations are shown. The scenarios that were used for these simulations were obtained by means of the link cost variation and a random initial distribution of the flows. The comparison was made between the scenarios where the SEHA (Social Election with Hidden Authorities Algorithm) was applied, in contrast with those in which the flows distribution was performed in a random manner.

As described 300 additional simulations were generated to verify the behavior of the SEHA (Social Election with Hidden Authorities Algorithm), varying the topology, node preferences and link costs. In addition, 300 simulations were generated by configuring BehaviorSpace following the same procedure, but the flow distribution was performed in a random manner, instead of using the SEHA (Social Election with Hidden Authorities Algorithm). The outcome using NetLogo Behavior Space with the initial values specified in Figure 19 is illustrated in Table 2.

The use of agent-based models to evaluate the performance of an algorithm is a suitable tool. The results shows that, although the variation in the costs greatly influences the random distribution, the affectation is minor when the SEHA (Social Election with Hidden Authorities Algorithm) is applied.

In Figure 20, the behavior of both simulated scenarios are shown, where it is observed that the outcome variation is smaller when the SEHA (Social Election with Hidden Authorities Algorithm) algorithm is applied than when it is not.

Figure 19. Initialization of Behavior Space of Netlogo.

Table 2. Results obtained after 600 simulations with Behavior Space of NetLogo.

Number of Runs	Flows Mean Arrival Rate	Cost	Social Election with Hidden Authorities Algorithm		Random Distribution	
			Mean of Congested Nodes	Standard Deviation	Mean of Congested Nodes	Standard Deviation
100	42	0.3	2.583	0.447	2.726	0.550
100	42	0.6	2.584	0.402	2.612	0.531
100	42	1.0	2.576	0.446	2.664	0.554

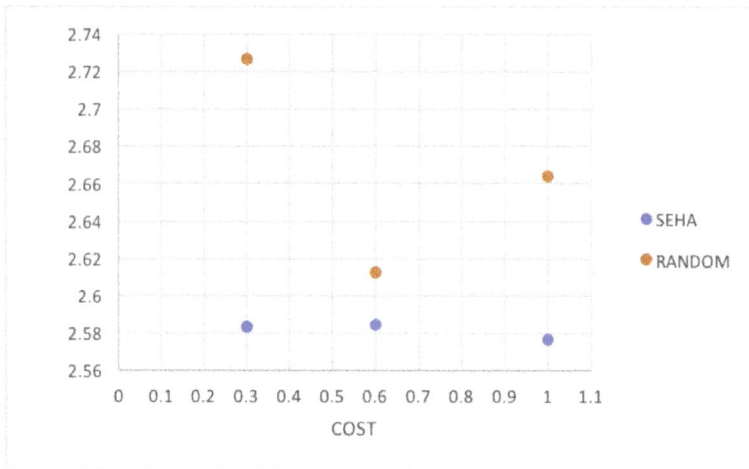

Figure 20. Comparation of performance using the SEHA (Social Election with Hidden Authorities Algorithm) and Random distribution of the flows.

6. Future Work

The integration of an agent that allows the automatic adaptation of network policies would bring us closer to the goal of automating the network-management tasks. This algorithm can be integrated into systems that require self-organization considering the connectivity between its components.

The proposed algorithm can be applied in a social network, where it is possible to identify those members who have a high hierarchy given their number of connections, in order to prioritize the messages that come from high hierarchy members.

7. Conclusions

The agent-based model allows the analysis of the results obtained to identify the patterns of behavior of a network. In addition, the model provides information about how the congestion of a network impacts the transmission of flows when classifying the nodes by their centrality.

Using the model to obtain the measures of complex networks of degree clustering and centrality by considering the network topology is a way to identify the hierarchy of each node. The identification of the nodes with greater centrality is one of the critical parameters of the SEHA Algorithm 1, which allows establishing a tiebreaker and a better flow distribution.

Flow transmission in the network are optimized according to the preferences of the community to which the congested node belongs, which exempts the network administrator from resolving the congestion failure by selecting manually the type of flow that must be transmitted according to its priority.

Considering that complex network metrics in a self-management system improves the network performance by decreasing congestion, this algorithm mainly considers the nodes with greater hierarchy in the network to transmit its flows because they have higher priority. The integration of intelligent algorithms that allows the autonomous negotiation of the resources improves the network performance.

Acknowledgments: This work was financially supported by CONACYT (Consejo Nacional de Ciencia y Tecnología) under the MYDCI (Mestría y Doctorado en Ciencias e Ingeniería) program at the Autonomous University of Baja California by the contract #101358.

Author Contributions: The corresponding authors Karina Raya-Díaz and Carelia Gaxiola-Pacheco proposed the research and drafted the manuscript. Karina Raya-Díaz was responsible for the development of the agent-based model. Manuel Castañón-Puga and Juan R. Castro gave helpful advice on the paper preparation and contributed towards experimental evaluation. Luis E. Palafox were involved in the network analysis and Dora-Luz Flores gave recommendations for the policies rules.

Conflicts of Interest: The authors declare no conflict of interest.

Appendix A

The cost of all links are random values, in the range of zero to one, and are listed in Table A1.

The preference values of each node are shown in Table A2, and these values are referenced in the Figure 4 of the Simulated Experiment and Results section. In Table A2, the flow type are Data = 0, Voice = 1 and Video = 2, and the order has an impact on the flow priority. The maximum size of queue in each node is set to 10 and the threshold set in the simulation are 20%. This means that, over this value, the bit of congestion ECG will be set to one, so a negotiation starts.

Table A4 provides the list of the agents' utilities in the second simulation using the SEHA algorithm and a network topology based on a degree distribution.

Table A1. Bidirectional links' costs.

Cost	Link	
	End1	End2
0.442	node 1	node 0
0.759	node 1	node 3
0.788	node 1	node 6
0.778	node 1	node 7
0.848	node 1	node 16
0.668	node 2	node 1
0.293	node 4	node 3
0.609	node 5	node 2
0.577	node 5	node 13
0.165	node 7	node 10
0.217	node 8	node 2
0.392	node 8	node 17
0.517	node 8	node 19
0.045	node 9	node 3
0.850	node 9	node 14
0.790	node 11	node 3
0.924	node 12	node 7
0.053	node 15	node 2
0.322	node 17	node 22
0.089	node 18	node 1
0.678	node 20	node 15
0.364	node 21	node 20
0.801	node 23	node 15

Table A2. Initialization values of nodes.

Source Node	Color	Preferences	Num. Flows	Type Flows	Destination Node
0	yellow	1 2 0	1	1	21
1	yellow	1 2 0	4	1,0,2,0	10,21,7,9
2	green	0 1 2	1	2	14
3	pink	0 2 1	2	1,1	22,6
4	green	0 1 2	3	1,2,2	22,3,21
5	green	0 1 2	5	1,2,2,0,0	8,7,21,15,7
6	green	0 1 2	3	1,0,0	22,10,21
7	yellow	1 2 0	1	0	6
8	red	2 1 0	1	1	20
9	pink	0 2 1	1	2	7
10	orange	1 0 2	1	1	17
11	orange	1 0 2	2	0,2	19,5
12	green	0 1 2	2	1,0	1,2
13	gray	2 0 1	2	1,0	21,8
14	orange	1 0 2	3	1,2,0	23,23,1
15	gray	2 0 1	1	2	20
16	pink	0 2 1	1	1	1
17	pink	0 2 1	4	1,2,0,1	11,20,7,4
18	red	2 1 0	2	2,1	23,20
19	orange	1 0 2	2	2,2	12,10
20	gray	2 0 1	2	1,0	5,19
21	gray	2 0 1	2	2,2	8,19
22	yellow	1 2 0	3	2,1,2	13,8,3
23	pink	0 2 1	0	-	-

Table A3. Agents' utilities after negotiation.

Ticks/Agent in Node	1	2	3	4	5	6	7	8
node 0	−1	−1	−1	−1	−1	−1	−1	−1
node 1	1	−1	−1	−1	−1	−1	1	1
node 2	−1	1	−1	−1	−1	−1	−1	−1
node 3	−1	−1	−1	−1	−1	1	1	1
node 4	−1	−1	−1	−1	−1	1	−1	−1
node 5	−1	−1	−1	−1	−1	−1	−1	−1
node 6	−1	−1	−1	1	−1	−1	1	1
node 7	−1	−1	−1	−1	−1	1	−1	−1
node 8	1	−1	1	−1	1	−1	−1	−1
node 9	−1	−1	−1	−1	−1	−1	1	1
node 10	1	−1	−1	−1	−1	1	1	1
node 11	−1	−1	−1	−1	−1	−1	−1	−1
node 12	−1	−1	−1	−1	−1	−1	−1	−1
node 13	−1	1	−1	−1	−1	−1	−1	−1
node 14	−1	−1	−1	−1	−1	−1	−1	−1
node 15	−1	1	−1	1	−1	−1	−1	−1
node 16	−1	−1	−1	−1	−1	−1	−1	−1
node 17	−1	−1	−1	−1	−1	−1	1	1
node 18	1	−1	1	−1	1	−1	−1	−1
node 19	−1	−1	−1	−1	−1	−1	−1	−1
node 20	−1	−1	−1	−1	1	1	1	1
node 21	−1	−1	−1	1	−1	−1	−1	−1
node 22	−1	−1	−1	−1	−1	−1	−1	−1
node 23	−1	−1	−1	−1	1	−1	1	1

Table A4. Agents' utilities.

Ticks /Agent in Node	1	2	3	4	5	6	7	8
node 0	−1	1	−1	−1	−1	−1	−1	−1
node 1	−1	−1	1	1	1	1	−1	−1
node 2	−1	−1	−1	−1	−1	−1	−1	−1
node 3	−1	1	1	1	−1	1	−1	−1
node 4	−1	−1	−1	1	−1	−1	−1	−1
node 5	−1	−1	−1	−1	−1	−1	−1	−1
node 6	1	1	−1	−1	−1	1	−1	−1
node 7	1	1	−1	−1	1	1	1	1
node 8	−1	1	1	1	1	1	1	1
node 9	−1	−1	−1	−1	−1	−1	−1	−1
node 10	1	−1	−1	−1	1	1	1	1
node 11	−1	−1	−1	−1	−1	−1	−1	−1
node 12	−1	−1	1	−1	−1	−1	−1	−1
node 13	−1	−1	1	−1	−1	1	−1	−1
node 14	−1	−1	−1	1	−1	−1	−1	−1
node 15	−1	1	−1	1	−1	−1	−1	−1
node 16	−1	1	1	1	1	−1	−1	−1
node 17	−1	−1	−1	−1	−1	−1	−1	−1
node 18	1	1	−1	−1	−1	−1	−1	−1
node 19	1	1	1	−1	−1	−1	−1	−1
node 20	1	−1	−1	1	−1	−1	−1	−1
node 21	−1	−1	−1	1	−1	1	−1	−1
node 22	−1	−1	−1	1	−1	−1	−1	−1
node 23	−1	−1	−1	−1	−1	−1	−1	−1

References

1. Samaan, N.; Karmouch, A. Towards autonomic network management: An analysis of current and future research directions. *IEEE Commun. Surv. Tutor.* **2009**, *11*, 22–36.
2. Mitchell, M. *Complexity: A Guided Tour*; Oxford University Press, Inc.: New York, NY, USA, 2009.
3. Toroczkai, Z. Complex networks the challenge of interaction topology. *Los Alamos Sci.* **2005**, *29*, 94–109.
4. Movahedi, Z.; Ayari, M.; Langar, R.; Pujolle, G. A survey of autonomic network architectures and evaluation criteria. *IEEE Commun. Surv. Tutor.* **2012**, *14*, 464–490.
5. International Business Machines Corporation. *An Architectural Blueprint for Autonomic Computing*; IBM Corporaton: Hawthorne, NY, USA, 2005.
6. Kalton, M.C.; Mobus, G.E. *Principles of Systems Science. Understanding Complex Systems*; Springer: New York, NY, USA, 2015.
7. Sayama, H. *Introduction to the Modeling and Analysis of Complex Systems*; Milne Library: Geneseo, NY, USA, 2015.
8. Newman, M. *Networks: An Introduction*; Oxford University Press, Inc.: New York, NY, USA, 2010.
9. Freeman, L.C. Centrality in social networks conceptual clarification. *Soc. Netw.* **1978**, *1*, 215–239.
10. Pavlopoulos, G.A.; Secrier, M.; Moschopoulos, C.N.; Soldatos, T.G.; Kossida, S.; Aerts, J.; Schneider, R.; Bagos, P.G. Using graph theory to analyze biological networks. *BioData Min.* **2011**, *4*, doi:10.1186/1756-0381-4-10.
11. Lada, A.A.; Rajan, M.L.; Huberman, B.A. Local Search in Unstructured Networks. In *Handbook of Graphs and Networks*; Wiley-VCH: Weinheim, Germany, 2003.
12. Gladwell, M. *The Tipping Point: How Little Things Can Make a Big Difference*; Little Brown: Boston, MA, USA, 2000.
13. Lempert, R. Agent-based modeling as organizational and public policy simulators. *Proc. Natl. Acad. Sci. USA* **2002**, *99* (Suppl. S3), 7195–7196.
14. *Introduction to Cisco IOS NetFlow—a technical overview.* Available online: http://www.cisco.com/en/US/prod/collateral/iosswrel/ps6537/ps6555/ps6601/prod_white_paper0900aecd80406232.html (accessed on 1 September 2017) .
15. Wooldridge, M. *An Introduction to MultiAgent Systems*; Wiley Publishing: Hoboken, NJ, USA, 2009.

16. Barabási, A.L. *Linked: How Everything Is Connected to Everything Else and What It Means for Business, Science, and Everyday Life*; Basic Books: New York, NY, USA, 2014.
17. Barabási, A.L. *Network Science*; Cambridge University Press: Cambridge, UK, 2017.
18. Weiler, R.; Engelbrecht, J.; Cadmus, J. The New Sciences of Networks & Complexity: A Short Introduction. *Cadmus* **2013**, *2*, 131–141.

applied
sciences

MDPI

Technical Note

On the Delayed Scaled Consensus Problems

Yilun Shang

School of Mathematical Sciences, Tongji University, Shanghai 200092, China; shylmath@hotmail.com

Academic Editors: Vicent Botti, Andrea Omicini, Stefano Mariani and Vicente Julian
Received: 28 May 2017; Accepted: 6 July 2017; Published: 11 July 2017

Abstract: In this note, we study the scaled consensus (tracking) problems, wherein all agents reach agreement, but with different assigned ratios in the asymptote. Based on the nearest neighbor-interaction rules, the scaled consensus processes are characterized with and without time delay. We consider both the signal transmission and signal processing delays and calculate the final scaled consensus values. When the underlying communication network contains a spanning tree, it is found that the scaled consensus can be achieved independent of the transmission delays while the specified consensus values in the asymptote depend on the initial history of the agents over a period of time. This phenomenon is in sharp contrast to the case of processing delays, where large delays are likely to jeopardize the consensus behavior, but the scaled consensus values once achieved are the same as the undelayed case.

Keywords: scaled consensus; delay; formation tracking; multi-agent system

1. Introduction

Multi-agent coordination of interconnected systems has found a diversity of applications in a number of fields, such as sensor networks, vehicle systems, social insects and cyber-physical systems. Consensus as a critical problem for multi-agent coordination aims to design appropriate protocols and strategies for reaching an agreement on a certain quantity of interest depending on the states of all agents. Different from traditional centralized controllers, the designed protocols take advantage of nearest-neighbor rules rendering the multi-agent systems in a distributed network framework governed by the graph Laplacians [1] since each agent can only interact with those within its local area due to limited communication capability. There has been an extensive literature on consensus problems using effective tools such as matrix theory, algebraic graph theory and system theory; see the recent surveys [2,3] on this wide ranging topics.

In many practical applications, the states of all agents may achieve consensus on a common quantity, but of their own scales due to the constraints of physical environments. Examples include water distribution systems, compartmental mass-action systems [4] and multiscale coordination control between spacecraft and their simulating vehicles on the ground [5]. As an extension to standard consensus, Roy [6] recently introduced a novel notion of scaled consensus, which permits prescribed ratios among the final convergent values of all of the agents. The scaled consensus offers a less conservative framework, which can be specialized to achieve standard consensus (with all ratios being one), cluster consensus [7], where agents in a subnetwork share a common value while there is no agreement between different subnetworks, and bipartite consensus [8] or signed consensus [9] by adopting appropriate scales.

Scaled consensus has been studied for a fixed strongly-connected topology in [6,10] and switching topologies in [11], where the agents are modeled by continuous-time single integrators. Scaled consensus can also be achieved through linear iterations [12,13]. However, time delay (especially distributed delay) has not been considered. Inspired by the stability analysis for delay systems in [14], we try to investigate the delayed scaled problems and distinguish between two main sources of time delays, to wit, signal transmission delays and signal processing delays. The contribution of this paper is two-fold. First, new scaled consensus protocols are proposed to accommodate discrete and distributed delays in networks containing spanning trees by extending the delayed complete consensus analysis in [14]. Second, for the undelayed scaled consensus process, we generalize some existing results on scaled consensus in [6,10,11]. In particular, we find that the ubiquitous spanning tree condition is both sufficient and necessary for scaled consensus, as well as for the related tracking formation problems on the scaled consensus manifold.

2. Preliminaries

The communication topology of a multi-agent system can often be characterized by a weighted directed graph [1] $\mathcal{G} = (\mathcal{V}, \mathcal{E}, \mathcal{A})$ with a node set $\mathcal{V} = \{1, 2, \cdots, n\}$ representing n agents, an edge set $\mathcal{E} \subseteq \mathcal{V} \times \mathcal{V}$ describing the information exchange among them and a nonnegative adjacency matrix $\mathcal{A} = (a_{ij}) \in \mathbb{R}^{n \times n}$. We assume $a_{ij} > 0$ if and only if $(j, i) \in \mathcal{E}$, representing the information flow from agent j to agent i. Because there are no self-loops in $\mathcal{G}$, we have $a_{ii} = 0$ for all $i \in \mathcal{V}$. The degree of i is defined as $d_i = \sum_{j=1}^{n} a_{ij}$. Given a sequence $i_1, i_2 \cdots, i_k$ of distinct nodes, a directed path in $\mathcal{G}$ from i_1 to i_k consists of a sequence of edges $(i_j, i_{j+1}) \in \mathcal{E}$ for $j = 1, \cdots, k-1$. If there is a node (called root) in $\mathcal{G}$ from which all other nodes can be reached along directed paths, then we say that it has a spanning tree. Furthermore, $\mathcal{G}$ is said to be strongly connected if between any pair of distinct nodes i, j, there exists a directed path from i to j. The graph Laplacian matrix of $\mathcal{A}$, $\mathcal{L} = (l_{ij}) \in \mathbb{R}^{n \times n}$, is defined by $l_{ii} = -\sum_{j \in \mathcal{V} \setminus \{i\}} l_{ij} = d_i$ and $l_{ij} = -a_{ij}$ for $i \neq j$. Clearly, $\mathcal{L}1_n = 0$, where $1_n \in \mathbb{R}^n$ represents the n-dimensional vector with elements being all ones. It is well known that $\mathcal{L}$ plays a significant role in the convergence analysis for consensus seeking [1]. By convention, the determinant, trace and adjugate of a square matrix are denoted by $\det(\cdot)$, $\mathrm{tr}(\cdot)$ and $\mathrm{adj}(\cdot)$, respectively.

Here, we consider a group of n agents, labeled from one to n, composing a fixed directed network $\mathcal{G} = (\mathcal{V}, \mathcal{E}, \mathcal{A})$. The dynamical behavior of each agent is described by a continuous-time system:

$$\dot{x}_i(t) = u_i(t), \quad i \in \mathcal{V}, \tag{1}$$

where $x_i(t) \in \mathbb{R}$ is the information state and $u_i(t) \in \mathbb{R}$ is the control input of the agent i at time t. In the vector form, we denote $x(t) = (x_1(t), \cdots, x_n(t))^{\mathrm{T}} \in \mathbb{R}^n$ and let $x(0) = (x_1(0), \cdots, x_n(0))^{\mathrm{T}} \in \mathbb{R}^n$ be the initial value. Given any scalar scale $\alpha_i \neq 0$ for the agent i, the system (1) is said to achieve scaled consensus to $(\alpha_1, \cdots, \alpha_n)$ if $\lim_{t \to \infty}(\alpha_i x_i(t) - \alpha_j x_j(t)) = 0$ for all $i, j \in \mathcal{V}$ and all initial conditions $x(0)$ [6].

Remark 1. *The notion of scaled consensus is a generalization of the standard consensus [2], cluster consensus [7] and bipartite consensus [8]. Note that the scales α_i associated with each individual i can be positive or negative, adding to the flexibility of the consensus-seeking process. Scaled consensus can also be defined in terms of the global asymptotic stability of the manifold defined by the above condition $\lim_{t \to \infty}(\alpha_i x_i(t) - \alpha_j x_j(t)) = 0$, both of which are known to be equivalent for linear processes [6].*

3. Undelayed Scaled Consensus Process

In this section, we first consider the undelayed scaled consensus problem. Given the scales $(\alpha_1, \cdots, \alpha_n)$ with nonzero α_i's, we apply the nearest-neighbor rules to propose the distributed strategy for agent i as:

$$u_i(t) = \mathrm{sgn}(\alpha_i) \sum_{j=1}^{n} a_{ij}(\alpha_j x_j(t) - \alpha_i x_i(t)), \tag{2}$$

where $\mathrm{sgn}(\cdot)$ is the signum function. Let $\mathcal{D} = \mathrm{diag}(d_1, \cdots, d_n) \in \mathbb{R}^{n \times n}$ be the degree diagonal matrix. Then, we have $\mathcal{L} = \mathcal{D} - \mathcal{A}$. Moreover, for ease of presentation, we define $\alpha := \mathrm{diag}(\alpha_1, \cdots, \alpha_n) \in \mathbb{R}^{n \times n}$, $|\alpha| := \mathrm{diag}(|\alpha_1|, \cdots, |\alpha_n|) \in \mathbb{R}^{n \times n}$, $\alpha^{-1} := \mathrm{diag}(\alpha_1^{-1}, \cdots, \alpha_n^{-1}) \in \mathbb{R}^{n \times n}$

and $\mathrm{sgn}(\alpha) := \mathrm{diag}(\mathrm{sgn}(\alpha_1), \cdots, \mathrm{sgn}(\alpha_n)) \in \mathbb{R}^{n \times n}$. With these notations, the switched multi-agent system (1) with Protocol (2) can be recast as:

$$\dot{x}(t) = -\mathrm{sgn}(\alpha)\mathcal{L}\alpha x(t). \tag{3}$$

The following result gives a concise characterization for the scaled consensus process governed by (3).

Theorem 1. *The multi-agent system* (3) *achieves scaled consensus to* $(\alpha_1, \cdots, \alpha_n)$ *if and only if* $\mathcal{G}$ *has a spanning tree.*

Proof. Sufficiency: It follows from (3) that $x(t) = \exp(-\mathrm{sgn}(\alpha)\mathcal{L}\alpha t)x(0)$ for $t \geq 0$ using the matrix exponential. We first consider the matrix $|\alpha|\mathcal{L}$. Clearly, it has zero row sums, and hence, zero is always an eigenvalue of $|\alpha|\mathcal{L}$. By applying the Gershgorin disk theorem [15], we see that all eigenvalues of $|\alpha|\mathcal{L}$ are in the right half plane, and its nonzero eigenvalues have strictly positive real parts (cf. [16]). Since $\mathcal{G}$ has a spanning tree, $\lambda_1 = 0$ is a simple eigenvalue of the Laplacian $\mathcal{L}$, as well as of $|\alpha|\mathcal{L}$ recalling that $|\alpha|$ is a positive definite diagonal matrix [17]. Noting that $\alpha^{-1}|\alpha|\mathcal{L}\alpha = \mathrm{sgn}(\alpha)\mathcal{L}\alpha$, we see that the state matrix $-\mathrm{sgn}(\alpha)\mathcal{L}\alpha$ is similar to $-|\alpha|\mathcal{L}$. Hence, all eigenvalues of $-\mathrm{sgn}(\alpha)\mathcal{L}\alpha$, denoted by $\lambda_1 = 0, \lambda_2, \cdots, \lambda_n$, are in the left half plane, and $\lambda_2, \cdots, \lambda_n$ have strictly negative real parts.

It is direct to see that $\exp(-\mathrm{sgn}(\alpha)\mathcal{L}\alpha t)$ is a normal matrix, and its eigenvalues are $e^{\lambda_1 t}, \cdots, e^{\lambda_n t}$. Therefore, $\exp(-\mathrm{sgn}(\alpha)\mathcal{L}\alpha t)$ possesses a complete set of eigenvectors, denoted by $\{v_1 = (a_1^{-1}, \cdots, a_n^{-1})^{\mathsf{T}}, v_2, \cdots, v_n\}$, which are exactly the corresponding eigenvectors of $-\mathrm{sgn}(\alpha)\mathcal{L}\alpha$ [15]. That is, $-\mathrm{sgn}(\alpha)\mathcal{L}\alpha v_i = \lambda_i v_i$ for $i = 1, \cdots, n$. Let $\{w_1, \cdots, w_n\}$ be the set of linearly independent eigenvectors of $(\exp(-\mathrm{sgn}(\alpha)\mathcal{L}\alpha t))^{\mathsf{T}}$, such that $w_i^{\mathsf{T}} v_j = \delta_{ij}$, where δ_{ij} is the Kronecker delta for $i, j = 1, \cdots, n$. Hence, $x(t) = \sum_{i=1}^{n}(w_i^{\mathsf{T}} x(t))v_i$, and furthermore,

$$x(t) = \exp(-\mathrm{sgn}(\alpha)\mathcal{L}\alpha t)x(0) = \sum_{i=1}^{n} w_i^{\mathsf{T}} \exp(-\mathrm{sgn}(\alpha)\mathcal{L}\alpha t)x(0)v_i$$

$$= \sum_{i=1}^{n} e^{\lambda_i t} w_i^{\mathsf{T}} x(0)v_i \to w_1^{\mathsf{T}} x(0)v_1 \tag{4}$$

as t tends to infinity. Recalling $v_1 = (\alpha_1^{-1}, \cdots, \alpha_n^{-1})^{\mathsf{T}}$, we see that the multi-agent system (3) achieves scaled consensus to $(\alpha_1, \cdots, \alpha_n)$.

Necessity: Suppose that $\mathcal{G}$ does not contain a spanning tree. Then, the zero eigenvalue is no longer a simple eigenvalue of $|\alpha|\mathcal{L}$ [17]. We consider the following two cases: (i) $\lambda_i = \lambda_1 = 0$ for some $i \neq 1$, and the geometric multiplicity of zero is equal to its algebraic multiplicity; and (ii) the geometric multiplicity of zero is less than the algebraic multiplicity.

In Case (i), say, $\lambda_2 = \lambda_1 = 0$, we have $\dot{x}(t) = 0$ for any $x(0) \in \mathrm{span}\{v_2\}$ by (3). However, for such initial values, scaled consensus to $(\alpha_1, \cdots, \alpha_n)$ is impossible because $v_2 \notin \mathrm{span}\{v_1\}$. Next, we consider Case (ii). In this situation, the state matrix $-\mathrm{sgn}(\alpha)\mathcal{L}\alpha$ is similar to a Jordan matrix $J \in \mathbb{R}^{n \times n}$. Hence, $\exp(-\mathrm{sgn}(\alpha)\mathcal{L}\alpha t) = P^{-1}e^{Jt}P$ for some invertible matrix P. For a $k \times k$ Jordan block ($k \geq 2$), we have:

$$\exp\left(\begin{pmatrix} 0 & 1 & \cdots & 0 \\ 0 & 0 & \ddots & \vdots \\ \vdots & \vdots & \ddots & 1 \\ 0 & 0 & \cdots & 0 \end{pmatrix} t\right) = \begin{pmatrix} 1 & t & \cdots & \frac{t^{k-1}}{(k-1)!} \\ 0 & 1 & \ddots & \vdots \\ \vdots & \vdots & \ddots & t \\ 0 & 0 & \cdots & 1 \end{pmatrix} \in \mathbb{R}^{k \times k},$$

which is divergent with respect to t. Similar reasoning as (4) implies that $x_i(t)/x_j(t)$ does not approach α_j/α_i for all i, j and generic initial conditions $x(0)$. Thus, scaled consensus to $(\alpha_1, \cdots, \alpha_n)$ is not achieved, and this completes the proof of necessity. $\square$

Remark 2. *The scaled consensus protocol specified by* (3) *is first studied in* [6]. *It is proven that* [6] *if G is strongly connected, then the scaled consensus can be reached. Theorem 1 improves the result by showing that a weaker spanning tree condition, actually, is not only sufficient, but necessary.*

Remark 3. *It is clear that the scaled consensus states are represented by* $w_1^T x(0) v_1 = w_1^T x(0) (\alpha_1^{-1}, \cdots, \alpha_n^{-1})^T$ *and the scaled states* $\alpha x(t) \to w_1^T x(0) 1_n$, *a common asymptote, as t grows. In fact, the quantity* $w_1^T x(t)$ *is conserved under the dynamics* (3) *since* $\frac{d}{dt}(w_1^T x(t)) = -w_1^T \text{sgn}(\alpha) \mathcal{L} \alpha x(t) = 0$.

Similarly as in [6], we may extend Theorem 1 by considering a tracking dynamics on the scaled consensus manifold as:

$$\dot{x}(t) = -\text{sgn}(\alpha) \mathcal{L} \alpha x(t) + \dot{f}(t) \alpha^{-1} 1_n, \tag{5}$$

where the forcing input $f : [0, +\infty) \to \mathbb{R}$ is continuously differentiable. Setting $y(t) = x(t) - f(t)\alpha^{-1} 1_n$, we can prove the following result using the same argument as in Theorem 1.

Corollary 1. *The multi-agent system* (5) *achieves scaled consensus to* $(\alpha_1, \cdots, \alpha_n)$ *if and only if G has a spanning tree. In particular,* $x(t)$ *converges to the time function* $(w_1^T x(0) + f(t)) v_1$ *as t grows.*

4. Scaled Consensus with Signal Transmission Delays

Signal transmission delay, also known as coupling delay, can be introduce to the control scheme (1) by considering:

$$u_i(t) = \text{sgn}(\alpha_i) \sum_{j=1}^{n} a_{ij}(\alpha_j x_j(t - \tau) - \alpha_i x_i(t)), \tag{6}$$

where $\tau \geq 0$ represents the time delay. Signal transmission delay often appears in networks of oscillators and has been investigated in, e.g., [14,18] for standard consensus problems. In fact, (6) can be generalized to accommodate a history dependence over the interval $[t - \tau, t]$ instead of a discrete past time instant $t - \tau$:

$$u_i(t) = \text{sgn}(\alpha_i) \sum_{j=1}^{n} a_{ij} \left(\int_{-\tau}^{0} \alpha_j x_j(t + s) d\eta(s) - \alpha_i x_i(t) \right), \tag{7}$$

where $\eta : [-\tau, 0] \to \mathbb{R}$ is a function of bounded variation describing the distributed delays. Assume that η is nondecreasing and appropriately normalized: $\int_{-\tau}^{0} d\eta(s) = 1$. η can be viewed as a probability distribution and admits scaled consensus solutions. In particular, when η is a Heaviside step function, the control input (7) readily reduces to (6).

The switched multi-agent system (1) with Protocol (7) can be recast in a compact form as:

$$\dot{x}(t) = -\text{sgn}(\alpha)\mathcal{D}\alpha x(t) + \text{sgn}(\alpha)\mathcal{A}\alpha \int_{-\tau}^{0} x(t + s) d\eta(s). \tag{8}$$

The formulation $x(t) = e^{\omega t} v$ $(v \in \mathbb{R}^n)$ yields the characteristic equation in $\omega \in \mathbb{C}$ as:

$$\chi(\omega) := \det(\omega I_n + \text{sgn}(\alpha)\mathcal{D}\alpha - G(\omega)\text{sgn}(\alpha)\mathcal{A}\alpha) = 0, \tag{9}$$

where $I_n \in \mathbb{R}^{n \times n}$ is the n-dimensional identity matrix and $G(\omega) = \int_{-\tau}^{0} e^{\omega s} d\eta(s)$. Let $\bar{\tau} := -\int_{-\tau}^{0} s d\eta(s) \geq 0$ represent the mean delay. It is clear that $G(0) = 1$ and $G'(0) = -\bar{\tau}$. Moreover, zero is always a characteristic value of (9) since $\chi(0) = \det(\text{sgn}(\alpha))\det(\mathcal{L})\det(\alpha) = 0$.

Our main result in this section shows that the stability of scaled consensus is independent of the transmission delays, generalizing the stability result of standard consensus [14] with all α_i being one.

Theorem 2. *If G has a spanning tree, the multi-agent system* (8) *achieves scaled consensus to* $(\alpha_1, \cdots, \alpha_n)$.

Proof. We first show that all nonzero characteristic values of (9) have negative real parts. In fact, assume that $\chi(\omega) = 0$ and $\mathrm{Re}(\omega) \geq 0$. It follows from (9) that $-\omega$ is an eigenvalue of $\mathrm{sgn}(\alpha)\mathcal{D}\alpha - G(\omega)\mathrm{sgn}(\alpha)\mathcal{A}\alpha$. Note that the matrix $\mathrm{sgn}(\alpha)\mathcal{D}\alpha - G(\omega)\mathrm{sgn}(\alpha)\mathcal{A}\alpha$ is similar to $|\alpha|\mathcal{D} - G(\omega)|\alpha|\mathcal{A}$. In light of the Gershgorin disk theorem, the eigenvalues of $\mathrm{sgn}(\alpha)\mathcal{D}\alpha - G(\omega)\mathrm{sgn}(\alpha)\mathcal{A}\alpha$ must lie in the union of n disks centered at $|\alpha_i|d_i$ with radius $|G(\omega)| \cdot |\alpha_i| \cdot \sum_{j=1}^n a_{ij} \leq |\alpha_i|d_i \cdot \int_{-\tau}^0 d\eta(s) \leq |\alpha_i|d_i$ for $i = 1, \cdots, n$. This implies that the nonzero characteristic values of (9) are contained in the set $\{\omega \in \mathbb{C} : \mathrm{Re}(\omega) > 0\} \cup \{0\}$, and thus, $\mathrm{Re}(-\omega) \leq 0$. Accordingly, we have $\omega = 0$, which completes the proof of the claim.

Next, we show that zero is a simple characteristic value of (9). It follows from (9) and some basic algebras [15] that:

$$\begin{aligned} \chi'(0) &= \mathrm{tr}\left(\mathrm{adj}(\mathrm{sgn}(\alpha)\mathcal{L}\alpha) \cdot (I_n + \bar{\tau}\mathrm{sgn}(\alpha)\mathcal{A}\alpha)\right) \\ &= \mathrm{tr}\left(\mathrm{adj}(\mathrm{sgn}(\alpha)\mathcal{L}\alpha) \cdot (I_n + \bar{\tau}\mathrm{sgn}(\alpha)(\mathcal{D} - \mathcal{L})\alpha)\right) \\ &= \mathrm{tr}\left(\mathrm{adj}(\mathrm{sgn}(\alpha)\mathcal{L}\alpha) \cdot (I_n + \bar{\tau}\mathrm{sgn}(\alpha)\mathcal{D}\alpha)\right), \end{aligned}$$

where in the last equality, we note that $\mathrm{adj}(\mathrm{sgn}(\alpha)\mathcal{L}\alpha) \cdot \mathrm{sgn}(\alpha)\mathcal{L}\alpha = \det(\mathrm{sgn}(\alpha)\ \mathcal{L}\alpha)I_n = \det(\mathrm{sgn}(\alpha))\det(\mathcal{L})\det(\alpha)I_n = 0$. Let $\tilde{\ell}_{ii}$ represent the diagonal entries of $\mathrm{adj}(\mathrm{sgn}(\alpha)\mathcal{L}\alpha)$. Then:

$$\chi'(0) = \sum_{i=1}^n \tilde{\ell}_{ii}(1 + \bar{\tau}|\alpha_i|d_i). \tag{10}$$

As in Theorem 1, we know that zero is a simple eigenvalue of $|\alpha|\mathcal{L}$ since $\mathcal{G}$ contains a spanning tree. Since zero is also a simple eigenvalue of $\mathrm{sgn}(\alpha)\mathcal{L}\alpha$ due to similarity, the matrix $\mathrm{sgn}(\alpha)\mathcal{L}\alpha$ has rank $n - 1$ and a one-dimensional kernel spanned by $v_1 = (\alpha_1^{-1}, \cdots, \alpha_n^{-1})^{\mathrm{T}}$. Therefore, $\mathrm{adj}(\mathrm{sgn}(\alpha)\mathcal{L}\alpha) \neq 0$. Moreover, we have $\mathrm{sgn}(\alpha)\mathcal{L}\alpha \cdot \mathrm{adj}(\mathrm{sgn}(\alpha)\mathcal{L}\alpha) = \det(\mathrm{sgn}(\alpha)\mathcal{L}\alpha)I_n = 0$, which implies that the columns of $\mathrm{adj}(\mathrm{sgn}(\alpha)\mathcal{L}\alpha)$ belong to the kernel of $\mathrm{sgn}(\alpha)\mathcal{L}\alpha$. Accordingly, $\mathrm{adj}(\mathrm{sgn}(\alpha)\mathcal{L}\alpha) = v_1 u$ for some row vector $u = (u_1, \cdots, u_n)$. Analogously, the equality $\mathrm{adj}(\mathrm{sgn}(\alpha)\mathcal{L}\alpha) \cdot \mathrm{sgn}(\alpha)\mathcal{L}\alpha = 0$ implies that the rows of $\mathrm{adj}(\mathrm{sgn}(\alpha)\mathcal{L}\alpha)$ belong to $\mathrm{span}\{w_1\}$, where $w_1 \in \mathbb{R}^n$ is defined in the proof of Theorem 1. Hence, for each $1 \leq i \leq n$, there exists some $c_i \in \mathbb{R}$ such that:

$$\left(\frac{u_1}{\alpha_i}, \cdots, \frac{u_n}{\alpha_i}\right) = c_i w_1^{\mathrm{T}}. \tag{11}$$

If $c_i = 0$, then the i-th row in the matrix (11) becomes zero, and $u_1 = \cdots = u_n = 0$. This indicates $\mathrm{adj}(\mathrm{sgn}(\alpha)\mathcal{L}\alpha) = 0$, which is a contradiction. Therefore, $c_i \neq 0$ for all i. We can show in the same way that $u_i \neq 0$ for all i. Therefore, there exists some $k \neq 0$ such that $c_i = \frac{k}{\alpha_i}$ and $\tilde{\ell}_{ii} = \frac{kw_{1i}}{\alpha_i}$ for all $1 \leq i \leq n$, where we write $w_1 = (w_{11}, \cdots, w_{1n})^{\mathrm{T}}$. In view of (10), we have:

$$\chi'(0) = k \sum_{i=1}^n \frac{w_{1i}}{\alpha_i}(1 + \bar{\tau}|\alpha_i|d_i). \tag{12}$$

By definition, $w_1^{\mathrm{T}}\mathrm{sgn}(\alpha)\mathcal{L}\alpha = 0$. Writing $w_1^{\mathrm{T}}\mathrm{sgn}(\alpha) := \tilde{w}_1^{\mathrm{T}} = (\tilde{w}_{11}, \cdots, \tilde{w}_{1n})$, we obtain $\tilde{w}_1^{\mathrm{T}}\mathcal{L} = 0$. For a sufficiently small $\varepsilon > 0$, $I_n - \varepsilon\mathcal{L}$ is a stochastic matrix, and $\tilde{w}_1^{\mathrm{T}}(I - \varepsilon\mathcal{L}) = \tilde{w}_1^{\mathrm{T}}$. Based on the Perron–Frobenius theory, we know that $\tilde{w}_1^{\mathrm{T}}$ (under an appropriate normalization) is the stationary distribution of the Markov chain described by the stochastic matrix. Hence, $\tilde{w}_1^{\mathrm{T}} \neq 0$ and has non-negative entries. Noting that $\frac{w_{1i}}{\alpha_i} = \frac{\tilde{w}_{1i}\mathrm{sgn}(\alpha_i)}{\alpha_i}$ and $1 + \bar{\tau}|\alpha_i|d_i > 0$ for all i, we see from (12) that $\chi'(0) \neq 0$. Consequently, zero is a simple characteristic value of (9).

Since all solutions of (8) involve a factor $e^{\omega t}$ with ω being a root of the characteristic Equation (9). Note that the root $\omega = 0$ corresponds to the kernel of $\mathrm{sgn}(\alpha)\mathcal{L}\alpha$, i.e., $\mathrm{span}\{v_1\}$. Moreover, the dynamics on the subspace span$\{v_1\}$ is constant, to wit, $\dot{x}(t) \equiv 0$. Combining the above discussions, we are led to the conclusion that:

$$\lim_{t\to\infty} x(t) = c v_1 = c(\alpha_1^{-1}, \cdots, \alpha_n^{-1})^{\mathrm{T}} \tag{13}$$

for some $c \in \mathbb{R}$. The proof is complete. $\square$

Remark 4. *The quantity c in (13) can be determined. In fact, noting that $w_1^{\mathrm{T}}\mathrm{sgn}(\alpha)\mathcal{L}\alpha = 0$, we obtain from (8) that:*

$$\frac{d}{dt}\left(w_1^{\mathrm{T}}\left(x(t) + \mathrm{sgn}(\alpha)\mathcal{D}\alpha \int_{-\tau}^0 \int_{t+s}^t x(t')dt'd\eta(s)\right)\right)$$

$$= w_1^{\mathrm{T}}\dot{x}(t) + w_1^{\mathrm{T}}\mathrm{sgn}(\alpha)\mathcal{D}\alpha \int_{-\tau}^0 (x(t) - x(t+s))d\eta(s)$$

$$= w_1^{\mathrm{T}}\mathrm{sgn}(\alpha)\mathcal{A}\alpha \int_{-\tau}^0 x(t+s)d\eta(s) - w_1^{\mathrm{T}}\mathrm{sgn}(\alpha)\mathcal{D}\alpha \int_{-\tau}^0 x(t+s)d\eta(s) = 0.$$

This means that the quantity $w_1^{\mathrm{T}}\left(x(t) + \mathrm{sgn}(\alpha)\mathcal{D}\alpha \int_{-\tau}^0 \int_{t+s}^t x(t')dt'd\eta(s)\right)$ remains unchanged with respect to time. By setting $t \to \infty$, we have:

$$w_1^{\mathrm{T}}\left(x(0) + \mathrm{sgn}(\alpha)\mathcal{D}\alpha \int_{-\tau}^0 \int_s^0 x(t)dtd\eta(s)\right)$$

$$= w_1^{\mathrm{T}}\left(cv_1 + \mathrm{sgn}(\alpha)\mathcal{D}\alpha \int_{-\tau}^0 \int_s^0 cv_1 dtd\eta(s)\right)$$

$$= c - cw_1^{\mathrm{T}}\mathrm{sgn}(\alpha)\mathcal{D}\alpha v_1 \int_{-\tau}^0 sd\eta(s)$$

$$= c\left(1 + \bar{\tau}w_1^{\mathrm{T}}\mathrm{sgn}(\alpha)\mathcal{D}1_n\right).$$

Hence, we have:

$$c = \frac{1}{1 + \bar{\tau}w_1^{\mathrm{T}}\mathrm{sgn}(\alpha)\mathcal{D}1_n}w_1^{\mathrm{T}}\left(x(0) + \mathrm{sgn}(\alpha)\mathcal{D}\alpha \int_{-\tau}^0 \int_s^0 x(t)dtd\eta(s)\right). \tag{14}$$

In the case of no delay, i.e., $\bar{\tau} = 0$, (14) reduces to $c = w_1^{\mathrm{T}}x(0)$, which agrees with Equation (4) in [6]. In the case of discrete delay delineated by (6), we have:

$$c = \frac{1}{1 + \tau w_1^{\mathrm{T}}\mathrm{sgn}(\alpha)\mathcal{D}1_n}w_1^{\mathrm{T}}\left(x(0) + \mathrm{sgn}(\alpha)\mathcal{D}\alpha \int_{-\tau}^0 x(t)dt\right).$$

Remark 5. *From (14), we observe that the scaled consensus values rely on the initial history of the agents' state x over the interval $[-\tau, 0]$ and that the scaled consensus is guaranteed independent of the magnitude of transmission delays. An important implication is that, in the event of additive measurement noise with zero temporal mean, the delayed system may have better performance by choosing a suitable τ to reduce the noise effect by averaging over $[-\tau, 0]$ via (14). In [18], a similar procedure for delayed standard consensus problems has been discussed.*

Finally, we consider a tracking dynamics on the scaled consensus manifold as:

$$\dot{x}(t) = -\mathrm{sgn}(\alpha)\mathcal{D}\alpha x(t) + \mathrm{sgn}(\alpha)\mathcal{A}\alpha \int_{-\tau}^0 x(t+s)d\eta(s) + \dot{f}(t)\alpha^{-1}1_n, \tag{15}$$

where the forcing input f is defined as in Section 3. We can similarly derive the following useful corollary.

Corollary 2. *If $\mathcal{G}$ has a spanning tree, then the multi-agent system* (15) *achieves scaled consensus to* $(\alpha_1, \cdots, \alpha_n)$. *In particular, $x(t)$ converges to the time function $(c + f(t))v_1$ as t grows, where c is given by* (14).

5. Scaled Consensus with Signal Processing Delays

In this section, we examine the signal processing delay, which is also known as internal delay. Signal processing delay has been much studied in the literature of consensus problems; see, e.g., [1–3]. In parallel with (6), the scaled consensus control scheme can be designed as:

$$u_i(t) = \text{sgn}(\alpha_i) \sum_{j=1}^{n} a_{ij}(\alpha_j x_j(t - \tau) - \alpha_i x_i(t - \tau)), \tag{16}$$

where $\tau \geq 0$ indicates the time delay. The distributed-delay version is:

$$u_i(t) = \text{sgn}(\alpha_i) \sum_{j=1}^{n} a_{ij} \int_{-\tau}^{0} (\alpha_j x_j(t + s) - \alpha_i x_i(t + s)) \, d\eta(s), \tag{17}$$

where $\eta : [-\tau, 0] \to \mathbb{R}$ is a function of bounded variation describing the distributed delays. In analogy to (8), the switched multi-agent system (1) with protocol (17) can be written as:

$$\dot{x}(t) = -\text{sgn}(\alpha)\mathcal{L}\alpha \int_{-\tau}^{0} x(t + s) d\eta(s). \tag{18}$$

The characteristic equation is given by:

$$\chi(\omega) := \det\left(\omega I_n + G(\omega)\text{sgn}(\alpha)\mathcal{L}\alpha\right) = 0, \tag{19}$$

where $\omega \in \mathbb{C}$ and $G(\omega) = \int_{-\tau}^{0} e^{\omega s} d\eta(s)$. Similarly as the transmission delay case, we have $G(0) = 1$, $\chi(0) = 0$ and $G'(0) = -\bar{\tau} := \int_{-\tau}^{0} s d\eta(s)$.

Here, we will only study the system (1) with discrete delay (16), which corresponds to a special form $G(\omega) = e^{-\omega\tau}$. We assume the following detailed balanced condition.

Assumption 1. $|\alpha_i|a_{ij} = |\alpha_j|a_{ji}$ holds for all $i, j = 1, \cdots, n$.

The detailed balanced condition has proven to be instrumental in studying coupled dynamics; see, e.g., [19,20] for details. Our main result in this section reads as follows.

Theorem 3. *Suppose that $\mathcal{G}$ has a spanning tree and that Assumption 1 holds. The multi-agent system* (1) *with Protocol* (16) *achieves scaled consensus to* $(\alpha_1, \cdots, \alpha_n)$ *if and only if* $0 \leq \tau < \frac{\pi}{2\max_{1 \leq i \leq n} \lambda_i}$, *where* $\{\lambda_i\}_{i=1}^{n}$ *are the eigenvalues of* $|\alpha|\mathcal{L}$.

Proof. Note that $|\alpha|\mathcal{L}$ is symmetric and positive semidefinite under Assumption 1. Let $\lambda_1 = 0$. The spanning tree condition implies $\lambda_i > 0$ for $i = 2, \cdots, n$.

We first show that zero is a simple characteristic value of (19). Indeed, we already see that $\chi(0) = 0$. Arguing similarly as in Theorem 2, we have:

$$\chi'(0) = \text{tr}\left(\text{adj}(\text{sgn}(\alpha)\mathcal{L}\alpha) \cdot (I_n - \tau\text{sgn}(\alpha)\mathcal{L}\alpha)\right)$$

$$= \text{tr}\left(\text{adj}(\text{sgn}(\alpha)\mathcal{L}\alpha)\right) = k \sum_{i=1}^{n} \frac{w_{1i}}{\alpha_i} \neq 0,$$

where $k \neq 0$ and $w_1 = (w_{11}, \cdots, w_{1n})^{\mathsf{T}}$ are interpreted in the same way as in (12). This means zero is a simple root of (19).

Next, we show that $0 < \lambda_i \tau < \frac{\pi}{2}$ for all $i \geq 2$ if and only if all nonzero characteristic values of (19) have negative real parts. Suppose that $\chi(\omega) = 0$ for some $\omega \in \mathbb{C}$. By (19), we know that $\frac{-\omega}{G(\omega)} = -\omega e^{\omega\tau}$ is an eigenvalue of $|\alpha|\mathcal{L}$. That is, $\omega e^{\omega\tau} = \lambda_i$ ($i = 2, \cdots, n$). Under Assumption 1, Theorem 1 in [21] implies that $\text{Re}(\omega) < 0$ for all solutions ω of the above system if and only if $0 < \lambda_i \tau < \frac{\pi}{2}$ for $i = 2, \cdots, n$.

It is easy to see that the sufficiency of Theorem 3 holds by the same argument at the end of Theorem 2. The necessity of Theorem 3 follows similarly from the necessity proof of Theorem 1. □

Remark 6. *It follows from the Gershgorin disk theorem that $0 \leq \lambda_i \leq 2\max_{1\leq i\leq n}|\alpha_i|d_i$. Therefore, the eigenvalue condition in Theorem 3 can be replaced by a more geometric (sufficient) condition $0 \leq \tau < \frac{\pi}{4\max_{1\leq i\leq n}|\alpha_i|d_i}$. We mention that a deep investigation regarding the conditions guaranteeing $\text{Re}(\lambda_i) < 0$ ($i \geq 2$) would be critical for tackling the general system (18).*

Remark 7. *Since $\frac{d}{dt}\left(w_1^T x(t)\right) = -w_1^T \text{sgn}(\alpha)\mathcal{L}\alpha x(t - \tau) = 0$, the quantity $w_1^T x(t)$ is conserved with respect to time t. Thus, setting $\lim_{t\to\infty} x(t) = cv_1$, we have $w_1^T x(0) = cw_1^T v_1 = c$, which is the same as the undelayed multi-agent system (3).*

Remark 8. *From Theorem 3, we find that introducing signal processing delay may prohibit the scaled consensus process, while the scaled consensus values are independent of the magnitude of the delay once scaled consensus is achieved, which is in sharp contrast to the effect of transmission delay.*

By considering the following tracking dynamics:

$$\dot{x}(t) = -\text{sgn}(\alpha)\mathcal{L}\alpha x(t - \tau) + \dot{f}(t)\alpha^{-1}\mathbf{1}_n, \tag{20}$$

we can similarly derive the following corollary.

Corollary 3. *Suppose that $\mathcal{G}$ has a spanning tree and Assumption 1 holds. The multi-agent system (20) achieves scaled consensus to $(\alpha_1, \cdots, \alpha_n)$ if and only if $0 \leq \tau < \frac{\pi}{2\max_{1\leq i\leq n}\lambda_i}$, where $\{\lambda_i\}_{i=1}^n$ are the eigenvalues of $|\alpha|\mathcal{L}$. In particular, $x(t)$ converges to the time function $(w_1^T x(0) + f(t))v_1$ as t grows.*

An example of delayed scaled consensus over a graph $\mathcal{G}$ of $n = 4$ agents with $\alpha = \text{diag}(1, -1, 1, 2)$

and $\mathcal{A} = \begin{pmatrix} 0 & 1 & 1 & 0 \\ 1 & 0 & 1 & 2 \\ 1 & 1 & 0 & 0 \\ 0 & 1 & 0 & 0 \end{pmatrix}$ is shown in Figure 1. Here, we take $\tau = 0.2$ and $x(-\tau) = (3, -2, -1, 1)^T$.

It is direct to check that the conditions of Theorems 2 and 3 are satisfied. As one would expect, the scaled consensus to $(1, -1, 1, 2)$ has been achieved under both signal transmission delays (Figure 2a) and signal processing delays (Figure 2b).

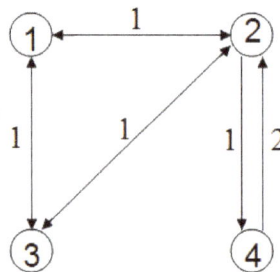

Figure 1. A communication network $\mathcal{G}$ over $n = 4$ agents.

(**a**)Transmission delay

(**b**)Processing delay

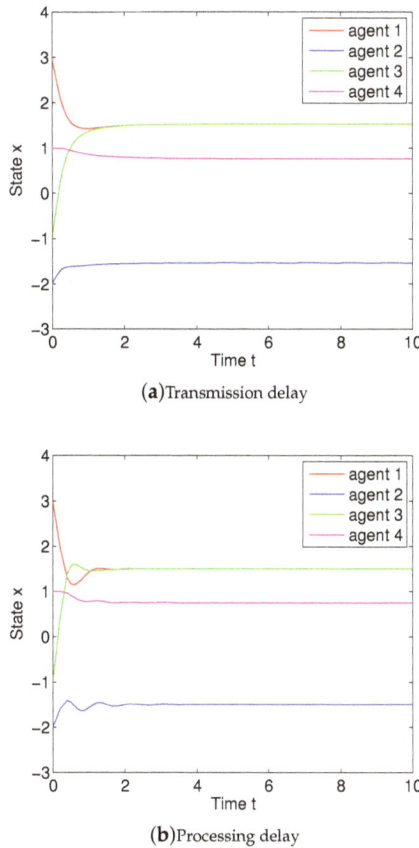

Figure 2. State evolution of agents under (**a**) signal transmission delays for system (1) with (6) and (**b**) signal processing delays for system (1) with (16).

6. Conclusions

Scaled consensus problems have found diverse applications in cooperative tasks in practical multi-agent systems. This paper discusses the stability of scaled consensus problems with and without time delays. We distinguish between signal transmission delays and signal processing delays by deriving sufficient and necessary scaled consensus conditions and calculating the final scaled consensus values in each case. These two types of delays are shown to possess distinct features on the scaled consensus processes, as well as the scaled consensus tracking on associated manifolds.

The scales encoded in α in this paper are assumed to be constant. It would be interesting to explore time-varying scales $\alpha = \alpha(t)$ (leading to a non-autonomous linear system) in the delayed system framework. It is noteworthy that our results assume a fixed communication topology and homogeneous agents. Therefore, delayed scaled consensus processes with switching/uncertain topologies and heterogeneously motivated agents (e.g., [22]) can be meaningful subjects of future work.

Acknowledgments: The author would like to thank the anonymous referees for their valuable suggestions that improved the presentation of the paper. The work is funded in part by the National Natural Science Foundation of China (11505127), the Shanghai Pujiang Program (15PJ1408300) and the Program for Young Excellent Talents in Tongji University (2014KJ036).

Conflicts of Interest: The author declares no conflict of interest.

References

1. Mesbahi, M.; Egerstedt, M. *Graph Theoretic Methods in Multiagent Networks*; Princeton University Press: Princeton, NJ, USA, 2010.
2. Olfati-Saber, R.; Fax, J.A.; Murray, R.M. Consensus and cooperation in networked multi-agent systems. *Proc. IEEE* **2007**, *95*, 215–233.
3. Cao, Y.; Yu, W.; Ren, W.; Chen, G. An overview of recent progress in the study of distributed multi-agent coordination. *IEEE Trans. Ind. Inf.* **2013**, *9*, 427–438.
4. Haddad, W.M.; Chellaboina, V.; Hui, Q. *Nonnegative and Compartmental Dynamical Systems*; Princeton University Press: Princeton, NJ, USA, 2010.
5. Guglieri, G.; Maroglio, F.; Pellegrino, P.; Torre, L. Design and development of guidance navigation and control algorithms for spacecraft rendezvous and docking experimentation. *Acta Astronaut.* **2014**, *94*, 395–408.
6. Roy, S. Scaled consensus. *Automatica* **2015**, *51*, 259–262.
7. Shang, Y. A combinatorial necessary and sufficient condition for cluster consensus. *Neurocomputing* **2016**, *216*, 611–616.
8. Altafini, C. Consensus problems on networks with antagonistic interactions. *IEEE Trans. Autom. Control* **2013**, *58*, 935–946.
9. Li, J.; Dong, W.; Xiao, H. Signed consensus problems on networks of agents with fixed and switching topologies. *Int. J. Control* **2017**, *90*, 148–160.
10. Meng, D.; Jia, Y. Robust consensus algorithms for multiscale coordination control of multivehicle systems with disturbances. *IEEE Trans. Ind. Electron.* **2016**, *63*, 1107–1119.
11. Meng, D.; Jia, Y. Scaled consensus problems on switching networks. *IEEE Trans. Autom. Control* **2016**, *61*, 1664–1669.
12. Hou, B.; Sun, F.; Li, H.; Chen, Y.; Liu, G. Scaled cluster consensus of discrete-time multi-agent systems with general directed topologies. *Int. J. Control* **2016**, *47*, 3839–3845.
13. Shang, Y. Finite-time scaled consensus through parametric linear iterations. *Int. J. Syst. Sci.* **2017**, *48*, 2033–2040.
14. Atay, F.M. On the duality between consensus problems and Markov processes, with application to delay systems. *Markov Process. Relat. Field.* **2016**, *22*, 537–553.
15. Horn, R.A.; Johnson, C.R. *Matrix Analysis*; Cambridge University Press: Cambridge, UK, 2012.
16. Ren, W.; Beard, R.W. Consensus seeking in multiagent systems under dynamically changing interaction topologies. *IEEE Trans. Autom. Control* **2005**, *50*, 655–661.
17. Merris, R. Laplacian matrices of graphs: A survey. *Linear Algebra Appl.* **1994**, *197–198*, 143–176.
18. Atay, F.M. The consensus problem in networks with transmission delays. *Phil. Trans. R. Soc. A* **2013**, *371*, doi:10.1098/rsta.2012.0460.
19. Haken, H. *Synergetics*; Springer: Berlin, Germany, 1978.
20. Chu, T.; Zhang, C.; Zhang, Z. Necessary and sufficient condition for absolute stability of normal neural networks. *Neural Netw.* **2003**, *16*, 1223–1227.
21. Hayes, H.D. Roots of the transcendental equation associated with a certain differential difference equation. *J. Lond. Math. Soc.* **1950**, *25*, 226–232.
22. Shang, Y. Consensus in averager-copier-voter networks of moving dynamical agents. *Chaos* **2017**, *27*, doi:10.1063/1.4976959.

applied
sciences

MDPI

Article

Artificial Neural Networks in Coordinated Control of Multiple Hovercrafts with Unmodeled Terms

Kairong Duan [1,*]**, Simon Fong** [1,*]**, Yan Zhuang** [1] **and Wei Song** [2]

[1]   Department of Computer and Information Science, University of Macau, Taipa 999078, Macau, China;
       syz@umac.mo
[2]   School of Computer Science, North China University of Technology, Beijing 100144, China; sw@ncut.edu.cn
*   Correspondence: yb67408@umac.mo (K.D.); ccfong@umac.mo (S.F.)

Received: 17 April 2018; Accepted: 22 May 2018; Published: 24 May 2018

check for
updates

Abstract: In this paper, the problem of coordinated control of multiple hovercrafts is addressed. For a single hovercraft, by using the backstepping technique, a nonlinear controller is proposed, where Radial Basis Function Neural Networks (RBFNNs) are adopted to approximate unmodeled terms. Despite the application of RBFNNs, integral terms are introduced, improving the robustness of controller. As a result, global uniformly ultimate boundedness is achieved. Regarding the communication topology, two different directed graphs are chosen under the assumption that there are no delays when they communicate with each other. In order to testify the performance of the proposed strategy, simulation results are presented, showing that vehicles can move forward in a specific formation pattern and RBFNNs are able to approximate unmodeled terms.

Keywords: surface vehicle; underactuated vehicle; RBFNNs; directed graph; coordinated control

---

1. Introduction

In recent years, Wireless Sensor Networks (WSNs) have attracted growing interests from researchers, because they have merits, compared with traditional networking solutions, such as reliability, flexibility, and an ease of deployment, that enable their use in a wide range of varied application scenarios [1]. They can be applied to track moving objects, to monitor special areas so as to trigger alarm systems when some dangerous signals are detected, etc. As the eyes and ears of the IoT, WSNs can work as bridges to build connections between the real-world and the digital-world. In light of this promising application scenario, this paper mainly focuses on a case study of mobile WSNs, where a group of hovercrafts equipped with specific sensors are chosen as test platforms. The objective is to enable them to move around and interact with the physical environment [2] and thus execute a mission of mapping, searching, and monitoring in a specific area.

Coordinated control of a fleet of hovercrafts is challenging, especially when we take into account their complex dynamic models. Until now, for a single surface vehicle, many research results have been reported. For example, a linear fuzzy-PID controller was proposed in [3]. Compared with the ordinary PID controller, the proposed controller therein performs better in term of improving settling time and reducing overshoot of the control signal. However, their works just consider the kinematic models of the vehicle without considering the dynamic models, which is not realistic in real operation scenarios. Another weakness of the linear controller is that it usually achieves local stability, e.g., [4], where velocity and position controllers were developed based on a linearized system, which is controllable only when the angular velocity is nonzero. Considering the limitations of linear controllers, in [5,6], nonlinear controllers for underactuated ships were designed, and global asymptotic stability is achieved. In [7], a nonlinear Lyapunov-based tracking controller was presented, and it was able to exponentially stabilize the position tracking error to a neighborhood of the origin that can be made arbitrarily small. A method of incorporating multiobjective controller selection into a closed-loop control system was presented in [8], where the authors designed three controllers so as to capture three "behaviors" representative of typical maneuvers that would be performed in a port environment. However, none of the works mentioned above consider disturbances and unmodeled terms of the vehicle. In order to ensure that the vehicle is robust to external disturbances, in [9], two controllers with application to a surface vehicle (named Qboat) and a hovercraft were proposed, and the authors designed disturbance estimators to estimate external constant disturbances. The disadvantage of this control strategy is that they did not consider unmodeled terms involved in the dynamic model of the vehicle. Considering this constraint, an estimator was developed in [10], where a fuzzy system was used to approximate unknown kinetics. A fault tolerant tracking controller was designed in [11] for a surface vessel. In addition, a self-constructing adaptive robust fuzzy neural control scheme for tracking surface vessels was proposed in [12], where simulation results were shown to testify the efficiency of the proposed method therein.

With respect to coordinated control strategy for multiple vehicles, many authors have presented their own approaches. In [13], a cooperative path following methodology was proposed under the assumption that the communication among a group of fully-actuated surface vehicles is undirected and continuous. A coordinated path following with a switching communication topology was designed in [14], while a null-space-based behavioral control technique was proposed in [15,16]. In [17–19], a leader–follower control strategy was presented. In [20], an adaptive coordinate tracking control problem for a fleet of nonholonomic chained systems was discussed under the assumption that the desired trajectory is available only to part of the neighbors. The reader is also referred to [21] for more results about multi-vehicle control approaches.

Inspired and motived by those works mentioned above, in this paper, we first develop a controller that is able to drive a single hovercraft to the neighborhood of a desired smooth path, where a Radial Basis Function Neural Network (RBFNN) is applied to approximate unmodeled dynamic terms of the vehicle while integral error terms are introduced, thus improving the robust performance of the controller. It is relevant to point out that all elements of the estimation weight matrix are always bounded through the use of a smooth projection function. We also derive a consensus strategy to make sure the desire paths progress in a specific formation. In order to validate the effectiveness of the proposed strategy, simulation results are presented.

The rest of the paper is organized as follows: Section 2 presents robot modeling, graph theory, RBFNNs, and coordinated control problem. A single controller is proposed in Section 3, while Section 4 devises a consensus strategy. Simulation results are given in Section 5 to validate the performance of the proposed approach herein. At last, Section 6 summarizes our work and describes the future work.

2. Problem Formulation

2.1. Vehicle Modeling

We first define a global coordinate frame $\{U\}$ and a body frame $\{B\}$ as shown in Figure 1. The kinematic equations of the vehicle are written as

$$\dot{\mathbf{p}} = \mathbf{R}\mathbf{v} \tag{1}$$

$$\dot{\psi} = \omega \tag{2}$$

where $\dot{\mathbf{p}} = [x, y]^T$ denotes the coordinates of its center of mass, $\mathbf{v} = [u, v]^T$ represents linear velocity, ψ is the orientation of the vehicle, and its angular velocity is represented by ω. Moreover, the rotation matrix $\mathbf{R}(\psi)$ is given by

$$\mathbf{R}(\psi) = \begin{bmatrix} \cos(\psi) & -\sin(\psi) \\ \sin(\psi) & \cos(\psi) \end{bmatrix}. \tag{3}$$

Its dynamic equations are

$$\dot{\mathbf{v}} = -\mathbf{S}(\omega)\mathbf{v} + m^{-1}T\mathbf{n}_1 + \Delta_v \tag{4}$$

$$\dot{\omega} = -J^{-1}\omega + J^{-1}\tau + \Delta_\omega \tag{5}$$

where $\mathbf{S}(\omega)$ is a skew symmetric matrix, given by

$$\mathbf{S}(\omega) = \begin{bmatrix} 0 & -\omega \\ \omega & 0 \end{bmatrix}. \tag{6}$$

$\mathbf{n}_1 = [1, 0]^T$, m and J denote the car's mass and rotational inertial, respectively. The force used to make the car move forward is denoted by T, and τ represents the torque that can steer the vehicle. Unmodeled dynamic terms are represented by Δ_v and Δ_ω. For more details about modeling surface vehicles, the reader is referred to [22].

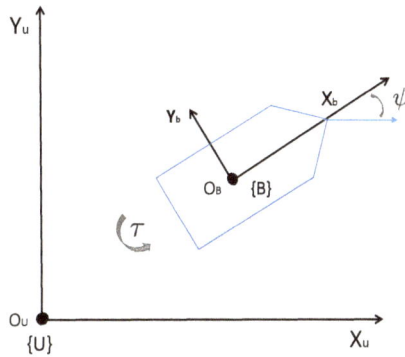

Figure 1. Simple Model of The Vehicle.

2.2. Graph Theory

In this paper, $\mathcal{G} = \mathcal{G}(\mathcal{V}, \mathcal{E})$ denotes a directed graph that can be used to model the interaction communication topology among mobile robots. The graph $\mathcal{G}$ consists of a finite set $\mathcal{V} = \{1, 2, ..., n\}$ of n vehicles and a finite set $\mathcal{E}$ of m pairs of vertices $\mathcal{V}_{ij} = \{i, j\} \in \mathcal{E}$. If $\mathcal{V}_{ij}$ belongs to $\mathcal{E}$, then i and j are said to be adjacent. A graph from i to j is a sequence of distinct vertices starting with i and ending with j such that consecutive vertices are adjacent. In this case, $\mathcal{V}_{ij}$ also represents a directional communication link from agent i to agent j. The adjacency matrix of the graph $\mathcal{G}$ is denoted by $\mathcal{A} = [a_{ij}] \in \mathbb{R}^{n \times n}$, which is a square matrix where a_{ij} equals to one if $\{j, i\} \in \mathcal{E}$ and zero otherwise. Moreover, the Laplacian matrix $\mathcal{L}$ is defined as $\mathcal{L} = \mathcal{D} - \mathcal{A}$, where the degree matrix $\mathcal{D} = [d_{ij}] \in \mathbb{R}^{n \times n}$ of the graph $\mathcal{G}$ is a diagonal matrix and d_{ij} equals the number of adjacent vertices of vertex i.

2.3. Radial Basis Function Neural Networks

Radial Basis Function Neural Networks (RBFNNs) can be used to approximate the unmodeled nonlinear dynamic terms due to their universal approximation capability [23]. For any unknown smooth function $f(\mathbf{x}) : \mathbb{R}^n \to \mathbb{R}^m$ can be approximated by RBFNNs in the following form, given by

$$\hat{f}(\mathbf{x}) = \mathbf{W}^T \sigma(\mathbf{x}) \tag{7}$$

where $x \in \Omega \subset \mathbb{R}^n$, Ω is a compact set. The adjustable weight matrix with n neurons is denoted by $W \in \mathbb{R}^{n \times m}$ under the assumption that it is a bounded matrix, that is

$$W \leq W_{\max}. \tag{8}$$

It is important to point out that here when we say matrix $x \in \mathbb{R}^{m \times n}$ is smaller than or equal to $x_{\max} \in \mathbb{R}^{m \times n}$, we mean all elements of x are smaller than or equal to their corresponding elements of $x_{\max} \in \mathbb{R}^{m \times n}$. Moreover, $\sigma(x)$ is the basis function vector and $\sigma_i(x_i) = \exp(-(x_i - \mu_i)^T(x_i - \mu_i)/c_i^2)$, $i = 1, 2, \ldots, n$ denotes its component, μ_i is the center of the receptive field, and c_i represents the width of the Gaussian function. Moreover, it is relevant to point out that, in order to achieve better approximate results, we should make the neuron number n large enough and choose the parameters properly. Going back to the smooth function $f(x)$ mentioned above, there is an ideal weight W_d such that

$$f(x) = W_d^T \sigma(x) + \epsilon(x) \tag{9}$$

where $\epsilon(x)$ denotes the approximation error and satisfies $||\epsilon(x)|| \leq \epsilon_{\max}$, where $\epsilon_{\max}$ is a positive number. It is noted that W_d is an "artificial" quantity for the purpose of mathematic analysis, in the process of controller design, we need to estimate it [24]. A simple RBFNNs is given by Figure 2.

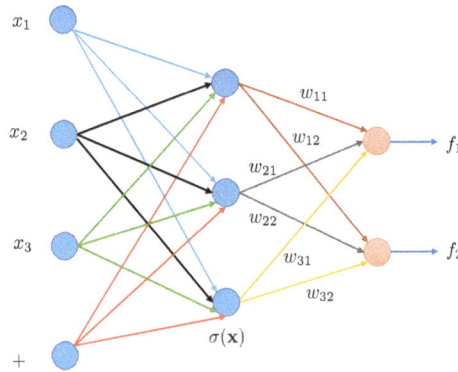

Figure 2. Simple example of RBFNNs.

2.4. Problem Statement

Now we can state our problem: Through designing a controller for each robot and proposing consensus strategies for their corresponding desired paths, we want a group of mobile robots to move forward in a specific formation pattern; that is,

(1) for an individual vehicle, $||p - p_d|| \to \delta$, where δ is an arbitrarily small constant value;
(2) for a group of n desired paths, $\gamma_i - \gamma_j \to 0$ and $\dot{\gamma}_i - \dot{\gamma}_d \to 0$, where agent j is the neighbor of agent i, $\dot{\gamma}_d$ represents the desired value of $\dot{\gamma}_i$, which is a known value.

3. Controller Design

Following the works of [7] and [9], we define the position error in the body frame as

$$e_1 = R^T(p - p_d) \tag{10}$$

where $\mathbf{R} = \mathbf{R}(\psi)$ denotes the rotation matrix. The time derivative of $\mathbf{e}_1$ yields

$$\dot{\mathbf{e}}_1 = -\mathbf{S}(\omega)\mathbf{e}_1 + \mathbf{v} - \mathbf{R}^T \dot{\mathbf{p}}_d. \tag{11}$$

Define our first Lyapunov function as

$$V_1 = \frac{1}{2}\mathbf{e}_1^T \mathbf{e}_1 \tag{12}$$

and compute its time derivative, we have

$$\dot{V}_1 = -\alpha_1 + \mathbf{e}_1^T (\mathbf{v} - \mathbf{R}^T \dot{\mathbf{p}}_d + k_1 \mathbf{e}_1) \tag{13}$$

where $\alpha_1 = k_1 \mathbf{e}_1^T \mathbf{e}_1$ is positive definite, and k_1 denotes gain, which is a positive value.

In order to continue to use the backstepping technique, a second error term is defined as

$$\mathbf{e}_2 = \mathbf{v} - \mathbf{R}^T \dot{\mathbf{p}}_d + k_1 \mathbf{e}_1 - \eta, \tag{14}$$

and its integral term

$$\mathbf{e}_{2n} = \int_0^t \mathbf{e}_2 dt \tag{15}$$

where $\eta = [\eta_1, \eta_2]^T$, $\eta_1 \neq 0$ is a constant vector. Define our second Lyapunov function as

$$V_2 = V_1 + \frac{1}{2}\mathbf{e}_2^T \mathbf{e}_2 + \frac{1}{2}\mathbf{e}_{2n}^T \mathbf{e}_{2n}, \tag{16}$$

and its time derivative yields

$$\dot{V}_2 = -\alpha_2 + \mathbf{e}_1^T \eta + \mathbf{e}_2^T \left(-\mathbf{S}(\omega)\eta + m^{-1}\mathbf{T}\mathbf{n}_1 - \mathbf{R}^T \ddot{\mathbf{p}}_d + k_1(\mathbf{e}_2 - k_1 \mathbf{e}_1 + \eta) \right.$$
$$\left. + k_2 \mathbf{e}_2 + \Delta_v + \mathbf{e}_{2n} \right) \tag{17}$$

where $\alpha_2 = k_1 \mathbf{e}_1^T \mathbf{e}_1 + k_2 \mathbf{e}^T \mathbf{e}_2$, which is positive definite, and k_2 is a positive number. It is relevant to point out that $\mathbf{e}_{2n}$ is introduced to eliminate the external slow-varying disturbances that act on the dynamic of the linear velocity $\mathbf{v}$. Moreover, notice that we do not know Δ_v, thereby we use RBFNNs mentioned before to approximate it, given by

$$\Delta_v = \mathbf{W}_{d1}^T \sigma(\mathbf{x}_1) + \epsilon_1(\mathbf{x}_1) \tag{18}$$

where $\mathbf{x}_1 = [1, \mathbf{v}^T]^T \in \mathbb{R}^3$. Moreover, notice that

$$-\mathbf{S}(\omega)\eta + m^{-1}\mathbf{T}\mathbf{n}_1 = \mathbf{N}\mathbf{I} \tag{19}$$

where

$$\mathbf{N} = \begin{bmatrix} m^{-1} & \eta_2 \\ 0 & -\eta_1 \end{bmatrix}. \quad \mathbf{I} = [T, \omega]^T.$$

Rewrite Equation (17), and we have

$$\dot{V}_2 = -\alpha_2 + \mathbf{e}_1^T \eta + \mathbf{e}_2^T \left(\mathbf{N}\mathbf{I} + \beta + \mathbf{W}_{d1}^T \sigma(\mathbf{x}_1) \right) + \mathbf{e}_2^T \epsilon_1(\mathbf{x}_1) \tag{20}$$

where $\beta = -\mathbf{R}^T \ddot{\mathbf{p}}_d + k_1(\mathbf{e}_2 - k_1 \mathbf{e}_1 + \eta) + k_2 \mathbf{e}_2 + \mathbf{e}_{2n}$.

Now, we can define our third Lyapunov function as

$$V_3 = V_2 + \frac{1}{2} \operatorname{tr} \left(\widetilde{\mathbf{W}}_{d1}^T \Gamma_1^{-1} \widetilde{\mathbf{W}}_{d1} \right) \tag{21}$$

where $\widetilde{\mathbf{W}}_{d1} = \mathbf{W}_d - \widehat{\mathbf{W}}_{d1}$ denotes the estimation error, $\Gamma_1 = \operatorname{diag}(\lambda_{11}, and\lambda_{12})$ is a matrix, where λ_{11} and λ_{12} are positive values. Compute the time derivative of V_3, and one obtains

$$\dot{V}_3 = -\alpha_2 + \mathbf{e}_1^T \eta + \mathbf{e}_2^T \epsilon_1(\mathbf{x}_1) + \mathbf{e}_2^T \left(\mathbf{NI} + \beta + \widehat{\mathbf{W}}_{d1}^T \sigma(\mathbf{x}_1) \right)$$
$$+ \mathbf{e}_2^T \widetilde{\mathbf{W}}_{d1}^T \sigma(\mathbf{x}_1) - \operatorname{tr} \left(\widetilde{\mathbf{W}}_{d1}^T \Gamma_1^{-1} \dot{\widehat{\mathbf{W}}}_{d1} \right). \tag{22}$$

Therefore, we choose our desired input $\mathbf{I}_d$ as

$$\mathbf{I}_d = -\mathbf{N}^{-1} \left(\beta + \widehat{\mathbf{W}}_{d1}^T \sigma(\mathbf{x}_1) \right), \tag{23}$$

and thereby our first controller T is chosen as

$$T = \mathbf{n}_1^T \mathbf{I}_d. \tag{24}$$

Correspondingly, the desired angular velocity is

$$\omega_d = \mathbf{n}_2^T \mathbf{I}_d \tag{25}$$

where $\mathbf{n}_2 = [0,1]^T$.

Notice that, if the updated law for $\widehat{\mathbf{W}}_{d1}$ is set as

$$\dot{\widehat{\mathbf{W}}}_{d1} = \Gamma_1 \sigma(\mathbf{x}_1) \mathbf{e}_2^T, \tag{26}$$

we cannot ensure that it is bounded by $\mathbf{W}_{\max}$. To solve this problem, a projection operator, which is Lipschitz continuous [25], is applied in our case, which is given by

$$\mathbf{proj}(\rho, \ell) = \begin{cases} \rho & \text{if } \Theta(\ell) \leq 0 \\ \rho & \text{if } \Theta(\ell) \geq 0 \text{ and } \Theta_{\ell}(\ell)\rho \leq 0 \\ (1 - \Theta(\ell))\rho & \text{if } \Theta(\ell) > 0 \text{ and } \Theta_{\ell}(\ell)\rho > 0 \end{cases} \tag{27}$$

where

$$\Theta(\ell) = \frac{\ell^2 - \ell_{\max}^2}{\epsilon^2 + 2\epsilon\ell_{\max}^2}, \quad \Theta_{\ell}(\ell) = \frac{\partial \Theta(\ell)}{\partial \ell}, \tag{28}$$

with the following condition: if $\dot{\ell} = proj(\rho, \ell)$ and $\ell(t_0) \leq \ell_{\max}$, then

(1) $\ell \leq \ell_{\max} + \epsilon, \forall 0 \leq t < \infty$;
(2) $\mathbf{proj}(\rho, \ell)$ is Lipschitz continuous;
(3) $|\mathbf{proj}(\rho, \ell)| \leq |\rho|$;
(4) $\tilde{\ell}\mathbf{proj}(\rho, \ell) \geq \tilde{\ell}\rho$, where $\tilde{\ell} = \ell - \hat{\ell}$.

Therefore, to make sure all elements of $\widehat{\mathbf{W}}_{d1}$ are upper-bounded, the update law for $\widehat{\mathbf{W}}_{d1}$ is finally set as

$$\dot{\widehat{\mathbf{W}}}_{d1} = \Gamma_1 \mathbf{proj} \left(\sigma(\mathbf{x}_1) \mathbf{e}_2, \widehat{\mathbf{W}}_{d1} \right). \tag{29}$$

To keep using the backstepping technique, we define a new error term

$$\mathbf{e}_3 = \omega - \omega_d, \tag{30}$$

and its corresponding integral term

$$\mathbf{e}_{3n} = \int_0^t \mathbf{e}_3 dt. \tag{31}$$

Then, we define a new Lyapunov function as

$$V_4 = V_3 + \frac{1}{2}\mathbf{e}_3^T \mathbf{e}_3 + \frac{1}{2}\mathbf{e}_{3n}^T \mathbf{e}_{3n}. \tag{32}$$

Compute its time derivatives, substitute Equation (29) into Equation (24), and combine the 4th property of projector ($\ell \mathbf{proj}(\rho, \ell) \geq \ell \rho$), and one obtains

$$\dot{V}_4 \leq -\alpha_3 + \mathbf{e}_1^T \boldsymbol{\eta} + \mathbf{e}_2^T \boldsymbol{\epsilon}_1(\mathbf{x}_1) + \mathbf{e}_3^T \left(\mathbf{n}_2^T \mathbf{N}^T \mathbf{e}_2 - \frac{1}{J}\boldsymbol{\tau} + \mathbf{n}_2^T \mathbf{N}^{-1}(\dot{\boldsymbol{\beta}} + \hat{\mathbf{W}}_{d1}\sigma(\mathbf{x}_1) \right.$$
$$\left. + \hat{\mathbf{W}}_{d1}\dot{\sigma}(\mathbf{x}_1)) + k_3 \mathbf{e}_3 + \mathbf{e}_{3n} + \Delta_\omega \right) \tag{33}$$

where $\alpha_3 = \alpha_2 + k_3 \mathbf{e}_3^T \mathbf{e}_3 \geq 0$, $k_3 > 0$. However, it is noted that both $\dot{\boldsymbol{\beta}}$ and $\dot{\sigma}(\mathbf{x}_1)$ contain unmodeled term Δ_v, so we need to separate Δ_v out from $\dot{\boldsymbol{\beta}}$ and $\dot{\sigma}(\mathbf{x}_1)$. After that, we can use RBFNNs to estimate it. Similar to Δ_ω, we also need to approximate it. This is given by,

$$\Delta_v = \mathbf{W}_{d2}^T \sigma(\mathbf{x}_2) + \boldsymbol{\epsilon}_2(\mathbf{x}_2) \tag{34}$$
$$\Delta_\omega = \mathbf{W}_{d3}^T \sigma(\mathbf{x}_3) + \boldsymbol{\epsilon}_3(\mathbf{x}_3) \tag{35}$$

where $\mathbf{x}_2 = [1, \mathbf{v}^T]^T \in \mathbb{R}^3$ and $\mathbf{x}_3 = [1, \mathbf{v}^T, \omega]^T \in \mathbb{R}^4$.

Now we define our last Lyapunov function as

$$V_5 = V_4 + \frac{1}{2}\operatorname{tr}\left(\tilde{\mathbf{W}}_{d2}^T \Gamma_2^{-1} \tilde{\mathbf{W}}_{d2}\right) + \frac{1}{2}\operatorname{tr}\left(\tilde{\mathbf{W}}_{d3}^T \Gamma_3^{-1} \tilde{\mathbf{W}}_{d3}\right) \tag{36}$$

where $\tilde{\mathbf{W}}_{d2} = \mathbf{W}_d - \hat{\mathbf{W}}_{d2}$, $\tilde{\mathbf{W}}_{d3} = \mathbf{W}_d - \hat{\mathbf{W}}_{d3}$, $\Gamma_2 = \operatorname{diag}(\lambda_{21}, \lambda_{22})$, and $\Gamma_3 = \operatorname{diag}(\lambda_{31}, \lambda_{32})$ are positive definite gain matrices. Then, we compute the time derivative of V_5 as

$$\dot{V}_5 \leq -\alpha_3 + \mathbf{e}_1^T \boldsymbol{\eta} + \mathbf{e}_2^T \boldsymbol{\epsilon}_1(\mathbf{x}_1) + \mathbf{e}_3^T (\mathbf{n}_2^T \mathbf{M}) \boldsymbol{\epsilon}_2(\mathbf{x}_2) + \mathbf{e}_3^T \boldsymbol{\epsilon}_3(\mathbf{x}_3) + \mathbf{e}_3^T \left(\mathbf{n}_2^T \mathbf{N}^T \mathbf{e}_2 + k_3 \mathbf{e}_3 \right.$$
$$\left. - \frac{\boldsymbol{\tau}}{J} - \mathbf{n}_2^T \hat{\mathbf{I}}_d - \mathbf{n}_2^T \mathbf{M}\hat{\mathbf{W}}_{2d}^T \sigma(\mathbf{x}_2) + \hat{\mathbf{W}}_{3d}^T \sigma(\mathbf{x}_3) + \mathbf{e}_{3n} \right) + \mathbf{e}_3^T \mathbf{n}_2^T \mathbf{M}\tilde{\mathbf{W}}_{2d}^T \sigma(\mathbf{x}_2) \tag{37}$$
$$+ \mathbf{e}_3^T \tilde{\mathbf{W}}_3^T \sigma(\mathbf{x}_3) - \operatorname{tr}\left(\tilde{\mathbf{W}}_{d2}^T \Gamma_2^{-1} \hat{\mathbf{W}}_{d2}\right) - \operatorname{tr}\left(\tilde{\mathbf{W}}_{d3}^T \Gamma_3^{-1} \hat{\mathbf{W}}_{d3}\right)$$

where

$$\hat{\mathbf{I}}_d = \dot{\hat{\boldsymbol{\beta}}} + \hat{\mathbf{W}}_{d1}\sigma(\mathbf{x}_1) + \hat{\mathbf{W}}_{d1}\dot{\sigma}(\mathbf{x}_1) \tag{38}$$

and

$$\mathbf{M} = \mathbf{N}^{-1}\hat{\mathbf{W}}_{d1}^T \mathbf{G} + (k_1 + k_2)\mathbf{N}^{-1} \tag{39}$$

with $\mathbf{G} = [\mathbf{0}_{2 \times 1}^T, \mathbf{n}_1 \sigma'(\mathbf{n}_1^T \mathbf{v}), and \mathbf{n}_1 \sigma'(\mathbf{n}_2^T \mathbf{v})]^T$. Then, we define our second control law, torque, as

$$\boldsymbol{\tau} = J(-\mathbf{n}_2^T \hat{\mathbf{I}}_d + \mathbf{n}_2^T \mathbf{N}^T \mathbf{e}_2 + k_3 \mathbf{e}_3 + \mathbf{n}_2^T \mathbf{M}\hat{\mathbf{W}}_{d2}^T \sigma(\mathbf{x}_2) + \hat{\mathbf{W}}_{d3}^T \sigma(\mathbf{x}_3) + \mathbf{e}_{3n}), \tag{40}$$

and estimate laws for $\hat{\mathbf{W}}_{d2}$ and $\hat{\mathbf{W}}_{d3}$

$$\dot{\hat{\mathbf{W}}}_{d2} = \Gamma_2 \mathbf{proj}\left(\sigma(\mathbf{x}_2)(\mathbf{n}_2^T \mathbf{M})\mathbf{e}_3^T \hat{\mathbf{W}}_{d2}\right), \tag{41}$$

$$\dot{\hat{\mathbf{W}}}_{d3} = \Gamma_3 \mathbf{proj}\left(\sigma(\mathbf{x}_3)\mathbf{e}_3^T, \hat{\mathbf{W}}_{d3}\right). \tag{42}$$

Substitute Equation (40)–(42) into Equation (37), one obtains

$$\dot{V}_5 \leq -\alpha_3 + \mathbf{e}_1^T \boldsymbol{\eta} + \mathbf{e}_2^T \boldsymbol{\epsilon}_1(\mathbf{x}_1) + \mathbf{e}_3^T(\mathbf{n}_2^T \mathbf{M})\boldsymbol{\epsilon}_2(\mathbf{x}_2) + \mathbf{e}_3^T \boldsymbol{\epsilon}_3(\mathbf{x}_3). \tag{43}$$

where $\alpha_3 = k_1 \mathbf{e}_1^T \mathbf{e}_1 + k_2 \mathbf{e}_2^T \mathbf{e}_2 + k_3 \mathbf{e}_3^T \mathbf{e}_3$.

3.1. Stability Analysis

Theorem 1. *For a single mobile robot, by applying control laws, Equations (24) and (40), and updated laws, Equation (29), Equation (41), and Equation (42), for any large initial position, the robot will converge to the neighborhood of its corresponding desired path $\mathbf{p}_d(\gamma)$, whose partial derivatives with respect to γ are all bounded. As a consequence, global uniformly ultimately boundedness is achieved.*

Proof. Let's go back to Equation (43). Rewrite it, and we obtain

$$\dot{V}_5 \leq -\mathbf{X}^T \mathbf{K} \mathbf{X} + \mathbf{X}^T \boldsymbol{\rho}$$
$$\leq |\mathbf{X}|^T\left(-k_{min}|\mathbf{X}| + |\boldsymbol{\rho}|\right) \tag{44}$$

where $\mathbf{X} = [\mathbf{e}_1^T, \mathbf{e}_2^T, \mathbf{e}_3^T]$, k_{min} is the smallest eigenvalue of $\mathbf{K} = \text{diag}(k_1, k_2, k_3)$, and $\boldsymbol{\rho} = [\boldsymbol{\eta}^T, \boldsymbol{\epsilon}_1(\mathbf{x}_1)^T, (\mathbf{n}_2^T \mathbf{M})\boldsymbol{\epsilon}_2(\mathbf{x}_2), \boldsymbol{\epsilon}_3(\mathbf{x}_3)]^T$, which is bounded due to the fact that $||\boldsymbol{\epsilon}_i(\mathbf{x})|| \leq \epsilon_{max}$, and the upper bound of $\boldsymbol{\rho}$ is

$$\boldsymbol{\rho}_{max} = [\boldsymbol{\eta}^T, \boldsymbol{\epsilon}_{1\,max}(\mathbf{x}_1)^T, (\mathbf{n}_2^T \mathbf{M})\boldsymbol{\epsilon}_{2\,max}(\mathbf{x}_2), \boldsymbol{\epsilon}_{3\,max}(\mathbf{x}_3)]^T. \tag{45}$$

Thereby, we can obtain that $\dot{V}_5$ is negative for $||\mathbf{X}|| \geq ||\boldsymbol{\rho}_{max}/k_{min}||$, which can be made as small as possible by tuning the value of k_{min}. As a result, the system is uniformly ultimate bounded, global uniformly ultimate boundedness is achieved. □

4. Consensus Strategy

Building upon the work of [14], the proposed solution is given by

$$\dot{\gamma}_i = v_d \mathbf{1} - a_1 \sum_{j \in N_i} (\gamma_i - \gamma_j) + z_i \tag{46}$$

$$\dot{z}_i = -a_2 z_i + a_3 \sum_{j \in N_i} (\gamma_i - \gamma_j) \tag{47}$$

where $a_1, a_2, and a_3$ are positive numbers, and v_d denotes the desired value of $\dot{\gamma}_i$. It is relevant to point out that z_i can be viewed as an auxiliary state that helps n paths to reach consensus.

4.1. Stability Analysis

Theorem 2. *For a group of n desired paths $\mathbf{p}_d(\gamma_i)$ ($i = 1, 2, \ldots, n$), by applying Equations (46) and (47), $\gamma_i - \gamma_j$ and $\dot{\gamma}_i - \dot{\gamma}_d$ converge to zero.*

Proof. We first choose Laplacian matrix $\mathcal{L}$, and define the coordinate error as

$$\Gamma_e = \mathcal{L}\Lambda \tag{48}$$

where $\Lambda = [\gamma_i]_{n \times 1}$. Rewrite Equations (46) and (47), and one obtains

$$\dot{\Gamma}_e = -A_1 \mathcal{L} \Gamma_e + \mathcal{L} Z, \tag{49}$$

$$\dot{Z} = -A_2 Z + A_3 \Gamma_e, \tag{50}$$

where $A_i = \mathrm{diag}(a_i), (i = 1,2,3)$ is a positive definite matrix. Define $x = [\Gamma_e, Z]^T$, and rewrite Equations (49) and (50), and we have

$$\dot{x} = Ax \tag{51}$$

where

$$A = \begin{bmatrix} -A_1 \mathcal{L} & \mathcal{L} \\ A_3 & -A_2 \end{bmatrix}. \tag{52}$$

In order to ensure that Equation (51) is stable, we need to guarantee that all the eigenvalues of A have negative or zero real parts and all the Jordan blocks corresponding to eigenvalues with zero real parts are 1×1. We consider n agents, where

$$\mathcal{L} = \begin{vmatrix} 0 & 0 & 0 & \cdots & 0 & 0 \\ -1 & 1 & 0 & \cdots & 0 & 0 \\ 0 & -1 & 1 & \cdots & 0 & 0 \\ \vdots & \vdots & \vdots & \vdots & \vdots & \vdots \\ 0 & 0 & 0 & \cdots & -1 & 1 \end{vmatrix}_{n \times n}. \tag{53}$$

For the sake of saving space, here we just present the eigenvalues of A directly, they are $\lambda_1, \lambda_2, \lambda_3$, and λ_4, and their corresponding multiplicities are $1, 1, n-1$, and $n-1$, with

$$\lambda_1 = 0 \tag{54}$$

$$\lambda_2 = -a_2 \tag{55}$$

$$\lambda_3 = -(a_1 + a_2)/2 + \sqrt{a_1^2 + a_2^2 - 2a_1 a_2 + 4a_3}/2 \tag{56}$$

$$\lambda_4 = -(a_1 + a_2)/2 - \sqrt{a_1^2 + a_2^2 - 2a_1 a_2 + 4a_3}/2. \tag{57}$$

Choosing $a_1, a_2,$ and a_3 properly, we can guarantee that λ_2, λ_3, and λ_4 are negative-definite. Moreover, we also have that the Jordan block corresponding to λ_1 is 1×1. As a consequence, Equation (51) is stable [26]. $\square$

To summarize, a fleet of n desired paths can progress in a specific formation, while each individual mobile robot converges to the neighborhood of its corresponding desired path. As a result, all those robots can move forward in a specific formation.

5. Simulation Results

In this section, we present simulation results with two different communication graphs including a cascade-directed communication graph (CDCG) and a parallel-directed communication graph (PDCG). Figures 3 and 4 show the sketches of the control blocks in Simulink/Matlab.

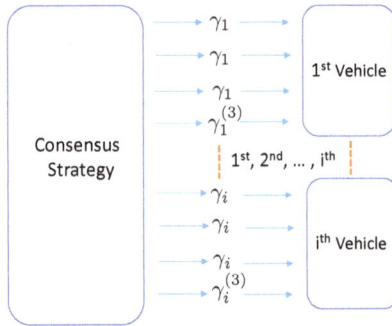

Figure 3. Coordinated control block in Simulink/Matlab.

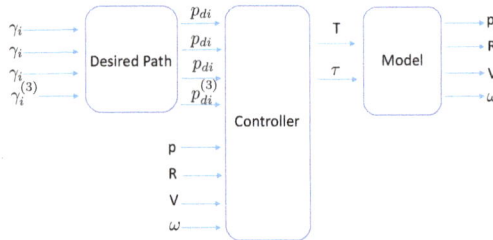

Figure 4. Control block in Simulink/Matlab for the *i*-th vehicle.

5.1. The Cascade-Directed Communication Graph

The CDCG used in this study is shown in Figure 5, where agent 1 can be viewed as the leader. Its corresponding Laplacian matrix $\mathscr{L}_1$ is

$$\mathscr{L}_1 = \begin{bmatrix} 0 & 0 & 0 & 0 & 0 \\ -1 & 1 & 0 & 0 & 0 \\ 0 & -1 & 1 & 0 & 0 \\ 0 & 0 & -1 & 1 & 0 \\ 0 & 0 & 0 & -1 & 1 \end{bmatrix}. \tag{58}$$

The desired paths are defined as

$$\mathbf{p}_{di}(\gamma_i) = R_i \begin{bmatrix} \cos(\gamma_i) \\ \sin(\gamma_i) \end{bmatrix} \text{ (m)} \tag{59}$$

where $R_i = 6 - i$, $(i = 1, 2, \ldots, 5)$ denotes the radius of the circle, and unmodeled terms Δ_V and Δ_ω are denoted by $[0.1u^2 + 0.01uv, 0.01uv + 0.1v^2]^T$ and $0.01uv + 0.05u\omega + 0.01v\omega$, respectively. The parameters used herein are as follows: $m = 0.6$, $J = 0.1$, $c_i = 2$, $a_1 = 1$, $a_2 = 2$, $a_3 = 1$, $k_1 = 6$, $k_2 = 3$, $k_3 = 2$, $\Gamma_1 = \text{diag}(0.6, 0.6)$, $\Gamma_2 = \text{diag}(0.01, 0.01)$, $\Gamma_3 = \text{diag}(0.022, 0.022)$, and $\eta = [0.2, 0]^T$.

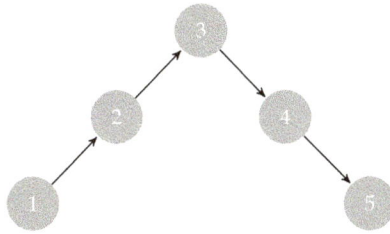

Figure 5. A cascade-directed communication graph (CDCG).

Figure 6a shows the actual trajectory of the mobile robots, and we can see that they move forward in a line formation. Moreover, Figures 7a,b display the convergence of $||\mathbf{e}_{1i}||$ and $||\mathbf{e}_{2i}||$, respectively, showing that all of them converge to the neighborhood of zero. The consensus performances are shown in Figures 8a,b, showing that $\gamma_{ij} = \gamma_i - \gamma_j$ converges to zero, where agent j is the neighbor of agent i. Moreover, we can also see that $\dot\gamma_i$ converges to the desired value $\dot\gamma_d = 0.5$. It is important to point out that, in this work, we chosen agent 1 as our leader, and this satisfies $\dot\gamma_1 = \dot\gamma_d$. The approximate performance of RBFNNs can be found in Figure 6b, where the blue lines denote the real values of $\boldsymbol{\Delta}_v$ and $\boldsymbol{\Delta}_\omega$, while the red lines represent their estimates $\hat{\boldsymbol{\Delta}}_v$ and $\hat{\boldsymbol{\Delta}}_\omega$. Thus, both estimates converge to their corresponding real values.

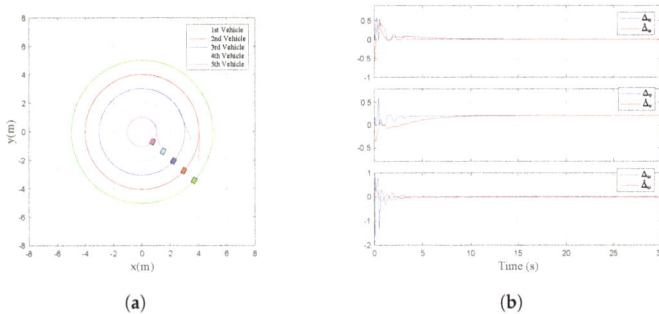

(a)   (b)

Figure 6. Norm of the position errors and the performance of unmodeled term approximation (CDCG). (a) Norm of the position errors. (b) Performance of unmodeled term approximation.

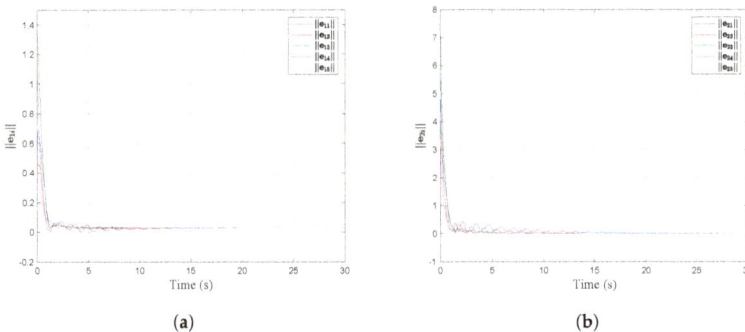

(a)   (b)

Figure 7. Norm of the position and linear velocity errors (CDCG). (a) Norm of the position errors. (b) Norm of the linear velocity errors.

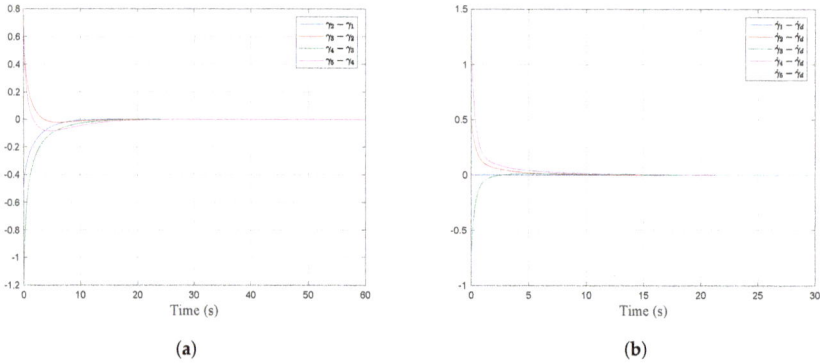

(a)

(b)

Figure 8. Performance of $\gamma_i - \gamma_j$ and $\dot{\gamma}_i - \dot{\gamma}_d$, where γ_j is the neighbor of γ_i (CDCG). (**a**) Performance of $\gamma_i - \gamma_j$. (**b**) Performance of $\dot{\gamma}_i - \dot{\gamma}_d$.

5.2. Parallel Communication Graph

The parallel communication graph is depicted in Figure 9.

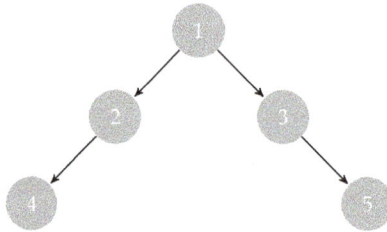

Figure 9. A parallel-directed communication graph (PDCG).

In this case, the Laplacian matrix $\mathscr{L}_2$ is

$$
\mathscr{L}_2 = \begin{bmatrix}
-1 & 1 & 0 & 0 & 0 \\
-1 & 0 & 1 & 0 & 0 \\
0 & -1 & 0 & 1 & 0 \\
0 & 0 & -1 & 0 & 1 \\
0 & 0 & 0 & 0 & 0
\end{bmatrix}. \tag{60}
$$

Moreover, it is noted that the graph presented in Figure 9 can be viewed as the combination of two cascade-directed graphs, where $1 \to 2 \to 4$ and $1 \to 3 \to 5$ and where agent **1** is the leader whose state is known. The desired paths are as follows:

(1) 1st vehicle:

$$
\mathbf{p}_{d1}(\gamma_1) = R_1 \begin{bmatrix} \cos(\gamma_1) \\ \sin(\gamma_1) \end{bmatrix} \ (\mathrm{m}) \tag{61}
$$

(2) 2nd vehicle:

$$
\mathbf{p}_{d2}(\gamma_2) = R_2 \begin{bmatrix} \cos(\gamma_2 - \pi/24) \\ \sin(\gamma_2) - \pi/24 \end{bmatrix} \ (\mathrm{m}) \tag{62}
$$

(3) 3rd vehicle:

$$\mathbf{P}_{d3}(\gamma_3) = R_3 \begin{bmatrix} \cos(\gamma_3 - \pi/12) \\ \sin(\gamma_3 - \pi/12) \end{bmatrix} \text{ (m)} \tag{63}$$

(4) 4th vehicle:

$$\mathbf{P}_{d4}(\gamma_4) = R_4 \begin{bmatrix} \cos(\gamma_4 - \pi/24) \\ \sin(\gamma_4 - \pi/24) \end{bmatrix} \text{ (m)} \tag{64}$$

(5) 5th vehicle:

$$\mathbf{P}_{d5}(\gamma_5) = R_5 \begin{bmatrix} \cos(\gamma_5 - \pi/12) \\ \sin(\gamma_5 - \pi/12) \end{bmatrix} \text{ (m)} \tag{65}$$

where $R_1 = 3$, $R_2 = 4$, $R_3 = 2$, $R_4 = 5$, and $R_5 = 1$. Figure 10a displays the actual paths of the robots. From Figure 11a,b, we can obtain that the norm of position errors and linear velocity errors converge to a ball centered at the origin. Moreover, Figure 12a,b show the performance of the consensus strategy introduced herein.

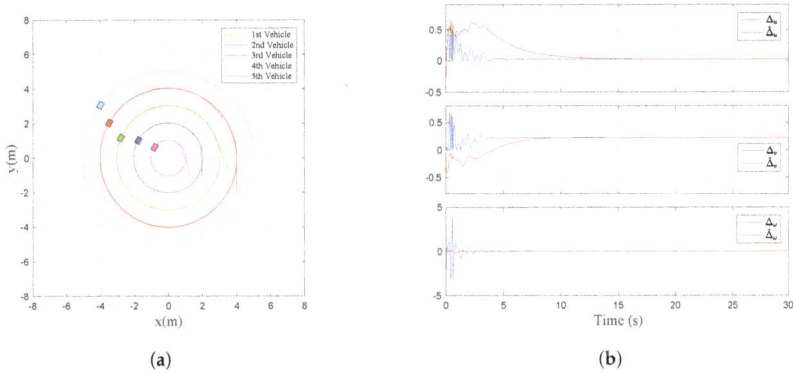

(a)   (b)

Figure 10. Norm of the position errors and performance of unmodeled term approximation (PDCG). (a) Norm of the position errors. (b) Performance of unmodeled term approximation.

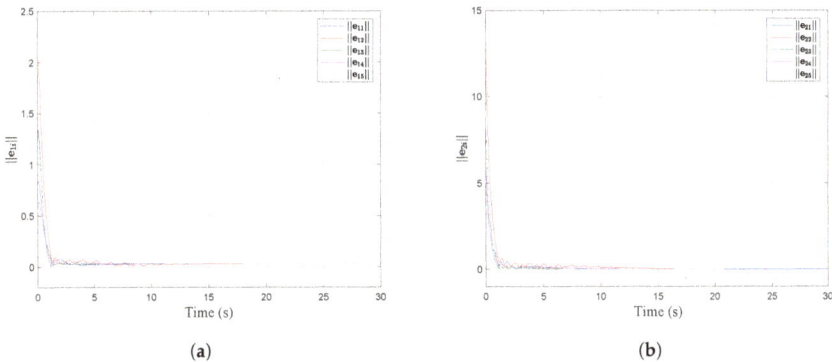

(a)   (b)

Figure 11. Norm of the position and linear velocity errors (PDCG). (a) Norm of the position errors. (b) Norm of the linear velocity errors.

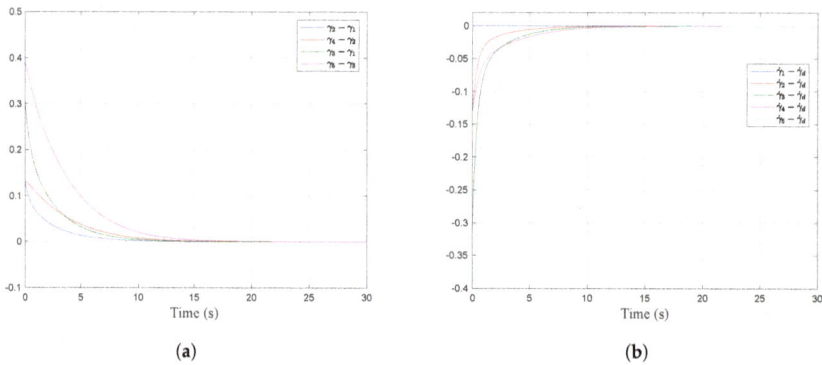

Figure 12. Performance of $\gamma_i - \gamma_j$ and $\dot{\gamma}_i - \dot{\gamma}_d$, where γ_j is the neighbor of γ_i (PDCG). (**a**) Performance of $\gamma_i - \gamma_j$. (**b**) Performance of $\dot{\gamma}_i - \dot{\gamma}_d$.

It is interesting to remark that, by using the proposed consensus strategy—Equations (46) and (47)—a consensus is researched if and only if the directed graph has a directed spanning tree [21]. However, in our case, the root must be the leader whose states are known beforehand.

6. Conclusions

In this paper, we mainly focus on designing coordinated control algorithms for multiple agents, where a group of underactuated hovercrafts were chosen as test platforms. In order to testify the efficiency of the devised control strategy, we implemented it by using Simulink/Matlab. Moreover, it is necessary to point out that agents can also be mobile robots, unmanned air vehicles, etc. For a single vehicle, we used RBFNNs to approximate unmodeled terms and introduce integral terms, which can improve the robustness of the controller. For multiple vehicles, we consider directed topology under the assumption that the communication among vehicles are continuous.

With respect to our future works, we plan to (i) use deep neural networks to estimate unmodeled terms so as to enhance the performance of approximation, (ii) build a mathematical model for external disturbance, such as winds, waves, or currents, (iii) take into account time-delays when we develop communication strategy for vehicles, and (iv) propose collision–avoidance algorithms so that we can ensure the operation is safe.

Author Contributions: K.D. and S.F. conceived and designed the experiments; K.D. performed the experiments and wrote the paper; Y.Z. and W.S. provided guidance for this paper.

Acknowledgments: The authors are grateful for financial support from the following research grants: (1) the 'Nature-Inspired Computing and Metaheuristics Algorithms for Optimizing Data Mining Performance' grant from the University of Macau (grant no. MYRG2016-00069-FST); the (2) 'Improving the Protein–Ligand Scoring Function for Molecular Docking by Fuzzy Rule-Based Machine Learning Approaches' grant from the University of Macau (grant no. MYRG2016-00217-FST), and (3) the 'A Scalable Data Stream Mining Methodology: Stream-based Holistic Analytics and Reasoning in Parallel' grant from the FDCT, the Macau Government (grant no. FDCT/126/2014/A3).

Conflicts of Interest: The authors declare no conflict of interest.

References

1. Rawat, P.; Singh, K.D.; Chaouchi, H.; Bonnin, J.M. Wireless sensor networks: a survey on recent developments and potential synergies. *J. Supercomput.* **2014**, *68*, 1–48. [CrossRef]
2. Yick, J.; Mukherjee, B.; Ghosal, D. Wireless sensor network survey. *Comput. Netw.* **2008**, *52*, 2292–2330. [CrossRef]
3. Majid, M.H.A.; Arshad, M.R. A Fuzzy Self-Adaptive PID Tracking Control of Autonomous Surface Vehicle. In Proceedings of the IEEE International Conference on Control System, Computing and Engineering, George Town, Malaysia, 27–29 November 2015.
4. Fantoni, I.; Lozano, R.; Mazenc, F.; Pettersen, K.Y. Stabilization of a nonlinear underactuated hovercraft. In Proceedings of the 38th Conference on Decision and Control, Phoenix, AZ, USA, 7–10 December 1999.
5. Do, K.D.; Jiang, Z.P.; Pan, J. Underactuated Ship Global Tracking Under Relaxed Conditions. *IEEE Trans. Autom. Control* **1999**, *47*, 1529–1536. [CrossRef]
6. Do, K.D.; Jiang, Z.P.; Pan, J. Universal controllers for stabilization and tracking of underactuated ships. *Syst. Control Lett.* **2002**, *47*, 299–317. [CrossRef]
7. Aguiar, A.P.; Cremean, L.; Hespanha, J.P. Position tracking for a nonlinear underactuated hovercraft: Controller design and experimental results. In Proceedings of the 42nd IEEE Conference on Decision and Control, Maui, HI, USA, 9–12 December 2003; Volume 4, pp. 3858–3863.
8. Bertaska, I.R.; Ellenrieder, K.D. Experimental Evaluation of Supervisory Switching Control for Unmanned Surface Vehicle. *IEEE J. Ocean. Eng.* **2017**. [CrossRef]
9. Xie, W. Robust Motion Control of Underactuated Autonomous Surface Craft. Master's Thesis, University of Macau, Macau, China, 2016.
10. Peng, Z.H.; Wang, J.; Wang, D. Distributed Maneuvering of Autonomous Surface Vehicles Based on Neurodynamic Optimization and Fuzzy Approximation. *IEEE Trans. Control Syst. Technol.* **2018**, *26*, 1083–1090. [CrossRef]
11. Chen, X.; Tan, W.W. Tracking control of surface vessel via fault tolerant adaptive backstepping interval type-2 fuzzy control. *Ocean Eng.* **2013**, *26*, 97–109. [CrossRef]
12. Wang, N.; Meng, J.E. Self-Constructing Adaptive Robust Fuzzy Neural Tracking Control of Surface Vehicles With Uncertainties and Unknown Disturbances. *IEEE Trans. Control Syst. Technol.* **2015**, *23*, 991–1002. [CrossRef]
13. Almeida, J.; Silvestre, C.; Pascoal, A. Cooperative control of multiple surface vessels in the presence of ocean currents and parametric model uncertainty. *Int. J. Robust Nonlinear Control* **2010**, *20*, 1549–1565. [CrossRef]
14. Ghabcheloo, R.; Aguiar, A.P.; Pascoal, A.; Silvestre, C.; Kaminer, I.; Hespanha, J.P. Coordinated path-following control of multiple underactuated autonomous vehicles in the presence of communication failures. In Proceedings of the 45th IEEE Conference on Decision Control, San Diego, CA, USA, 13–15 December 2006.
15. Arrichiello, F.; Chiaverini, S.; Fossen, T.I. Formation control of marine surface vessels using the null-space-based behavioral control. In *Group Coordination and Cooperative Control. Lecture Notes in Control and Information Science*; Springer: Berlin, Germany, 2006; Volume 336.
16. Balch, T.; Arkin, R.C. Behavior-based formation control for multirobot teams. *IEEE Trans. Robot. Autom.* **1999**, *14*, 926–939. [CrossRef]
17. Shojaei, K. Leader–follower formation control of underactuated autonomous marine surface vehicles with limited torque. *Ocean Eng.* **2015**, *105*, 196–205. [CrossRef]
18. Peng, Z.H.; Wang, D.; Chen, Z.Y.; Hu, X.J.; Lan, W.Y. Adaptive Dynamic Surface Control for Formations of Autonomous Surface Vehicles with Uncertain Dynamics. *IEEE Trans. Control Syst. Technol.* **2013**, *21*, 513–520. [CrossRef]
19. Luca, C.; Fabio, M.; Domenico, P.; Mario, T. Leader-follower formation control of nonholonomic mobile robots with input constraints. *Automatica* **2008**, *44*, 1343–1349.
20. Wang, Q.; Chen, Z.; Yi, Y. Adaptive coordinated tracking control for multi-robot system with directed communication topology. *Int. J. Adv. Rob. Syst.* **2017**, *14*. [CrossRef]
21. Ren, W.; Beard, R. *Distributed Consensus in Multi-Vehicle Cooperative Control-Theory and Applications*; Springer: Berlin, Germany, 2007.
22. Fossen, T.I. Modeling of Marine Vehicles. In *Guidance and Control of Ocean Vehicles*; John Wiley & Sons Ltd.: Chichester, UK, 1994; pp. 6–54, ISBN 0-471-94113-1.

23. Wen, G.X.; Chen, P.C.L.; Liu, Y.J.; Liu, Z. Neural-network-based adaptive leader-following consensus control of second-order nonlinear multi-agent systems. *IET Control Theory Appl.* **2015**, *9*, 1927–1934. [CrossRef]

24. Wen, G.X.; Chen, P.C.L.; Liu, Y.J.; Liu, Z. Neural Network-Based Adaptive Leader-Following Consensus Control for a Class of Nonlinear Multiagent State-Delay Systems. *IEEE Trans. Cybern.* **2017**, *47*, 2151–2160. [CrossRef] [PubMed]

25. Do, K.D. Robust adaptive path following of underactuated ships. *Automatic* **2004**, *40*, 929–944. [CrossRef]

26. Hespanha, J.P. Internal or Lyapunov Stability. In *Linear Systems Theory*; Princeton University Press: Princeton, NJ, USA, 2009; pp. 63–78, ISBN 978-0-691-14021-6.

applied
sciences

MDPI

Article

A Multi-Agent System for the Dynamic Emplacement of Electric Vehicle Charging Stations

Jaume Jordán *,†, Javier Palanca †, Elena del Val †, Vicente Julian † and Vicente Botti †

Departamento de Sistemas Informáticos y Computación, Universitat Politècnica de València,
Camino de Vera s/n, 46022 Valencia, Spain; jpalanca@dsic.upv.es (J.P.), edelval@dsic.upv.es (E.d.V.);
vinglada@dsic.upv.es (V.J.); vbotti@dsic.upv.es (V.B.)
* Correspondence: jjordan@dsic.upv.es; Tel.: +34-963-877-350
† These authors contributed equally to this work.

Received: 30 December 2017; Accepted: 14 February 2018; Published: 23 February 2018

Abstract: One of the main current challenges of electric vehicles (EVs) is the creation of a reliable, accessible and comfortable charging infrastructure for citizens in order to enhance demand. In this paper, a multi-agent system (MAS) is proposed to facilitate the analysis of different placement configurations for EV charging stations. The proposed MAS integrates information from heterogeneous data sources as a starting point to characterize the areas where charging stations could potentially be placed. Through a genetic algorithm, the MAS is able to analyze a large number of possible configurations, taking into account a set of criteria to be optimized. Finally, the MAS returns a configuration with the areas of the city that are considered most appropriate for the establishment of charging stations according to the specified criteria.

Keywords: multi-agent systems; electric vehicles; charging stations; genetic algorithm

1. Introduction

European governments are focused on greatly reducing the transport sector's carbon emissions. An element that plays a key role is the electric vehicle (EV) and the expected uptake of the EV market [1]. The electric vehicle, powered in whole or in part by electricity from electricity grids, is more efficient and environmentally friendly (i.e., results in lower emissions of gases and noise) when compared to other current propulsion technologies and current vehicles. Among the modes of transportation in cities that need solutions for their migration from emitting technologies to non-polluting alternatives such as electricity, we must take into account public vehicles (i.e., buses, taxis, postal service vehicles, rental vehicles) as well as as private vehicles (i.e., cars, motorcycles, bicycles, mototaxis).

Currently, consumers can choose from an wide number of electric vehicle models that provide many environmental benefits. According to the European Environment Agency, although the number of electric vehicle sales has increased rapidly over past years, they represented just 1.2% of all new cars sold in the EU in 2015 [2].

The causes that can hinder the introduction of EV use in cities include the following: limited information and technological uncertainty (in comparison with the more familiar conventional vehicle technologies), the limitations of battery life and charging times of EVs, and the lack a charging infrastructure that covers the potential demand for the EVs [3]. The last cause is closely related to range anxiety that is considered a significant obstacle to market acceptance of EVs [4]. Range anxiety is the fear that the vehicle has insufficient range to reach the destination [5]. An option to deal with this problem is through the deployment of an efficient public charging infrastructure. However, since infrastructure development is expensive, there is a need to direct investments towards the establishment of refueling facilities in areas with maximum impact (i.e., maximum coverage at a low cost).

In this paper, the aim is to provide a multi-agent system (MAS) for the planning of efficiently located infrastructures for electric vehicle charging stations for the public and private sectors in a city. For this purpose, we present an MAS that integrates the collection of information from heterogeneous data sources and the optimization of the charging station locations using artificial intelligence algorithms. In this way, the proposed MAS allows the evaluation of a set of possible configurations of charging station locations. The MAS also considers different configurable criteria and determines which is the most advisable configuration of charging station locations.

This proposal will contribute in the research and technology transfer of new analysis systems for the strategic planning and distribution of urban elements, such as charging stations for EVs. The proposed system will make it possible to: (1) guarantee the supply of electricity for charging EVs in the city; and (2) optimize investment in infrastructures and charging stations, which results in greater sustainability.

The problem lies in where should the charging stations should be established and how many charging piles should be established in each charging station in order to provide service to as many EV users as possible.

The article is structured in the following sections. Section 2 presents previous works related to the proposal presented. Section 3 describes the main aspects of the proposed MAS and details the *Emplacement Optimizer Agent*, which is the core of the system. Section 4 presents a case study for the city of Valencia, and Section 5 shows some experimental results over the proposed case study. Finally, Section 6 provides some conclusions and information on future works.

2. Related Work

From the user's point of view, there are several commercial tools related to charging stations. Most of them focus on the user orientation, trying to help them to find nearby stations or plan a route while taking into account charging needs during the journey [6–9].

From the manager's point of view (i.e., governments, administrations, town councils, etc.), there are initiatives that try to provide support, through implementation guidelines, to the selection problem of the most appropriate placement of charging stations in a city [10,11]. A large part of the research work related to charging stations focuses on the scenario where N EVs must be charged along T units of time [12] or how to reduce the impact of EVs on the grid of the electric supply [13].

Other works are more focused on the analysis of placement configurations of charging stations (see Table 1). Some works analyze the distribution of charging stations considering vehicle travel range constraints. Shukla et al. [14] propose mathematical programming for determining the best locations for establishing alternative transportation fuel stations. The goal of this proposal was to site the refueling stations at locations that maximize the number of vehicles served, while staying within budget constraints. The proposed model is a modification of the flow interception facility location model. Nie et al. [15] present conceptual optimization model to analyze travel by EVs along a long corridor. The objective of the model is to select the charging power at each station and the number of stations needed along the corridor to meet a given level of service in such a way that the total social cost is minimized.

Other studies analyzed the distribution of electric stations in cities. Wood et al. [16] focus on the estimated number of charging stations needed to substantially increase the utility of the vehicle and how the stations can be strategically located to maximize their potential benefit to future EV owners. This approach uses travel profiles, driver behavior, vehicle performance, battery attributes, environmental conditions, and charging infrastructure to optimize the performance of the EV and the charging station.

Table 1. Comparison of approaches that deal with the electric vehicle (EV) charging station placement. PoI: point of interest.

	Traffic, Frequent Routes	Social Data	Population	Time Spent in a PoI	Cost Per Station	Demand Per Station
Shukla et al.	✓				✓	
Nie et al.	✓				✓	
Wood et al.	✓			✓		✓
Wagner et al.	✓			✓		✓
Wei et al.	✓					✓
Dong et al.	✓				✓	✓
Li et al.	✓			✓	✓	✓
Sweda et al.	✓	✓				
Proposal	✓	✓	✓	✓	✓	✓

Lacey et al. [17] present a work that focuses primarily on a tool that performs the voltage analysis needed at charging stations and not on the planning of charging stations. The tool uses Excel to allow the analysis of the effects of typical loads and the load of EV on the distribution network.

Wagner et al. [18] take into account the EV users' travel destinations (i.e., restaurants, shops or banks). These destinations are considered as points of interest (PoIs). The authors propose a model that ranks the PoIs according to their attraction for EV users. To solve the problem of where to place a station, they propose two approaches: (1) a method based on obtaining maximum demand coverage while at the same time calculating the most optimal location of charging points; and (2) an iterative method that penalizes a PoI if it is close to an existing charging point. The main drawback is that it only uses information about the journeys of EV users.

Wei et al. [19] propose a tool that models the demand for taxis, stations, and electric taxis. The objective is to maximize the service of the electric taxi and the service of charging them using a genetic algorithm. For this purpose, the authors take into account the range that a taxi can travel, the charging time, and the capacity of the EV stations.

Li et al. [20] transform the location problem of the EV charging stations into a problem of maximum coverage in a weighted network where the weight of the arcs is the number of cars going from the origin to the destination. Its aim is to maximize demand coverage.

Dong et al. [5] determine the location and type of public charging stations using an optimization model. This model is based on a genetic algorithm that minimizes lost trips taking into account budget constraints. The authors consider a grid where grid cells are sorted by the number of trips ending in the cell. The 500 most popular destinations are selected as possible locations for public charging stations.

Sweda et al. [21] propose an agent-based decision support system to identify patterns in residential EV ownership and driving activities to enable the strategic deployment of new charging infrastructure. The proposed model incorporates road network data to permit micro-level analyses of the market for EVs. However, other factors relevant for an effective charging infrastructure such as geography as well as to demographics are not considered.

Our proposal is based on an MAS that aims to facilitate decision-making processes on the location of EV charging stations in a city. This system is based on the integration of a set of heterogeneous sources of data such as open data web portals to obtain data about traffic, populations in different places of a city, data from Google applications that provide information about the average time spent in PoIs, or social networks to collect geo-located information about users activity. All these data sources are the input of an Artificial Intelligence (AI) algorithm that estimates a near-optimal solution for the most appropriate EV charging station location according to a utility function. This functionality can be configurable by the user of the system to adapt it to specific circumstances of the city.

3. Proposal and Methodology

Several studies in the literature analyze users charging behaviors [22,23]. One of the main causes that is currently limiting EV adoption is what is called range anxiety,which, among other factors, is determined by the infrastructure of the charging stations [24]. In order to deal with this problem, it is necessary to evaluate the set of possible configurations in order to offer a distribution of charging stations that satisfies the users and enhances the use of the EV.

3.1. Multi-Agent System Design

In this paper, we propose a multi-agent system (MAS) that integrates a genetic algorithm to obtain configurations for the localization of charging stations according to a utility function. The MAS is composed of a set of agents that offer services and provide flexibility, scalability, and reuse in other municipalities. The agents that participate in the MAS are the following (see Figure 1):

- The Urban Agent: This agent is responsible for obtaining information about the amount of population in the different neighborhoods or blocks of the city under study. It extracts the information from census sections of Open Data portals.
- The Traffic Agent: The agent's responsibility is to collect traffic information. With the information collected, the agent is able to answer queries about how much traffic there is on average in a certain defined area of the city.
- The PoI Agent: The aim of the PoI Agent is to detect and classify points of interest (PoIs) for the installation of future charging stations in the city based on its urban development plan. In addition, the agent carries out a clustering process to eliminate points of interest that are too close and to define zones of influence of the point with a minimum area.
- The Popularity Agent: The task of the Popularity Agent is to determine the popularity of a point of interest based on the number of people who visit it and how much time those people spend in the area. To do this, it uses third party services, through an exhaustive search on the network, to locate this data. One example is Google's own service, which can be used to make a reverse resolution of coordinates to entities of interest on the map. Then, based on this information, the agent uses the results of the search engine to consult the estimated time spent by visitors in that area.
- The Social Networks Agent: This agent retrieves geolocated information from social networks (Twitter, Instagram, ...) to measure the popularity of a PoI based on the amount of activity that occurs through social networks in that area.
- The Data Processing Agent: This module is in charge of aggregating all the information obtained by the previous collecting agents. The above mentioned collecting agents are each specialized in a specific type of information, while the Data Processing Agent is in charge of combining and completing all this information to send it to the next agent (the Emplacement Optimizer Agent) so that data analysis can begin. This agent could collect information from various collecting agents based on the information available in each city.
- The Emplacement Optimizer Agent: This agent applies a genetic algorithm to determine the set of points of interest which are more appropriate according to the criteria that the user wants to optimize.
- The User Interface Agent: this agent consists of a dashboard that will offer an interface for the criteria specification to be optimized, and information sources that are going to be taken into account by the Emplacement Optimizer Agent. This agent provides the visualization of the results of the Emplacement Optimizer Agent.
- The system uses also a centralized database where all the persistence is done. This way, the Data Processing Agent and the Emplacement Optimizer Agent can share the computed data and the results, which are shown in the User Interface.

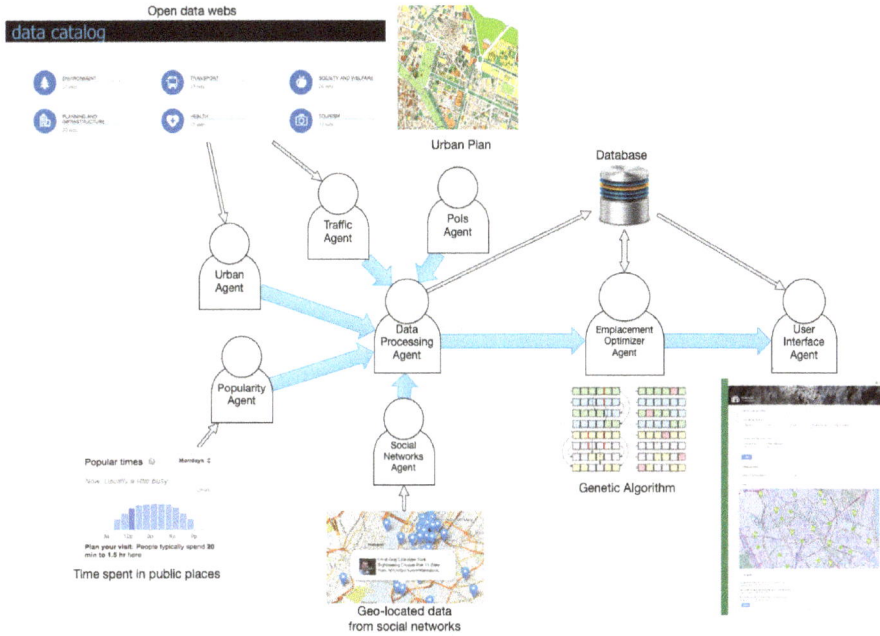

Figure 1. Modules that comprise the tool. PoIs: points of interest.

The proposed MAS follows the next phases: (1) extraction of the set of points of interest (PoIs) of the city P and their characterization considering different sources of information; (2) application of the genetic algorithm according to the conditions defined by the problem to be treated; and (3) visualization of the results.

In the first phase, the PoI Agent determines the PoIs P for the location of a charge station s_i. Initially, the agent considers public parking and garages to be PoIs. These PoIs are visited by a high number of users, usually the duration of stay is enough to charge a EV, and the power contracted in the installation is suitable for EV charging stations [25]. The consideration of these initial locations will not only increase the utilization but also increase the visibility, which might help to relieve range anxiety and promote EV acceptance. In order to be able to determine these PoIs (i.e., areas with collective car parkings and garages), we used the General Urban Development Plan and the land uses of the city. This set of PoIs was considered as a starting point for the possible locations of the charging stations.

To determine the area of influence of each PoI, a Voronoi diagram was created. Some (rectangular) boundaries are defined to limit the area of the city, and then the Voronoi diagram is calculated using the PoIs. Since a Voronoi diagram divides the whole area into regions based on the distance to points in each specific subset, each of these regions or polygons represents the influence area of the corresponding PoI. This allows us to consider an area rather than a single point to provide more flexibility in locating a charging station. The polygons around the PoIs designate their area of influence.

Once the polygons containing the PoIs were identified, they were characterized from the extraction of information from different sources: geo-positioned activity in the city from social networks [26–28], census sections, traffic status, traffic intensity per section, existing charge stations, tourist areas, and time spent in areas where there may be collective vehicle parking (i.e., shopping malls, work areas, etc.).

In a second phase, the Emplacement Optimizer Agent takes the polygons, the information that characterizes each of them, and the constraints of the problem (e.g., the number of stations to be installed, time limit to obtain a solution, etc.), and initiates the search for solutions by means of the previously described genetic algorithm. Finally, once a configuration has been obtained, the user can visualize the location of the stations on the city map.

3.2. The Emplacement Optimizer Agent

This subsection describes the main aspects of the *Emplacement Optimizer Agent* which is in charge of determining the more appropriated set of locations for a set of charging stations. First, we describe how the problem has been modeled, and secondly, we describe how the best configuration is found through a genetic algorithm.

3.2.1. Problem Description

The problem consists of the location of a set of charging stations starting from a set of PoIs of the city under study. In this way, $P = \{p_1, \ldots, p_n\}$ is a set of possible locations for charging stations (PoIs), and $S = \{s_1, \ldots, s_n\}; 0 < s_i \leq max\_chargers\_per\_poi$ is the set of charging stations that are going to be finally deployed in the city, with values ranging from 0 to a constant value $max\_chargers\_per\_poi$ of charging stations per PoI. A PoI p_i is characterized by a set of attributes that define it, $p_i = \{a_1, a_2, \ldots, a_n\}$:

- Population in the area around p_i
- Average traffic in the area
- Average time spent by citizens in public places in the area
- Geo-located social networking activity in the area
- Cost depending on the area covered by the stations
- Cost per charging station

The goal of the application is to find a configuration of charging stations at some of the pre-defined PoIs. A configuration is composed of a set of points p_i where one or more stations s_i are located, $C_i = \{\langle p_1, s_i \rangle, \langle p_2, s_j \rangle, \ldots, \langle p_k, s_n \rangle\}$. Each configuration C has a value associated which has been assigned by a utility function $V(C_i) : 2^E \rightarrow R$. This function V is a lineal combination of a set of factors that are intended to be maximized or minimized. The final goal is to determine which is the optimal configuration (or a solution close to the optimal one) for the placement of the EV charging stations $argmax\ V(C_i)$. To deal with this goal we propose the use of a genetic algorithm.

3.2.2. Genetic Algorithm

Genetic algorithms consist of general algorithms of optimization and learning based on evolutionary processes present in nature. This type of algorithm gradually converges towards high-quality solutions through the application of a set of operators. Another characteristic of genetic algorithms is their ability to solve problems with near-optimal solutions in solution spaces where a brute force algorithm would take too long. In the case of searching for the best EV charging station configuration, if we had 100 potential locations (PoIs) where we could place a EV charging station and we would like to select 20, we would have to analyze $100!/(20!(100 - 20)!) = 5.36 \times 10^{20}$ possible configurations. Given this scenario, the fact of using a genetic algorithm allows us to use a heuristic dedicated to the stochastic search that reaches near optimal solutions.

To solve the problem of the configuration of stations, we proposed a genetic algorithm that generates sets of solutions (i.e., generations), where each generation inherits the properties of the best solutions (i.e., configurations) of the previous one. Initially, the algorithm creates an initial random population of individuals N. Each individual is a solution to the problem (i.e., configuration C_i) (see Figure 2). In the scenario of EV charging stations, the size of the chromosome is the number of possible locations, i.e., P. The value of a gene within a chromosome indicates whether that location (p_i) will be used to locate one or more charging stations (s_i) or none.

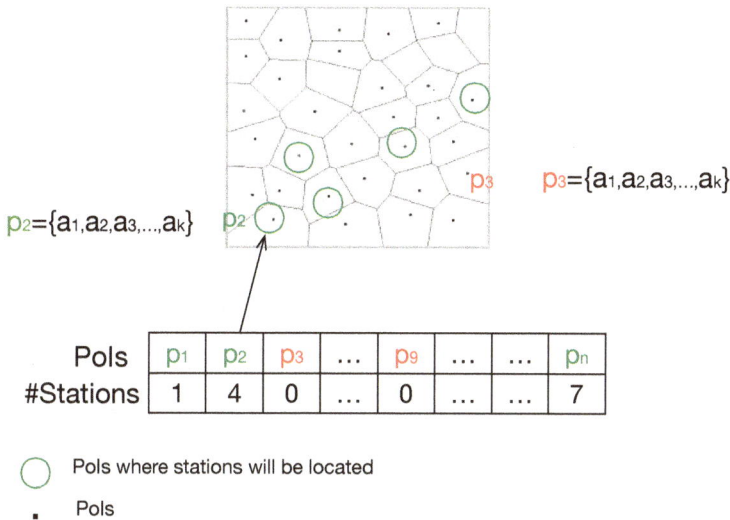

Figure 2. The encoding of an individual.

The fitness function evaluates the quality of the solutions, that is, the quality of the individual (i.e., configurations C_i). In our problem, the fitness function corresponds to the usefulness of placing charging stations in the selected points. This function considers different attributes described above ($p_i = \{a_1, a_2, \ldots, a_n\}$) balanced by weights.

Genetic operators are applied to existing individuals to generate new individuals. The proposed genetic algorithm considers the following genetic operators:

- Selection. The best individuals according to the fitness measure are selected to pass their genes to the next generation of the algorithm. This is done before applying the crossover operator. In our implementation, we used the tournament selection method, which creates several random groups of individuals, called tournaments, and selects the best one of each group.

- Crossover. Two parents are selected and a new individual child is created by combining the genes of the parents. Among the different crossover methods that exists, we used the cross uniform, which swaps genes from parents taking a uniform number of genes from them.

- Mutation. A new individual is generated by mutating some of the genes of a selected individual. This operator is used to maintain genetic diversity from different generations. Mutation is applied during evolution according to a probability which must be set low.

The proposed genetic algorithm performs as follows (see Algorithm 1). Initially, a population is randomly generated. This population consists of a set of possible configurations of EV charging stations. During each iteration (i.e., generation) of the algorithm, a randomly selected genetic operator is applied to each individual in the population. Then, the fitness value is calculated for each individual according to the equation:

$$V(C_i) = \sum_{\forall p_i \in C} ((\omega_p \cdot a_{population} + \omega_{tr} \cdot a_{traffic} + \omega_t \cdot a_{time} + \omega_s \cdot a_{social}) - (\omega_a \cdot cost\_area + \omega_c \cdot cost\_per\_charger \cdot |s_i|)), \quad (1)$$

where $a_{population}$ refers to the population of the area covered by the charging stations located in p_i; $a_{traffic}$ refers to the traffic generated in the area covered by the charging stations located in p_i; a_{time} refers to the average time citizens spend in public/commercial places in the area covered by the charging stations located in p_i; a_{social} refers to the average social activity in the area covered by the charging stations located in p_i; $cost\_area$ refers to the cost of locating stations in p_i that covers the demand of that specific area; and *cost_per_charger* is a constant cost per each charger ($|s_i|$) located in p_i. The value of these parameters ranges in the interval $[0, 1]$. Each parameter has associated a weight value ω established by the user of the system. In this way, the users of the system can tune the importance of each parameter depending on the shape of the city where the stations have to be placed.

Algorithm 1 The Genetic Algorithm

Generate an initial population of N random individuals
Evaluate the fitness of each individual of the population N
Select the best solution s
Number of generations $k = 0$
Number of generations without improving the solution $q = 0$
Temporal constraint $t = 0$
while ($k < max\_gen \wedge q < max\_gen$ without improving) **do**

  for ($j = 0; j < N; j + +$) **do**

    randomly apply one of the genetic operators over individual j
    evaluate the fitness value of j and j'
    insert j and j' in the new generation
  end for
  selection of N best individuals
  selection of the best individual s'
  if ($s' \leq s$) **then**

    q++
  end if
  k++
end while

Once the operators have been applied to the population of a generation, the new individuals are inserted into the new generation. The best N individuals remain in the new generation and the others are removed. The process ends when at least one of these situations occurs: (1) the number of generations exceeds a number established by the system; (2) when there are a certain number of generations where there is no individual in the new generation who has a fitness value higher than the best individual in previous generations; and (3) when the algorithm exceeds a time limit established by the system.

At the end of the process, the *Emplacement Optimizer Agent* sends the obtained results to the *User Interface Agent*.

4. Case Study: Valencia

This section describes a case study of the proposed system using data from the city of Valencia. At the European level, the European Commission has produced the *White Paper on Transport* [29], which sets targets for 2050 for the elimination of conventional fuel cars in cities. Although there are many initiatives and programs to implement hybrid and electric vehicles by the International Energy Agency, European Battery Manufacturers Association, and forums for global cooperation for the development and deployment of EV, there is a need to focus on the problems from the user's point of view to define and implement solutions at the municipal level. The municipalities should incorporate strategy and development plans of infrastructures, regulations, and planning, in order to face an imminent acceleration of the deployment of EVs in cities, taking into account that technology is in continuous evolution. Therefore, one of the main challenges of the electric vehicle is to create a reliable, accessible and convenient charging infrastructure for the citizen to boost the demand of EV.

Currently, there are 76 charging points in the province of Valencia, according to [30], and 24 of these are located in the city of Valencia. The Valencia City Council has carried out various initiatives aimed at improving infrastructure to facilitate the introduction of EV. Among these initiatives, studies have been carried out for the installation of EV charging points.

Although actions have been taken to facilitate the integration of EV through charging stations, we believe it would be interesting to provide the proposed MAS that allows a global study to be carried out analyzing the different configurations of charging stations according to different criteria to be optimized. Providing good planning and distribution of charging stations could drive demand for EVs among users considering the use and/or acquisition of an EV and who ultimately do not carry it out due to lack of charging points or poorly located EVs [31].

In the first phase, the MAS determines the P potential PoIs for the location of a charge station s_i taking into account data from the General Urban Development Plan. In particular, the points shown in Table 2 were selected. To determine the area of influence of each of the PoIs, a Voronoi diagram is created around the selected zones (see Figure 3).

Table 2. Areas selected as potential points to locate EV charge stations.

	Charge Mode	Charge Schedule	Stay Time	Connection Property
Shopping Mall	Fast charge, slow charge	Weekdays 19:00–22:00 h and weekends	1.2 h	Public
Workplace	Slow charge	Weekdays 07:00–19:00 h and weekends	9 h	Public/Private
Parking	Slow charge	24 h	2 h	Public
Public thoroughfare	Fast charge, slow charge	24 h	1–12 h	Public
Neighborhood community	Slow charge	08:00–20:00 h	12 h	Private
Private garage	Slow charge	24 h	12 h	Private
Refueling stations	Fast charge, battery change	24 h	10 min	Public
Vehicle fleet parking	Fast charge, slow charge, battery change	24 h	15 min–12 h	Private

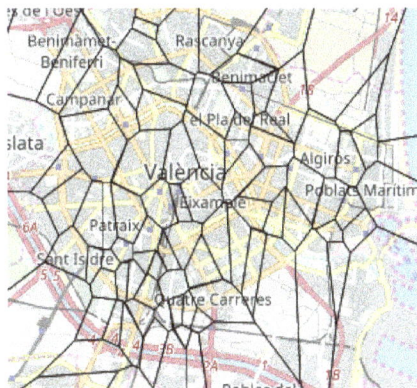

Figure 3. Voronoi diagram from the selected zones as potential points to locate charging stations.

In the second phase, the MAS collects data about the city of Valencia. Specifically, the MAS collects data about traffic (see Figure 4 left), population (Figure 4 center), average time spent in commercial spaces, and geo-located social activity from social networks (Figure 4 right) for each of the polygons around a PoI. This data is collected and aggregated using the different proposed agents described in Section 3.1 and represents the input of the *Emplacement Optimizer Agent* that determines the solution.

Figure 4. Maps that show the characteristics considered for each charging station.

In the third phase, polygons and their characteristics represented by the extracted data are considered in order to obtain a solution by means of the *Emplacement Optimizer Agent* through the proposed genetic algorithm. The number of stations and the time needed to find a solution are constraints of the problem. The weights that we use to balance out the attributes of each PoI in Valencia are: 0.4 for the population around the PoI; 0.3 for the average traffic around the PoI; 0.2 for the average time spent in public places of the area; and 0.1 for the geo-located social networking activity in the area. Once the genetic algorithm finishes, the best individual (i.e., configuration of EV charging stations) in the population is provided based on the value provided by the function of *fitness* presented in Equation (1). An example of a solution given by the *User Interface Agent* in which the locations where each charge point would be located is shown in Figure 5.

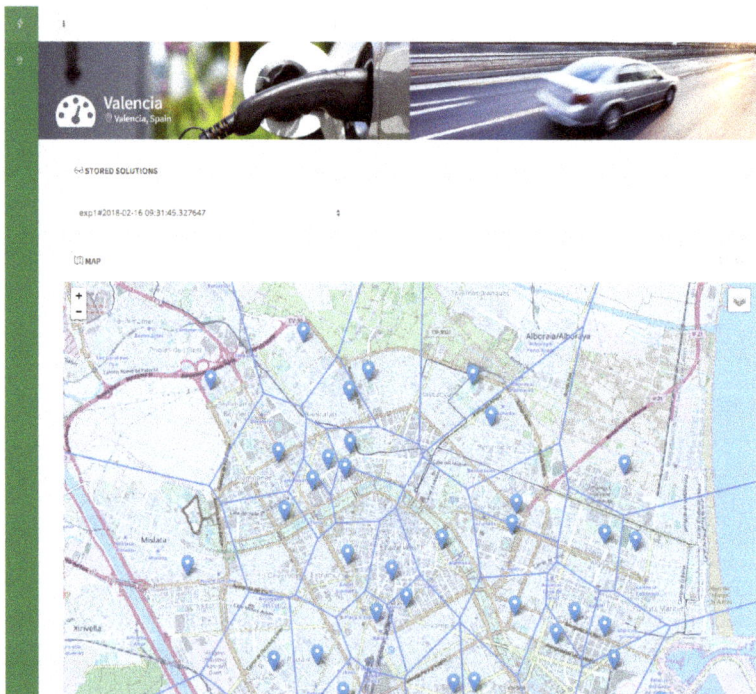

Figure 5. Configuration of the location of the charging stations displayed by the *User Interface Agent*.

5. Experimental Results

In this section, we present a series of experiments in which we analyze the behavior of the genetic algorithm that the Emplacement Optimizer Agent uses in order to find out the best parameters for our scenario. However, the entire MAS platform was used for the experiments carried out. Moreover, prior to the execution of the proposed experiments, unit tests were carried out on the entire system. These tests allowed us to validate that the system performed as expected.

Our genetic algorithm is implemented using the *deap* (urlhttps://github.com/DEAP/deap) library of *Python*. For our experiments, the PoI Agent considers a set of 103 points of interest of the city of Valencia. We assume a maximum of five chargers per PoI. Hence, depending on the final result of the genetic algorithm, for each PoI, a number of chargers between 0 and 5 will be installed.

The graph of Figure 6 represents the evolution of the fitness function of the genetic algorithm with different initial populations from 0 to 200 generations. Each line is the mean of five different executions with the same corresponding parameter. The results of Figure 6 show that the more individuals in the initial population, the higher the maximum value of fitness, as it would be expected. This is caused by the fact that with more individuals it is easier to have high fitness values reached by the big population which may be rich genetically. Nevertheless, when the initial population is 1500 or more, there is no significant difference when the algorithm reaches more than 50 generations. The reason behind this is that a near-optimal solution is found if there is enough diversity in the population. Hence, there is no need for using a bigger initial population that does not contribute to finding better solutions. Therefore, it seems that in this particular scenario, an initial population of 2000 would be enough, while values of 3000 and 4000 may be considered if computation time is not a problem.

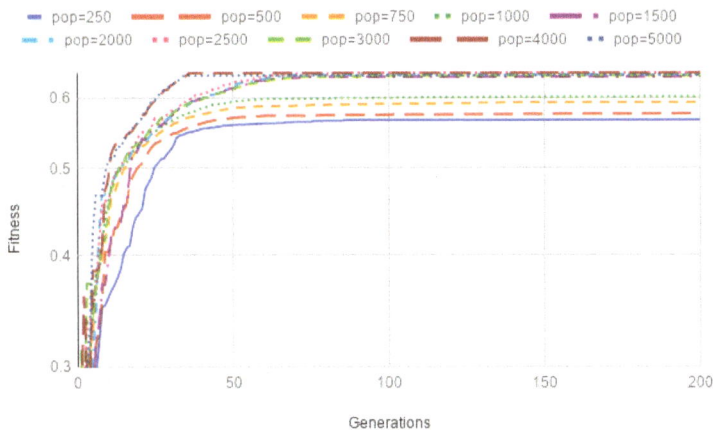

Figure 6. Evolution of maximum fitness values for different initial populations.

We also tested different crossover techniques of the genetic algorithm. Specifically, we tested the single point technique in which a single crossover point is selected and all data of the parents is swapped from that point; the two points technique, where two points of the chromosomes are selected and everything between these two points is swapped between the parents; and the cross uniform technique, that uses a fixed mixing ratio between the two parents and parents' genes are swapped instead of segments of genes.

Figure 7 represents the value of fitness (the mean value of five executions) for each of the aforementioned crossover techniques with initial populations of 2000 individuals and 100 generations. The cross uniform technique has the best results in our scenario, followed by the two points technique. The reason for the better performance of the cross uniform technique is that it produces new individuals by mixing the parents at a gene level, which generates a richer population genetically. The lower performance of single point and two points techniques is caused by the crossover at segment level of genes instead of gene by gene, which yields in a lower rate of mixing of parents through different generations and greater difficulty in varying the population.

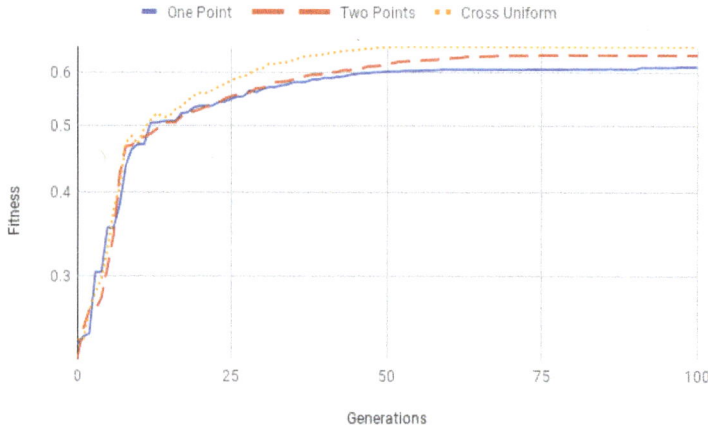

Figure 7. Maximum fitness values for different crossover techniques with an initial population of 2000.

Figure 8 presents the computation time in seconds (all the tests were conducted on a single machine with an Intel Core i7-3770 CPU at 3.40 GHz and 8 GB RAM) for different initial populations of the genetic algorithm for 100 and 200 generations in our scenario. Each bar is calculated with the mean of five different executions of the genetic algorithm with the specified parameters. The computation time increases linearly with the population and generations. In this way, the computation times for 200 generations are approximately the double of the computation times for 100 generations.

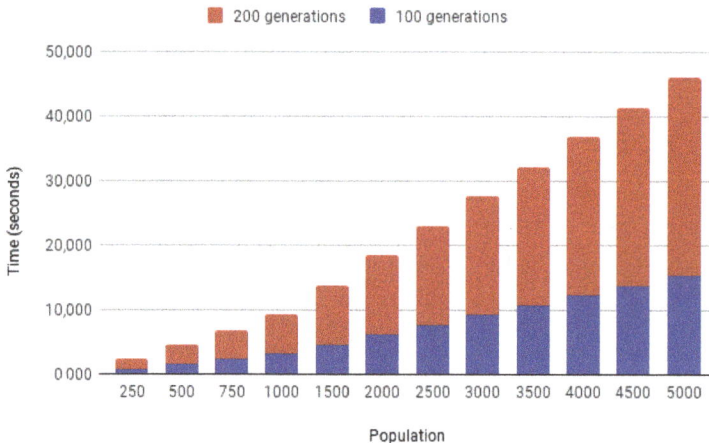

Figure 8. Computation time of different executions of the genetic algorithm.

Considering the complexity of this task, the computation times are tractable for initial populations around 2000 to 3000, which proved to be enough to find near-optimal solutions in the test of Figure 6. We provide a reminder here that the goal of our MAS is to find a solution to emplace charging stations in the city. Hence, computation times of the Emplacement Optimizer Agent of hours or even days can be acceptable for the users of our system.

Finally, we show two specific results of the genetic algorithm for the particular case study of Valencia in order to compare solutions of different quality. Our goal is to analyze how accurate a solution with a high fitness value is compared to another with lower value. In this way, Figure 9 represents two computed solutions with our genetic algorithm. Figure 9a is a solution computed with an initial population of 250 and a fitness value of 0.563. The solution of Figure 9b is computed with an initial population of 4000 that yields a fitness value of 0.639. Both fitness values are close. However, as we showed in the analysis of Figure 6, there is a significant difference between the fitness values obtained.

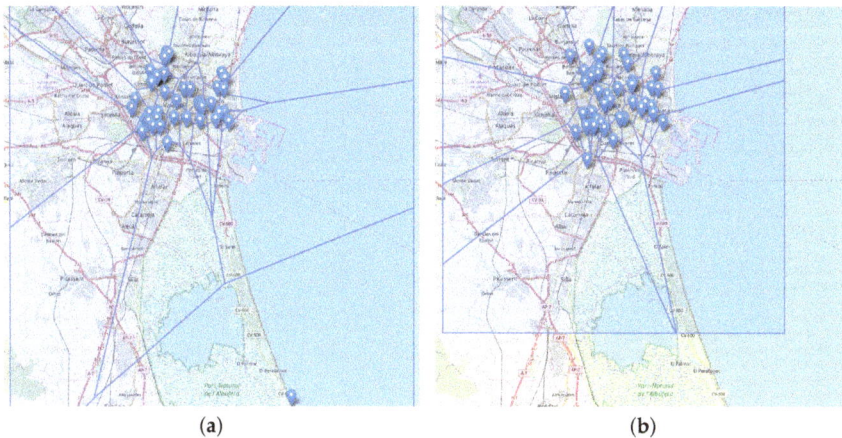

(a)   (b)

Figure 9. Computed solutions by the genetic algorithm for the city of Valencia. (**a**) Solution with a fitness value of 0.563; (**b**) Solution with a fitness value of 0.639.

The solution of Figure 9a has 40 charging stations, while the solution of Figure 9b uses 42 charging stations. In this way, both solutions are pretty similar, so the costs of implementing them would be almost the same (fixed cost by charging station is almost equal). There is a difference in the areas that covers the solution of Figure 9a, which considers bigger areas in the south of the city that will increase the cost of installing the stations. This is why solution of Figure 9b has a very similar cost even when placing more charging stations, but it better optimizes the areas covered by them. Therefore, the locations of the charging stations in both solutions also determine the quality of them. For instance, Figure 9a places a charging station in the far south of the city because there is some activity there. However, this activity is not significant enough and it would be a waste of resources to place a charging station so far out of the city because it is an area that does not need to be covered (in our scenario). The solution of Figure 9b places the charging stations more uniformly in the city, covering the full area where the main activity occurs, that is, the more populated and crowded areas. Concretely, there are several charging stations covering the center and north of the city which are not present in Figure 9a. Therefore, the solution of Figure 9b is more accurate according to the data showed in Figure 4, which explains the slightly higher fitness value that makes the solution better to implement in the particular case of Valencia.

6. Conclusions

This paper has presented a multi-agent system for which the goal is the planning of efficient placement of infrastructures for electric vehicle charging stations for public and private users in a city.

The main advantages of the proposed system are the integration of information from different sources, and the modeling and location analysis of charging stations through the use of optimization techniques. The proposed solution allows us to obtain real-time data and different contexts with respect to the activity in a city. This information is relevant for detecting relevant points/areas of the city in order to place charging points. The proposed system has been implemented in the city of Valencia where different experiments have been done in order to validate the implementation. Results have shown how the genetic algorithm behaves with real input data from the city of Valencia.

In future work, the system can be extended for a more detailed analysis of the activity and mobility of people in the city. Specifically, the proposal can be applied for the detection of alternative routes to locate charging stations by the citizens, in order to facilitate accessibility between different charging stations or to identify more mobility patterns that affect the charging point infrastructure. We also plan to include information about the availability of dedicated parking spaces so as to potentially install charging points in the model.

Acknowledgments: This work was partially supported by the MINECO/FEDER TIN2015-65515-C4-1-R and MOVINDECI projects of the Spanish government.

Author Contributions: All authors contributed equally to this work.

Conflicts of Interest: The authors declare no conflict of interest. The founding sponsors had no role in the design of the study; in the collection, analyses, or interpretation of data; in the writing of the manuscript, and in the decision to publish the results.

References

1. Wolfram, P.; Lutsey, N. *Electric Vehicles: Literature Review of Technology Costs and Carbon Emissions*; Technical Report; The International Council on Clean Transportation: Washington, DC, USA, 2016.
2. *Electric Vehicles in Europe*; Technical Report; European Environment Agency: Copenhagen, Denmark, 2016.
3. Klabjan, D.; Sweda, T. The nascent industry of electric vehicles. *Wiley Encycl. Oper. Res. Manag. Sci.* **2011**, doi:10.1002/9780470400531.eorms0997.
4. Skippon, S.; Garwood, M. Responses to battery electric vehicles: UK consumer attitudes and attributions of symbolic meaning following direct experience to reduce psychological distance. *Trans. Res. Part D Trans. Environ.* **2011**, *16*, 525–531.
5. Dong, J.; Liu, C.; Lin, Z. Charging infrastructure planning for promoting battery electric vehicles: An activity-based approach using multiday travel data. *Trans. Res. Part C Emerg. Technol.* **2014**, *38*, 44–55.
6. Alternative Fueling Station Locator. 2017. Available online: http://www.afdc.energy.gov/locator/stations/ (accessed on 30 December 2017).
7. How to Use the PlugShare EV Charging Station Tool. 2017. Available online: http://www.plugincars.com/how-to-use-plugshare-guide.html (accessed on 30 December 2017).
8. AAA Adds EV Charging Station Locations to Mapping Tools. 2017. Available online: http://newsroom.aaa.com/2012/03/aaa-adds-ev-charging-station-locations-to-mapping-tools/ (accessed on 30 December 2017).
9. Electric Vehicle Station Locator. 2017. Available online: https://www.nyserda.ny.gov/Researchers-and-Policymakers/Electric-Vehicles/Tools/Electric-Vehicle-Station-Locator (accessed on 30 December 2017).
10. EV Infrastructure Corridor Development Toolkit. 2017. Available online: http://altfueltoolkit.org/ev-infrastructure-corridor-development-toolkit/ (accessed on 30 December 2017).
11. Association, C.E. Planning for Electric Vehicle Charging Infrastructure: A Toolkit. 2013. Available online: https://www.zap-map.com/live/ (accessed on 30 December 2017).
12. Gan, L.; Topcu, U.; Low, S.H. Optimal decentralized protocol for electric vehicle charging. *IEEE Transp. Power Syst.* **2013**, *28*, 940–951.
13. Ma, T.; Mohammed, O.A. Optimal charging of plug-in electric vehicles for a car-park infrastructure. *IEEE Transp. Ind. Appl.* **2014**, *50*, 2323–2330.
14. Shukla, A.; Pekny, J.; Venkatasubramanian, V. An optimization framework for cost effective design of refueling station infrastructure for alternative fuel vehicles. *Comput. Chem. Eng.* **2011**, *35*, 1431–1438.
15. Nie, Y.M.; Ghamami, M. A corridor-centric approach to planning electric vehicle charging infrastructure. *Transp. Res. Part B Methodol.* **2013**, *57*, 172–190.

16. Wood, E.; Neubauer, J.S.; Burton, E. *Measuring the Benefits of Public Chargers and Improving Infrastructure Deployments Using Advanced Simulation Tools*; Technical Report; SAE Technical Paper; National Renewable Energy Lab. (NREL): Golden, CO, USA, 2015.

17. Lacey, G.; Putrus, G.; Bentley, E.; Johnston, D.; Walker, S.; Jiang, T. A modelling tool to investigate the effect of electric vehicle charging on low voltage networks. In Proceedings of the 2013 World Electric Vehicle Symposium and Exhibition (EVS27), Barcelona, Spain, 17–20 November 2013; pp. 1–7.

18. Wagner, S.; Götzinger, M.; Neumann, D. *Optimal Location of Charging Stations in Smart Cities: A Points of Interest Based Approach*; ICIS: Milan, Italy, 2013.

19. Tu, W.; Li, Q.; Fang, Z.; Shaw, S.l.; Zhou, B.; Chang, X. Optimizing the locations of electric taxi charging stations: A spatial-temporal demand coverage approach. *Transp. Res. Part C Emerg. Technol.* **2016**, *65*, 172–189.

20. Li, Z.; Cui, X. Research on Location Problem of Electric Vehicle Charging Station. *J. Appl. Sci. Eng. Innov.* **2015**, *2*, 495–498.

21. Sweda, T.; Klabjan, D. An agent-based decision support system for electric vehicle charging infrastructure deployment. In Proceedings of the 2011 IEEE Vehicle Power and Propulsion Conference, Chicago, IL, USA, 6–9 September 2011; pp. 1–5.

22. Franke, T.; Krems, J.F. Understanding charging behaviour of electric vehicle users. *Transp. Res. Part F Traffic Psychol. Behav.* **2013**, *21*, 75–89.

23. Franke, T.; Günther, M.; Trantow, M.; Krems, J.F.; Vilimek, R.; Keinath, A. Examining user-range interaction in battery electric vehicles—A field study approach. In *Advances in Human Aspects of Transportation Part II*; AHFE Conference: Krakow, Poland, 2014; pp. 334–344.

24. Needell, Z.A.; McNerney, J.; Chang, M.T.; Trancik, J.E. Potential for widespread electrification of personal vehicle travel in the United States. *Nat. Energy* **2016**, *1*, 16112.

25. Sedano Franco, J.; Portal García, M.; Hernández Arauzo, A.; Villar Flecha, J.R.; Puente Peinador, J.; Varela Arias, J.R. Sistema inteligente de recarga de vehículos eléctricos: Diseño y operación. *Dyna* **2013**, *88*, 644–651.

26. Vivanco, E.; Palanca, J.; del Val, E.; Rebollo, M.; Botti, V. Using geo-tagged sentiment to better understand social interactions. In *Advances in Practical Applications of Cyber-Physical Multi-Agent Systems: The PAAMS Collection*; Springer: Cham, Switzerland, 2017; pp. 369–372.

27. del Val, E.; Palanca, J.; Rebollo, M. U-Tool: A Urban-Toolkit for enhancing city maps through citizens' activity. In *Advances in Practical Applications of Scalable Multi-agent Systems: The PAAMS Collection*; Springer: Cham, Switzerland, 2016; pp. 243–246.

28. del Val, E.; Martínez, C.; Botti, V. Analyzing users' activity in online social networks over time through a multi-agent framework. *Soft Comput.* **2016**, *20*, 4331–4345.

29. Transporte, D.G.d.M. Hoja De Ruta Hacia Un Espacio único Europeo De Transporte: Por Una Política De Transportes Competitiva Y Sostenible. 2011. Available online: https://ec.europa.eu/transport/sites/transport/files/themes/strategies/doc/2011_white_paper/white-paper-illustrated-brochure_es.pdf (accessed on 30 December 2017).

30. Electromaps. Electromaps: Puntos De Recarga en Valencia. 2017. Available online: https://www.electromaps.com/puntos-de-recarga/espana/valencia (accessed on 30 December 2017).

31. Levante. Grezzi Anuncia Más Puntos De Recarga Para Los Coches Eléctricos. 2017. Available online: http://www.levante-emv.com/valencia/2017/04/23/grezzi-anuncia-puntos-recarga-coches/1557495.html (accessed on 30 December 2017).

applied
sciences

MDPI

Article

Multi-Agent System for Demand Prediction and Trip Visualization in Bike Sharing Systems

Álvaro Lozano [1,*], Juan F. De Paz [1], Gabriel Villarrubia González [1], Daniel H. De La Iglesia [1] and Javier Bajo [2]

[1]   Faculty of Science, University of Salamanca, Plaza de la Merced s/n, 37002 Salamanca, Spain;
      fcofds@usal.es (J.F.D.P.); gvg@usal.es (G.V.G); danihiglesias@usal.es (D.H.D.L.I.)
[2]   Department of Artificial Intelligence, Polytechnic University of Madrid, Campus Montegancedo s/n,
      Boadilla del Monte, 28660 Madrid, Spain; jbajo@fi.upm.es
*   Correspondence: loza@usal.es; Tel.: +34-646-891-785

Received: 15 December 2017; Accepted: 2 January 2018; Published: 5 January 2018

Featured Application: The main application of this work is the analysis and prediction of the demand in bike sharing systems using their open/private data. A multi agent system is proposed and a case study is conducted using the data of a bicycle sharing system from a middle size city.

Abstract: This paper proposes a multi agent system that provides visualization and prediction tools for bike sharing systems (BSS). The presented multi-agent system includes an agent that performs data collection and cleaning processes, it is also capable of creating demand forecasting models for each bicycle station. Moreover, the architecture offers API (Application Programming Interface) services and provides a web application for visualization and forecasting. This work aims to make the system generic enough for it to be able to integrate data from different types of bike sharing systems. Thus, in future studies it will be possible to employ the proposed system in different types of bike sharing systems. This article contains a literature review, a section on the process of developing the system and the built-in prediction models. Moreover, a case study which validates the proposed system by implementing it in a public bicycle sharing system in Salamanca, called SalenBici. It also includes an outline of the results and conclusions, a discussion on the challenges encountered in this domain, as well as possibilities for future work.

Keywords: bike sharing systems (BSS); regression models; open data; data visualization; multi agent systems; organizations and institutions; socio-technical systems

---

1. Introduction

There is a consensus in the literature [1,2] which states that bicycles are one of the most sustainable modes of urban transport and they are suitable for both short trips and medium distance trips. Riding a bicycle does not have any negative impact on the environment [3], it promotes physical activity and improves health. Furthermore, its use is cost-effective from the perspective of users and infrastructure.

Moreover, due to the increased CO_2 levels, the European Union and other states are taking measures to reduce greenhouse gas emissions in every sector of the economy [4].

These facts explain the growing popularity of sustainable means of transport such as bike sharing systems. From 1965 when they came into use in Amsterdam to 2001, there were only few systems around the world. Bike sharing systems (BSS) began to spread in 2012, when their number increased to over 400 [5]. By 2014 this number had doubled [6] and nowadays there are approximately 1175 cities, municipalities or district jurisdictions in 63 different countries where these systems are in active use, according to BikeSharingMap [7].

Bike sharing systems allow users to travel in the city at a low cost or even for free. They can pick up a bicycle at one of the stations distributed across the city and leave it at another. These systems have evolved over time [8] and today the vast majority include sensors that provide information on the interaction of users with the system. However, the management of these systems and the data collected by them, is often poor and as a result the numbers of bicycles available at stations are not sufficient.

These are the reasons as to why bike sharing systems should be improved with data produced by the systems themselves. They should include predictive models for user behaviour and demand, which will notify the system administrator of the stations where more bicycles are required for satisfying user demand. This will also allow to set up new stations in places where the demand is high or, on the contrary, to close down the stations at which the demand is too low.

This article presents a multi-agent system which collects bike sharing system data together with other useful data. The system uses these data to create demand prediction models and to offer services through an Application Programming Interface (API), used as a prediction and visualization tool. This application has two main functions: (1) it visualizes historical data including the flow of bicycles between stations and (2) it predicts the demand for each station in the system. In order to perform the prediction of demands, the system includes different regressor systems based on previously collected data. The system has been tested with data provided by Salamanca's bike sharing system (SalenBici) getting successful results regarding to the demand prediction. This will provide the administrator with more information about stations and users in the system as well as useful data for the bike reallocation process.

This work is structured as follows: Section 2 overviews the state of the art of current works and techniques involved in bike sharing systems and prediction models and points to the current issues in this field. Section 3 describes the proposed system architecture and provides a concise description of every part. Section 4 presents a specific case study: we test with the bike sharing system called SalenBici of Salamanca to validate the system proposed. Finally, Section 5 provides insights about the results and conclusions of this case study and points out research challenges for future works.

2. Background

2.1. Bike Sharing Systems

All bike sharing systems operate on the basis of a common philosophy, their principle is simple: individuals use bicycles on an "as-needed" basis without the costs and responsibilities that owning a bicycle normally entails [9]. Figure 1 shows a classical bike station, where there are free docks and available bikes to rent within a station map of Madrid's Bike Sharing System called BiciMad. As shown on the map, the bike sharing system offers a wide list of stations located across the centre of the city.

Peter Midgley indicates in his study [3] that bike sharing systems can be categorized into 4 generations depending on their features: (1) First generation bike sharing systems: the first generation of bike sharing systems was introduced in Amsterdam (1965), La Rochelle (1976) and Cambridge (1993). These systems provided totally free bicycles which could be picked and returned in any location. The vast majority of these systems were closed due to vandalism. (2) Second generation: this generation tries to solve the drawbacks of the previous one. In this case, the systems had a coin deposit (like the supermarket trolleys) but they still suffered from thefts due to the anonymity of the users. (3) Third generation: this generation uses high tech solutions including electronic locking docks, smart cards, mobile applications, built-in GPS devices in the bikes and totem applications. (4) Fourth generation: it is still in the process of development, this generation includes movable docking stations, solar-powered docking stations, electric bikes and real-time system data.

This last generation of systems includes a new scheme named dock-less bike sharing solution which reuses the first-generation philosophy: free bicycles around the city. However, in this case, they are equipped with a GPS tracker and they can be found and rented through a mobile application. This kind of systems are quite spread nowadays in China and they are arriving to cities like Singapore, Cambridge and Seattle through enterprises as OFO [10], Mobike [11] and Bluegogo. Nevertheless, these bike sharing systems may suffer from vandalism and other problems already presented in first-generation. Some of these big enterprises—such as Bluegogo [12] in China—has recently gone bankrupt due to financial problems.

Furthermore, these bike sharing systems present other kind of operating diagram that varies from the one based on stations. In consequence, there are huge differences regarding to the data processing. This work focuses on systems based on stations which will be deeply discussed.

Concerning Third and Fourth generations of BSS, there are some common issues that will be described in detail in the following section.

Figure 1. One of BiciMad System's bike docking stations (Madrid, Spain) with electrical bikes.

2.2. Bike Sharing Main Issues

Current literature lists factors that influence the success of bike sharing systems, these include the built environment, psychological factors related to the natural environment, as well as the utility theory. This last reason focuses on providing users with the best quality of service, by addressing any issues encountered in the bike sharing systems. Vogel et al. [9] describes three main issues in bike sharing systems; their design, management and operation. The authors distinguish three categories:

- Network design and redesign issues: These decisions address issues related to the initial design of the system, they consider the topography, traffic and equity of stations located across the city [10]. The amount of docks and the number of vehicles at each station are important factors in this category. These issues occur not only in the initial design but also in the subsequent use of the system: long term corrections of stations could be done thanks to the insights provided by the data analysis tools created with the information generated using the system.
- Incentivizing users to balance the system: These decisions are related to operational matters and they aim to mitigate the main operational problem: the shortage event [13] that occurs when a customer wants to pick a bike from an empty station or return it to a station whose docks are occupied. Users are incentivized to occupy the stations with available bikes and docks. These incentives could involve changes in the pricing policies or even a reservation station system that ensure availability at the final station. This kind of incentives could be implemented in

systems that operate on a pay-per-use basis but they are more difficult to implement in systems with a fixed quota per year.

- Operational reallocation issues: This category also focuses on an operational aspect, known as the commutation pattern in bike sharing systems. There are specific bike station usage patterns, for example, in the mornings the stations located on the outskirts of cities are empty because many people travel to the city to work. On the other hand, the stations in the city tend to be full in the mornings [14]. As we mentioned previously, the empty and full stations must be balanced by system administrators and this reallocation must follow a specific strategy that is based on the demand predicted for each station provided by prediction demand models. They must additionally use a reallocation algorithm which minimizes the total cost of managing bike fleets. In this case, there are two more problems: the prediction engine and the optimization algorithm used to reallocate the bikes. This work addresses the prediction problem but does not discuss the reallocation strategy.

Now that we have identified the main issues that should be considered in these systems, in the next paragraphs it will be described how a visualization tool is used to solve the problems related to the first category. We will also focus on the third category, especially on the demand prediction models. Moreover, we will show works in the literature that regard demand prediction models and the kind of data have been employed to build them.

2.3. Bike Demand Prediction Models

In the literature, we can currently find numerous works on station demand prediction. Some address the problem from a data mining point of view [15] with an exploratory data approach [16]. Afterwards, works like Kaltenbrunner et al. [17] show how to implement prediction models based on time series analysis methods like Auto-Regressive Moving Average (ARMA).

Subsequently, in May 2014, Kaggle [18], a known machine learning competition platform, launched a competition about the Washington D.C. BSS. In this competition, participants were asked to combine historical usage patterns with weather data in order to forecast bike rental demand in the Capital Bikeshare [19] program in Washington, D.C. Due to this competition, a lot of researchers began to work in prediction models to obtain the forecast of bike demand in stations.

Romain et al. [20] use the data described above to in the analysis of several regression algorithms from state of the art. They applied these algorithms to predict the global use of the bike sharing system. They establish baseline predictors as references of performance and apply Ridge Regression, Adaboost Regression, Support Vector Regression, Random Forest Regression and Gradient Tree Boosting Regression. They perform predictions up to 24 h ahead and their study covers the whole system instead of each station independently. Patil et al. [21] also present prediction models about the total number of bikes rented per hours in the system in their work. The authors compare Random Forest, Conditional inference trees and Gradient Boosted Machines. Wen Wang [22] perform a great comparison of the approaches proposed by other authors, while others like Lee et al. [23], Du et al. [24], Yin et al. [25] among others in their dissertation. This work offers a great overview of past research works on this topic.

Previous works are interesting if we want to look at total demand in the entire bike sharing system but nowadays it is possible to obtain station data from Open Data services [26,27] and private services [28] which can be used to build prediction models for each station and forecast the demand. This will allow to provide useful information for a further bike reallocation strategy and avoid shortages.

Diogo et al. [29] present an interesting work where they achieve the previous two objectives: prediction models for each station and a rebalancing algorithm. The data employed in this work provided information about Chicago BSS [30] and Washington [31] bike systems and it was related to the trips travelled in the system. In the system proposal section of this work it will be explained what kind of information is contained in trip data. The authors propose the use of Gradient Boosting

Machines, Poissson Regression and Random Forest to build prediction models and they use a variant of the Vehicle Routing Problem (VRP) with the framework Jsprit to perform the reallocation strategy. The authors do not provide any open source code nor an in-depth explanation.

In this work, we will employ some well-known methods which have also been used in the described literature, to predict the demand at each station and provide useful information for the subsequent reallocation task, which will be studied in a future work.

3. Proposed System

This section details the proposed multi agent system (MAS) and presents a general diagram of the system's architecture which is shown in Figure 2. Literature shows that multi agent systems have previously been employed in similar tasks like taxi fleet coordination [32] where they offer an ideal solution to abstract away issues of the different existing platforms and communication protocols. The system is divided into the following groups of agents: bike sharing data agents, weather data collector agents, geographical information agents, data persistence agents, demand prediction agents and API agents. Several multi agent platforms such as SPADE [33,34], JADE [35], PANGEA [36,37], AIOMAS [38] and osBrain [39] were evaluated and osBrain was finally selected for this system because of its ease of use. Furthermore, it is implemented in Python (like SPADE and AIOMAS) and it is in continuous development nowadays.

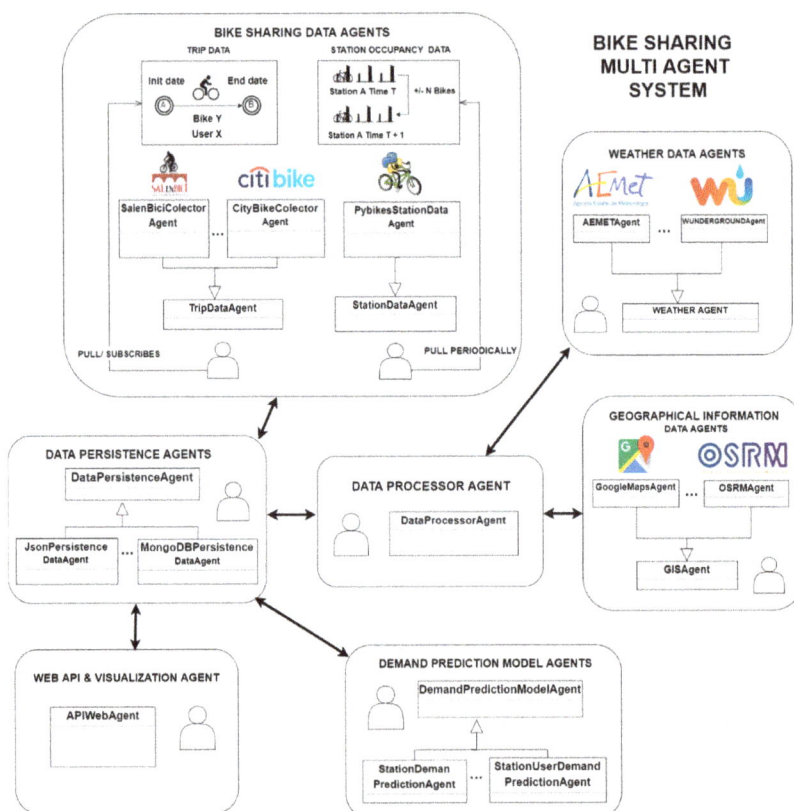

Figure 2. General diagram of the proposed multi agent system for bike sharing systems.

Each agent group will be described in detail in following sections, including issues encountered, techniques, developed processes and decisions taken.

3.1. Bike Data Sharing Agents

Bike Data Sharing Agents are responsible for obtaining data from the BSS, which usually offers two kinds of data:

- Station information: It indicates the total number of available docks and bicycles at each station at a given time, that is, the current situation at the station. This data is usually available and provided to the end users in order to keep them informed about the stations with available bicycles. If this data is periodically polled, pickups and returns at each station can be calculated regularly. This kind of information only provides us with knowledge about variations at each station without specifying the flow of bikes from one station to another or data regarding the users involved in these changes. It is usually provided through an API in real time by a great amount of BSS, so much so that there is a project called PyBikes [27,40] which unifies the access to this information through a common API for a large number of current BSS. The multi agent system presented in this paper includes a station data agent that periodically collects this information for the target BSS.
- Trip information: This sort of information is not commonly published but it is usually available in csv files offered every month as open data [41]. It is related to the trips recorded by the BSS: a user picks a bike at a specific station at a given time and later returns it at another station. This kind of data discloses more details as it provides insights on the flow of bikes between stations and the users who perform the trips recorded in the system. This information will be especially useful for the visualization tool and demand prediction in BSS with registered users (occasional users are not included). There is a trip data agent in the system that asynchronously gets data from a public or private data source.

This group of agents sends the collected information to the Data Persistence Agent, which is in charge of storing the raw data that will later be processed by the Processor Agent.

3.2. Weather & Environment Data Agents

Weather & Environment Data Agents obtain the weather data requested by the Processor Agent, they provide historical weather data for the requested dates and locations which will later be merged with trip or station data. These agents will also provide weather forecasts to Prediction Model Agents; they need this information when making predictions for a specific station at a given time.

The weather data obtained by these agents is daily data on: minimum, maximum and average temperature, wind speed, rain and minimum and maximum pressure. There is a common base agent to provide this information and specific data agents responsible for implementing that functionality for each weather provider such as Wunderground [42] or Aemet [43].

3.3. Geographic Information Data Agents

The Geographic Information Data Agents obtain the geographic location and the distance between stations as well as the altitude of each station. A base agent offers services related to this information and a specific agent implements this functionality with different data providers such as Open Source Routing Machine (OSRM) or Google Maps. This information will be stored and employed later by API agents in the visualization tool. This information will also be useful for future works where a reallocation bike strategy must be calculated.

3.4. Data Processor Agents

Data Processor Agent processes the data obtained by Bike Sharing Data agents. Additionally, this agent will request the weather agents to provide information regarding the bike sharing data dates and it will clean and merge that information with the previously obtained bike sharing data. This process will be executed periodically in order to obtain all the recent information that is pending to be processed.

Since different kinds of information are obtained by the bike sharing data agents, the data processor agent is capable of perform two different data processing workflows. The selected workflow will depend on the type of input information: trip data information or station data information. As it has been pointed out previously, the output data obtained by processing either kind of input information will be different and will be employed differently depending on the case study.

Figure 3 shows the flowchart of the data processor agent in detail. This agent will perform different data processing methods based on the input information and then it will send the results to the data persistence agent, for storage. Every workflow and their included sub processes are explained in-depth below.

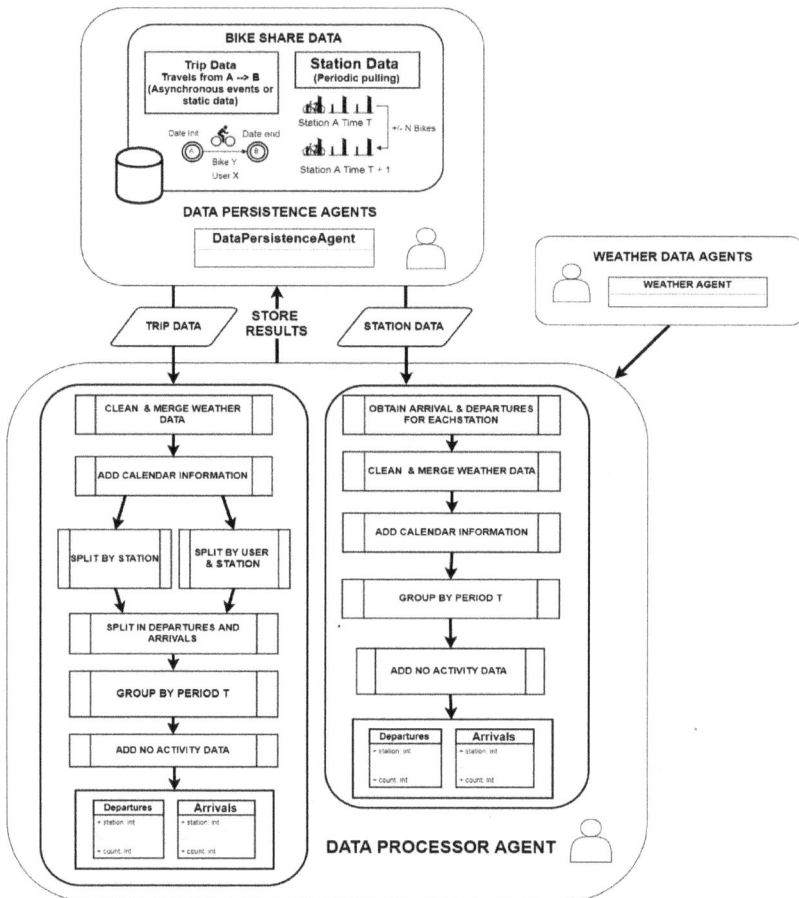

Figure 3. Data processing flow chart of the processes performed by the data processor agent.

3.4.1. Trip Data Workflow

First of all, trip information is received as input data. A process of dropping outliers is performed for Trip Data. Trips that lasted less than 2 min are dropped out because they are considered an attempt of using a bike which is malfunctioning, so it is taken and immediately returned to the same station.

Besides that, the processor data agent requests the weather data agent to provide weather data in the range of dates of the input data. The information provided by the weather agent is cleaned and if there is no weather data for data records, a lineal interpolation is performed in the data. This information is merged with the input information. Then new features are added to the current dataset. These features divide dates into years, months and days of the week, day of the month, hours and other derivative features as seasons, weekends, holidays (obtained for each country with the library Workalendar library [44]). At this point, the received information is cleaned and merged with extra information. The next step involves splitting the information by station in the system and creating departures and arrivals. For each station in the system, that had the same station of origin are converted into departures and all trips with the same station as their destination are converted into arrivals.

Once this is executed in every arrivals/departures dataset, the information will be grouped by a period T and the total number of arrivals/departures in each period T is summed. An additional step is performed, this is necessary in order to generate records for the periods for which no information on trips is available, periods of the day when the number of trips is zero. This process is called *adding no activity data*. Finally, the information is sent to the persistence data agent that will store it for further utilization by the model prediction agent.

3.4.2. Station Data Workflow

This process is slightly different from the previous one but they share common steps in the sub processes. The information received is related to each individual station in the system, the difference between consecutive records will show the arrivals or departures of bikes in the system. From the beginning the data is split into arrivals and departures at each station that is part of the system. Then the previously mentioned process of cleaning and merging weather information is performed. Next, the process of adding information on dates is also performed. Finally, the data is grouped by a period T, no activity data is added and the information is sent to the persistence data agent, as in the previous data process. Once the data is grouped and cleaned it is ready to be used by the Predictor Engine in the making of demand forecasts for each of the stations in the system.

3.5. Demand Prediction Agents

These agents are responsible for carrying out two main tasks: generating the prediction models for each station as well as providing demand forecasting for a specific station, at a concrete time. These agents will run periodically generating new models if new data is available. The demand prediction agent will request the cleaned data to the data persistence agent and it will generate the following prediction models using the machine learning library Scikit-learn [45].

3.5.1. Regression Models Employed

Regression models such as Random Forest Regressor [46], Gradient Boosting Regressor [47] and Extra Tree Regressor have been employed as they have obtained the best results in previous studies [22,48–50]. In the case study, the tested algorithms and their results are described in-depth. In order to compare the quality of the estimators a dummy regressor has been employed as a baseline regressor, which uses the mean as a strategy for performing predictions.

3.5.2. Model Selection and Employed Evaluation Metric

The evaluation metric selected in this work is the Root Mean Square Logarithmic Error (RMSLE) Equation (1). Although there are many common methods for comparing the predicted value with the validation value, such as Root Mean Square Error (RMSE) or Mean Absolute Error (MAE). However, the metric employed in this problem is RMSLE as suggested by the Kaggle competition.

$$\epsilon = \sqrt{\frac{1}{n} \sum_{i=1}^{n} (\log(p_i + 1) - \log(a_i + 1))^2} \tag{1}$$

where ϵ is the RMSLE value (score), n is the total number of observations in the dataset, p_i is the prediction and a_i is the actual response for i, $\log(x)$ is the natural logarithm of x.

The fundamental reason behind this kind of error is that using the logarithmic-based calculation ensures that the errors during peak hours do not dominate the errors made during off-peak hours. Kang [51] discusses why this error is employed in the competition.

There are many Kaggle competitions where this metric is employed such us Grupo Bimbo Inventory Demand [52] and Sberbank Russian Housing [53]. This metric is not included in Scikit-learn, so we had to implement it in order to use it as a scoring function.

3.6. WEBAPI Agent: Visualization and Forecast API and Web Application

This agent will offer access to historical data as well as demand prediction services through a Web Application Programming Interface (WebAPI). Likewise, this agent will provide a web application that will use the previous services to offer information to the BSS operator. The web application will be compound by two main sections:

- Historical data visualization: This section will provide information on the BSS's historical data. It also offers visualizations of trip data information, permitting the user that is filtering historical data to explore user behaviour between stations.
- Forecasting & Status: this section will show the current status of the system and the demand forecasts for the selected station.

4. Case Study

4.1. BSS SalenBici

The SalenBici System is a bike sharing system located in Salamanca, a medium city with a population of 144.949 according the last 2016 census [11]. Today, this system has 29 stations and 176 bikes throughout the city. The system working hours are: work days from 7:00–22:00 and weekends and holidays from 10:00 to 22:00. Figure 4b shows the station map of SalenBici BSS.

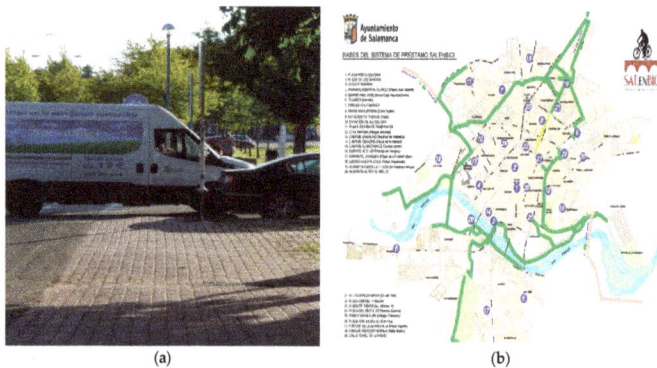

(a)　　　　　　　　　(b)

Figure 4. (a) One of the trucks of the SalenBici system operators; (b) Map of the stations provided by the council of Salamanca.

The system operators employ trucks as it is showed in Figure 4a to perform reallocation tasks in the evening, at noon and in the afternoon, thus extra information on the expected demand in the mornings and evenings could be useful for performing the reallocation tasks.

4.2. Data Collection Process

The data collected regards all trips travelled in the system from a station of origin to a final station. In this case the information provided by SalenBici company is trip data, as mentioned previously and it has the following information in the original dataset: (1) Time Start: timestamp of the beginning of the trip, (2) Time End: timestamp of the end of the trip, (3) Bicycle ID: unique bike identifier, (4) Origin Station: origin station name, (5) End Station: destination station name, (6) Origin dock: origin dock identifier, (7) End dock: destination dock identifier, (8) User ID: user unique identifiers.

The SalenBici company provided us this information in the form of csv files, one for each year. To process this information, the files were given to the multi-agent system. The trip agent loads all the data from the provided data source, the processor agent is notified by the persistence agent because new raw data is added to the system. The processor agent checks the data and request geographic and weather data from the Geographic Information System (GIS) Agent and Weather Agents respectively. Before these agents respond to the request, they check the availability of data with the persistence agent, if the data is available it is sent immediately. On the other hand, if information is not available it is first collected, saved and then sent to the processor agent. Figure 5 shows a sequence diagram of the activities performed by each agent, the communication between the GIS and weather agents and the persistence agent is omitted for clarity.

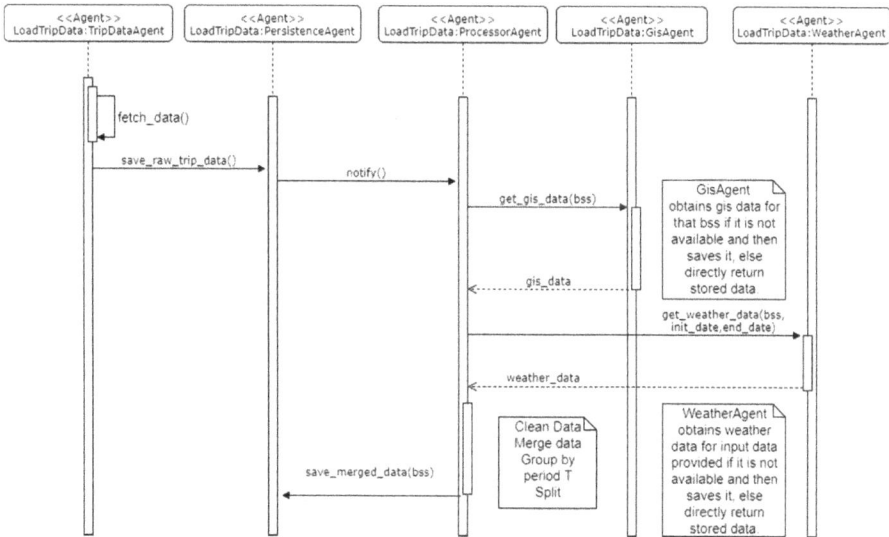

Figure 5. Sequence diagram of trip data loading by multiagent system.

After this process, the stored output data contains the following information on arrivals and departures: Year, Month, Day of the Week, Day of the month, week of the year, hour, season (winter, spring, summer or autumn), weekends and holidays as time related information. The information that is available in relation to the weather on a particular day, is the following: the maximum, minimum and average temperature in degrees Celsius, the average wind speed in km/h, the minimum and maximum atmospheric pressure in millibars and the amount of rainfall. We have the total amount of arrivals or departures in the grouped period T is a dependent variable.

Before spawning the predictor data agent in the multi agent system, an exploratory analysis of the processed data was performed. This analysis determined what model should be included in the predictor data agent.

Figure 6 shows the total number of trips loaded in the system, which are divided into arrivals and departures for each of the stations in the system. This data dates from January 2013 until March 2017 a total of 1520 days and we can see a clear difference between the use of stations in the system. There are a lot of stations where the mean of the bike trips is less than 2 events (arrivals or departures) per day. The station with the highest mean is called "Plaza Poeta Iglesias" near *Plaza Mayor* (the most touristic place in the city of Salamanca), with 15 events per day (arrivals or departures).

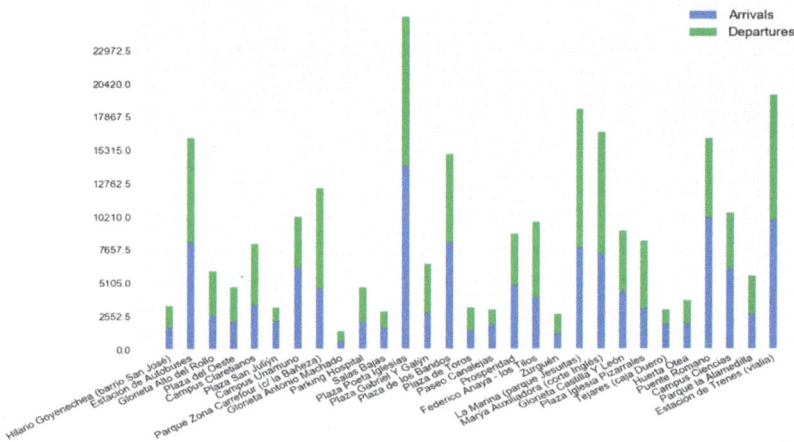

Figure 6. Total number of events (arrivals or departures) performed in Bike sharing systems (BSS) SalenBici from January 2013 until March 2017.

In addition, Figure 7 shows the time period from which each of the stations in the system is active. When reading the total number of trips from the previous figure, it is important to take into account that not all stations have been working since January 2013.

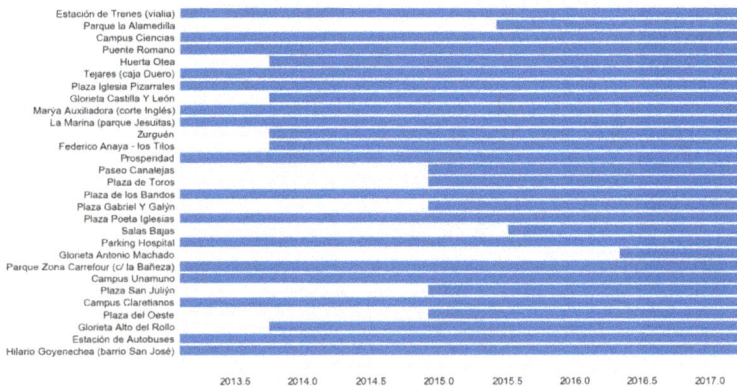

Figure 7. Periods of working time of each station in the BSS SalenBici.

4.3. Model Selection

This section describes how the BSS dataset was split and the methodology that was used in order to select the prediction models that will be included in the predictor agent. Like in Kaggle competitions [17], the data have been split into two datasets; a training dataset and a validation dataset. Figure 8 shows schematically how the available data were employed in the training, selection and validation of the models used. In the upper part of the figure, the green part represents the entire dataset. Like in Kaggle competition, a validation dataset is initially extracted and it is formed from the 20th to the end of each month, in the diagram it is represented in blue. The rest of the dataset, (those from the 1st to the 19th of each month), will be used as training data, represented in violet in the diagram. These data will be one of the inputs of the hyperparameter search technique: GridSearchCV [50]. This technique will use the following as inputs: (1) regression algorithms with their corresponding parameter grid; (2) a scoring function, in order to evaluate the input models, in this case RMSLE and R^2; finally; (3) a cross validation method, in this case *TimeSeriesSplit*, a method that is intended specifically for time series and which resembles the usual functioning of a system in production. This method makes it possible to progressively use data from the past for training and use future data for validation, a diagram showing its functioning is situated in the lower part under GridSearchCV, in Figure 8.

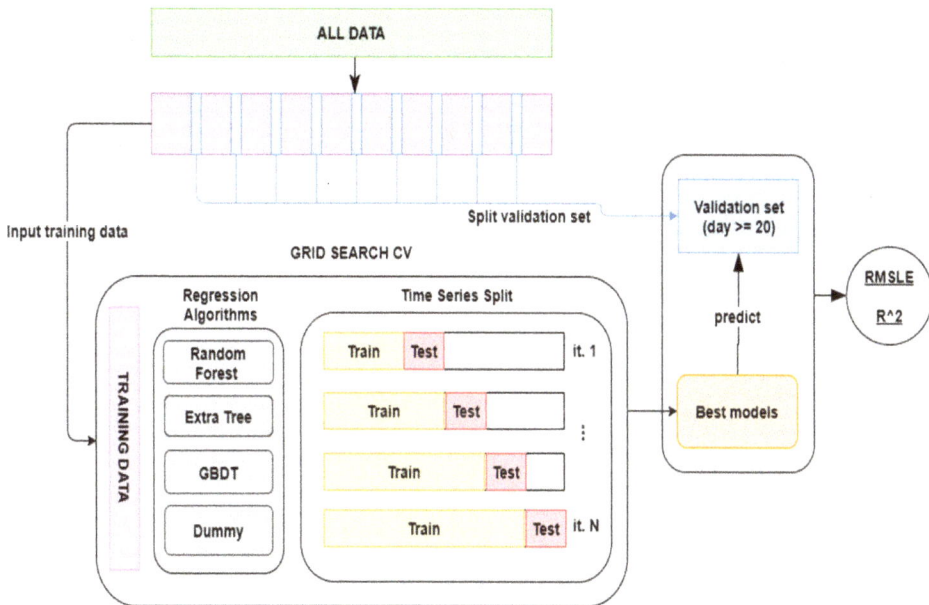

Figure 8. Data splitting for model training and selection.

The GridSearch method will make all the possible combinations for each algorithm with the provided grid parameters; this will be done by employing the cross-validation method (Time Series Split) and evaluating the trained methods with the provided scoring functions. As the output of this method, models for each algorithm with the best results will be obtained and these will be evaluated with the validation dataset that had been split at the beginning, on the right-hand side of Figure 8. A Dummy Regressor has been added to the models used and was established as their prediction strategy in order to continually predict the average. The regression algorithms as well as the following parameter girds have been trained using GridSearchCV:

- Extra Tree Regressor: learning rate: [0.1, 0.01, 0.001], subsample: [1.0, 0.9, 0.8], max depth: [3, 5, 7], min samples leaf: [1, 3, 5]
- Random Forest Regressor: criterion: [mae, mse], number estimators: [10, 100, 1000], max features: [auto, sqrt, log2]
- Gradient Boosting Regressor: learning rate: [0.1, 0.01, 0.001], subsample: [1.0, 0.9, 0.8], max depth: [3, 5, 7], min samples leaf: [1, 3, 5]

Once the model is selected, the results of the best selected model are validated with the validation dataset that had been split previously. Additionally, the coefficient of determination R^2 of the models selected for each station has been calculated.

4.4. Results and Discussion

Next, the results of the RMSLE validation dataset are outlined. The results of the best models selected by the GridSearchCV for each station, are described. Figure 9 shows the results for arrivals and Figure 10 shows results for departures. From these graphs, we can see that regressors tend to have better results than the established baseline. The Random Forest Regressor and the Extra Tree Regressor have the best results.

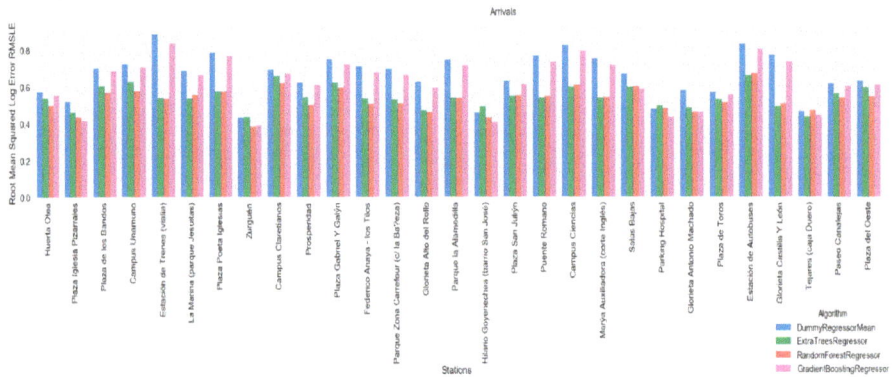

Figure 9. Root Mean Square Logarithmic Error (RMSLE) provided by each algorithm for each station in the arrivals data.

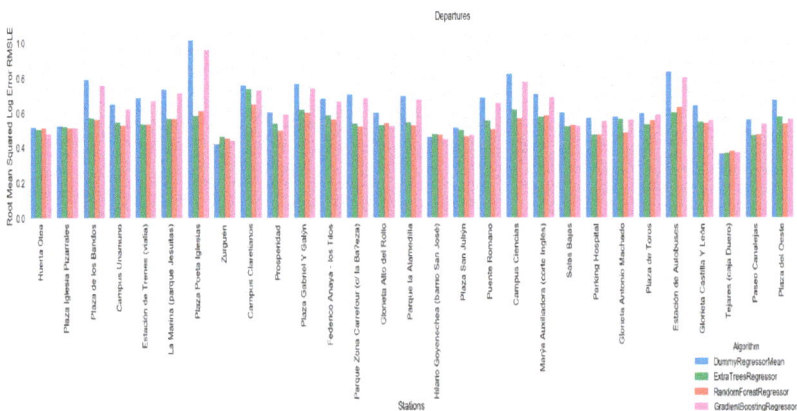

Figure 10. RMSLE provided by each algorithm for each station in the departures data.

Additionally, the coefficient of determination R^2 has been calculated for each of the algorithms, with the aim of using it to select a model that will be included in the predictor agent. The results of the coefficient of determination are shown for each of the algorithms at each of the stations, for arrivals in Figure 11 and departures in Figure 12. Graphically, that the method that performs the best with R^2 is Random Forest Regressor, however results will be analysed to see if this difference is significant.

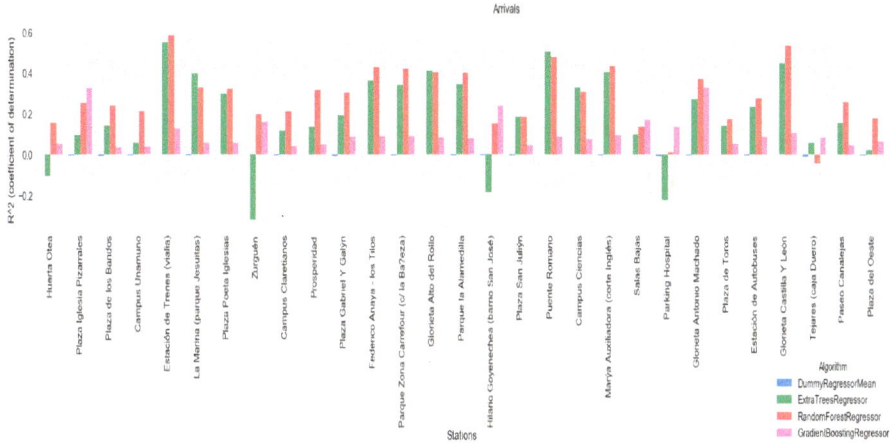

Figure 11. Coefficient of determination of models for the arrivals at each station.

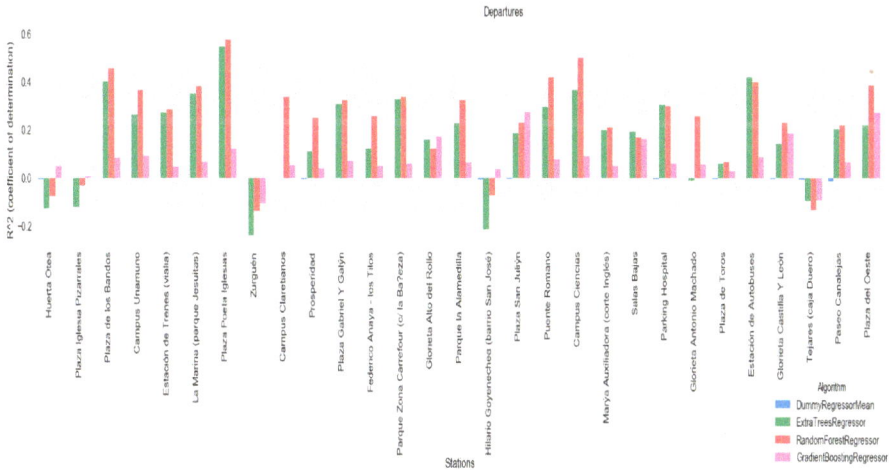

Figure 12. Coefficient of determination of models for departures from each station.

In order to select the best algorithm for all the stations in the system, a statistical Mann-Whitney U test was used. First, the test was done in order to check if one of the algorithms used, functions differently to the rest. The following confirmatory data analysis was applied: H_0 considers the median of two equal methods while H_1 considers the median of two different methods. As we can see in Figure 13, in the case of both departure and arrival models, the Random Forest Regressor Algorithm has a median that is different to the rest, thus, it could be considered that there is a significant statistical difference. The *p*-value obtained for the Random Forest Regressor and Extra Tree pair, surpasses slightly the test's 0.05 significance level, thus it cannot be assumed that there is a significant statistical difference. However, by calculating the median of both methods we can see that the median for Random Forest Regressor is greater in comparison to that of ExtraTreesRegressor.

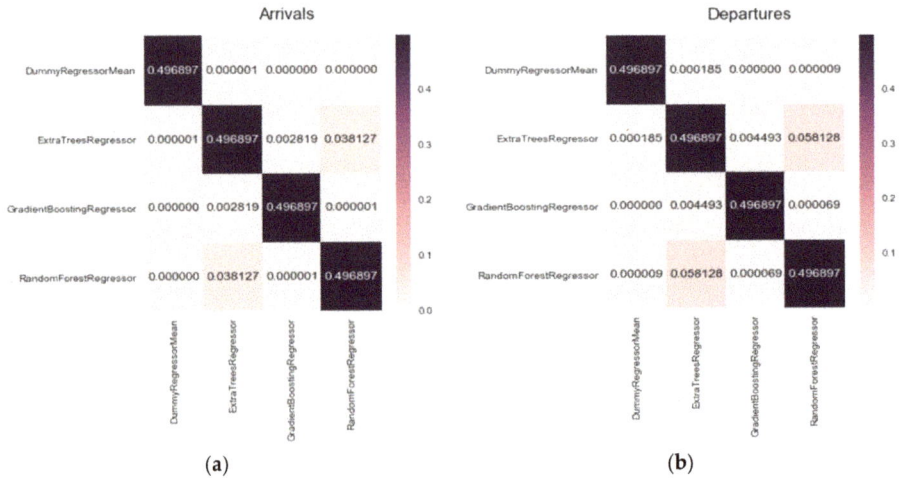

Figure 13. Mann-Whitney U test two sided for (**a**) arrivals and (**b**) departures.

Once this test is completed, it is necessary to repeat it, in the case significant statistical differences are detected, we would proceed to determining whether the median is smaller or greater. In this case, the defined H_1 states that the median of the classifier from the row is greater than the median of the classifier form the column. Figure 14 verifies that Random Forest Regressor has a greater median than the rest of the algorithms for all the stations.

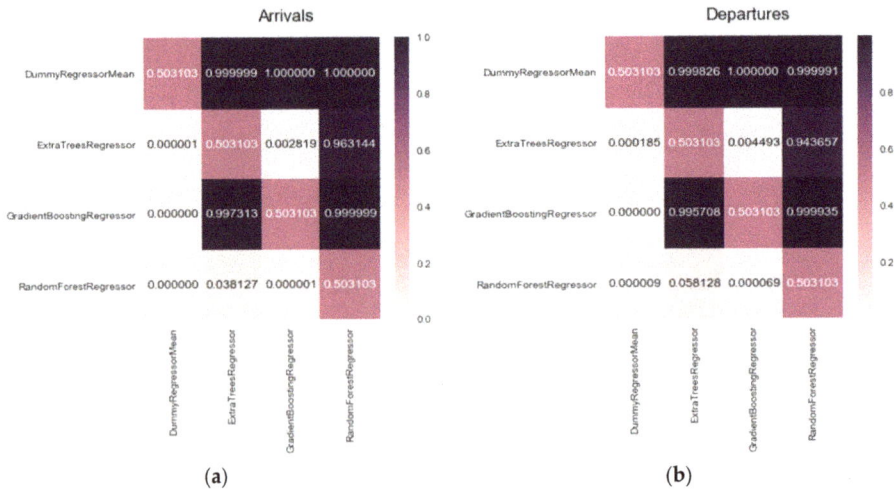

Figure 14. Mann-Whitney U test greater for (**a**) arrivals and (**b**) departures.

After seeing these results, Random Forest Regressor has been selected for inclusion in the predictor agent; it will generate models and store them to subsequently provide predictions through the WebAPI agent. In Figure 15, the sequence diagram for performing the predictions is shown. The predictor agent periodically trains and saves the model for each station in the system. Later these models are used by the WebAPI agent in order to show predictions on the web application and the API REST.

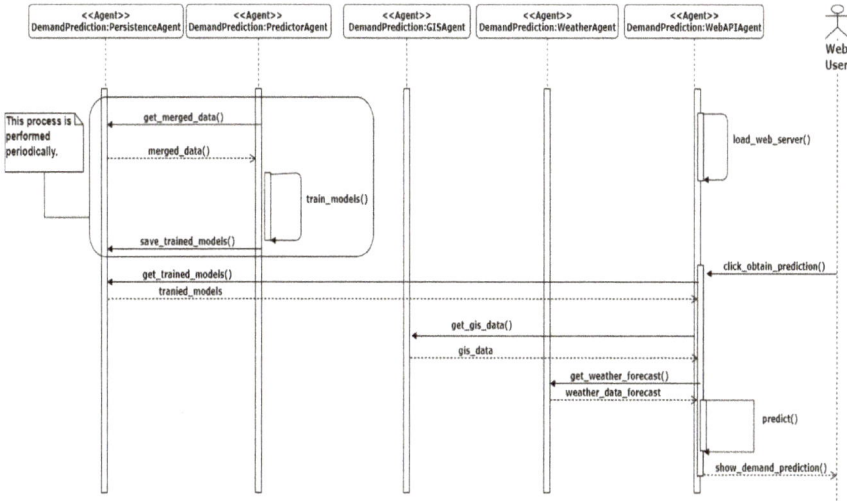

Figure 15. Sequence diagram of trip data loading by multiagent system.

The WebAPI agent provides a web application where users can request predictions for a selected station. This agent will communicate with the weather agent, the GIS agent and the persistence agent in order to obtain the information it needs for making a prediction that was requested by a user. In the next section, the web application will be explained in detail.

4.5. Web Application

The WebAPI agent offers an API REST for third party applications which can obtain the data processed in the MAS and ask for demand predictions of the stations in the system. In addition, there is a web application where users can make visualization and prediction tasks. The web application consists of two sections; in Figure 16 we can see an initial view of the visualization section. In this section, it is possible to view the trips travelled in the system by year and month.

Figure 16. Web application served by the WebAPI agent. Visualization of the historical data section.

The visualization feature has various sections where data can be looked up by introducing a particular station, time (year, month and day), trip (arrivals or departures) and user. These sections act as filters and thus allow to visualize a series of specific data. In Figure 17, we can see the map with departures from the selected station, within the time period indicated in the left corner of the image. The number of trips at the selected station is represented by an arrow, its colour scale and width is proportional to the number of rides made between the station of origin and the destination.

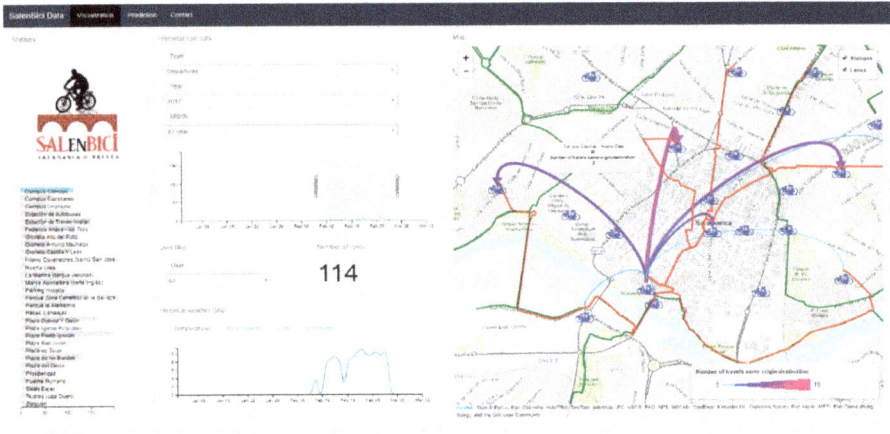

Figure 17. Visualization section. Where a station is selected for a period of time and the number of trips (origin-destination) is visualized.

Trips between the station of origin and the rest of stations are represented with arrows, whose colour and width is proportional to the number of rides made between a particular station and the rest. These visualizations allow the BSS administrator to analyse the history of the system and find the stations with the greatest numbers of arrivals and departures. Figure 18 shows the demand prediction section which employs the selected models to predict demand at each station.

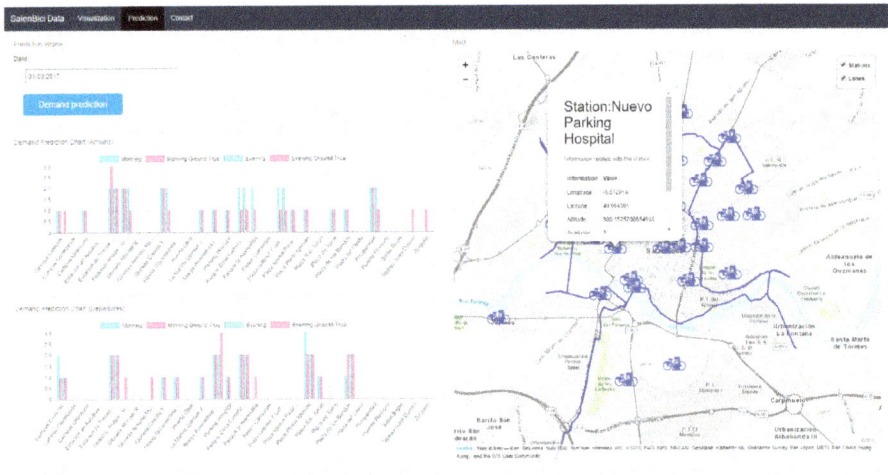

Figure 18. Prediction tool section.

It is possible to select both, an arrow from the past and an arrow for the future. An arrow from the past, apart of showing the data collected by the system, will also visualize the prediction made by the model and the real data that were collected for that station, within the indicated date. These data will be useful to BSS administrators at the time of relocating the bicycles at the stations.

5. Conclusions and Future Works

This work proposed a multi agent system for bike sharing systems. As part of this work, a case study was conducted to test the feasibility of the system, using the data of a BSS located in Salamanca, a middle-sized city. The inclusion of agents in charge of collecting data from heterogeneous sources as well as its cleaning and fusion has eased these tasks. This agent architecture will also make it easier to obtain data from different bike sharing systems.

Moreover, the information collected by the other system agents was employed by the agents that perform demand forecasting. The use of these agents has allowed us to offer services to third party applications.

Different regression algorithms have also been employed in the process of bike demand prediction. Additionally, a statistical analysis has been performed in order to show the differences in their performance and to determine the relevance of results. Random Forest Regressor is the regressor algorithm that outperformed the rest of the algorithms.

With regards to the visualization of the data collected by the agents, a web application has been developed where it is possible to analyse and filter the previously processed data. Thanks to filters in this application the BSS operator can view the desired information thanks and have an insight on the behaviour of the users in the system.

This web application includes a prediction module where the BSS operator can request predictions for future days and make decisions in matters of bike reallocation strategies.

In future works we will employ this multi agent system to collect, process and analyse the data of BSSs in bigger cities such as Madrid.

We will also evaluate the different prediction models with that data in order to compare the performance of the system in middle-sized and big cities. Furthermore, in order to employ the output demand prediction of this system, future work will concern reallocation bike strategies in BSS and the obtaining of an optimal route.

In future works we will analyse how to include the new dock-less bike sharing systems into the developed multi agent system. We will analyse the new additional data that these kinds of systems provide and how to include them into our system. In dock-less systems, the deployed fleet of bicycles could be grouped, creating clusters which act as virtual stations in the city. This data could be used to analyse the bicycle demand and flow of dock-less bike sharing systems.

Acknowledgments: Álvaro Lozano is supported by the pre-doctoral fellowship from the University of Salamanca and Banco Santander. This work was supported by the Spanish Ministry, Ministerio de Economía y Competitividad and FEDER funds. Project. SURF: Intelligent System for integrated and sustainable management of urban fleets TIN2015-65515-C4-3-R. We would also like to show our gratitude to the SalenBici bike share system for sharing with us the data for the current work.

Author Contributions: Álvaro Lozano and Juan F. De Paz conceived and designed the multi agent system while Daniel H. De La Iglesia and Gabriel Villarrubia González contributed to the development and implementation of the system; Álvaro Lozano carried out the experimentation, the model selection and implementation and wrote the paper. Javier Bajo reviewed and supervised the whole work and provided his great expertise in the field.

Conflicts of Interest: The authors declare no conflict of interest.

References

1. Pucher, J.; Buehler, R. Cycling towards a more sustainable transport future. *Transp. Rev.* **2017**, 1–6. [CrossRef]
2. Mátrai, T.; Tóth, J. Comparative assessment of public bike sharing systems. *Transp. Res. Procedia* **2016**, *14*, 2344–2351. [CrossRef]

3. Midgley, P. Bicycle-sharing schemes: Enhancing sustainable mobility in urban areas. *Commun. Sustain. Dev.* **2011**, *18*, 1–12.

4. European Commission—PRESS RELEASES—Press Release—Energy Union and Climate Action: Driving Europe's Transition to a Low-Carbon Economy. Available online: http://europa.eu/rapid/press-release_IP-16-2545_en.htm (accessed on 28 June 2017).

5. Plan B Updates—112: Bike-Sharing Programs Hit the Streets in over 500 Cities Worldwide | EPI. Available online: http://www.earth-policy.org/plan_b_updates/2013/update112 (accessed on 28 June 2017).

6. Ricci, M. Bike sharing: A review of evidence on impacts and processes of implementation and operation. *Res. Transp. Bus. Manag.* **2015**, *15*, 28–38. [CrossRef]

7. The Bike-Sharing World Map. Available online: https://www.google.com/maps/d/u/0/viewer?ll=-3.81666561775622e-14%2C-42.25341696875&spn=143.80149%2C154.6875&hl=en&msa=0&z=1&source=embed&ie=UTF8&om=1&mid=1UxYw9YrwT_R3SGsktJU3D-2GpMU (accessed on 28 June 2017).

8. Dhingra, C.; Kodukula, S. Public Bicycle Schemes: Applying the Concept in Developing Cities. 2010. Available online: http://sutp.org/files/contents/documents/resources/B_Technical-Documents/GIZ_SUTP_TD3_Public-Bicycle-Schemes_EN.pdf (accessed on 15 December 2017).

9. Shaheen, S.A.; Guzman, S.; Zhang, H. Bikesharing in Europe, the Americas, and Asia Past, Present, and Future. *Transp. Res. Rec. J. Transp. Res. Board* **2010**, *2143*, 159–167. [CrossRef]

10. Station-Free Bike Sharing | ofo. Available online: https://www.ofo.com/us/en (accessed on 29 December 2017).

11. The World's First & Amp; Largest Smart Bike Share | Mobike. Available online: https://mobike.com/global/ (accessed on 29 December 2017).

12. Share Bike Bubble Claims First Big Casualty as Bluegogo Reportedly Goes Bankrupt. Available online: http://www.smh.com.au/world/share-bike-bubble-claims-first-big-casualty-as-bluegogo-reportedly-goes-bankrupt-20171116-gzn0k9.html (accessed on 29 December 2017).

13. Raviv, T.; Tzur, M.; Forma, I.A. Static repositioning in a bike-sharing system: Models and solution approaches. *EURO J. Transp. Logist.* **2013**, *2*, 187–229. [CrossRef]

14. Di Gaspero, L.; Rendl, A.; Urli, T. Balancing bike sharing systems with constraint programming. *Constraints* **2016**, *21*, 318–348. [CrossRef]

15. Vogel, P.; Greiser, T.; Mattfeld, D.C. Understanding bike-sharing systems using Data Mining: Exploring activity patterns. *Procedia Soc. Behav. Sci.* **2011**, *20*, 514–523. [CrossRef]

16. Froehlich, J.; Neumann, J.; Oliver, N. Measuring the pulse of the city through shared bicycle programs. In Proceedings of the UrbanSense08, Raleigh, NC, USA, 4 November 2008; pp. 16–21.

17. Kaltenbrunner, A.; Meza, R.; Grivolla, J.; Codina, J.; Banchs, R. Urban cycles and mobility patterns: Exploring and predicting trends in a bicycle-based public transport system. *Pervasive Mob. Comput.* **2010**, *6*, 455–466. [CrossRef]

18. Kaggle INC Bike Sharing Demand | Kaggle. Available online: https://www.kaggle.com/c/bike-sharing-demand (accessed on 27 July 2017).

19. Capital Bikeshare Capital Bikeshare: Metro DC's Bikeshare Service | Capital Bikeshare. Available online: https://www.capitalbikeshare.com/ (accessed on 27 July 2017).

20. Giot, R.; Cherrier, R. Predicting bikeshare system usage up to one day ahead. *IEEE Symp. Ser. Comput. Intell.* **2014**, 1–8. [CrossRef]

21. Patil, A.; Musale, K.; Rao, B.P. Bike share demand prediction using RandomForests. *IJISET Int. J. Innov. Sci. Eng. Technol.* **2015**, *2*. Available online: http://ijiset.com/vol2/v2s4/IJISET_V2_I4_195.pdf (accessed on 15 December 2017).

22. Wang, W.; Wang, W.; Curley, A. Forecasting Bike Rental Demand Using New York Citi Bike Data. Master's Thesis, Dublin Institute of Technology, Dublin, Ireland, 2016.

23. Lee, C.; Wang, D.; Wong, A. Forecasting Utilization in City Bike-Share Program (Vol. 254). Technical Report, CS 229 2014 Project. 2014. Available online: http://cs229.stanford.edu/proj2014/Christina%20Lee,%20David%20Wang,%20Adeline%20Wong,%20Forecasting%20Utilization%20in%20City%20Bike-Share%20Program.pdf (accessed on 15 December 2017).

24. Du, J.; He, R.; Zhechev, Z. Forecasting Bike Rental Demand. 2014. Available online: http://cs229.stanford.edu/proj2014/Jimmy%20Du,%20Rolland%20He,%20Zhivko%20Zhechev,%20Forecasting%20Bike%20Rental%20Demand.pdf (accessed on 15 December 2017).

25. Yin, Y.-C.; Lee, C.-S.; Wong, Y.-P. Demand Prediction of Bicycle Sharing Systems. 2012. Available online: http://cs229.stanford.edu/proj2014/Yu-chun%20Yin,%20Chi-Shuen%20Lee,%20Yu-Po%20Wong,%20Demand%20Prediction%20of%20Bicycle%20Sharing%20Systems.pdf (accessed on 15 December 2017).

26. NYC. Citi Bike Station Map. Available online: https://member.citibikenyc.com/map/ (accessed on 27 July 2017).

27. Lluís Esquerda CityBikes: Bike Sharing Networks around the World. Available online: https://citybik.es/ (accessed on 27 July 2017).

28. SalEnBici Salenbici. Sistema de Préstamo de Bicicletas de Salamanca. Salenbici. Available online: http://www.salamancasalenbici.com/ (accessed on 27 July 2017).

29. Matos, D.M.; Lopes, B.; Bento Dei, C.; Machado, E.R. *An Intelligent Bike-Sharing Rebalancing System*; Universidade de Coimbra: Coimbra, Portugal, 2015.

30. Chicago Bike Share System Divvy: Chicago's Bike Share Program | Divvy Bikes. Available online: https://www.divvybikes.com/ (accessed on 27 July 2017).

31. Capital Bike Share Whasington System Data | Capital Bikeshare. Available online: https://www.capitalbikeshare.com/system-data (accessed on 27 July 2017).

32. Billhardt, H.; Fernandez, A.; Lujak, M.; Ossowski, S.; Julian, V.; De Paz, J.F.; Hernandez, J.Z. Coordinating open fleets. A taxi assignment example. *AI Commun.* **2017**, *30*, 37–52. [CrossRef]

33. Gregori, M.E.; Cámara, J.P.; Bada, G.A. A jabber-based multi-agent system platform. In Proceedings of the Fifth International Joint Conference on Autonomous Agents and Multiagent Systems—AAMAS '06, Hakodate, Japan, 8–12 May 2006; ACM Press: New York, NY, USA, 2006; p. 1282.

34. Criado, N.; Argente, E.; Julian, V.; Botti, V. Organizational services for the spade agent platform. *IEEE Lat. Am. Trans.* **2008**, *6*, 550–555. [CrossRef]

35. Bellifemine, F.; Poggi, A.; Rimassa, G. JADE—A FIPA-compliant agent framework. *Proc. PAAM* **1999**, *99*, 33.

36. Zato, C.; Villarrubia, G.; Sánchez, A.; Barri, I.; Rubión, E.; Fernández, A.; Rebate, C.; Cabo, J.A.; Álamos, T.; Sanz, J.; et al. *PANGEA—Platform for Automatic Construction of Organizations of Intelligent Agents*; Springer: Berlin/Heidelberg, Germany, 2012; pp. 229–239.

37. Villarrubia, G.; Paz, J.F.; De Iglesia, D.H.D.; La Bajo, J. Combining multi-agent systems and wireless sensor networks for monitoring crop irrigation. *Sensors* **2017**, *17*, 1775. [CrossRef] [PubMed]

38. Aiomas—Aiomas 1.0.3 Documentation. Available online: http://aiomas.readthedocs.io/en/latest/ (accessed on 14 November 2017).

39. osBrain—0.5.0—osBrain 0.5.0 Documentation. Available online: http://osbrain.readthedocs.io/en/stable/ (accessed on 14 November 2017).

40. Esquerda, L. Documentation | CityBikes API. Available online: https://api.citybik.es/v2/ (accessed on 4 October 2017).

41. Citi Bike System Data | Citi Bike NYC. Available online: https://www.citibikenyc.com/system-data (accessed on 4 October 2017).

42. Weather Forecast & Amp; Reports—Long Range & Amp; Local | Wunderground | Weather Underground. Available online: https://www.wunderground.com/ (accessed on 4 October 2017).

43. Agencia Estatal de Meteorología—AEMET. *Gobierno de España*; AEMET: Madrid, Spain, 2017.

44. Workalendar. Available online: https://github.com/novafloss/workalendar (accessed on 5 October 2017).

45. Pedregosa Fabianpedregosa, F.; Alexandre Gramfort, N.; Michel, V.; Thirion Bertrandthirion, B.; Grisel, O.; Blondel, M.; Prettenhofer Peterprettenhofer, P.; Weiss, R.; Dubourg, V.; Vanderplas Vanderplas, J.; et al. Scikit-learn: Machine Learning in Python Gaël Varoquaux. *J. Mach. Learn. Res.* **2011**, *12*, 2825–2830.

46. Breiman, L. Random Forests. *Mach. Learn.* **2001**, *45*, 5–32. [CrossRef]

47. Friedman, J.H.; Friedman, J.H. Greedy function approximation: A gradient boosting machine. *Ann. Stat.* **2000**, *29*, 1189–1232. [CrossRef]

48. Regue, R.; Recker, W. Using gradient boosting machines to predict bikesharing station states. In Proceedings of the 93rd Annual Meeting of Transportation Research Board, Washington, DC, USA, 12–16 January 2014.

49. Malani, J.; Sinha, N.; Prasad, N.; Lokesh, V. Forecasting Bike Sharing Demand. Available online: https://cseweb.ucsd.edu/classes/wi17/cse258-a/reports/a050.pdf (accessed on 15 December 2017).

50. Prakash Nekkanti, O. Prediction of Rental Demand for a Bike-Share Program. 2017. Available online: https://library.ndsu.edu/ir/handle/10365/25949 (accessed on 15 December 2017).

51. Kang, S.C.; Otani, T.W. *Learning to Predict Demand in a Transport-Resource Sharing Task*; Naval Postgraduate School: Monterey, CA, USA, 2015.

52. Grupo Bimbo Inventory Demand | Kaggle. Available online: https://www.kaggle.com/c/grupo-bimbo-inventory-demand (accessed on 15 November 2017).

53. Sberbank Russian Housing Market | Kaggle. Available online: https://www.kaggle.com/c/sberbank-russian-housing-market (accessed on 15 November 2017).

applied
sciences

MDPI

Article

An Organizational-Based Model and Agent-Based Simulation for Co-Traveling at an Aggregate Level

Iftikhar Hussain [1,*], Muhammad Arsalan Khan [1], Syed Fazal Abbas Baqueri [1],
Syyed Adnan Raheel Shah [1,2], Muhammad Khawar Bashir [3], Mudasser Muneer Khan [4]
and Israr Ali Khan [5]

[1] Instituut Voor Mobiliteit (IMOB), Hasselt University, Wetenschapspark 5 Bus 6, 3590 Diepenbeek, Belgium;
    Muhammadarsalan.khan@uhasselt.be (M.A.K.); syed.fazalabbasbaqueri@uhasselt.be (S.F.A.B.);
    Syyed.adnanraheelshah@uhasselt.be (S.A.R.S.)
[2] Department of Civil Engineering, Pakistan Institute of Engineering & Technology, Multan 66000, Pakistan
[3] Department of Statistics and Computer Science, University of Veterinary and Animal Sciences,
    Lahore 54000, Pakistan; mkbashir@uvas.edu.pk
[4] Department of Civil Engineering, University College of Engineering & Technology,
    Bahauddin Zakariya University, Multan 66000, Pakistan; mudasserkhan@bzu.edu.pk
[5] Department of Electrical Engineering, Namal College, Talagang Road, Mianwali 42250, Pakistan;
    Israr.khan@namal.edu.pk
*   Correspondence: iftikhar.hussain@uhasselt.be; Tel.: +32-(0)486-207-755

Received: 3 November 2017; Accepted: 21 November 2017; Published: 27 November 2017

Abstract: Carpooling is an environmentally friendly and sustainable emerging traveling mode that enables commuters to save travel time and travel expenses. In order to co-travel, individuals or agents need to communicate, interpret information, and negotiate to achieve co-operation to find matching partners. This paper offers the scheme of a carpooling model for a set of candidate carpoolers. The model is interpreted using an agent-based simulation to analyze several effects of agents' interaction and behavior adaptations. Through communication and negotiation processes, agents can reach dynamic contracts in an iterative manner. The start of the negotiation process relies on the agents' intention to emit an invitation for carpooling. The realization of the negotiation process depends significantly on the departure time choice, on the agents' profile, and on route optimization. The schedule or agenda adaptation relies on the preferences among the realistic schedules of the agents and usually depends on both the participation of the trip and on the time of day. From the considerations, it is possible to reveal the actual representation of the possible carpoolers during the simulated period. Experiments demonstrate the nearly-polynomial relationship between computation time and the number of agents.

Keywords: agent technology; organizational model; agent behavior; travel behavior; commuting; carpooling

1. Introduction

At present, various research fields including *transport behavior* require investigations and simulations of the dynamic behavior of travelers. In recent studies, it has become increasingly important to establish the interaction between agents. Traditional modeling applications have encountered difficulties in dealing with the complication of the interaction and negotiation of individuals (agents) that are essential to simulating the carpool travel demand [1]. The method that is more suitable for autonomous entity interaction is based on agent-based modeling (ABM). ABM is a distributed and personal-centric approach that allows agents to recognize the interaction of physical elements and describes many issues arising in research fields such as astronomy, biology, ecology, and social sciences. It has also been used in a wide range of topics within the field of transportation science, including vehicle or pedestrian flow simulations, route selection modeling, vehicle tracking and lane change models, and traffic simulations.

Carpooling is an environmentally friendly and sustainable emerging traveling mode that aids travelers (carpoolers) to save travel time and travel expenses (fuel, charges, and parking fees). It also cuts emissions and reduces traffic jams. Change in some factors such as the rise in fuel prices, charged parking, or the introduction of new traffic policies may lead to a carpool initiative for an individual. In addition, the social economic characteristics (SEC) such as gender, age, relationships, income, education, job, and vehicle and driving license ownership, may play a dynamic role in finding the right persons to co-travel. In order to co-travel, individuals need to negotiate, coordinate, and in most cases, can adapt their daily schedules to enable co-operation [2,3]. The agent's daily schedule or agenda is a group of sequential trips and activities. Each activity and trip of the daily schedule is specified by the start time and duration. Each negotiation consists of a small number of agents and their daily schedule can be interrelated through co-operation [2,3].

This research presents an organizational-based model and agent-based simulation for long-term commuting at an aggregate level (Travel Analysis Zones (TAZ)) based on work trips (home-to-work *HW* and work-to-home *WH*) that is designed to establish a framework and a carpooling social network of the candidates. The purpose of this research is to examine the effects of time limitations and models on how people adjust their own schedule to co-operate. Carpooling is a concrete example of co-operation between people. The focus is on the mechanisms that simulate human behavior when making collaborative decisions. The agent-based simulations can analyze the aggregation consequences of individual behavioral changes to enable carpooling. The model shall support research on both topics in the arena of large scale agent-based modeling and in the field of co-operation and rescheduling in activity-based models.

This paper presents the model and simulation for co-traveling which generalize the conception of interaction, negotiation, and co-ordination of a multi-trips negotiation model. It takes the option of flexible activity planning into account. The model is based on an agent-based and organizational meta-model [4]. The role and organization are the foremost entities in the organizational meta-model. In order to co-travel, agents search for other participants to cooperate with while performing their periodic trips and explore the carpooling social network (CPSN) by emitting carpool invitations and sharing their agendas within a carpooling social group (CPSG). During negotiation, agents may reach complex contracts depending on the "negotiation model" (used to match with the carpooling participants) and also on the travel behavior of the individuals. The agents negotiate about trip timings, and on the driver and vehicle selection, and may adjust their daily agenda to enable co-operation. After finding matching partners during the negotiation procedure, agents may decide to carpool for a specified time period. Carpoolers may re-negotiate and/or re-schedule their daily schedule when someone joins the carpool group or leave it permanently. For implementation, the Janus (http://www.janus-project.org/Home) [5], multi-agent based platform is used which offers an efficient implementation of agent-based and organizational-based concepts.

This paper is organized as follows. After the detailed related work presented in Section 2, an organizational-based model for carpooling is described in Section 3. It is based on three layers (from bottom to top): (1) problem layer; (2) organizational layer; and (3) the agent layer. In depth results and the discussion are presented in Section 4. Finally, the conclusion and ideas for future work are presented in Section 5.

2. Related Work

The related work on carpooling negotiation models, negotiation, and co-operation on multi-agent environments, agent-based studies in other fields (i.e., tourism), and activity rescheduling is presented in this section.

It starts by explaining some of the agent-based models covering *coordination and negotiation* in carpooling. Hussain et al. [6] present a negotiation model for carpooling based on multiple trips (*HW* and *WH*) that is derived from the existing departure time studies presented in [7]. The authors designated the direct communication between agents using a state-transition machine. It also restricted the interaction to within the small groups and the formation of groups is based on the home and Work Travel Analysis Zones (TAZs). The same authors in [8] offer the agents' interaction mechanism by making an independent group of individuals living in different areas but having the same work TAZ. Only the individuals within a group are permitted to negotiate with each other. The negotiation model on trip departure times and on driver selection is also presented in literature. For instance, Garland et al. [9] propose the conceptual design of an ABM for carpooling applications by simulating autonomous agents and by analyzing the impact of changes in infrastructure, individuals' behavior, and cost factors. The model uses agent information and social networks to start communications, and then uses routing algorithms and utility functions to trigger negotiation processes. This study is based on [10], which presents the conceptual design of ABM for carpool applications.

Bellemans et al. [11] introduce an agent-based simulation model to support the carpooling of large manufacturing plants. The authors describe the following services: (a) agent-based simulations are used to study the opportunities and inhibitors; and (b) online matching can be used to match the commuter profile. They claim that the complex consultations between agents is necessary for fruitful carpooling because of re-routing and re-scheduling parameters. In [12], an adaptive genetic algorithm based on multi-agent approach is proposed, which can effectively solve the problem of long-term carpooling with a limited search space. The system is a combination of a multi-agent system and genetic paradigm, and by the collective learning process dynamically modified super-heuristic guidance. In [13], the authors used NetLogo to design and present a multi-agent based Dynamic Carpooling System (DCS). The DCS optimizes the use of transportation by sharing the same route. The authors claim that their system provides intelligent matching services and intelligent routing engines that can use real-time information (i.e., weather and traffic conditions).

In terms of travel requirements, co-operation is appropriate for the implementation of supportive activity and trip accomplishment. Ronald et al. [14] presents an agent-based model that emphases the negotiation scheme of cooperative activities. The proposed scheme comprises a precise and structured interactive protocol: a combination of transport and social dimensions. A utility method based on personal and composite attributes is presented. Agents negotiate on the type of activity, destination location, and the departure time of their communal activities. Chun and Wong [15] presented a generalized model to schedule events dynamically that uses a negotiation mechanism. It is based on an agent-based approach. The authors define a group and a negotiating protocol to establish agreements on agendas. Each agent is supposed to state its most preferred choices first. It also identified new proposals in a non-increasing order of preference.

The authors in [16] applied game theory in environments where the agents are autonomous, but may follow some agreed-upon protocols. This revealed that classical mechanic models are valuable for task distribution in large sets of co-operative agents. The queueing networks for task distribution among a comparatively small set of co-operative agents are also applied. The authors also used the less formal social science models of co-operation when there were no strict protocols. Furthermore, they verified that concepts drawn from philosophy can be the basis for the development of shared plans among agents. The authors in [17] present the first complete-implementation of a multi-agent based Delphi process for the document relevance domain. The Delphi method is a procedure that promises a new way of dealing with the co-ordination of agents. Intuitively, the specialists should engage in dialogue, interchange ideas, and change their mind as the discussion progresses. The authors solve the document relevance evaluation problem with a case study where a community of specialists decide whether a document is relevant or not. The domain-specific portion is summarized in certain objects and external modules.

In the research presented in [18], the authors established the Jiuzhaigou tourist shunt model based on the multi-agent simulation platform of NetLogo. It determines the allocation probability according to the space-time load based on the Logit model. The performance is evaluated using two indexes of tourist balanced distribution including the variance model and the Gini–Simpson model. The research study of tourist route selection during the peak travel period illustrates further exploration based on previous studies. The authors in [19] presented an Agent-based Simulator for Tourist Urban Routes (ABSTUR) that provides the number of tourist people signed up for each route. The simulation considered both the characteristics of routes and the types of tourists with their preferences. ABSTUR has been experienced by assisting a group of tourism experts in designing a set of tourist routes in the historic center of Madrid. In this manner, they were able to avoid both overcrowded tourist and non-profitable routes.

A large body of literature (e.g., Nijland, et al. [20] and Guo [12]) has been published about the idea of rescheduling activities. In the context of negotiation to co-operate, this considered agenda adaptation to surprising events which are opposite to the rescheduling activities. Knapen et al. [21] present a framework at a large scale to investigate algorithms for rescheduling. The exchange of information between traffic information applications and commuters is aided by explicit modeling. It chains macroscopic traffic assignment with the microscopic simulation of the agents. The Aurora model developed by Joh et al. [22] provides agenda generation and the decisions of the dynamic activity travel rescheduling. Arentze et al. [23] present a complete description of the Aurora activity-based model for schedule generation and adaptation. A comprehensive model has been indicated by describing the following operations: insertion, shifting, deletion, and replacement of activities, as well as changing locations, trip chaining options, and transportation modes. Prototypes of this level of detail are a prerequisite to incorporate co-operation concepts in carpooling. Gupta and Vovsha [24] describe a hybrid discrete choice-duration framework for work activity by modeling interactions between workers in a multiple-worker household. The important feature is to introduce the intra-household interactions through the worker's agenda synchronization techniques.

Xia et al. [25] present matching services for carpooling by applying both optimal and heuristic approaches to test and find solutions. It is verified that a new formulation and related solution processes can authorize the purpose of the optimal carpool teams and their routes. Martinez et al. [26] present an agent-based simulation for shared taxis to identify a set of rules for matching in terms of space and time. The authors considered that the client is only eager to agree a maximum deviation from his or her straight path. Knapen et al. [2] present an automated advisory service to match travelling trips for carpooling. The probability for fruitful negotiation is measured by means of a learning mechanism. The matching tool needs to deal with dynamically changing the graph with regards to the topology and edge weights. The problem of finding an optimal tree structured route for carpooling is studied by the same authors in [27], in case some participants leave their car at a carpool parking area. In this research, the authors proposed an algorithm to find the optimal solution for the join tree.

None of the reported research effectively analyses the effect of agents' behavior and the negotiated schedule adaptation of a set of candidate carpoolers required for co-traveling. In this paper, we propose and present a model to examine the problem.

3. Agent-Based Model for Co-Traveling

An agent-based framework for cooperative travelling is presented to investigate the behavior of carpooling participants. The aim is to simulate how individuals decide to carpool by adjusting their daily agendas, and how actual trip execution activity is performed in the carpooling process. Agents can communicate autonomously to find matching partners in order to co-travel. Agents will carpool for numerous days with the same carpool-group. They can also carpool for several consecutive carpooling periods in different CPSG, one at a time. The negotiation procedure integrates a constant preference scheme for the trip departure times. The travelling trips, HW and WH, in the daily agendas are specifically comprehensive and discussed in the long-term perspective. The *negotiation model* is used to find the matching partners: by adapting the trip departure times and by selecting the vehicle (car) and the driver of the carpool-group. The driver and vehicle selection is achieved by inspecting the individual's profiles during the *negotiation process*. The presented carpooling model is established on the movement of vehicles between traffic analysis zones (TAZ). It is anticipated that people board at home, alight at work TAZ, and vice versa; participants with similar trips can carpool with each other.

3.1. Agent-Based Modeling

An agent-based approach is used for evaluating the effects of an agent's behavior and for simulating the communications of autonomous entities. It is individual-centric and essentially distributed, and suitable for systems that display complex behavior. The proposed carpooling model is based on the Capacity, Role, Interaction, and Organization (CRIO) meta-model [4]. It provides organizational views for demonstrating dynamic structures in terms of their role and relations. This meta-model offers the transformation from organizational views to ones that are processed for generating an agent-based simulation. Accepting an organizational approach permits agents to dynamically adapt their actions without altering their core structure.

According to the CRIO meta-model, the definition of an organization is "a cluster of roles that participate in structured logical institutionalized designs of interactions with the other roles in a collective perspective. This perspective involves common knowledge, common rules or standards and common feelings, etc. and is considered as an ontology. The goal of an organization is to accomplish some requirements". A group is used for splitting organizations. The group is an organizational object in which all associates are able to cooperate according to pre-defined interaction designations and rules. A role is defined as "a predictable behavior where a set of role jobs are well-ordered by a strategy, and a set of privileges and responsibilities in the organization context." Each role contributes to the completion of the necessities of the organization within which it is defined.

According to [28,29], the CRIO meta-model is suitable because the carpoolers are dynamically altering their roles in the carpooling social network. For implementation, the Janus [5] multi-agent based platform is used which offers the well-organized application of agent-based and organizational-based models. Janus is built upon the CRIO organizational meta-model in which the concepts of organization and role are special entities. Each agent has the ability to play its role inside an instance (CPSG) of the organization. The organizational-based modeling permits the situations to be defined in an organized way. It provides the capability to determine where the associations between agents occur and how these associations influence the outcomes [4,30].

3.2. Framework Structure

This section presents the setup of an organizational-based framework and the associated agent-based simulation model for the carpooling problem. The presented framework is based on three layers (see Figure 1 from bottom to top): (1) the problem layer; (2) organizational layer; and (3) agent layer. In the problem layer of the framework, different activities of the carpooling process performed by the agents (carpooling and non-carpooling) are described. The negotiation model on trip departure times and on the selection of the driver for multiple trips is presented. The organizational layer of the proposed model emphases the organizations of the carpooling scheme. It also focuses on the activities in terms of role behaviors of the agents. Subsequently, the design of the day-switching mechanism is revealed by the use of organizational concepts. The agent layer section presents the agent model of the framework where the agent's behavior is demonstrated and discussed using state-machines.

The *agent* or *individual* (carpooler and non-carpooler) is someone who belongs to the region where (s)he executes a daily agenda to fulfil personal needs. The agent's personal daily schedule is a group of sequential trips and activities. Each activity and trip of the daily schedule is specified by the start time and duration. The demonstrating construction claims that the agents perform their daily activities and also spend their time on traveling between those activities. The schedule adjustments of the participants depends on personal preferences of their feasible schedule. Agents who belong to the same carpooling social group may cooperate while performing their periodic trips by exploring the carpooling social network.

Different layers of the agent-based framework for co-traveling are illustrated in Figure 1. In what follows, each of these layers from bottom to top are described in more detail.

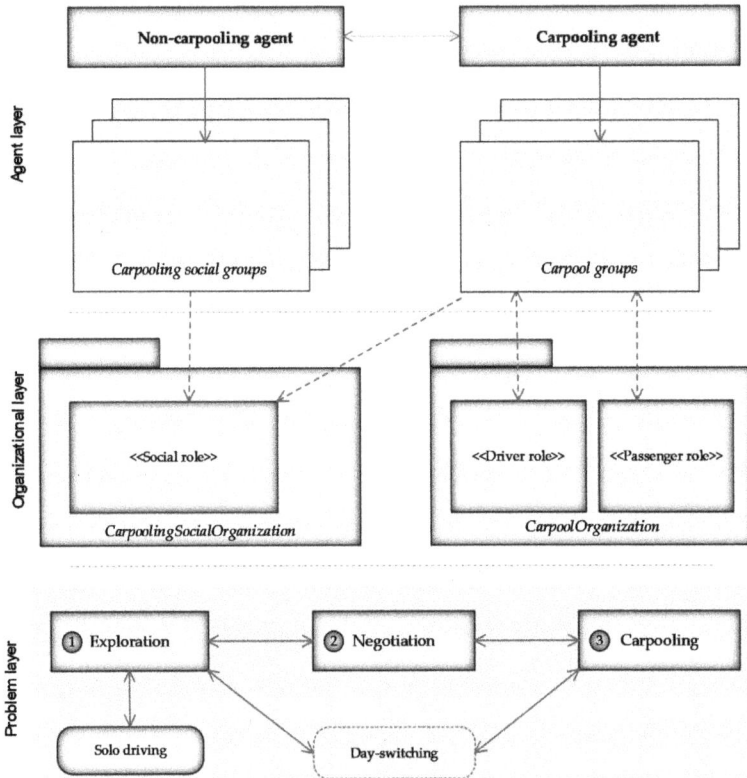

Figure 1. The agent-based framework for co-traveling based on organizational concepts.

3.3. Problem Layer

This section presents the problem domain of the carpooling model. It is based on three main activities which are iterative and performed by the agents: (1) exploration; (2) negotiation; and (3) carpooling, and each of these activity is described in more detail.

3.3.1. Exploration

Each non-carpooling agent explores the carpooling social network for carpool partners to cooperate with while performing their daily trips. The presented framework is established on the movement of vehicles between Traffic Analysis Zones (TAZ) as different to agents' street or home addresses. The model considers that people board at home and alight at work TAZ locations. People agree to choose carpool companions from the group of individuals segmented on the basis of similar home and work TAZ locations.

The agents, once they decide to carpool, may start searching for carpool partners from the carpooling social network. In case they do not want to carpool, they continue traveling solo by using their own car. The agents who reside in the same carpooling social group (grouped on the basis of same home and work TAZs) may interact by exchanging text messages. Through the interaction process, the agents may search for the ideal candidate for carpooling and may reach a coordinated solution by negotiating on the departure times of both the trips, on the selection of the driver, and hence on the choice of vehicle. Agents may continue looking for participants in the exploration process throughout the simulated period (in case they are unable to find a carpool partner).

For a carpooling agent, the time intervals, path, and personal profile are considered as the relationship information to compare with other agents. In general, every agent has a basic set of properties for interaction such as mutual interests and requirements. The common interest (intention to invite someone for carpooling) and requirements (the similarity values: path, route, time intervals, and some personal attributes) for the corresponding individuals need to match satisfactorily well for the interaction process.

Each agent (the sender) may look for a carpooling partner by sending him a carpool invitation message. This emission of a carpool invitation depends on the personal intension to invite someone with similar attributes i.e., the similar home and work TAZ. Some people do not like to carpool so they will not emit any carpool invitation. Within a day, an individual can search for carpool partners multiple times in the carpooling social network. The receiver agent may accept the sender or inviter as a carpooling partner when the negotiation succeeds. The negotiation process is described in the following section.

3.3.2. Negotiation

The final decision to form a carpool group is taken in the negotiation process where the actual matching is applied using similarity constraints. The negotiation among agents is taken into account for the selection of departure times of the multiple trips (morning and evening) and also to select the driver of the carpool group. The schedule is adapted based on the preferences amongst the feasible schedules of the agents. The negotiation will become successful only when the driver is available and the preferred trip departure times for both the trips are commonly well-matched within the carpool group. Compatible indicators (described in more detail below) are used in the negotiation process.

Driver and Vehicle Selection

During the negotiation process, the agent's personal profile is used for the driver and vehicle selection. The agent having the car and the driving-license ownership may be selected as a driver when carpooling. The driver in the carpool is responsible for picking other carpool participants from their home-locations. Hence the first agent of the candidate sequence shall be the driver. The candidate sequence can be considered infeasible and dropped immediately, where the first agent cannot act as the driver.

Time Intervals Similarity

In this study, every moment or interval between the earliest and latest possible trip departure time in the time windows as indicated by the candidate carpoolers is supposed to be equal. The time preference function is the same for each participant over the time and supposed to be constant. For matching agents, time windows equipped with a constant preference function, i.e., $\forall t \in I : P(t) = 1$ are used. For each trip (either HW or WH) of an agent, two time windows (departure and arrival) are used.

Let A be a set of all individuals or agents. For an agent a_i, the earliest and latest departure time intervals for the trip are $T_{a_i}^b$, $T_{a_i}^e$ and the preferred trip start time of a_i is t_{a_i}. The time windows of carpooling participants at the specific TAZ-location l_i are represented by $T_{carpool,l_i}$. The departure and arrival time windows for the carpool trips are the intersection of the departure and arrival time windows of the particular participants. The earliest $T_{carpool,\ l_i}^b$ and latest $T_{carpool,\ l_i}^e$ intervals of the intersection of the time windows can be calculated as specified in Equation (1); the indices used for the maximum and minimum range are max() and min() functions, respectively, over the set of carpooling candidates. The available departure and arrival time intervals of trips for the carpool group are given by Equation (1), where j recognizes the candidate carpooler.

$$
\begin{aligned}
T^b_{carpool,\ l_i} &= \underset{j=1...N}{max}\ (T^b_{a_j}) \\
T^e_{carpool,\ l_i} &= \underset{j=1...N}{min}\ (T^e_{a_j})
\end{aligned}
= \bigcap_{j=1}^{N} T_{a_j\ l_i}
\tag{1}
$$

For some l_i, when the time windows $T_{carpool,\ l_i}$ of the individuals participating in the negotiation process are empty (the $T_{carpool,\ l_i}$ do not intersect), the case is considered as infeasible (given by the Equation (2)). Similarly when the time windows of the trip departure times intersect, then the case is feasible (given by the Equation (2)) and the negotiation among participants on the trip departure time succeeds.

$$
\forall_h :\ \begin{array}{l} T_{carpool,\ l_i} = 0\ \text{infeasible case} \\ T_{carpool,\ l_i} \neq 0\ \text{feasible case} \end{array}
\tag{2}
$$

Hence, a carpooling participant may join the carpool group if and only if the desired trip departure times are within the suitable intervals and can be achieved by Equation (3).

$$
\bigcap_{j=1}^{N} T_{HW,\ a_j\ l_i} \cdot \bigcap_{j=1}^{N} T_{WH,\ a_j\ l_i} \neq 0
\tag{3}
$$

Path Similarity

Detour or maximum excess time duration integrates the path similarity concepts which are assigned to each individual of the ordered pair that specifies the OD (Origin-Destination) pairs are involved in the particular trips. An individual with a solo trip duration $d_{solo,\ a_i}$ has an upper limit $d_{detour,\ a_i}$ for the detour delay in the trip: from home-to-work and from work-to-home TAZs. The maximum excess duration depends on the personal preference of the agent and it may be specified by the individuals.

For negotiation success, the carpool duration for an a_i must be less than or equal to the individual's detour travel or maximum excess duration $d_{detour,\ a_i}$.

The negotiation becomes successful when the above described constraints are satisfied and the carpooling participants adapt their daily schedules according to the output of the negotiation. During the schedule adaptation process, the carpoolers agree on pickup and drop-off times, pickup and drop-off orders, and also how long (number of days/months) everyone may carpool with this carpooling group.

When the negotiation succeeds and everyone agrees on the negotiated agendas then the actual trips are executed. The invited agent (the first agent in the sequence of the participants) who is able to drive starts playing his role as driver in the carpooling group. The rest of the participants continue carpooling by playing the passenger role in the carpooling group.

3.3.3. Carpooling

The actual trips (HW and WH) execution activity of the carpoolers is operational for multiple days (for long-term). The presented carpooling model considers that the travel times are oblivious to the actual carpooling trips. The carpooling trips do not expressively decrease jamming on the roads. The pre-computed travel times between TAZs are expected to be time independent. Travel times in the presented model are to be advanced by making it time dependent (by the use of real-time travel times). Since the carpooling model is presented for the long-term, it is considered that the day-to-day schedule of individuals remains the same for all the simulated working days.

During the carpooling trips, the carpoolers may receive carpool invitations from the non-carpooling agents and will reply with either accept or reject messages. The actual negotiation process presented in Section 3.3.2 will be repeated again by considering the car capacity. New trip departure times for both the trips and the pick-up and drop-off orders of the carpoolers must be assigned before a new carpooling candidate can be accepted.

The potential carpoolers may terminate carpooling when their agreed carpool participation period expires. When the driver leaves the carpool group and the carpool group exceeds one, the remaining passengers will re-negotiate the driver selection and decide the pick-up, drop-off order of the carpool group. In this case, the driver first transfers the driver-charge to the newly selected driver from the rest of passengers of the same carpool group and then leaves the carpool group. An agent who once left the group can immediately start exploring the carpooling social networks again and be part of the same or any other active carpool group at a later instance. The same agent can also establish a new carpool group with the carpool participants of his or her interest group. The destruction of the carpool group occurs when only one agent is left in the group or if no persons with a car and a driving license are available.

3.4. Organizational Layer

In an organization layer, organizational concepts are used solely to divide organizations into groups. A group is an organizational body in which all the participants are capable of interacting according to pre-defined interaction definitions and protocols. In groups, sets of roles are stated jointly and specify the communal norms for the roles. Within a group of an organization, every agent has the capability to play a role. Organizational-based modeling permits the situations to be well-defined in an organized way. It offers the aptitude to determine where the associations exist between the agents and how it influences the outcomes [4,30].

In this model, the following organizations are introduced in the carpooling simulation: (1) Social Organization; and (2) the Carpool Organization.

3.4.1. Carpooling Social Organization

In the above presented organizational model, *Carpooling social organization* is used to divide the social network into segments, namely *carpooling social groups* (*CPSG*). This limits the communication requirements to individuals with similar characteristics (see Figure 2a) as the negotiators for carpooling are members of the same *CPSG*. In this model, agents are segmented on the basis of the same origin and destination TAZs. Within the *CPSG*, the agents play their respective *Carpooling social role*. The agents (who are playing the respective roles) may communicate with each other by exchanging text messages within their communal CPSG. To achieve the required tasks, agents should remain in their own *CPSG* throughout the simulation period.

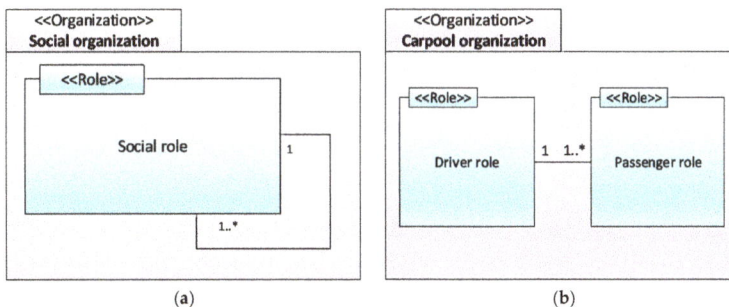

(a) (b)

Figure 2. (a) The individuals are negotiating together, are member of the *Carpooling Social Organization* by playing *Carpooling Social Role*; (b) The carpoolers are members of *Carpool Organization* by playing either *Driver role* or *Passenger role*.

3.4.2. Carpool Organization

The carpool organization contains the carpooling agents. When the negotiation succeeds, the receiver agent creates a *Carpool group* of the *Carpool organization* and starts his/her role either as *Driver role* or *Passenger role* depending on the negotiation outcome. Figure 2b shows the *Carpool organization* that supports the trip execution or co-traveling of the carpoolers. All carpooling agents must play a role (driver or passenger) in the carpool group of this *Carpool organization*. The driver of the carpool group will play the *Driver role*, while the passenger will play the *Passenger role* in the *Carpool group* throughout the carpooling period. When the driver's carpool period expires, he or she will allocate the driving tasks to one of the passengers who owns the vehicle and driving-license and has been playing the passenger role for the longest period. He or she will immediately leave the *Carpool group* by leaving the *Driver role*. On the other hand, the passenger who is selected as the driver will start playing the *Driver role* and will leave the *Passenger role* immediately. During the lifetime of the carpool group, agents can also interact with the other agents who have the intention to join. An agent can be part of one or more carpool groups throughout the simulation period, but only one at a time. When the carpool period of the particular agent expires, the agent will simply leave the respective role of the carpool group which they belong to. The same agent can later be part of it again or may join or create a new *Carpool group.*

3.4.3. Day-Switching Organization

In the presented simulation, the synchronization of a day using a time resolution is sufficient and it is a complex task. The day switching mechanism is compulsory to familiarize the concept of synchronized time for each day among agents as the organizational-based concept that is used exclusively for simulated time synchronization. The *Day-switching organization* is used to create instances for each simulated day. As soon as the agent has completed their daily actions, they need to join a *Day-switching group* (instance of *Day-switching organization*). For each day, the first agent will create the day-switching group and joins it by playing the day-switching role. The following agents, after finishing their daily activities, will join the same day-switching group. Every agent joining such a group instantly starts playing the *day-switching role*. They will wait for other agents to finish their daily actions and to join the *Day-switching group* by playing their particular role.

The joining of the last individual to the *Day-switching group* triggers the start of the next-day activities. Different day-switching groups will be created for each day. The day-switching process is repeated over and over until the termination of the simulation.

3.5. Agent Layer

The agent layer is committed to an agent-oriented model (see class-diagram in Figure 3) that is a solution to the problem defined in the problem layer. Agents denote people in the population whose particular properties and communal associations are modeled at the discrete level. Each agent acts independently and therefore is considered an autonomous entity.

In this simulation model, the agents live and execute their own daily agenda in the carpooling social network, and the agent environment is therefore established. The simulation starts the execution by loading personal profiles of the agents and the network information (OD-based travel time matrix). The carpooling social network is segmented and each agent becomes a member of a CPSG as determined by its home and work TAZs. Each agent once in its lifetime becomes part of such a social group (CPSG). The simulator comprises at most one carpool social group for each home-work TAZ combination. As soon as the agent becomes a member of the CPSG, it starts playing the role (social role).

A finite state-transition machine is used to model the behavior of the agent. It is composed of four states: *"Exploring"*, *"Waiting-for"*, *"Driving"*, and *"As-passenger"* (see Figure 4). The transition of the agents between different states depends on the interaction mechanism. The interaction between agents is modeled by the use of the text messages. The agents can exchange text messages with the other agents who belong to the same CPSG. The *Carpool invitation*, *Accept*, and *Reject* messages are used for interaction.

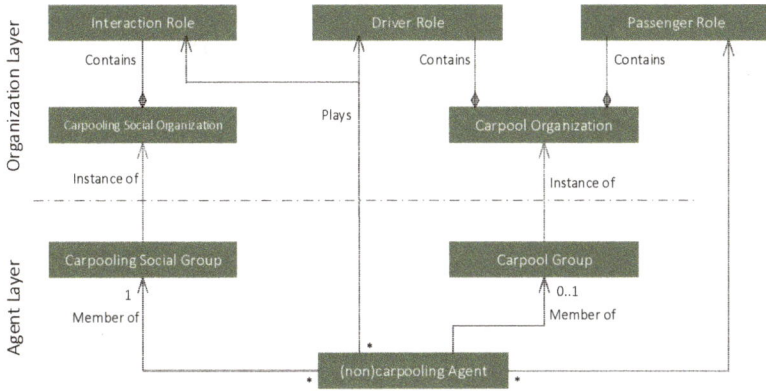

Figure 3. Class-diagram of the organizational model for co-traveling that is mapped to the agent-oriented model.

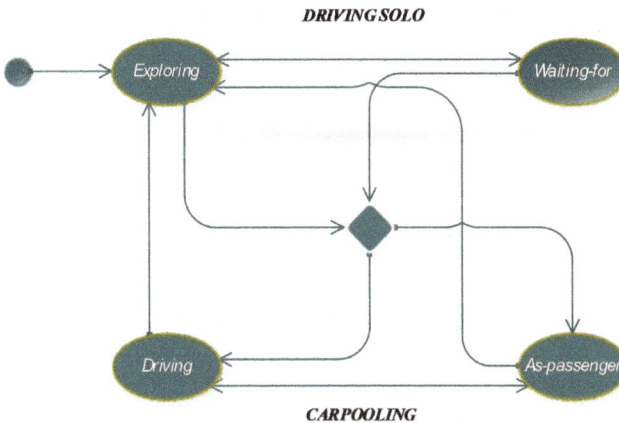

Figure 4. The state-machine is used to model the interaction behavior of the agent.

The state-machine is described in more detail in the following sub-sections.

3.5.1. Exploring

Exploration in the carpooling social network is the action of searching for people involved in co-operation. In the *Exploring* state, the sender agent may search for a partner by sending a Carpool invitation message to a randomly chosen agent, of the same CPSG. An invitation can be emitted on the basis of an agent's properties and interests. A variable *probToInvite* is used to model these properties and interests. If the value of *probToInvite* for someone is set to 30%, it implies that on a given day an agent has a probability of 0.3 to invite someone for carpooling. Once an invitation has been released, the sender enters into the WAIT state, and waits for the receiver's reply.

An agent can receive a Carpool invitation message from another agent in the *Exploring* state and will take the cooperative decision based on the negotiation model, discussed in Section 3.3.2. In case there is a feasible invitation for the receiver agent, he or she will reply with the Accept message and will change his or her state to either *Driving* or *As-passenger*. When the invitation is not decent, then he or she will reply with the Reject message and will continue the exploration process to find the ideal carpool partners in the same state.

When the negotiation succeeds between agents, the receiver agent creates a carpool group which is an instance of the Carpool organization. The receiver agent registers himself and starts playing the suitable role, either Driver role or Passenger role. Within a day, an agent can search CPSG for the participants multiple times sequentially. During the interaction process, the agents can switch multiple times between *Exploring* and *Waiting-for* states.

3.5.2. Waiting-For

In the *Waiting-for* state, the sender agent is waiting for the receiver's reply. If the sender receives an Accept message then the sender agent joins the same Carpool group that the receiver belongs to. The sender agent changes its state to either *Driving* or *As-passenger* depending on the receiver's reply. If a driver exists in the carpool group, then the sender agent will join as a passenger and in the case where the sender agent is selected as the driver then he or she will join the carpool group as the driver. When the receiver agent's reply is a Reject message, the sender agent changes its state to *Exploring* and again starts to find other carpool partners. In the *Waiting-for* state, if the sender agent receives an invalid (Carpool invitation or other irrelevant) message, then he or she simply replies with a Reject message and remains in the same state.

The Accept message specifies the role (Driver role or Passenger role) to play in the carpool group. The agent (1) leaves the *Waiting-for* state; (2) joins the Carpool group; and (3) starts playing the coordinated role of either Driver or Passenger. If the response is a Reject message, the sender agent changes its state to *Exploring* and tries to find a carpool partner again.

While in the *Waiting-for* state, the agent rejects any incoming invitation (simply by replying with a Reject message).

3.5.3. Driving

In the *Driving* state, the driver agent plays the Driver role in the Carpool group and remains in this state throughout his or her carpooling period. In this state, the driver agent can receive Carpool invitation messages and negotiate and coordinate with the sender agent. The driver agent can reply with either Accept or Reject messages on the basis of a sender's profile and the remaining seats in the vehicle. The driver leaves this carpool group when his or her carpooling period expires by leaving the Driver role. In this case, if the carpool group size still exceeds one, the remaining passengers will renegotiate for the driver and vehicle selection, and pick-up and drop-off orders of the participants. The driver agent who leaves the carpool group will change its state to *Exploring*. The driver agent destroys the carpool group when he or she is the only one left in the group after all passengers have quit.

3.5.4. As-Passenger

The passenger agent's behavior with regard to carpool association and co-operation while present in the *As-passenger* state is identical to the driver agent in the *Driving* state. Except, when the driver agent's carpooling period deceases and the remaining passengers re-negotiate to choose a driver. The selected driver agent will continue carpooling by starting the Driver role and by leaving the Passenger role of the same Carpool group.

The agents, once they finish their daily activities, will move to the following day immediately. Because of the combination of subsequent reasons: (1) agents will carpool on a long-term basis, for a specified time period; (2) agents are associated with carpool groups sequentially; (3) a carpool calendar is not followed by the agents; and (4) the newly accepted carpooling agents can join a carpool group on any day.

4. Results

For the experimentations, a *carpooling social network* was recognized by generating a population. FEATHERS [31], an operative activity-based traffic demand model for Flanders (Belgium region), is used to establish the social network for carpooling. There are about six million inhabitants living in the Flanders region. The area is subdivided into 2386 TAZs and on average a TAZ covers 5 [km^2] only. The daily schedule for each participant of the synthetic population is generated. The individuals allocate the day taking part in activities and traveling between them. The detailed commuting trips (HW and WH) for co-traveling are taken and discussed relative to the long term. The set of constraining activities, i.e., pick-up and drop-off, shopping, etc., is considered because the induced timing constraints can also have an effect on the commuting trips. Note that only individuals with the same home and work TAZs can carpool together.

To examine the behavior of the carpoolers, the presented framework is simulated for 150 working days and the constraints shown in Table 1 are applied to accomplish the simulation results.

Table 1. Important constraints and their values for the experiments.

Constraints	Values
No. of persons invited	At most 5 people/simulated day.
No. of agents	18,000 individuals whose transportation mode is car.
Destination TAZs	Twenty two TAZs from the Brussels region of Belgium.
Time windows' length	5 min, 10 min, 15 min, 20 min, 25 min, and 30 min.
Maximum car capacity	5 persons (driver included).
Carpool period	[30 to 60 days] by sampling from a uniform distribution.
Constraining activities	Enabled: pick-up and drop-off, shopping.

As a proof-of-concept, some experiments were conducted and are explained as follows:

Figure 5 demonstrates the development of the number of active carpool groups over the simulated period. The horizontal axis indicates the simulated working days and the vertical axis specifies the number of active carpool groups for each day. The graph in Figure 5 is based on six curves. Each curve indicates the length of the time windows. As shown in Table 1, the experiments were conducted by taking different time window lengths ranging from 5 min to 30 min with an interval of 5 min (i.e., 5 min, 10 min, 15 min, etc.). As per intuition, a larger time acceptance window allows for more carpooling. During the initial period of the simulation, also known as the warm-up period, the number of active carpool groups monotonically increase with the passage of time since the shortest possible carpooling period lasts for 30 days only. In the warm-up period, the active carpool groups are placed in a realistic condition and this means that the initial transient has passed and the simulation output is in a steady state. Figure 5 shows a visual representation of active carpool groups over the simulated period. After the warm-up period, each curve shows a decline because the destruction rate of carpool groups is higher than creating the new carpool groups. The new carpoolers give the impression of joining existing carpool groups more frequently than creating new ones and joining them. It appears to be convenient to join an existing carpool group rather than to create a new one. In this case, the number of active carpool groups decreases, but the number of active carpoolers does not decrease. The curves become stable with a linear decrease after 60 days because the possibility to join existing carpool groups becomes lower due to the car saturation effect.

The graph in Figure 6 expresses the number of active carpoolers over the simulated period of 150 days. The *x*-axis displays the working days while the *y*-axis shows the number of active carpoolers. It is perceived that, on average, a larger time acceptance window permits more carpooling. The curves represent the active carpoolers for each length of the time window ranging from 5 min to 30 min with an interval of 5 min (i.e., 5 min, 10 min, 15 min, etc.). The number of active carpoolers quickly rises during the warm-up period of the simulation and the shortest carpooling period continues up to 30 days. Note that the warm-up period is 30 days where the individuals are only joining the carpool groups. After the warm-up period, the increase rate is lower because of the joining and leaving the carpool groups, respectively. The curves show the decrease of the number of active carpoolers after the warm-up period because the rate of joining by the new carpoolers is lower compared to the rate of leaving the carpool periods by the existing carpoolers. The share of carpooling individuals appears to have become saturated after 100 simulated working days. The results show that when the length of the time window is larger, the chances for negotiation success are greater than when using the smaller one and the behavior of the carpoolers evolves over time due to their carpooling periods and the car capacity parameters.

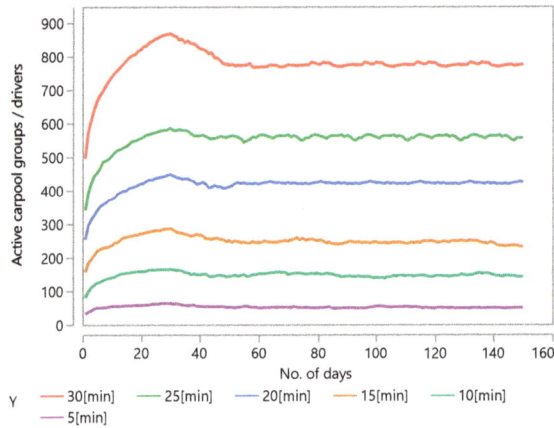

Figure 5. Number of active carpool groups (cars/drivers) over the simulated period.

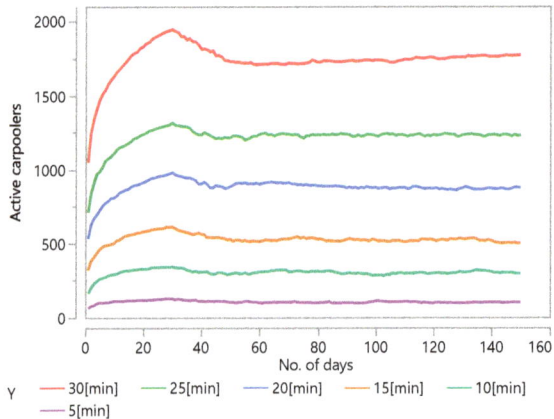

Figure 6. Number of active carpoolers over the simulated period.

The graph in the Figure 7 illustrates the average carpool group size as a function of time. The carpool size gradually increases during the warm-up period up to 30 days of the simulation. After the warm-up period, the carpool group size curve almost shows spikes. This is enlightened as follows: when the carpooling period of the carpoolers expires and they leave the existing carpool group, there is an enhanced likelihood that they will join another (existing or the same) carpool group instead of creating a new one. After a few days (nearly 80-days), the carpool group size increases at an aggregate level because the rate of the passengers joining groups is higher than the new drivers joined. During this period, the passengers join the existing carpool groups rather than create new ones.

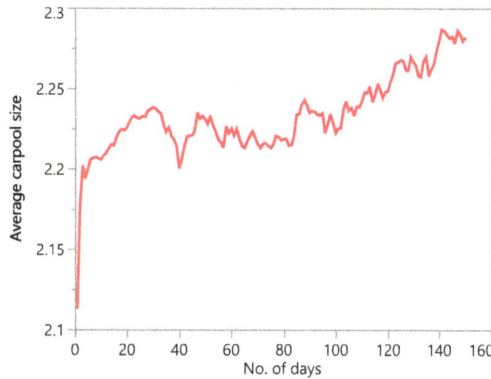

Figure 7. Carpool groups and carpoolers over the simulated period.

The measurement of the computation time of the agent-based simulation is also one of the goals of this study. This is done to analyze the effect of the interaction between individuals. In addition, it aims to determine whether optimization is required to iterate reality and precisely predict carpooling negotiation consequences for the whole population of the study area. The graph in Figure 8 indicates the average computation time of the presented simulation. The experiment was conducted on an Intel® Xeon® CPU E5-2643 v2@3.50GHz 3.50GHz (2-processors) machine, with 128 GB of RAM and 64-bits operating system. The population scale is achieved by taking 1000 to 10,000 agents (i.e., 1000, 2000, 3000, up to 10,000) for various simulation runs. Each time, the simulation is executed for 150 work-days. To perform this experiment, a constant time window of 30 min is used for each run. The rest of the constraints are the same as shown in the Table 1 except the number of agents. The curve in the graph shows that the processing time increases in a nearly polynomial way with different numbers of agents. It happens due a greater execution of the interaction and the negotiation mechanisms.

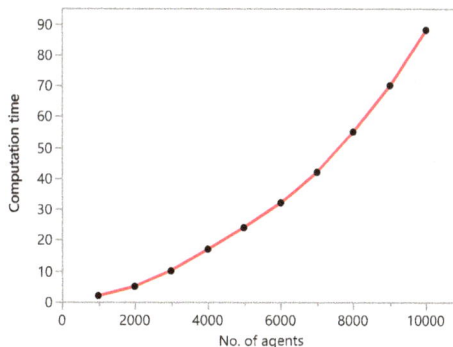

Figure 8. Computation time in minutes by running different numbers of agents.

The model presented in this paper requires that all carpooling individuals share the source *travel analysis zone*, as well as the destination *travel analysis zone*. This constraint will be considered in future research, which will allow the route of one participant to be a subpath of the route of another participant. The car trip depends on the selection of the *driver* and the pick-up and drop-off orders of the passengers. This information is used to select the optimal driver.

5. Conclusions and Future Work

An agent-based simulation for long-term carpooling has been developed and presented to simulate human behavior when decisions for co-operation are to be made. It setups the framework and also the social network of the carpooling candidates. The model covers the co-operation of individuals' matching for commuting trips. The negotiation model is based on the constant preference function of the trip start times and the personal profile of the driver and the vehicle selection. The implementation also applies constraining activities by considering the personal daily schedule of each individual. Sensitivity analysis of the simulation model was also conducted during this research. Experiments showed that when the time window was larger, the chances for a successful negotiation were greater. The computation time of the simulation increased in a nearly polynomial manner with the increase of the number of agents. The simulation model requires a large amount of accurate input data, and has scalability issues that are still to be solved. Apart from scalability issues, future research will mainly focus on the effect of schedule adaptation and enhancing the mechanisms for communication and negotiation between agents. To make this simulation operational, multi-zonal interaction and negotiation will be taken into account. In addition, other features, including feasible pick-up and drop-off orders of the carpooling agents of the same carpool groups, will be added to increase the volume of the potential carpoolers.

Acknowledgments: This research was partially supported by the Higher Education Commission, Pakistan by the award letter No. HRDI-UESTPs/Batch-I/Belgium/2013.

Author Contributions: Iftikhar Hussain, Muhammad Arsalan Khan, and Syed Fazal Abbas Baqueri conceived and designed the concept; Syyed Adnan Raheel Shah and Israr Ali Khan performed the literature review; Mudasser Muneer Khan and Muhammad Khawar Bashir reviewed the manuscript; Iftikhar Hussain implemented the model and also wrote the manuscript.

Conflicts of Interest: The authors declare no conflict of interest.

References

1. Galland, S.; Ansar-Ul-Haque Yasar, L.K.; Knapen, L.; Gaud, N.; Bellemans, T.; Janssens, D. Simulation of Carpooling Agents with the Janus Platform. *JUSPN* **2014**, *5*, 9–15.
2. Knapen, L.; Yasar, A.; Cho, S.; Keren, D.; Dbai, A.A.; Bellemans, T.; Janssens, D.; Wets, G.; Schuster, A.; Sharfman, I.; et al. Exploiting graph-theoretic tools for matching in carpooling applications. *J. Ambient Intell. Humaniz. Comput.* **2014**, *5*, 393–407. [CrossRef]
3. Horvitz, E.; Apacible, J.; Sarin, R.; Liao, L. Prediction, Expectation, and Surprise: Methods, Designs, and Study of a Deployed Traffic Forecasting Service. In Proceedings of the Twenty-First Conference on Uncertainty in Artificial Intelligence (UAI'05), Edinburgh, UK, 26–29 July 2005; AUAI Press: Arlington, VA, USA, 2005; pp. 275–283.
4. Cossentino, M.; Gaud, N.; Hilaire, V.; Galland, S.; Koukam, A. ASPECS: An agent-oriented software process for engineering complex systems. *Auton. Agents Multi-Agent Syst.* **2010**, *20*, 260–304. [CrossRef]
5. Gaud, N.; Galland, S.; Hilaire, V.; Koukam, A. An Organisational Platform for Holonic and Multiagent Systems. In Proceedings of the International Workshop on Programming Multi-Agent Systems (ProMAS), Estoril, Portugal, 13 May 2008; Hindriks, K.V., Pokahr, A., Sardina, S., Eds.; Springer: Berlin/Heidelberg, Germany, 2009; pp. 104–119.
6. Hussain, I.; Knapen, L.; Yasar, A.-U.-H.; Bellemans, T.; Janssens, D.; Wets, G. Negotiation and Coordination in Carpooling. *Transp. Res. Rec. J. Transp. Res. Board* **2016**, *2542*, 92–101. [CrossRef]

7. Hendrickson, C.; Plank, E. The flexibility of departure times for work trips. *Transp. Res. Part Gen.* **1984**, *18*, 25–36. [CrossRef]
8. Hussain, I.; Knapen, L.; Galland, S.; Yasar, A.-U.-H.; Bellemans, T.; Janssens, D.; Wets, G. Organizational-based model and agent-based simulation for long-term carpooling. *Futur. Gener. Comput. Syst.* **2016**, *64*, 125–139. [CrossRef]
9. Galland, S.; Knapen, L.; Yasar, A.-U.-H.; Gaud, N.; Janssens, D.; Lamotte, O.; Koukam, A.; Wets, G. Multi-agent simulation of individual mobility behavior in carpooling. *Transp. Res. Part C Emerg. Technol.* **2014**, *45*, 83–98. [CrossRef]
10. Cho, S.; Yasar, A.-U.-H.; Knapen, L.; Bellemans, T.; Janssens, D.; Wets, G. A Conceptual Design of an Agent-based Interaction Model for the Carpooling Application. *Procedia Comput. Sci.* **2012**, *10*, 801–807. [CrossRef]
11. Bellemans, T.; Bothe, S.; Cho, S.; Giannotti, F.; Janssens, D.; Knapen, L.; Körner, C.; May, M.; Nanni, M.; Pedreschi, D.; et al. An Agent-Based Model to Evaluate Carpooling at Large Manufacturing Plants. *Procedia Comput. Sci.* **2012**, *10*, 1221–1227. [CrossRef]
12. Guo, Y.; Goncalves, G.; Hsu, T. RETRACTED ARTICLE: A Multi-agent Based Self-adaptive Genetic Algorithm for the Long-term Car Pooling Problem. *J. Math. Model. Algorithms Oper. Res.* **2013**, *12*, 45–66. [CrossRef]
13. Armendáriz, M.; Burguillo, J.C.; Peleteiro-Ramallo, A.; Arnould, G.; Khadraoui, D. Carpooling: A Multi-Agent Simulation in Netlogo. *Eur. Counc. Model. Simul.* **2011**, 61–67. [CrossRef]
14. Ronald, N.; Arentze, T.; Timmermans, H. Modeling social interactions between individuals for joint activity scheduling. *Transp. Res. Part B Methodol.* **2012**, *46*, 276–290. [CrossRef]
15. Chun, H.W.; Wong, R.Y.M. N∗—An agent-based negotiation algorithm for dynamic scheduling and rescheduling. *Adv. Eng. Inform.* **2003**, *17*, 1–22. [CrossRef]
16. Kraus, S. Negotiation and cooperation in multi-agent environments. *Artif. Intell.* **1997**, *94*, 79–97. [CrossRef]
17. García-Magariño, I.; Gómez-Sanz, J.J.; Pérez-Agüera, J.R. A Multi-agent Based Implementation of a Delphi Process. In Proceedings of the 7th International Joint Conference on Autonomous Agents and Multiagent Systems—(AAMAS'08), Estoril, Portugal, 12–16 May 2008; International Foundation for Autonomous Agents and Multiagent Systems: Richland, SC, USA, 2008; Volume 3, pp. 1543–1546.
18. Du, S.; Guo, C.; Jin, M. Agent-based simulation on tourists' congestion control during peak travel period using Logit model. *Nonlinear Dyn. Complex* **2016**, *89*, 187–194. [CrossRef]
19. García-Magariño, I. ABSTUR: An Agent-based Simulator for Tourist Urban Routes. *Expert Syst. Appl.* **2015**, *42*, 5287–5302. [CrossRef]
20. Nijland, E.W.L.; Arentze, T.A.; Borgers, A.W.J.; Timmermans, H.J.P. Individuals' Activity–Travel Rescheduling Behaviour: Experiment and Model-Based Analysis. *Environ. Plan. A* **2009**, *41*, 1511–1522. [CrossRef]
21. Knapen, L.; Bellemans, T.; Usman, M.; Janssens, D.; Wets, G. Within day rescheduling microsimulation combined with macrosimulated traffic. *Transp. Res. Part C Emerg. Technol.* **2014**, *45*, 99–118. [CrossRef]
22. Joh, C.-H.; Timmermans, H.; Arentze, T. Measuring and predicting adaptation behavior in multidimensional activity-travel patterns. *Transportmetrica* **2006**, *2*, 153–173. [CrossRef]
23. Arentze, T.; Pelizaro, C.; Timmermans, H. An agent-based micro-simulation framework for modelling of dynamic activity–travel rescheduling decisions. *Int. J. Geogr. Inf. Sci.* **2010**, *24*, 1149–1170. [CrossRef]
24. Gupta, S.; Vovsha, P. A model for work activity schedules with synchronization for multiple-worker households. *Transportation* **2013**, *40*, 827–845. [CrossRef]
25. Xia, J.; Curtin, K.M.; Li, W.; Zhao, Y. A New Model for a Carpool Matching Service. *PLoS ONE* **2015**, *10*, e0129257. [CrossRef] [PubMed]
26. Martinez, L.M.; Correia, G.H.A.; Viegas, J.M. An agent-based simulation model to assess the impacts of introducing a shared-taxi system: An application to Lisbon (Portugal). *J. Adv. Transp.* **2015**, *49*, 475–495. [CrossRef]
27. Knapen, L.; Keren, D.; Yasar, A.-U.-H.; Cho, S.; Bellemans, T.; Janssens, D.; Wets, G. Analysis of the Co-routing Problem in Agent-based Carpooling Simulation. *Procedia Comput. Sci.* **2012**, *10*, 821–826. [CrossRef]
28. Jennings, N.R. On agent-based software engineering. *Artif. Intell.* **2000**, *117*, 277–296. [CrossRef]

29. Ferber, J.; Gutknecht, O.; Michel, F. From Agents to Organizations: An Organizational View of Multi-agent Systems. In Proceedings of the 4th International Workshop on Agent-Oriented Software Engineering IV (AOSE), Melbourne, Australia, 15 July 2003; Giorgini, P., Müller, J.P., Odell, J., Eds.; Springer: Berlin/Heidelberg, Germany, 2004; pp. 214–230.

30. Cossentino, M.; Galland, S.; Gaud, N.; Hilaire, V.; Koukam, A. An Organisational Approach to Engineer Emergence within Holarchies. *Int. J. Agent-Oriented Softw. Eng.* **2010**, *4*, 304–329. [CrossRef]

31. Bellemans, T.; Kochan, B.; Janssens, D.; Wets, G.; Arentze, T.; Timmermans, H. Implementation Framework and Development Trajectory of FEATHERS Activity-Based Simulation Platform. *Transp. Res. Rec. J. Transp. Res. Board* **2010**, *2175*, 111–119. [CrossRef]

*applied
sciences*

MDPI

Article

Collision Avoidance from Multiple Passive Agents with Partially Predictable Behavior

Khalil Muhammad Zuhaib [1], Abdul Manan Khan[2], Junaid Iqbal [1], Mian Ashfaq Ali [3],
Muhammad Usman [1], Ahmad Ali [1], Sheraz Yaqub [1], Ji Yeong Lee [2] and Changsoo Han [2,*]

[1] Department of Mechatronics Engineering, Hanyang University ERICA Campus, Ansan 15588, Korea;
    kmzuhaib@gmail.com (K.M.Z.); jibssp@gmail.com (J.I.); musman@hanyang.ac.kr (M.U.);
    ahmadali@hanyang.ac.kr (A.A.); sheraz.yaqub@yahoo.com (S.Y.)
[2] Department of Robot Engineering, Hanyang University ERICA Campus, Ansan 15588, Korea;
    kam@hanyang.ac.kr (A.M.K.); jiyeongl@hanyang.ac.kr (J.Y.L.)
[3] School of Mechanical and Manufacturing Engineering (SMME), National University of Science and
    Techonology (NUST), Islamabad 44000, Pakistan; ishfaqaries@gmail.com
*   Correspondence: cshan@hanyang.ac.kr; Tel.: +82-31-400-4062

Received: 14 July 2017; Accepted: 30 August 2017; Published: 4 September 2017

Abstract: Navigating a robot in a dynamic environment is a challenging task, especially when
the behavior of other agents such as pedestrians, is only partially predictable. Also, the kinodynamic
constraints on robot motion add an extra challenge. This paper proposes a novel navigational strategy
for collision avoidance of a kinodynamically constrained robot from multiple moving passive agents
with partially predictable behavior. Specifically, this paper presents a new approach to identify the set
of control inputs to the robot, named control obstacle, which leads it towards a collision with a passive
agent moving along an arbitrary path. The proposed method is developed by generalizing the concept
of nonlinear velocity obstacle (NLVO), which is used to avoid collision with a passive agent, and takes
into account the kinodynamic constraints on robot motion. Further, it formulates the navigational
problem as an optimization problem, which allows the robot to make a safe decision in the presence
of various sources of unmodelled uncertainties. Finally, the performance of the algorithm is
evaluated for different parameters and is compared to existing velocity obstacle-based approaches.
The simulated experiments show the excellent performance of the proposed approach in term
of computation time and success rate.

Keywords: collision avoidance; multiple passive agents; Mobile Robot Navigation; pedestrian
environment; kinodynamic planning; velocity obstacle

1. Introduction

Motion planning in dynamic environments has become central to the operations of robots.
Most modern applications require navigation of robots among humans, vehicles and other robots.
Almost all of the mobile robots in the real world applications are subjected to kinodynamic constraints
like differential driven or car-like robots [1,2]. Many different kinds of motion planning algorithms
have been developed for such robots facing static environment, and then they were further extended
for dynamic environments. In [3], a motion planning approach for the car-like robots was presented,
and it was proved that the path for holonomic robots lying in an open configuration space could
be transformed into an equally useful path for nonholonomic robots. Particularly, an algorithm
was proposed to generate a useful path for nonholonomic systems based on the path obtained for
holonomic robots. However, that transformation was not smooth for car-like robots. It was first
proposed in [4] and their idea was further improved in [5]. However, instead of using a steering
function, authors presented a method based on computing clothoid curves. Although their method

improved transition and smoothness, authors did not address complete trajectory planning for dynamic environments. For this propose many authors have proposed different algorithms focusing on complete trajectory planning , such as [6–8]. In some of these works, the problem of robot motion planning in a dynamic environment was decomposed into hatching an achievable path for nonholonomic systems, and designing a velocity profile by which such vehicles could maneuver safely [9–11]. However, these approaches do not perform well for navigating a robot in a dynamic environment with multiple moving agents, like pedestrian environment, due to following reasons. (1) These approaches do not take into account the future prediction of obstacle motion thus robot being blinded to a potential collision. (2) When planning in such dynamic environment the time available to compute solution is limited, it is the function of nature and dynamicity of the environment. Therefore, in a highly dynamic environment there is a high probability that a complete path to the goal cannot be computed in the available time.

In [12,13], the problem of kinodynamic motion planning for a robot in a dynamic environment was addressed. The proposed approaches explore the state-time space of the robot to find a collision free path to the goal, while it was assumed that the entire path of the passive agents is known. However, this assumption restricts their application when extending them to pedestrian or multi-agent environment. In most of the applications of navigating an agent, the moving obstacles has free will, and their future behavior is only partially predictable (if at all). When facing such situation, obstacle's path must be predicted using prediction techniques such as the ones presented in [14–17]. When the on-line path prediction is used to plan motion, it is likely that the model of the future that is obtained will have limited time duration. In addition, such on-line prediction is noisy. Therefore, a planner is required that takes into account the validity duration of the model of the environment and allowable time for computing a solution. Also, it is necessary to consider various sources of uncertainties present, e.g., passive agents unpredictability, uncertainty in resulting state under a given control action, and localization error.

Some of the principal work that considers the future behavior of pedestrians or other kind of agents is focused on the concept of velocity obstacle (VO) [18]. The original formulation of VO was designed for an agent with simple-agent dynamics to avoid a collision with a passive agent moving along a known straight path. In [19], authors proposed an optimal reciprocal collision avoidance strategy (ORCA) for multiple active agents, considering similar behavior for all agents. More specifically, the work assumes that each agent employs a similar collision avoidance strategy. Instead of complete motion planning, their approach was to plan local motion directed towards the next (sub) goal extracted from a global way point plan. Several efforts have been made to extend the concept of ORCA to more complex dynamic systems ranging from the single integrator, differential driven, car-like robot and arbitrary linear equation of motion in [20–25]. However, these approaches require every agent in a collision to run a similar collision avoidance algorithm. In [26], authors examined the issue of navigating car-like agent in the dynamic environment with multiple passive agents. They extended the concept of VO to consider the constraint of the kinematic car-like agent to avoid collision with passive agents moving along the linear path. The approach was designed for specific agent dynamics and cannot be simply extended for avoiding passive agents moving along an arbitrary path, probably nonlinear.

1.1. Contribution

This paper addresses the problem of navigating a kinodynamically constrained robot, among multiple passive agents with partially predictable behavior. In order to solve the problem, this paper develops the following two contributions.

- First, it generalizes the concept of Nonlinear velocity obstacle (NLVO) [27] to develop a new approach to identify the set of control inputs to robot that will lead the robot towards collision, named control obstacle. It seeks to address the issue of navigating a robot with kinodynamic constraints while considering that a passive agent is moving along an arbitrary path, probably non-linear. The original approach of NLVO does not consider the robot model, and thus is limited to the agent with dynamics as of single integrator.

- Secondly, a novel collision avoidance strategy is proposed, that allows the robot to make a safe decision for avoiding a collision with the passive agents. The safe navigation decision is based on the concept of minimum safety margin which is the measure of how safe the path is in the presence of various sources of unmodeled uncertainties.

This paper presents the implementation of the proposed approach for a car-like agent and a double integrator. The performance of the proposed algorithm is evaluated for dynamic environment considered in [26], where the predicted trajectory of passive agents change frequently. The simulated experiments show better performance of the proposed approach compared to current VO based approaches, in terms of computation time and success rate.

1.2. Limitations

As this work extends the concept of velocity obstacle that increases its applicability, it inherits some limitations of VO method. First, it requires the passive agents to be of circular shape. It is a logical assumption for pedestrians or mobile robots, as it simplifies the problem. For a non-circular passive agent, collision avoidance is achieved by selecting a circle enclosing geometry of a passive agent. Second, this work proposes a local navigation approach. Thus it requires a global motion planner to converge to goal in the presence of the large static obstacles. The paper presents one possible implementation of the global navigational plan in Section 5, which results in model predictive partial motion planning framework.

Finally, the presented approach is probabilistic in nature. It is possible that in some cases the solution may not be found even if one exists. However, simulation results show that the presented approach has much higher success rate compared to existing VO based approaches.

1.3. Organization

The rest of this paper is organized as follows. Section 2 discusses the previous work done relating to VO based navigation. Section 3 reviews the nonlinear velocity obstacle and the issue that appears when applying it to navigate a kinodynamically constrained agent. Section 4 introduces the idea of a control obstacle and describes the safe margin input space. Section 5 presents the proposed navigation approach. Section 6 presents the implementation of the proposed approach for robot dynamics as of the double integrator and a car-like robot. Further, this section evaluates the performance of the algorithm for a set of parameters and compares the performance of the proposed approach with current approaches. Finally, Section 7 concludes the paper.

2. Previous Work

One of the early development in collision avoidance is of velocity obstacle (VO) [18]. VO is a cone in the velocity space of agent, which represents the set of velocities that will lead an active agent towards a collision with a passive agent. To avoid the collision, active agent has to select its velocity outside VO. The early approach was developed for simple agent dynamics to avoid collision with a passive agent which is moving along a straight path with constant velocity. In [27], authors proposed NLVO algorithm, which expands the concept of VO to allow an agent with linear equation of motion to avoid collision with a passive agent with known, possibly nonlinear trajectories. The generalized velocity obstacle (GVO) algorithm proposed in [26] does principally the opposite of NLVO. It considers a problem of car-like agent avoiding passive agents moving along linear paths.

In contrast of avoiding passive agents in [19], authors examined the issue of collision avoidance among active agents and proposed the concept of reciprocal collision avoidance. The approach was based on dividing the responsibility for collision avoidance among agents involved in a collision. Their approach generates collision free piecewise paths for active agents. Efforts have been made to extend the applicability of this approach to consider the kinodynamic constraints of agents. In [23], authors proposed the concept of Acceleration-Velocity Obstacle, (AVO), that takes into account the acceleration constraints of the robot. In [24], authors proposed the idea of continuous control obstacle, to generate a continuous collision free path rather than piecewise linear path. All these reciprocal collision based approaches require that all dynamic objects should employ the same algorithm to avoid a collision. Thus, these approaches cannot easily be extended for avoiding a collision with passive agents moving along an arbitrary path, possibly nonlinear and only partially predictable.

3. Background

This section will review the concept of nonlinear velocity obstacle proposed in [27] and discuss its application for a robot with kinodynamic constraints.

3.1. Nonlinear Velocity Obstacle

For a disc-shaped robot R_a and moving obstacle O_b of radii r_a and r_b, respectively, the nonlinear velocity obstacle (NLVO) induced for R_a is a set of its velocities that will result in its collision with O_b at some time in the future. NLVO is defined in terms of its temporary components. Let at current time $t = 0$, the center of the robot represented by $p_a(0)$ is at origin. In a robot configuration space, let set B represent the obstacle's circular geometry with radius equal to the sum of radii r_a and r_b, and the robot geometry is represented by a point, as shown in Figure 1a. For an obstacle following a generalized trajectory $c(t)$, the temporary velocity obstacle induced for a robot due to obstacle's position at time instant t_{i+n} is as follows:

$$NLVO(t_{i+n}) = \frac{c(t_{i+n}) \oplus B}{t_{i+n}} \tag{1}$$

where $c(t_{i+n}) \oplus B$ denotes the Minkowski sum of vector $c(t_{i+n})$ and set B. $NLVO(t_{i+n})$ is a set of all absolute velocities of a robot that will result in a collision of p_a with a point in B at time instant t_{i+n}. The set of robot velocities that will result in collision of p_a with B within time horizon $[0, \tau]$ can be defined in terms of $NLVO(t_{i+n})$ as follows:

$$NLVO^\tau = \bigcup_{0 < t \leq \tau} NLVO(t) \tag{2}$$

Geometrically, $NLVO^\tau$ is a warped cone, such as the one shown in Figure 1b.

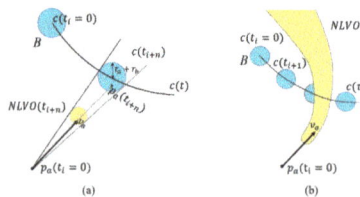

Figure 1. For a robot with current position $p_a(0)$ at origin and obstacle following trajectory $c(t)$, (a) shows the temporary velocity obstacle, $NLVO(t_{i+n})$ induced for a robot due to the position of an obstacle at time t_{i+n}. It is a disc of radius $(r_a + r_b)/t_{i+n}$ and its center is at $c(t_{i+n})/t_{i+n}$ in the robot velocity space. If the velocity of a robot is such that $v_a \in NLVO(t_{i+n})$, then $p_a(t_{i+n}) \in c(t_{i+n}) \oplus B$. In (b), $NLVO^\tau$ is shown as the union of its temporary components over time horizon $[0, \tau]$. If a robot velocity v_a is in $NLVO^\tau$, then the collision will occur within time horizon $[0, \tau]$. NLVO: Nonlinear Velocity Obstacle.

Collision avoidance is then achieved as follow: Robot R_a selects its new velocity at time $t = 0$ such that $v_a \notin NLVO^\tau$ to remain safe for at least τ seconds into the future. The new velocity is usually selected that minimize the Euclidean distance to velocity v_{pref}, which in turn points to next sub goal extracted from global waypoint plan. v_a is then applied for short time horizon until a next control loop begins.

3.2. Robot with Kinodynamic Constraints

For a known, possibly nonlinear trajectory of a passive agent, NLVO defines the set of velocities that may lead the robot towards a collision with a passive agent into the future. For an arbitrary starting point, a new velocity outside $NLVO^\tau$ cannot be attained instantly by a mobile robot under kinodynamic constraints. For example, the robot with car-like dynamics, as shown in Figure 2, has feasible velocity in a single direction, specifically in the direction of rear wheels. If the current velocity of robot is in $NLVO^\tau$, a new velocity outside $NLVO^\tau$ can only be attained over some finite time, and there is no guarantee that a robot can attain that new velocity into the future without having a collision.

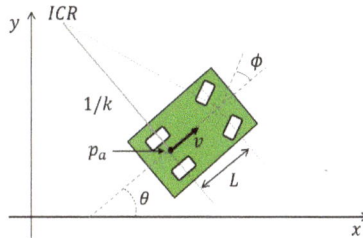

Figure 2. Kinematic model of the car-like robot. States are given by the center position of rear wheel axle p_a, the orientation angle θ, the steering angle ϕ. L represents the wheelbase of the car. Robot velocity v is in the direction of rear wheels. ICR: Instantaneous Center of Rotation.

4. Safe Margin Control Space

This section presents a new concept of temporary control obstacle. It is the generalization of temporary velocity obstacle. It seeks to address the problem of computing collision free motion by taking the kinodynamic constraints of the robot into account. Secondly, it introduces the idea of safe margined control space to select the safest trajectory in the presence of various sources of uncertainties.

4.1. Passive Agent Representation

This work assumes that there is a system other than a robot, like the one presented in [16,17], that tracks passive agents, predicts their future behaviors, and presents them in a general format, used in this paper. We are given a list of the passive agents and each passive agent is considered to be a disc or sphere having a radius and a trajectory. The future behavior of a passive agent is represented in the form of set points that represent predicted future states of the passive agent, specifying its position and time.

4.2. Notations and Assumptions

Let the state space of robot A be $X_a \subset \mathbb{R}^N$. The dimension of robot workspace is typically either $d = 2$ or $d = 3$. It is assumed that the position of the robot p_a in configuration space can be obtained from its states $x_a(t)$, potentially by some nonlinear projection function $f : X_a \longrightarrow \mathbb{R}^d$.

$$p_a(t) = f(x_a(t)) \tag{3}$$

Appl. Sci. **2017**, *7*, 903

Let the control space of robot has the same dimension as the workspace. A constraint c defined on the robot states by constraint function $q_c(x_a(t), t)$ bounded in $[C_-, C_+]$ $(C_-, C_+ \in \mathbb{R})$ will constraint the allowable control inputs in the robot control space. It is assumed that this constrained control region is convex. In the case of multiple constraints defined on robot states an admissible control space represented by U_{ad} is obtainable by taking the intersection of allowable control regions induced by each constraint.

Lets the continuous time state transition is given by some potential nonlinear function $g: X_a \times U_{ad} \longrightarrow \mathbb{R}^N$

$$\dot{x}_a(t) = g(x_a(t), u(t)) \tag{4}$$

where $x_a(t)$ is the state of the robot and $u(t)$ is the control input given to it. For the current state $x_a(t) = x_a(0)$ and constant control $u(t) = u(0)$, the state of the robot at $t > 0$ is given by:

$$x_a(t) = h(x_a, u, t) \tag{5}$$

where $h : X_a \times U_{ad} \times \mathbb{R} \longrightarrow X_a$ is the solution of (4) which is considered to be obtainable by its integration.

4.3. Control Obstacle

Consider a mobile robot that shares its workspace with multiple other passive agents. Let A and B_j are the geometry of the robot and the jth passive agent respectively. The robot and passive agent geometries are considered as their bounding circles, similarly as in the original formulation of VO [18]. Let O_j is the Minkowski sum of robot's and jth passive agent's geometries, $O_j = A \oplus B_j$. The current position of passive agent is $c_j(0)$ and its position at $t > 0$ is given by $c_j(t)$. To avoid collision with a passive agent within time horizon τ, their relative position should remain outside the Minkowski sum of their geometries.

$$p_a(t) - c_j(t) \notin O_j, \forall t \in [0, \tau] \tag{6}$$

Therefore, the temporary control obstacle induced due to the state of the jth passive agent at a future time instant $t_{i+n} \in [0, \tau]$, denoted by $UO_j(t_{i+n})$, is defined as follows:

Definition 1. *(Temporary Control Obstacle) It is a set of control inputs for which the relative position vector is inside O_j at time t_{i+n}*

$$UO_j(t_{i+n}) = \{u | f(h(x_a, u, t_{i+n})) - c_j(t_{i+n}) \in O_j\} \tag{7}$$

where $f(h(x_a, u, t_{i+n})) - c_j(t_{i+n})$ is a relative position vector at time t_{i+n}, for control input u given to a robot.

Now, the control obstacle induced for a robot due to the predicted motion of jth passive agent over the time horizon $[0, \tau]$ can be defined as follows:

Definition 2. *(Control Obstacle) The control obstacle induced due to the states of a passive agent over time horizon $[0, \tau]$ is the union of temporary control obstacles over that horizon.*

$$UO_j^\tau = \bigcup_{0 < t_{i+n} \leq \tau} UO_j(t_{i+n}) \tag{8}$$

UO_j^τ is the set of control inputs to a robot, that will lead it towards collision with a passive agent in time horizon $[0, \tau]$ into the future. In another words, the collision will not occur between the robot and passive agent B_j with in time horizon $[0, \tau]$ into the future, if robot selects its control input such that $u \notin UO_j^\tau$.

In case of avoiding collision from multiple passive agents, we can extend the given approach as follows. Let $N = \{1, ..., m\}$ is the index of passive agents to be avoided. Then the control obstacle can be given as:

$$UO^\tau = \bigcup_{j \in N} UO_j^\tau \tag{9}$$

Robot selecting its control input outside UO^τ will lead it to move without collision with m passive agents within time horizon $[0, \tau]$ into the future.

4.4. Safe Control Inputs

All control inputs in admissible control space that are not in control obstacle are considered as safe control inputs. The definition of safe control space is as follows:

Definition 3. *(Safe control space) It is the relative complement of UO^τ in U_{ad}.*

$$U_{safe}^\tau = U_{ad} \setminus UO^\tau \tag{10}$$

Although all control inputs in U_{safe}^τ are considered safe, but each control input has a different level of proximity to the control obstacle. One possible metric defining the closeness of a safe control input $u_s \in U_{safe}^\tau$ to a control obstacle is defined in below definition.

Definition 4. *(Margin) The margin of safe control input u_s is its minimum weighted distance to the control obstacle.*

$$mrg(u_s, UO^\tau) = \min_{u \in UO^\tau} \sqrt{(u - u_s)^T M (u - u_s)} \tag{11}$$

where M is a positive definite diagonal weighted matrix whose values are assigned based on the importance of jth dimension of control space.

The maximum margined control input is considered to be the safest control input to a robot, considering the various sources of unmodeled uncertainties.

4.5. Example: Robot as Single-Integrator

To illustrate the concept of safe control space, take an example of a simple robot R_A that was considered in [27]. The position of the center of a disc-shaped robot R_A, at time t, for a given control input u, is given as

$$p_a(t, u) = p_a(0) + tu \tag{12}$$

where $p_a(0)$ is the current position of robot. Similarly to [19], consider the constraint on robot states as follow:

$$\sqrt{\dot{x}(t)^2 + \dot{y}(t)^2} \in [0, v_m] \tag{13}$$

where v_m is the maximum speed of the robot. For system in (12), the control input directly corresponds to the velocity of the system $\dot{p}_a(t, u) = u$; therefore, U_{ad} is a set of control inputs to a robot such that $||u|| \leq v_m$. Geometrically, U_{ad} is a disc of the radius v_m with its center at the origin in the control space.

The temporary control obstacle $UO_j(t_{i+n})$ is the set of control inputs for which $p_a(t_{i+n}, u) - c(t_{i+n}) \in B$. For the single integrator, temporary control obstacle is equivalent to

temporary velocity obastcle, $NLVO(t_{i+n})$ defined in [27], and is shown in Figure 1a. Geometrically, it is a disc of the radius $(r_a + r_a)/t_{i+n}$ with its center at $(c(t_{i+n}) - p_a(0))/t_{i+n}$ in robot control space .
  Consider the following situation of the robot:

- current position at $p_a(0) = (0,0)$, radius $r_a = 0.4$, and maximum linear speed limit of 1unit/s.
- disc shaped static obstacle of radius $r_b = 0.4$ centered at $(2,0)$.
- temporary control obstacle considered at every time step of 0.1 s up to time horizon of $\tau = 5$ s.

  Figure 3 shows the obtained safe velocity space and the path undertaken by a robot for selected velocities with different margins.

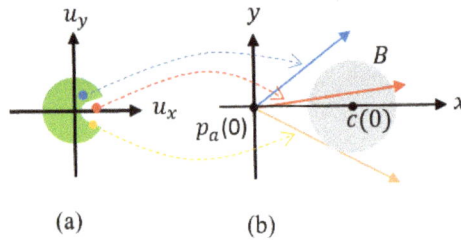

(a)   (b)

Figure 3. The green region in (**a**) is a safe control space $U_{safe}^{\tau=5}$ obtained by taking the complement of control obstacle $UO^{\tau=5}$ in U_{ad}. In (**b**), the paths undertaken by a robot for selected control inputs in $U_{safe}^{\tau=5}$ are shown. The velocity in red lies is in control obstacle and its associated path leads robot to penetrate into obstacle within 5 s. Velocity in yellow lies on the boundary of control obstacle thus, have margin zero. It leads the robot to graze the obstacle within 5 s. Velocity in blue has margin greater than zero and it leads the robot to navigate from a distance to the obstacle.

5. Navigational Approach

  This section will address the problem of navigating a robot among multiple passive agents. It further discusses a possible global navigation plan for navigating a robot among large static obstacles.

5.1. Avoiding Multiple Passive Agents

  We will present the optimization procedure to navigate a robot among multiple passive agents, from a random start point to next (sub) goal p_g, in an open environment. This paper refers the goal as a point extracted from some global way point plan. For the considered environment, the algorithm should be able to quickly re-plan motion for an arbitrary starting point. Instead of conforming to any specific path, it requires the robot to avoid collision with passive agents while moving towards the goal. The proposed approach is based on computing optimal control input that brings robot closer to goal, while its margin should be greater than or equal to some allowable minimum safety margin, represented by β.

  The control input u_s^* that is actually given to a robot has margin greater than or equal to β, (14) and it minimizes the cost in (15) :

$$mrg(u_s^*, UO^\tau) \geq \beta \tag{14}$$

$$u_s^* = arg \min_{u_s \in U_{safe}^\tau} ||f(h(x_a, u_s, \tau)) - p_g|| \tag{15}$$

  That is, the navigation problem can be formulated as the problem of finding $u_s \in U_{safe}^\tau$ for which position $f(h(x_a, u, \tau))$ is closest to p_g in terms of Euclidean distance, and the margin $mrg(u_s^*, UO^\tau)$ is greater than or equal to minimum safety margin β. In case, if no control input in U_{safe}^τ satisfies (14) then control input with the biggest margin is selected, thus giving priority to safety. In this optimization problem, the interaction time horizon τ can be set equal to time horizon over which the behavior of obstacles is predicted, that is typically in the range of 3 to 10 s. The computed control input is applied for a short time step unless a new control loop begins. It results in continuous sense-plan-act navigation framework. The rest of the section will present the procedure of solving the optimization problem.

Explicit construction of control obstacle and safe control space is computationally challenging. We can earn computational savings by adopting a sampling based optimization procedure. Algorithm 1 summarizes the procedure for obtaining a set of safe control inputs and a set of inputs in the control obstacle. The sampling function on line 4 generates uniformly distributed samples in admissible control space for simplicity. However, more intelligent sampling procedure can be used. Each sampled control input is tested against temporary control obstacle at discrete time steps to obtain a set of safe control inputs and a set of control inputs in the control obstacle.

Algorithm 1 Sample Control Space

1: $U^\tau_{safe} \longleftarrow \varnothing$
2: $UO^\tau \longleftarrow \varnothing$
3: **for** $i = 0$ to n **do**
4:     $u_i \longleftarrow$ Sample(U_{ad})
5:     **for** $t = 0 : \delta t : \tau$ **do**
6:         **if** $u_i \in \bigcup_{j\in N} UO_j(t)$ **then**
7:             $UO^\tau = UO^\tau \bigcup u_i$
8:             break loop
9:         **end if**
10:     **end for**
11:     **if** $u_i \notin UO^\tau$ **then**
12:         $U^\tau_{safe} = U^\tau_{safe} \bigcup u_i$
13:     **end if**
14: **end for**

Algorithm 2 summarizes the procedure used for selecting a best safe control input to be given to a robot. A sampling based margin of u_s can be computed, which will be the weighted Euclidean distance of u_s to the nearest sampled control input within control obstacle. It is the overestimate of the actual margin, as shown in Figure 4. Margin function on line 4 returns overestimated margin if it is less than β, else it returns β as the margin of tested control input. For computing margin of u_s, only those samples are visited that are within a distance β from u_s in control space. In line 6, the function $max(m)$ returns largest margin found for the tested control inputs. Lines 7 to 17 mention the procedure of finding control input, that minimize the cost given in (15), and have overestimated margin greater than or equal to β. If all the tested control inputs have margin less than β then the one with the biggest margin is returned.

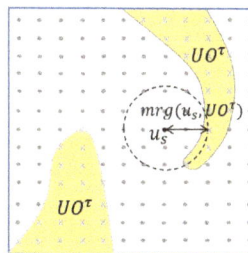

Figure 4. A sampling-based estimate of u_s margin , $mrg(u_s, UO^\tau)$ is the weighted Euclidean distance of u_s to the nearest tested control in control obstacle. It is the overestimate of actual margin of u_s.

Algorithm 2 Find Best Safe Control Input

1: $\beta \longleftarrow$ minimum margin for inputs
2: $min \longleftarrow \infty$
3: **for all** $u_i \in U_{safe}^\tau$ **do**
4:     $m[i] \longleftarrow mrg(u_i, UO^\tau, \beta)$
5: **end for**
6: $m_a = m_b = max(m)$
7: **if** $max(m) \geq \beta$ **then**
8:     $m_a = max(m); m_b = \beta$
9: **end if**
10: **for all** $u_i \in U_{safe}^\tau$ **do**
11:     **if** $m[i] \in [m_a, m_b]$ **then**
12:        **if** $||f(h(x_a, u_i, \tau) - p_g))|| < min$ **then**
13:           $min \longleftarrow ||f(h(x_a, u_i, \tau) - p_g)||$
14:           $argmin \longleftarrow u_i$
15:        **end if**
16:     **end if**
17: **end for**

5.2. Global Navigation

As other VO based approaches, the above presented approach is also subjected to a local minimum in the vicinity of large static concave obstacles. In such environment, the proposed framework can be incorporated into a global navigation plan. The key insight is that the admissible control space represents system compliant trajectories. By considering these trajectories up to time horizon τ, a tree of depth one can be obtained. This tree can be searched and expanded based on margin of segments using a graph search method as one presented in [28]. During the search, all those segments that lie in control obstacle or are in collision with a static obstacle are considered as forbidden control inputs and are discarded. By running these components in a loop, a model predictive partial motion planning framework is obtained.

6. Implementation and Results

The previous sections have presented the framework of defining safe control space for a generalized agent dynamics. This section applies the framework to two types of kinodynamic model of the robot, namely the double integrator, and a car-like robot. Further, this section evaluates the performance of the proposed navigational approach for different parameters and presents its comparison with current approaches.

6.1. Considered Kinodynamic Model of Robot

6.1.1. Car-Like Robot

As illustrated in Figure 2, the states of the car-like robot can be given by the center position of the rear wheel axle $p_a = [x_a \ y_a]^T$, its orientation θ and steering angle ϕ. Its state-transition equations are given by:

$$\dot{x}_a(t) = v_s \cos \theta(t)$$

$$\dot{y}_a(t) = v_s \sin \theta(t) \tag{16}$$

$$\dot{\theta}(t) = v_s k$$

where v_s is the speed control input, k is the curvature control input. As in [26], curvature is directly taken as a control input, and the steering angle ϕ is computed as $\phi = \tan^{-1}(kL)$, where L represents the wheelbase of the car. We will denote the control input vector $[v_s \ k]^T$ by u. The states of the car-like robot are constrained as follow:

$$\sqrt{\dot{x}_a(t)^2 + \dot{y}_a(t)^2} \in [0, v_m] \tag{17}$$

$$\dot{\theta}(t) / \sqrt{\dot{x}_a(t)^2 + \dot{y}_a(t)^2} \in [-k_m, k_m] \tag{18}$$

where v_m and k_m are the maximum velocity and curvature constraints on robot states, respectively. For these constraint, U_{ad} will be a rectangular region in control space such that $|k| < k_m$ and $|v_s| < v_m$. The expression for the position of the robot at a time t is obtained by integrating (16) under the assumption that the control inputs will remain constant over the time horizon t, and it is given as follows:

- if $k \neq 0$,

$$p_a(t, u) = p_a + R(\theta_a) * \frac{1}{k} \begin{bmatrix} \sin(v_s k t) \\ 1 - \cos(v_s k t) \end{bmatrix} \tag{19}$$

- if $k = 0$,

$$p_a(t, u) = p_a(0) + R(\theta_a)[t v_s \ 0]^T \tag{20}$$

where p_a and θ_a are the current position and orientation of the robot, respectively and $R(\theta_a)$ is a rotation matrix equal to $(\cos\theta_a \ -\sin\theta_a; \sin\theta_a \ \cos\theta_a)$.

6.1.2. Double Integrator

Consider a robot with dynamics as of double integrator and its states are constrained as follows:

$$\sqrt{\dot{x}_a(t)^2 + \dot{y}_a(t)^2} \in [0, v_m] \tag{21}$$

$$\sqrt{\ddot{x}_a(t)^2 + \ddot{y}_a(t)^2} \in [0, a_m] \tag{22}$$

where v_m and a_m are the maximum velocity and acceleration constraints of robot, respectively. Similar to work in [23], we let the robot to choose a velocity u instead of acceleration. Due to constraints on its states, the new velocity cannot be adopted instantaneously. A proportional control for acceleration is used for the robot, that is the acceleration applied at time t is equal to the difference between new velocity u and velocity $\dot{p}_a(t)$ at that time.

$$\ddot{p}_a(t, u) = \frac{u - \dot{p}_a(t)}{\eta} \tag{23}$$

where η is a control parameter whose unit is time. By integrating (23) we obtain:

$$\dot{p}_a(t, u) = u - e^{-t/\eta}(u - \dot{p}_a(0)) \tag{24}$$

where $\dot{p}_a(0)$ is the current velocity of the robot. By integrating (24) we obtain:

$$p_a(t, u) = p_a(0) + t u + \eta(e^{-t/\eta} - 1)(u - \dot{p}_a(0)) \tag{25}$$

where $p_a(0)$ is the current position of a robot. The admissible velocities space due to acceleration constraints is a disc in velocity space, its radius is equal to ηa_m and its center is at $\dot{p}_a(0)$. The admissible velocities space due to speed constraints is a disc in velocity space, its radius is equal to v_m and its center is at the origin. The intersection of these two convex spaces will give us an admissible velocity space U_{ad}.

6.2. Implementation Details and Simulation Setup

This section will describe the implementation of the proposed algorithm and discuss its performance based on a set of simulated experiments. The experiments are conducted for a car-like robot and a double integrator, governed by the kinodynamic model described in Section 6.1. For each experiment, the interaction time horizon τ is set to 3.5 s. The weighted matrix M defined in Definition 04 is set to identity matrix. The following constraints are considered on robot states in the simulations:

- *Car-like robot:* Maximum velocity $v_m = 1.5$ unit/s, Maximum curvature $k_m = 1.5$ unit^{-1}.
- *Double integrator:* Maximum velocity $v_m = 2$ unit/s, Maximum acceleration $a_m = 1$ unit/s^2.

The algorithm is implemented in C++, on a computer running Windows 10. The timing results are generated on Intel i5-3550 PC with 4-GB RAM. Although it is possible to build U_{safe}^{τ} and UO^{τ} using multiple cores, we use single-core to produce timing results.

Figure 5 shows the trajectory undertaken by the car-like robot to reach the goal while avoiding a collision with a static passive agent for different selected β. Figure 6 shows the same for the double integrator.

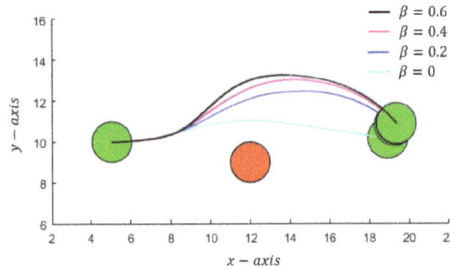

Figure 5. It shows the paths taken by a car-like robot to reach the goal at (20,10) from its initial position at (5,10), for selected β, while avoiding collision with a static passive agent at (12,9).

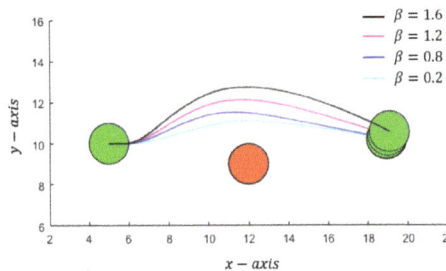

Figure 6. It shows the paths taken by a double integrator to reach the goal at (20,10) from its initial position at (5,10), for selected β, while avoiding collision with a static passive agent at (12,9). The control parameter η was set to 3.

6.3. Collision Avoidance with Multiple Passive Agents

Each experiment is conducted in an open environment. Robot initial position is set to $(5, 10)$, and it has to move towards a goal at $(20, 20)$. Several circular obstacles representing passive agents are randomly distributed in a square region bounded by $(0, 0)$ and $(22, 22)$. All passive agents and robot have radii equal to 1 unit. Similar to [26], passive agents are assigned arbitrary velocities and the maximum upper limit on their speeds is set to ± 1 unit/s. The probability that the passive agent will change its velocity within 1 s is 0.2. Figure 7 shows multiple snapshots at different time instances of the car-like robot, in green, avoiding collision with multiple passive agents, in red, while moving towards the goal marked in yellow.

An empirical test showing the performance of the algorithm is presented in the following subsections, where each experimental value is the mean of 10 trials.

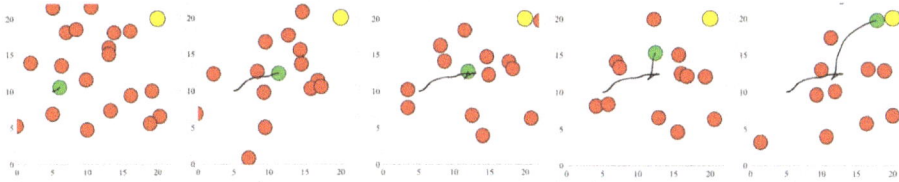

Figure 7. Avoiding multiple passive agents. Multiple snapshots at different time instances show a car-like robot, in green, navigates among multiple passive agents, in red, while moving towards a goal p_g, in yellow. The passive agents are moving with arbitrary velocities. The probability that a passive agent will change its velocity within one second is 0.2. {$\beta = 0.4$ is set for the simulation}.

6.3.1. Performance Results

In Figure 8, the performance of the proposed algorithm for a car-like agent is presented for the set of three parameters, the number of passive agents present, minimum saftey margin β, and the length of the time step. The effect on computation time, success rate, and elapsed time are investigated. The computation time is the time required to compute optimal control input. A successful run is a trial in which agent successfully reaches the goal without a collision. A collision might occur when a passive agent changes its direction and traps the robot. In a successful trial, elapsed time is the amount of time passed, starting from the point in time when the robot was at its initial position, to the point in time when robot reaches the goal position. Figure 8a shows an average computation time taken when the number of obstacles present grows. The computation time increases approximately linearly with the increase in a number of passive agents, while Figure 8b shows the success rate drops with the growth in the number of passive agents. For these experiments the step time was set to 5 ms, the margin was set to 0.4, and 256 samples was taken in U_{ad}. In Figure 8c, the success rate is shown for selected values of β, when 20 passive agents are present. It shows that the success rate increases with the increase in β, while Figure 8d shows that the elapsed time also increases with the increase in β. For larger value of β, the robot takes comparatively longer time to reach the goal because it selects a more safe path to the goal rather than a direct path to it. Finally, Figure 8e shows the success rate when β is set equal to 0, and 0.4, for varying step time. It shows that the increase in step time will lead to a drop in success rate, whereas the success rate is comparatively higher for $\beta = 0.4$ than that for $\beta = 0$.

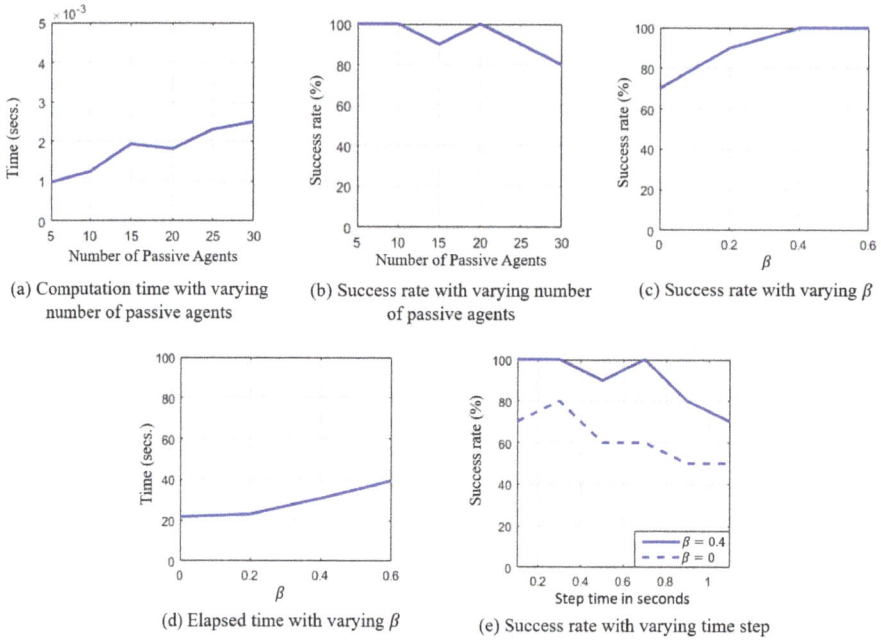

Figure 8. In this graph, the performance of the proposed algorithm for a car-like robot is presented for the set of three parameters, the number of passive agents present, minimum saftey margin β, and the length of the time step. The effects on computation time, success rate, and elapsed time, are shown.

The results in Figure 8c,d show that the car-like robot reaches the goal with relatively high success rate when $\beta = 0.4$ is selected, while the elapsed time is comparatively reasonable. The next subsection will compare the performance of the proposed approach (U_{safe}) with $\beta = 0.4$ to that of GVO.

6.3.2. Comparison with GVO

In Figure 9, a comparison of performance is shown for the proposed U_{safe} approach and GVO approach. The comparison is made regarding the effect on computation time, success rate, and elapsed time, for varying number of passive agents present. In this comparison 256 control inputs are sampled in admissible control space. In Figure 9, (a) shows average computation time for the two approaches, where computation time for GVO is much higher compared to U_{safe}. (b) shows the comparison of average success rate. The success rate of GVO is much lower compared to U_{safe}, as for GVO approach the robot is frequently trapped by the passive agents. Finally, Figure 9c compares the elapsed time for the two approaches. For U_{safe}, the robot takes a safer path rather than a direct path to reach the goal. Thus, the robot takes a comparatively longer time to reach the goal. In summary, we can conclude that the performance of U_{safe} is much better than GVO.

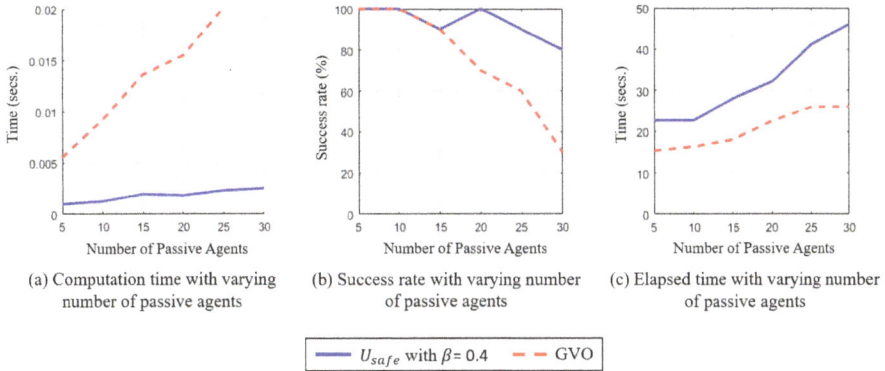

(a) Computation time with varying number of passive agents

(b) Success rate with varying number of passive agents

(c) Elapsed time with varying number of passive agents

— U_{safe} with $\beta = 0.4$ - - GVO

Figure 9. It shows the performance of GVO algorithm and the proposed U_{safe} approach. The effect on average computation time, success rate, and elapsed time, are shown against varying number of passive agents. GVO: Generalized Velocity Obstacle.

6.3.3. Comparison with AVO

In order to make a comparison with AVO [23], the robot with kinodynamic model as of double integrator is considered. We have modified AVO by removing its reciprocal collision avoidance aspect and keeping the linear programming optimization. The control parameter η is set to 3 for these experiments. For U_{safe}, β is set to 1.2. The graphs in Figure 10 presents the performance of proposed U_{safe} approach for selected values of β. The experiment results in Figure 11 show the performance of AVO algorithm and the proposed U_{safe} approach. It shows the effect on average computation time, success rate, and time elapsed, against varying number of passive agents. In Figure 11, (a) shows the average computation time for the two approaches, where AVO takes slightly lesser time on average for computing the solution. (b) shows the comparison of average success rate, the success rate for AVO is much lower as compared to U_{safe}. (c) shows that the elapsed time is almost same for the two approaches.

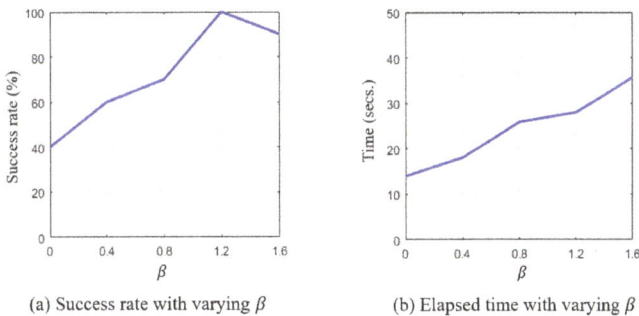

(a) Success rate with varying β

(b) Elapsed time with varying β

Figure 10. In this graph, the performance of the proposed algorithm U_{safe} for double integrator is presented for the scenario when 20 passive agents are present. (**a**) show the success rate for selected values of β. (**b**) show the elapsed time for different values of β. The robot takes comparatively longer time to reach goal when the larger value for β is set, as robot take the safer path to the goal rather than a direct path to it. Results show that robot reaches the goal with relatively high success rate when $\beta = 1.2$ is selected, while the elapsed time is comparatively reasonable.

Under AVO approach, the velocity obstacle induced due to each passive agent is approximated by a half plane in velocity space, this leads to a computed collision free control space which is frequently empty, thus give rise to the number of the failures in computing the solution.

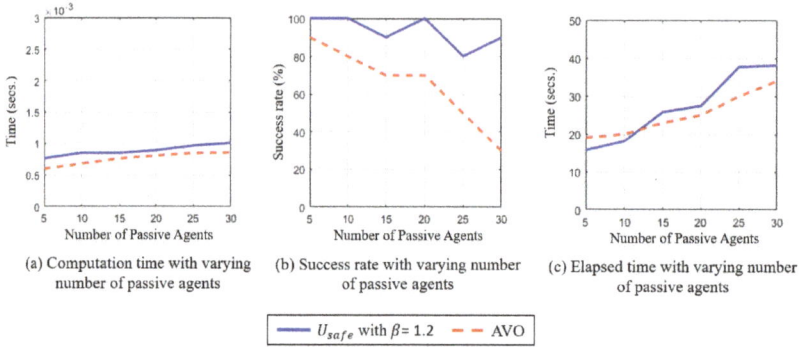

(a) Computation time with varying number of passive agents

(b) Success rate with varying number of passive agents

(c) Elapsed time with varying number of passive agents

U_{safe} with $\beta = 1.2$ — — AVO

Figure 11. It shows the performance of AVO algorithm and the proposed U_{safe} approach. The effect on average computation time, success rate, and elapsed time, are shown against varying number of passive agents. AVO: Acceleration-Velocity Obstacle.

6.4. Global Navigation

The real world environment usually has static obstacles in addition to passive agents. These static obstacles are usually concave, and the VO based approaches are subjected to a local minimum in such environments. The presented concept of safe control space allows us to integrate the control space in a graph search fashion. Figure 12 shows the framework where multiple snapshots at different time instances are shown. In this simulation, the interaction time horizon τ is set to 4 s and 256 control inputs are sampled in an admissible control space. All samples that lie in the control obstacle or collide with static obstacles are discarded to obtain a tree depth of one. In this experiment, the graph search terminates at the depth of one, the point at which $f(h(x_a, u, \tau))$ connects to Dijkstra heuristic function. The trajectory that is executed for one control cycle is the one that minimize the Dijkstra heuristic function and has a margin greater than or equal to 1.2. As a result, we obtain a global optimization criteria for navigating the agent. However, in case if all tested samples has margin less then β, than the sample with the biggest margin is selected.

Figure 12. Global navigation plan. Robot employs a heuristic graph search strategy. The color gradient shows the Dijkstra heuristic field for the maze. The interaction horizon is $\tau = 4$ s. The safe control inputs are computed, which are neither in control obstacle nor did they lead to a collision with the static obstacle. At the same time, Dijkstra heuristic function computes the cost for reaching the goal. A safe control input is executed which minimizes the cost and has margin greater than or equal to $\beta = 1.2$.

7. Conclusions

This paper presented a fast navigation approach for a dynamic environment where the behavior of passive agents is partially predictable. The concept of safe control space is presented by generalizing the nonlinear velocity obstacle, which makes it possible to consider the robot model for computing the system compliant collision-free trajectories. The safety margin defined for safe control inputs allows the robot to make a safe decision in the environment where the predicted trajectories of passive agents change frequently. The ability of the proposed algorithm to compute a safe decision in real-time is demonstrated. The comparative study validates that under the proposed approach the robot reaches the goal with high success rate. The proposed approach can be incorporated into a waypoint plan for global navigation. One possible implementation of a global navigation plan is presented in the paper.

In future, authors will attempt the problem of dynamically adapting the minimum safety margin parameter based on the circumstances. Also, an interesting direction for future work will be the extension of the proposed approach to the environment with multiple active and passive agents. In addition to this, the authors are interested in planning for situations with dynamic obstacles have deterministic behavior.

Acknowledgments: This research was partially supported by the Higher Education Commission of Pakistan by the award letter No. HRDI-UESTPs/Batch-II/South Korea/2012.

Author Contributions: K.M.Z. proposed the idea, implemented the simulated experiments and wrote the manuscript; C.H. supervised the study and the manuscript writing process; J.Y.L. co-supervised the study. All the authors discussed the results and contributed to the final manuscript.

Conflicts of Interest: The authors declare no conflict of interest.

Abbreviations

The following abbreviations are used in this manuscript:

VO	Velocity Obstacle
NLVO	Nonlinear Velocity Obstacle
AVO	Accleration Velocity Obstacle
GVO	Generalized Velocity Obstacle

References

1. Prassler, E.; Scholz, J.; Fiorini, P. A robotic wheelchair for crowded public environments. *IEEE Robot. Autom. Mag.* **2001**, *8*, 38–45.
2. Breitenmoser, A.; Tâche, F.; Caprari, G.; Siegwart, R.; Moser, R. MagneBike: Toward multi climbing robots for power plant inspection. In Proceedings of the 9th International Conference on Autonomous Agents and Multiagent Systems: Industry track, Toronto, ON, Canada, 10–14 May 2010; International Foundation for Autonomous Agents and Multiagent Systems: Singapore, 2010; pp. 1713–1720.
3. Laumond, J.P.; Jacobs, P.; Taïx, M.; Murray, R. A Motion Planner for Nonholonomic Mobile Robots. *IEEE Trans. Robot. Autom.* **1994**, *10*, 577–593.
4. Scheuer, A.; Fraichard, T. Continuous-curvature path planning for car-like vehicles. In Proceedings of the IEEE International Conference on Intelligent Robots and Systems, Grenoble, France, 7–11 September 1997; Volume 2, pp. 997–1003.
5. Lamiraux, F.; Laumond, J.P. Smooth motion planning for car-like vehicles. *IEEE Trans. Robot. Autom.* **2001**, *17*, 498–502.
6. Frazzoli, E.; Dahleh, M.; Feron, E. Real-time motion planning for agile autonomous vehicles. *J. Guid. Control Dyn.* **2002**, *25*, 116–129.
7. Hsu, D.; Kindel, R.; Latombe, J.C.; Rock, S. Randomized kinodynamic motion planning with moving obstacles. *Int. J. Robot. Res.* **2002**, *21*, 233–255.
8. Zucker, M.; Kuffner, J.; Branicky, M. Multipartite RRTs for rapid replanning in dynamic environments. In Proceedings of the IEEE International Conference on Robotics and Automation, Roma, Italy, 10–14 April 2007; pp. 1603–1609.
9. Kant, K.; Zucker, S. Toward Efficient Trajectory Planning: The Path-Velocity Decomposition. *Int. J. Robot. Res.* **1986**, *5*, 72–89.
10. Peng, J.; Akella, S. Coordinating multiple Robots with kinodynamic constraints along specified paths. *Int. J. Robot. Res.* **2005**, *24*, 295–310.

11. Van Den Berg, J.; Overmars, M. Kinodynamic motion planning on roadmaps in dynamic environments. In Proceedings of the IEEE International Conference on Intelligent Robots and Systems, San Diego, CA, USA, 29 October–2 November 2007; pp. 4253–4258.

12. Gaillard, F.; Soulignac, M.; Dinont, C.; Mathieu, P. Deterministic kinodynamic planning with hardware demonstrations. In Proceedings of the IEEE/RSJ International Conference on Intelligent Robots and Systems (IROS), San Francisco, CA, USA, 25–30 September 2011; pp. 3519–3525.

13. Chen, C.; Rickert, M.; Knoll, A. Kinodynamic motion planning with space-time exploration guided heuristic search for car-like robots in dynamic environments. In Proceedings of the IEEE/RSJ International Conference Intelligent Robots and Systems (IROS), Hamburg, Germany, 28 September–2 October 2015; pp. 2666–2671.

14. Bennewitz, M.; Burgard, W.; Thrun, S. Learning motion patterns of persons for mobile service robots. In Proceedings of the IEEE International Conference on Robotics and Automation, Washington, DC, USA, 11–15 May 2002; Volume 4, pp. 3601–3606.

15. Vasquez, D.; Fraichard, T. Motion prediction for moving objects: A statistical approach. In Proceedings of the IEEE International Conference on Robotics and Automation, New Orleans, LA, USA, 26 April–1 May 2004; Volume 4, pp. 3931–3936.

16. Kim, S.; Guy, S.; Liu, W.; Wilkie, D.; Lau, R.; Lin, M.; Manocha, D. BRVO: Predicting pedestrian trajectories using velocity-space reasoning. *Int. J. Robot. Res.* **2015**, *34*, 201–217.

17. Bera, A.; Kim, S.; Randhavane, T.; Pratapa, S.; Manocha, D. GLMP-realtime pedestrian path prediction using global and local movement patterns. In Proceedings of the IEEE International Conference on Robotics and Automation (ICRA), Stockholm, Sweden, 16–21 May 2016; pp. 5528–5535.

18. Fiorini, P.; Shiller, Z. Motion planning in dynamic environments using velocity obstacles. *Int. J. Robot. Res.* **1998**, *17*, 760–772.

19. Van Den Berg, J.; Guy, S.J.; Lin, M.; Manocha, D. Reciprocal n-body collision avoidance. In *Robotics Research*; Pradalier, C., Siegwart, R., Hirzinger, G., Eds.; Springer: Berlin, Germany, 2011; pp. 3–19.

20. Best, A.; Narang, S.; Manocha, D. Real-time reciprocal collision avoidance with elliptical agents. In Proceedings of the IEEE International Conference on Robotics and Automation (ICRA), Stockholm, Sweden, 16–21 May 2016; pp. 298–305.

21. Alonso-Mora, J.; Breitenmoser, A.; Rufli, M.; Beardsley, P.; Siegwart, R. Optimal reciprocal collision avoidance for multiple non-holonomic robots. In *Distributed Autonomous Robotic Systems*; Springer: Berlin, Germany, 2013; pp. 203–216.

22. Alonso-Mora, J.; Breitenmoser, A.; Beardsley, P.; Siegwart, R. Reciprocal collision avoidance for multiple car-like robots. In Proceedings of the IEEE International Conference on Robotics and Automation (ICRA), Saint Paul, MN, USA, 14–18 May 2012; pp. 360–366.

23. Van Den Berg, J.; Snape, J.; Guy, S.J.; Manocha, D. Reciprocal collision avoidance with acceleration-velocity obstacles. In Proceedings of the IEEE International Conference on Robotics and Automation (ICRA), Shanghai, China, 9–13 May 2011; pp. 3475–3482.

24. Rufli, M.; Alonso-Mora, J.; Siegwart, R. Reciprocal collision avoidance with motion continuity constraints. *IEEE Trans. Robot.* **2013**, *29*, 899–912.

25. Bareiss, D.; Van Den Berg, J. Generalized reciprocal collision avoidance. *Int. J. Robot. Res.* **2015**, *34*, 1501–1514.

26. Wilkie, D.; Van Den Berg, J.; Manocha, D. Generalized velocity obstacles. In Proceedings of the IEEE/RSJ International Conference on Intelligent Robots and Systems (IROS), St. Louis, MO, USA, 10–15 October 2009; pp. 5573–5578.

27. Shiller, Z.; Large, F.; Sekhavat, S. Motion planning in dynamic environments: Obstacles moving along arbitrary trajectories. In Proceedings of the IEEE International Conference on Robotics and Automation, Seoul, Korea, 21–26 May 2001; Volume 4, pp. 3716–3721.

28. Park, J.; Iagnemma, K. Sampling-based planning for maximum margin input space obstacle avoidance. In Proceedings of the IEEE/RSJ International Conference on Intelligent Robots and Systems (IROS), Hamburg, Germany, 28 September–2 October 2015; pp. 2064–2071.

*applied
sciences*

MDPI

Article

Agreement Technologies for Coordination in Smart Cities

Holger Billhardt [1], Alberto Fernández [1], Marin Lujak [2] and Sascha Ossowski [1,*]

[1] Centre for Intelligent Information Technologies (CETINIA), University Rey Juan Carlos, 28933 Madrid, Spain; holger.billhardt@urjc.es (H.B.); alberto.fernandez@urjc.es (A.F.)
[2] IMT Lille Douai, 59508 Douai, France; marin.lujak@imt-lille-douai.fr
* Correspondence: sascha.ossowski@urjc.es; Tel.: +34-916-647-485

Received: 16 March 2018; Accepted: 14 May 2018; Published: 18 May 2018

check for updates

Abstract: Many challenges in today's society can be tackled by distributed open systems. This is particularly true for domains that are commonly perceived under the umbrella of smart cities, such as intelligent transportation, smart energy grids, or participative governance. When designing computer applications for these domains, it is necessary to account for the fact that the elements of such systems, often called software agents, are usually made by different designers and act on behalf of particular stakeholders. Furthermore, it is unknown at design time when such agents will enter or leave the system, and what interests new agents will represent. To instil coordination in such systems is particularly demanding, as usually only part of them can be directly controlled at runtime. Agreement technologies refer to a sandbox of tools and mechanisms for the development of such open multiagent systems, which are based on the notion of agreement. In this paper, we argue that agreement technologies are a suitable means for achieving coordination in smart city domains, and back our claim through examples of several real-world applications.

Keywords: agreement technologies; coordination models; multiagent systems; smart cities

1. Introduction

The transactions and interactions among people in modern societies are increasingly mediated by computers. From email, over social networks, to virtual worlds, the way people work and enjoy their free time is changing dramatically. The resulting networks are usually large in scale, involving huge numbers of interactions, and are open for the interacting entities to join or leave at will. People are often supported by software components of different complexity to which some of the corresponding tasks can be delegated. In practice, such systems cannot be built and managed based on rigid, centralised client-server architectures, but call for more flexible and decentralised means of interaction.

The field of agreement technologies (AT) [1] envisions next-generation open distributed systems, where interactions between software components are based on the concept of agreement, and which enact two key mechanisms: a means to specify the "space" of agreements that the agents can possibly reach, and an interaction model by means of which agreements can be effectively reached. Autonomy, interaction, mobility and openness are key characteristics that are tackled from a theoretical and practical perspective.

Coordination in distributed systems is often seen as governing the interaction among distributed processes, with the aim of "gluing together" their behaviour so that the resulting ensemble shows desired characteristics or functionalities [2]. This notion has also been applied to distributed systems made up of software agents. Initially, the main purpose of such multiagent systems was to efficiently perform problem-solving in a distributed manner: both the agents and their rules of interaction were designed together, often in a top-down manner and applying a divide-and-conquer strategy to solve the problem at hand [3]. However, many recent applications of multiagent systems refer to domains where agents, possibly built by different designers and representing different interests, may join and leave the system at a pace that is unknown at design time. It is apparent that coordination in such open multiagent systems requires a different, extended stance on coordination [3].

Application areas that fall under the umbrella of *smart cities* have recently gained momentum [4]. Intelligent transportation systems, smart energy grids, or participative governance are just some examples of domains where an improved efficiency of the use of shared urban resources (both physical and informational) can lead to a better quality of life for the citizens. It thus seems evident that new applications in the context of smart cities have the potential for achieving significant socioeconomic impact.

We believe that applying AT to the domain of smart cities may enable the development of novel applications, both with regard to functionality for stakeholders, as well as with respect to the level of sustainability of smart city services. In particular, in this article, we discuss how coordination can be achieved in practical applications of multiagent systems, with different levels of openness, by making use of techniques from the sandbox of AT. Section 2 briefly introduces the fields of AT, coordination models, and smart cities, and relates them to each other. Section 3 describes several real-world applications, related to the field of smart cities, that illustrate how coordination models can be tailored to each particular case and its degree of openness. Section 4 summarises the lessons learnt from this enterprise.

2. Background

In this section, we introduce the fields of agreement technologies and coordination models and relate them to each other. We then briefly characterise the field of smart cities and argue that agreement technologies are a promising candidate to instil coordination in smart city applications.

2.1. Agreement Technologies

Agreement technologies (AT) [1] address next-generation open distributed systems, where interactions between software processes are based on the concept of agreement. AT-based systems are endowed with means to specify the "space" of agreements that can be reached, as well as interaction models for reaching agreement and monitoring agreement execution. In the context of AT, the elements of open distributed systems are usually conceived as software agents. There is still no consensus where to draw the border between programs or objects on the one hand and software agents on the other, but the latter are usually characterised by four key characteristics, namely, *autonomy*, *social ability*, *responsiveness* and *proactiveness* [5]. The interactions of a software agent with its environment (and with other agents) are guided by a reasonably complex program, capable of rather sophisticated activities such as reasoning, learning, or planning. Two main ingredients are essential for such multiagent systems based on AT: firstly, a normative model that defines the "rules of the game" that software agents and their interactions must comply with; and secondly, an interaction model where agreements are first established and then enacted. AT can then be conceived as a sandbox of methods, platforms, and tools to define, specify, and verify such systems.

The basic elements of the AT sandbox are related to the challenges outlined by Sierra et al. for the domain of agreement computing [6], covering the fields of semantics, norms, organisations, argumentation and negotiation, as well as trust and reputation. Still, when dealing with open distributed systems made up of software agents, more sophisticated and computationally expensive models and mechanisms can be applied [7].

The key elements of the field of AT can be conceived of in a tower structure, where each level provides functionality to the levels above, as depicted in Figure 1.

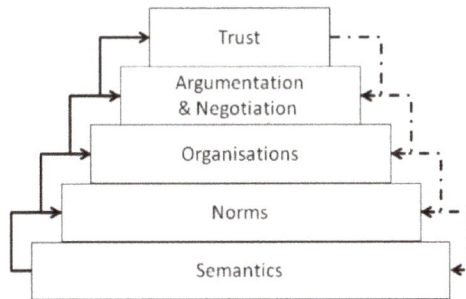

Figure 1. Agreement technologies (AT) tower.

Semantic technologies provide solutions to semantic mismatches through the alignment of ontologies, so agents can reach a common understanding on the elements of agreements. In this manner, a shared multifaceted "space" of agreements can be conceived, providing essential information to the remaining layers. The next level is concerned with the definition of *norms* determining constraints that the agreements, and the processes leading to them, should satisfy. Thus, norms can be conceived of as a means of "shaping" the space of valid agreements. *Organisations* further restrict the way agreements are reached by imposing organisational structures on the agents. They thus provide a way to efficiently design and evolve the space of valid agreements, possibly based on normative concepts. The *argumentation and negotiation* layer provides methods for reaching agreements that respect the constraints that norms and organisations impose over the agents. This can be seen as choosing certain points in the space of valid agreements. Finally, the *trust and reputation* layer keeps track of whether the agreements reached, and their executions, respect the constraints put forward by norms and organisations. So, it complements the other techniques that shape the "agreement space" by relying on social mechanisms that interpret the behaviour of agents.

Even though one can clearly see the main flow of information from the bottom towards the top layers, results of upper layers can also produce useful feedback that can be exploited at lower levels. For instance, as mentioned above, norms and trust can be conceived as a priori and a posteriori approaches, respectively, to security [6]. Therefore, in an open and dynamic world it will certainly make sense for the results of trust models to have a certain impact on the evolution of norms. Some techniques and tools are orthogonal to the AT tower structure. The topics of environments [8] and infrastructures [9], for instance, pervade all layers. In much the same way, coordination models and mechanisms are not just relevant to the third layer of Figure 1, but cross-cut the other parts of the AT tower as well [10]. We will elaborate on this matter in the next subsection.

2.2. Coordination Models

The notion of coordination is central to many disciplines. Sociologists observe the behaviour of groups of people, identify particular coordination mechanisms, and explain how and why they emerge. Economists are concerned with the structure and dynamics of the market as a particular coordination mechanism; they attempt to build coordination market models to predict its behaviour. Biologists observe societies of simple animals demonstrating coordination without central coordinators; coordination mechanisms inspired from biology have proven useful to various scientific disciplines. In organisational theory, the emphasis is on predicting future behaviour and performance of an organisation, assuming the validity of a certain coordination mechanism. From a computer science point of view, the challenge is to *design* mechanisms that "glue together" the activities of distributed actors in some efficient manner. However, beyond such high-level conceptions, within the computer science field, and even among researchers working on multiagent systems, there is no commonly agreed definition for the concept of coordination. An important reason for this is the different *interests* of the designers in coordination mechanisms (micro- and/or macro-level properties), as well as different levels of control that designers have over the elements of the distributed intelligent system (the degree of *openness* of the system), as we will argue in the following [3].

Early work on coordination in multiagent systems (MAS) focused essentially on (cooperative) *distributed problem solving*. In this field, it is assumed that a system is constructed (usually from the scratch) out of several intelligent components, and that there is a single designer with *full control* over these agents. In particular, this implies that agents are *benevolent* (as instrumental local goals can be designed into them) and, by consequence, that the designer is capable of imposing whatever interaction patterns are deemed necessary to achieve efficient coordination within the system. Efficiency in this context usually refers to a trade-off between the system's resource consumption and the quality of the solution provided by the system: agents necessarily have only partial, and maybe even inconsistent views of the global state of the problem-solving process, so they need to exchange just enough information to be able to locally make good decisions (i.e., choices that are instrumental with respect to the overall system functionality). Resource consumption is not only measured in terms of computation but also of communication load.

From a *qualitative* perspective, coordination in distributed problem-solving systems can be conceived as a *distributed constraint problem* (see [11] for an example). Agents locally determine individual actions that comply with the constraints (dependencies) that affect them, so as to give rise to "good" global solutions. Alternatively, in *quantitative* approaches, the structure of the coordination problem is hidden in the shape of a shared global multi-attribute utility function. An agent has control over only some of the function's attributes, and the global utility may increase/decrease in case there is a positive/negative dependency with an attribute governed by another agent, but these dependencies are hidden in the algorithm that computes the utility function and are thus not declaratively modelled. Quantitative approaches to coordination can be understood in terms a of a *distributed optimisation problem*.

More recent research in the field of MAS has been shifting the focus towards *open* systems, where the assumption of a central designer with full control over the system components no longer holds. This raises interoperability problems that need to be addressed. In addition, the benevolence assumption of distributed problem-solving agents needs to be dropped: coordination mechanisms now have to deal with *autonomous*, self-interested behaviour—an aspect that is usually out of the scope of models from the field of distributed computing. Approaching agent design in open systems from a *micro-level* perspective means designing an intelligent software entity capable of successful autonomous action in potentially hostile (multiagent) environments. In this context, coordination can be defined as "a way of adapting to the environment" [12]: adjusting one's decisions and actions to the presence of other agents, assuming that they show some sort of *rationality*. If the scenario is modelled within a quantitative framework, we are still concerned with multi-attribute utility functions, where only some attributes are controlled by a particular agent, but now there are different utility functions for each agent. The most popular way of characterising a problem of these characteristics is through (nonconstant sum) *games* [13]. Coordination from a micro-level perspective boils down to agents applying some sort of "best response" strategy, and potentially leads to some notion of (Nash) equilibrium. From a macro-level perspective, coordination is about designing "rules of the game" such that, assuming that agents act rationally and comply with these rules, some desired properties or functionalities are instilled. In the field of game theory, this is termed *mechanism design* [13]. In practice, it implies designing potentially complex interaction protocols among the agents, which shape their "legal" action alternatives at each point in time, as well as institutions or infrastructures that make agents abide by the rules [9]. From this perspective, instilling coordination in an open multiagent system can be conceived as an act of *governing interaction* within the system.

If the environment is such that agents can credibly commit to mutually binding agreements, coordinating with others comes down to negotiating the terms of such commitments. This is where the link to AT becomes evident. *Norms* and *organisations* define and structure the interactions that may take place among agents. The shape of these interactions depends on the particular case, but often they can be conceived as *negotiating* an agreement for a particular outcome of coordination. In addition, depending on the agents' interests, information and structured *arguments* can be provided to make agents converge on such an agreement. Norms and trust can be seen as a priori and a posteriori measures, respectively, that make agents comply with the constraints imposed by norms and organisations.

2.3. Smart Cities

There is a broad variety of domains where the potential of AT becomes apparent (see Part VII of [1]). In these domains, the choices and actions of a large number autonomous stakeholders need to be coordinated, and interactions can be regulated, by some sort of intelligent computing infrastructure [9], through some sort of institutions and institutional agents [14], or simply by strategically providing information in an environment with a significant level of uncertainty [15]. The advent of intelligent road infrastructures, with support for vehicle-to-vehicle and vehicle-to-infrastructure communications, make *smart transportation* a challenging field of application for AT, as it allows for a decentralised coordination of individually rational commuters. However, also the infrastructure of the electricity grid is evolving, allowing for bidirectional communication among energy producers and consumers. Therefore, in the near future, large numbers of households could coordinate and adapt their aggregate energy demand to the supply offered by utilities. AT can also be applied to the domain of *smart energy* in order to integrate large numbers of small-scale producers of renewable energy into the grid infrastructure. In much the same way, *smart governance* can make use of electronic institutions that support citizens, for instance, in the process of dispute resolution.

The above are only a few examples of applications and domains that are often referred to under the umbrella of *smart cities*. Even though many definitions of that term exist [3], there is still no commonly agreed conception of a smart city. Still, we believe that authors tend to concur that a key challenge of smart cities is to improve the efficiency of the use of shared urban resources (both physical and informational) through the use of Information and communications technology (ICT), so as to improve the quality of life of citizens (see, e.g., [16–18]). Most of the world's urban areas have a limited space to expand, congestion and contamination seriously affect people's well-being, and a constant and reliable supply of energy is essential for almost all aspects of urban life. Therefore, ICT-based solutions can help to adequately disseminate information and effectively coordinate the urban services and supplies so as to make urban life more comfortable and efficient.

While initially smart city research had a strong focus on ICT and "smartness", more recently, impact indicators of environmental, economic, or social *sustainability* have also gained importance [19], so in present days the term *smart sustainable city* is commonly used [20]. This notion underlines that, on the tack to making our cities smarter, preserving the "needs of present and future generations with respect to economic, social and environmental aspects" is of foremost importance [21].

The Internet of Things (IoT) is often considered as crucial in the development of smart cities [22]. It is usually conceived as a global infrastructure, enabling advanced services by interconnecting (physical and virtual) devices based on ICT. Recently, it has been moving from interesting proofs of concept to systemic support for urban processes that generates efficiency *at scale* (see, e.g., [23]). With increasing connectivity between people, data, and things based on IoT, the challenge is how to manage and *coordinate* the decisions of a myriad of decision makers in real time, considering the scarcity of resources and stochasticity in demand.

We believe that there is a significant potential in applying AT outlined in Section 2.1, targeting methods and tools that support the formation and execution of agreements in large-scale open systems, in order to progress towards the vision of smart and sustainable cities mentioned above. In much the same way, it seems straightforward that the efficient discovery, orchestration, and maintenance of services, based largely on data from heterogeneous sensors and all sorts of embedded devices, calls for the application of both scalable and tailorable coordination models. In the following sections, we will focus on different types of assignment problems in the context of sustainable smart cities: we provide examples of how AT-based coordination services mediate the use of scarce resources to the benefit of citizens.

3. Applications

In this section, we show how the AT paradigm can be applied to achieve coordination in various real-world problems. Depending on the structure and characteristics of each domain, different technologies from the AT sandbox need to be selected and combined so as to meet the requirements for each case. Section 3.1 highlights the use of techniques related to norms and organisations (in particular, auction protocols and market-based control) in an open domain, where flows of autonomous vehicles, controlled by individually rational driver agents, are coordinated through a network of intelligent interactions. Section 3.2 is dedicated to the problem of evacuation guidance in smart buildings, where evacuees, suffering from significant levels of uncertainty concerning the state of an emergency, are provided with individualised route recommendations in a coordinated manner. In this context, issues related to situation-awareness and semantics play a major role. Section 3.3 addresses the coordination of fleets of ambulance vehicles. Even though this is a primarily closed scenario, we address it with techniques from the field of AT, applying an algorithm that *simulates* multiple concurrent computational auctions. Section 3.4 focuses on coordination of emergency medical services for angioplasty patients—a problem similar to the previous one, even though its internal structure (different types of agents, etc.) leads to a more complex coordination mechanism. Finally, Section 3.5 also addresses a coordination problem related to fleet management, but applied to the field of taxi services. Here, we again have a higher degree of openness, as taxis are conceived of as autonomous agents, so coordination needs to be induced by incentives, targeted at influencing the choices of drivers whose actions are not fully determined by organisational rules and protocols.

3.1. Coordination of Traffic Flows through Intelligent Intersections

Removing the human driver from the control loop through the use of autonomous vehicles integrated with an intelligent road infrastructure can be considered as the ultimate, long-term goal of the set of systems and technologies grouped under the name of "intelligent transportation systems" (ITS). Autonomous vehicles are already a reality. For instance, in the *DARPA Grand Challenges* (https://en.wikipedia.org/wiki/DARPA_Grand_Challenge), different teams competed to build the best autonomous vehicles, capable of driving in traffic and performing complex manoeuvres such as merging, passing, parking, and negotiating with intersections. The results have shown that autonomous vehicles can successfully interact with both manned and unmanned vehicular traffic in an urban environment. In line with this vision, the *IEEE Connected Vehicles Initiative* (http://sites.ieee.org/connected-vehicles/ieee-connected-vechicles/ieee-cv-initiative/) promotes technologies that link road vehicles to each other and to their physical surroundings (i.e., by vehicle-to-infrastructure and vehicle-to-vehicle wireless communications). The advantages of such an integration span from improved road safety to a more efficient operational use of the transportation network. For instance, vehicles can exchange critical safety information with the infrastructure so as to recognise high-risk situations in advance and therefore to alert drivers. Furthermore, traffic signal systems can communicate signal phase and timing information to vehicles to enhance the use of the transportation network.

In this regard, some authors have recently paid attention to the potential of a tighter integration of autonomous vehicles with the road infrastructure for future urban traffic management. In the *reservation-based* control system [24], an intersection is regulated by a software agent, called *intersection manager* agent, which assigns reservations of space and time to each autonomous vehicle intending to cross the intersection. Each vehicle is operated by another software agent, called *driver* agent. When a vehicle approaches an intersection, the driver requests that the intersection manager reserves the necessary space–time slots to safely cross the intersection. The intersection manager, provided with data such as vehicle ID, vehicle size, arrival time, arrival speed, type of turn, and other variables, simulates the vehicle's trajectory inside the intersection and informs the driver whether its request is in conflict with the already confirmed reservations. If such a conflict does not exist, the driver stores the reservation details and tries to meet them; otherwise, it may try again at a later time. The authors show through simulations that in situations of balanced traffic, if all vehicles are autonomous, their delays at the intersection are drastically reduced compared to traditional traffic lights.

In this section, we report on our efforts to use different elements of the sandbox of AT to further improve the effectiveness and applicability of Dresner and Stone's approach, assuming a future infrastructure where all vehicles are autonomous and capable of interacting with the regulating traffic infrastructure. We extend the reservation-based model for intersection control at two different levels.

- *Single Intersection*: our objective is to elaborate a new policy for the allocation of reservations to vehicles that takes into account the drivers' different attitudes regarding their travel times.
- *Network of Intersections*: we built a computational market where drivers must acquire the right to pass through the intersections of the urban road network, implementing the intersection managers as competitive suppliers of reservations which selfishly adapt the prices to match the actual demand, and combine the competitive strategy for traffic assignment with the auction-based control policy at the intersection level into an adaptive, market-inspired mechanism for traffic management of reservation-based intersections.

3.1.1. Mechanism for Single Intersection

For a single reservation-based intersection, the problem that an intersection manager has to solve comes down to allocating reservations among a set of drivers in such a way that a specific objective is maximised. This objective can be, for instance, minimising the average delay caused by the presence of the regulated intersection. In this case, the simplest policy to adopt is allocating a reservation to the first agent that requests it, as occurs with the first-come first-served (FCFS) policy proposed by Dresner and Stone in their original work. Another work in line with this objective takes inspiration from adversarial queuing theory for the definition of several alternative control policies that aim at minimising the average delay [25]. However, these policies ignore the fact that in the real world, depending on people's interests and the specific situation that they are in, the relevance of travel time may be judged differently: a business person on his or her way to a meeting, for instance, is likely to be more sensitive to delays than a student cruising for leisure. Since processing the incoming requests to grant the associated reservations can be considered as a process of assigning resources to agents that request them, one may be interested in an intersection manager that allocates the disputed resources to the agents that value them the most. In the sequel, we designed an auction-based policy for this purpose. In line with approaches from mechanism design, we assumed that the more a human driver is willing to pay for the desired set of space–time slots, the more they value the good. Therefore, our policy for the allocation of resources relies on *auctions*.

The first step is to define the resources (or items) to be allocated. In our scenario, the auctioned good is the use of the space inside the intersection at a given time. We modelled an intersection as a discrete matrix of space slots. Let S be the set of the intersection space slots, and T the set of future time-steps, then the set of items that a bidder can bid for is $I = S \times T$. Therefore, differing from other auction-based approaches for intersection management (e.g., [26]), our model of the problem calls for a *combinatorial auction*, as a bidder is only interested in bundles of items over the set I. As Figure 2 illustrates, in the absence of acceleration in the intersection, a reservation request implicitly defines which space slots at which time the driver needs in order to pass through the intersection.

The bidding rules define the form of a valid bid accepted by the auctioneer. In our scenario, a bid over a bundle of items is implicitly defined by the reservation request. Given the parameters *arrival time*, *arrival speed*, *lane*, and *type of turn*, the auctioneer (i.e., the intersection manager) is able to determine which space slots are needed at which time. Thus, the additional parameter that a driver must include in its reservation request is the *value of its bid*, that is, the amount of money that it is willing to pay for the requested reservation. A bidder is allowed to withdraw its bid and to submit a new one. This may happen, for instance, when a driver that submitted a bid b, estimating to be at the intersection at time t, realises that, due to changing traffic conditions, it will more likely be at the intersection at time $t' > t$, thus making the submitted bid b useless for the driver. The rational thing to do in this case, as the driver would not want to risk being involved in a car accident, is resubmit the bid with the updated arrival time. However, we require the new bid to be greater than or equal to the value of the previous one. This constraint avoids the situation whereby a bidder "blocks" one or several slots for itself, by acquiring them early and with overpriced bids.

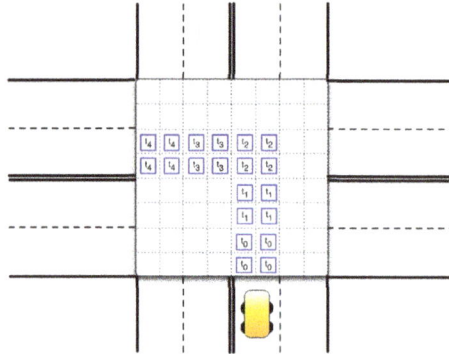

Figure 2. Bundle of items for a reservation request.

Figure 3 shows the interaction protocol used in our approach. It starts with the auctioneer waiting for bids for a certain amount of time. Once the new bids are collected, they constitute the bid set. Then, the auctioneer executes the algorithm for the winner determination problem (WDP), and the winner set is built, containing the bids whose reservation requests have been accepted. During the WDP algorithm execution, the auctioneer still accepts incoming bids, but they will only be included in the bid set of the next round. The auctioneer sends a *CONFIRMATION* message to all bidders that submitted the bids contained in the winner set, while a *REJECTION* message is sent to the bidders that submitted the remaining bids. Then, a new round begins, and the auctioneer collects new incoming bids for a certain amount of time.

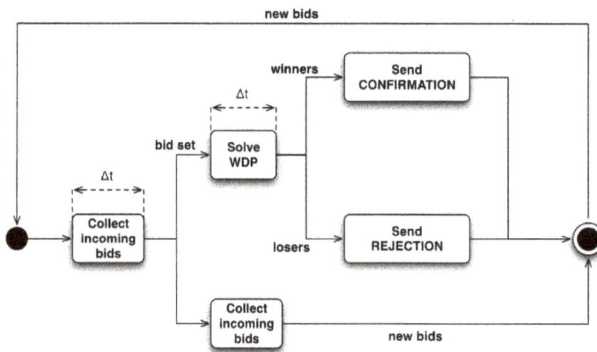

Figure 3. Auction policy.

Notice that the auction must be performed in real time, so both the bid collection and the winner determination phase must occur within a specific time window. This implies that optimal and complete algorithms for the WDP are not suitable. Therefore, we used an approximation algorithm with *anytime* properties (i.e., the longer the algorithm keeps executing, the better the solution it finds) [27].

We expect our policy based on combinatorial auction (CA) to enforce an inverse relation between the amounts spent by the bidders and their *delay* (the increase in travel time due to the presence of the intersection). That is, the more money a driver is willing to spend for crossing the intersection, the faster will be its transit through it. For this purpose, we designed a custom, microscopic, time-and-space-discrete simulator, with simple rules for acceleration and deceleration [27]. The origin O and destination D of each simulated vehicle are generated randomly. The destination implies the type of turn that the vehicle will perform at the intersection, as well as the lane it will use to travel. We created different traffic demands by varying the expected number of vehicles λ that, for every O–D pair, are spawned in an interval of 60 s, using a Poisson distribution. The bid that a driver is willing to submit is drawn from a normal distribution with mean 100 cents and variance 25 cents, so the agents are not homogeneous in the sense that the amount of money that they are offering differs from one to another.

Figure 4 plots (in logarithmic scale) the relation between travel time and bid value for values of $\lambda = 20$, with error bars denoting 9% confidence intervals. It clearly shows a sensible decrease of the delay experienced by the drivers that bid from 100 to 150 cents. The delay reduction tends to settle for drivers that bid more than 1000 cents. Similar results are achieved with lower and higher densities [27]. Notice that even with a theoretically infinite amount of money, a driver cannot experience zero delay when approaching an intersection, as the travel time is influenced by slower potentially "poorer" vehicles in front of it. Extensions to our mechanism that address this problem are subject to future work.

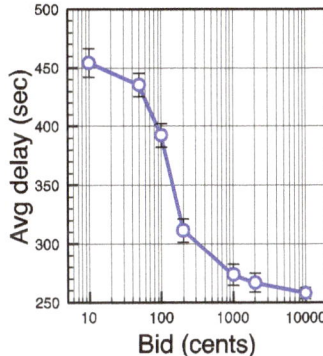

Figure 4. Bid-delay relation.

We also analysed the impact that such a policy has on the intersection's *average* delay, comparing it to the FCFS strategy. Figure 5 shows that when traffic demand is low, the performance of the CA policy and the FCFS is approximately the same but, as demand grows, there is a noticeable increase of the average delay when the intersection manager applies CA. The reason is that the CA policy aims at granting a reservation to the driver that values it the most, rather than maximising the number of granted requests. Thus, a bid b whose value is greater than the sum of n bids that share some items with b is likely to be selected in the winner set. If so, only one vehicle will be allowed to transit, while n other vehicles will have to slow down and try again. When extending the CA mechanism to multiple intersections, we tried to reduce this "social cost" of giving preference to drivers with a high valuation of time.

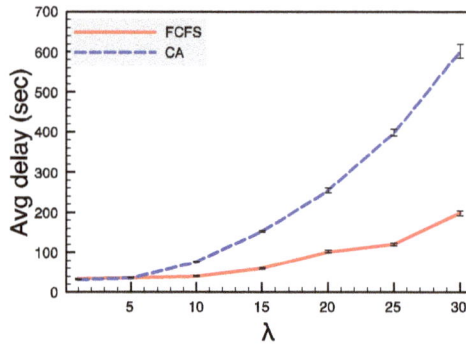

Figure 5. Average delay.

3.1.2. Mechanisms for Multiple Intersections

In the previous section, we analysed the performance of an auction-based policy for the allocation of reservations in the *single* intersection scenario. A driver is modelled as a simple agent that selects the preferred value for the bid that will be submitted to the auctioneer. The decision space of a driver in an urban road network with *multiple* intersections is much broader: complex and mutually dependent decisions must be taken, such as route choice and departure time selection. Therefore, this scenario opens up new possibilities for intersection managers to affect the behaviour of drivers. For example, an intersection manager may be interested in influencing the collective route choice performed by the drivers, using variable message signs, information broadcast, or individual route guidance systems, so as to evenly distribute the traffic over the network. This problem is called *traffic assignment*. In the following, we first evaluate how market-inspired methods can be used within a traffic assignment strategy for networks of reservation-based intersections (*competitive traffic assignment* (CTA) strategy). Then, we combine this traffic assignment strategy with the auction-based control policy into an integrated mechanism for traffic management of urban road networks (CTA–CA strategy). Finally, the performance of the different approaches is evaluated.

The complexity of the problem puts limits to coordination approaches based on cooperative multiagent learning [28]. Therefore, our CTA models each intersection manager as a provider of the resources (in this case, the reservations of the intersection it manages). Each intersection manager is free to establish a price for the reservations it provides. On the other side of the market, each driver is modelled as a buyer of these resources. Provided with the current prices of the reservations, it chooses the route according to its personal preferences about travel times and monetary costs. Each intersection manager is modelled so as to compete with all others for the supply of the reservations that are traded. Therefore, our goal as market designers is making the intersection managers adapt their prices towards a price vector that accounts for an efficient allocation of the resources.

In CTA, for each incoming link l, an intersection manager defines the following variables:

- Current price $p^t(l)$: the price applied by the intersection manager to the reservations sold to the drivers that come from the incoming link l.
- Total demand $d^t(l)$: the total demand of reservations from the incoming link l that the intersection manager observes at time t, given the current price $p^t(l)$, i.e., the number of vehicles that intend to cross the intersection coming from link l at time t.
- Supply $s(l)$: the reservations supplied by the intersection manager for the incoming link l. It is a constant and represents the number of vehicles that cross the intersection coming from link l that the intersection manager is willing to serve.
- Excess demand $z^t(l)$: the difference between total demand at time t and supply, i.e., $d^t(l) - s(l)$.

We define the price vector p^t as a vector that comprises all prices at time t (i.e., the prices applied by all intersection manager to each of its controlled links). In particular, we say that a price vector p^t maps the supply with the demand if the excess demand $z^t(l)$ is 0 for all links l of the network. This price vector, which corresponds to the market equilibrium price, can be computed through a Walrasian auction [29], where each buyer (i.e., driver) communicates to the suppliers (i.e., intersection managers) the route that it is willing to choose, given the current price vector p^t. With this information, each intersection manager computes the demand $d^t(l)$, as well as the excess demand $z^t(l)$, for each of its controlled links. Then, each intersection manager adjusts the prices $p^t(l)$ for all the incoming links, lowering them if there is excess supply ($z^t(l) < 0$) and raising them if there is excess demand ($z^t(l) > 0$). The new price vector p^{t+1} is communicated to the drivers that iteratively choose their new desired route on this basis. Once the equilibrium price is computed, the trading transactions take place and each driver buys the required reservations at the intersections that lay on its route.

In order to adapt the Walrasian auction to the traffic domain, we implemented a pricing strategy that aims at reaching the equilibrium price but works on a continuous basis, with drivers that leave and join the market dynamically, and with transactions that take place continuously. To reach general equilibrium, each intersection manager applies the following *price update rule*: at time t, it independently computes the excess demand $z^t(l)$ and updates the price $p^t(l)$ as follows:

$$p^{t+1}(l) = max\left[\delta, p^t(l) + p^t(l)\frac{z^t(l)}{s(l)}\right]$$

where δ is the minimum price that the intersection manager charges for the reservations that it sells. As drivers that travel through road network links with low demand shall not incur any costs, for the CTA strategy, we choose $\delta = 0$.

The *integrated mechanism for traffic management* (CA–CTA) combines the competitive traffic assignment strategy (CTA) with the auction-based policy (CA). Since the intersection manager is the supplier of the reservations that are allocated through the combinatorial auction, it may control the *reserve price* of the auctioned reservations, that is, the minimum price at which the intersection manager is willing to sell. At time t, for each link l, CTA–CA simply sets this reserve price to the price $p^t(l)$ computed by the price update rule of the CTA strategy.

The experimental evaluation of the strategies was performed on a hybrid mesoscopic-microscopic simulator, where traffic flow on road segments was modelled at mesoscopic level, while traffic flow inside intersections was modelled at microscopic level. Although our work does not depend on the underlying road network, we chose a topology inspired by the urban road network of the city of Madrid for our empirical evaluation (see Figure 6). The network is characterised by several freeways that connect the city centre with the surroundings and a ring road. Each large dark vertex in Figure 6, if it connects three or more links, is modelled as a reservation-based intersection. In the experiments, we recreated a typical high-load situation (i.e., the central, worst part of a morning peak), with more than 11,000 vehicles departing within a time window of 50 min (see [27] for details).

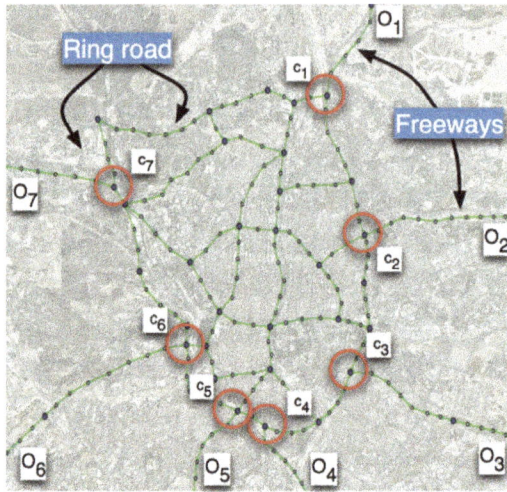

Figure 6. Urban road network.

We aimed at comparing the performance of FCFS, CTA, and CTA–CA. In FCFS, each intersection manager performs combinatorial auctions (without reserve price) in isolation. In this case, the drivers' route choice model simply selects the route with minimum expected travel time at free flow, since there is no notion of price. For the other strategies, we assumed that drivers choose the most preferred route they can afford. Since the prices of links are changing dynamically, a driver continuously evaluates the utility of the route it is following and, in the case that a different route becomes more attractive, it may react and change on-the-fly how to reach its destination, selecting a route different from the original one.

To assess the social cost incurred by CA–CTA at the global level, we measured the moving average of the travel time, that is, how the average travel time of the entire population of drivers, computed over all the O–D pairs, evolves during the simulation. The results, with 95% confidence interval error bars, are plotted in Figure 7. In the beginning, the average travel time is similar for all the scenarios, but as the number of drivers that populate the network (i.e., its load) increases, it grows significantly faster with FCFS than with the CA–CTA policy. In terms of average travel times, CTA is the best performing policy. CA–CTA has a slightly inferior performance, but it can be shown that it enforces an inverse relationship between bid value and delay, similar to the results presented in the previous section [27]. The fact that both CA–CTA and CTA outperform FCFS is an indication that, in general, a traffic assignment strategy (the "CTA" component of both policies) improves travel time. In fact, FCFS drivers always select the shortest route, which in some cases is not the best route choice. Furthermore, granting reservations through an auction (the "CA" component of the CA–CTA policy) ensures that bid value and delay reduction are correlated.

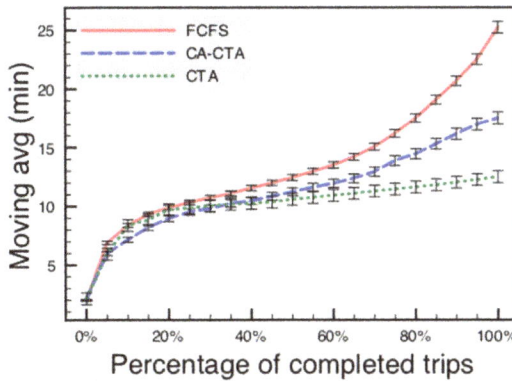

Figure 7. Moving average of travel times.

3.2. Evacuation Coordination in Smart Building

The objective of an evacuation is to relocate evacuees from hazardous to safe areas while providing them with safe routes. Present building evacuation approaches are mostly static and preassigned (e.g., [30]). Frequently, no coordination is available except for predefined evacuation maps. Still, due to the lack of overall evacuation network information, there might be casualties caused by too slow an evacuation on hazardous routes. *Real-time* route guidance systems, which dynamically determine evacuation routes in inner spaces based on the imminent or ongoing emergency, can help reduce those risks. Chen and Feng in [31] proposed two heuristic flow control algorithms for a real-time building evacuation with multiple narrow doors: with no limitation on the number of evacuation paths and k required evacuation paths, respectively. Filippoupolitis and Gelenbe in [32] proposed a distributed system for the computation of shortest evacuation routes in real time. The routes are computed by decision nodes and are communicated to the evacuees located in their vicinity. However, this approach considers only the physical distance and the hazard present in each link and does not take into consideration crowd congestion on the routes.

A dynamic, *context-sensitive* notion of route *safety* is a key factor for such recommendations, particularly because herding and stampeding behaviours may occur at potential bottlenecks, depending on (among other factors) the amount of people who intend to pass through them. Furthermore, smart devices allow guidance to be *personalised*, taking into account, for instance, the specific circumstance of the elderly, disabled persons, or families. In such settings, an adequate notion of *fairness* of evacuation route recommendations is of utmost importance to assure the trustworthiness of the system from the standpoint of its users [33]: the guidance should not only achieve good overall performance of the evacuation process but must also generate proposals for each of its users that each of them perceive as efficient. Finally, large groups of people may need to be evacuated, so *scalability* plays a key role.

Our proposal concentrates on real-time situation-aware evacuation guidance in smart buildings. The system aims at assigning efficient evacuation paths to individuals based on their mobility limitations, initial positions, respecting individual privacy, and other evacuation requirements. In our approach, a network of smart building agents calculates individual routes in a decentralised fashion. Complex event processing, semantic technologies, and distributed optimisation techniques are used to address this problem. In addition, we use the notion of agility to determine *robust* routes, in the sense that they are not only fast but also allow finding acceptable alternatives in case of upcoming contingencies.

We rely on the existence of a rather extensive set of possible evacuation routes, which may be determined by evacuation experts or through some automated online or offline process. The different evacuation routes are stored in an emergency ontology that, together with an ontology describing the topological structure of the building, specifies the a priori knowledge of our system. In addition, situational knowledge about the current situation in each moment of the building and of the evacuees is generated in real time through a network of sensors. This dynamic knowledge is merged with the static knowledge about the infrastructure. In an emergency situation, semantic inference is used to select the most appropriate agile evacuation route for each individual in the building. Furthermore, real-time monitoring allows the system to reroute evacuees in case of contingencies and, thus, to propose evacuation routes that are adaptive to unpredictable safety drops in the evacuation network.

3.2.1. Distributed Architecture

The objective of the evacuation route guidance architecture (ERGA) is to provide individualised route guidance to evacuees over an app on their smartphones based on the evacuation information received from connected smartphones within the building and the building sensor network. However, even if an evacuee does not have a smartphone available, she could still receive information on relevant evacuation directions, for example, through LED displays on the walls of a smart building.

ERGA (Figure 8) consists of user agents (UAs) and a network of smart building (SB) agents.

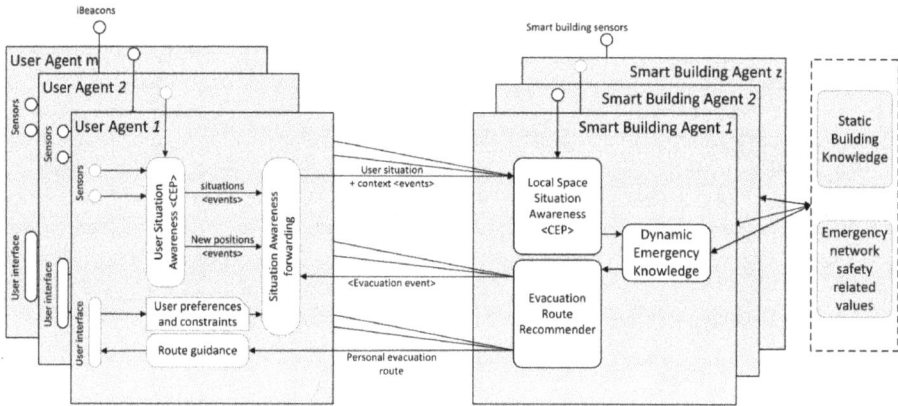

Figure 8. Situation-aware real-time distributed evacuation route guidance architecture (ERGA). User agents 1, 2, and m are located in the physical space of smart building (SB) Agent 1 so that they are given route recommendations by SB Agent 1.

User agents. The user agent is associated with the application on a smartphone of an evacuee. It manages and stores all the information that is related to a specific evacuee in the building. Here, we assume that people that enter the building own a smartphone with the evacuation app installed, or they have been provided with some smartphone-like device that runs the app when they start to evacuate. The user agent contains three modules: (i) user preferences and constraints, (ii) user situation awareness module, and (iii) route guidance module.

The *user preferences and constraints* module allows defining constraints such as disabilities (e.g., the use of wheelchair or vision impairment), as well as evacuation-related behavioural disorders (e.g., agoraphobia, social phobia, etc.), while the preferences include the affiliate ties with other users of the building. The *user situation awareness* module exploits sensor data (from the smartphone and building) and reasons about the behaviour and location of the user. The presence of an evacuee, together with the information derived from the situation awareness module and the individual preferences and constraints, are passed to the closest SB agent. In order to assure privacy, only certain basic data about the user's situation should be forwarded to the SB agent (e.g., location, running events). In case of an emergency evacuation, the *user interface* provides the user with personalised navigation guidelines for evacuation.

Smart building agents. Situation awareness and decision making are distributed in the network of SB agents such that each agent is responsible for the semantic reasoning concerning the safety of its assigned physical space, as well as the evacuation route computation for the evacuees positioned in its physical space. We assumed that each SB agent has at its disposal the information regarding the evacuation network's layout, topology, and safety.

A single SB agent controls only its own physical space. The number and location of SB agents is defined when the system is installed. Each SB agent has a corresponding region (Voronoi cell) consisting of all user agents closer to that SB agent than to any other SB agent. Each SB agent contains a *local space situation awareness* module that perceives the safety conditions of the physical space it controls through combining and analysing the events provided by the sensors and individual user agents located within the smart space it controls. Moreover, each SB agent communicates with its neighbouring SB agents and with the user agents present within its physical space.

The *local space situation awareness* module functions in cycles. At the first phase, the local building sensor data is fused with the data from the locally present user agents. Then, the safety value is deduced. This data is sent to a blackboard or similar globally shared data structure containing the overall network safety values and is visible to all agents. When an SB agent detects an emergency situation, it sends the updated safety value of its physical space to the shared blackboard. This allows, on the one hand, to monitor the real-time situation of the building and, on the other hand, to trigger an evacuation process and to execute control actions in such a process. SB agent's *evacuation route recommender* module computes optimised evacuation routes for each locally present user agent by distributed computation and communication with the rest of the SB agents in a multi-hop fashion. In this process, the algorithm uses: (i) data regarding the building topology, (ii) general knowledge about emergency and evacuation scenarios (e.g., facts that people with strong affiliate ties should always be evacuated together, the appropriateness of certain routes for people with limited mobility, and the influence of certain events like fire and smoke on the security level), and (iii) the current physical space situation awareness of the SB agent itself, as well as regarding the evacuees that are currently in its space and evacuation network's safety values.

During evacuation, the global safety situation of the building is dynamically updated in real time and each SB agent recalculates the evacuation routes if necessary.

3.2.2. Situation Awareness

We assumed the existence of data provided by a smart infrastructure, as well as by the users currently in the building. In particular, we required information for identifying the location of each user in the building.

There are various technological techniques to localise people in buildings. Measuring the strength of the signal of several Wi-Fi access points could be used to calculate a person's location via trilateration. However, the signal strength is easily affected by the environment (obstacles, users, etc.) making it very difficult to obtain accurate positions. Another option is using Radio-frequency identification (RFID) technology, but a lot of expensive readers would need to be installed in the building, and there are also similar trilateration problems to those for Wi-Fi. In addition, it would require providing an RFID tag to each person in the building. We opted for using beacons, a recent technology to support indoor navigation. Beacons are cheap devices that emit Bluetooth signals, which can be read by beacon readers, in particular, smartphones. Beacons send, among other information, a unique ID that allows identifying the specific sensor the user is near to, thus providing accurate user location.

Besides user location, other infrastructure sensors provide different measures, such as temperature, smoke, fire, and so on. In addition, the users' smartphones built-in sensors provide information that allows detecting their activity (e.g., if the person carrying the phone is running).

Sensor events (each piece of information forwarded by or read from a sensor) are processed using complex event processing (CEP), a software technology to extract the information value from event streams [34]. CEP analyses continuous streams of incoming events in order to identify the presence of complex sequences of events (event patterns). Event stream processing systems employ "sliding windows" and temporal operators to specify temporal relations between the events in the stream. The core part of CEP is a declarative event processing language (EPL) to express event processing rules. An event processing rule contains two parts: a condition part describing the requirements for firing the rule, and an action part that is performed if the condition matches. An event processing engine analyses the stream of incoming events and executes the matching rules.

UAs exploit sensor data and infer the location and behaviour of their user. For example, data read from beacons is introduced as events of type *beaconEvent(beaconID)*. Then, the subsequent CEP rule creates *enteringSection* and *leavingSection* events, meaning that the user is entering and leaving a certain space, respectively. The rule describes the situation that a new *beaconEvent b2* has been read in the phone, where the beacon ID has changed. The symbol "->" indicates that event *b1* occurs before event *b2*.

> CONDITION:
>
> beaconEvent AS b1 -> beaconEvent AS b2 ∧
>
> b1.id <> b2.id
>
> ACTION:
>
> CREATE enteringSection(userID, b2)
>
> CREATE leavingSection(userID, b1)

enteringSection and *leavingSection* events, as well as others, like *runningEvent* (generated by a CEP rule that checks if the average velocity of the user is higher than 5 km/h for the last 10 s) are forwarded to the SB agent monitoring the user's location area.

SB agents receive processed events from the UA in their area. That information, as well as that obtained from smart building sensors, is incorporated into a stream of events. Again, the event stream is processed by the CEP engine generating more abstract and relevant situation awareness information. For instance, a panic event can be inferred if more than 40% of persons in a certain section of the building emit a running event.

Finally, situation awareness information, in the form of events, is then transformed into a semantic representation, namely RDF facts. Afterwards, the situation information is ready to be consumed by semantic inference engines. We used OWL ontologies to represent information semantically in our system (user preferences, building topology, emergency knowledge, building situation). Semantic representations provide the means to easily obtain inferred knowledge. For example, if we define a class *DisabledPerson* to represent people with at least one disability, then we can infer disabled people even though they have not been explicitly described as instances of that class. For more complex reasoning tasks, we use rules on top of our OWL ontologies, which typically add new inferred knowledge. In particular, we use rules to determine the accessibility of certain sections in the building, and to select possible evacuation routes.

3.2.3. Personalised Route Recommendation

Our aim is to safely evacuate all the evacuees (or at least as many as possible) within an allotted upper time limit. This limit is usually given by the authorities in charge of evacuation.

Initially, we rely on the existence of a set of *predefined evacuation routes*. This set is independent of user constraints. The set of routes is analysed with the objective of generating *personalised efficient evacuation routes*, that is, sets of alternative routes for each particular user, considering the current situation of the building and user constraints (e.g., wheelchair, blind, kids, etc.). This is carried out in two steps. First, those routes that are not *time-efficient* (e.g., their expected evacuation times are not within the time limit) are filtered out. Next, using a rule-based system, *safe personalised plans* for each user are created. These routes only include traversing sections that are accessible for that particular user (e.g., avoiding paths through staircases if the person uses a wheelchair). Semantic rules and OWL reasoning are used in this task. For example, the following Jena (http://jena.apache.org) rule identifies staircase sections that are not accessible for people in a wheelchair:

(*?user:hasDisability:Wheelchair*)

(*?section rdf:type:Staircase*)

->

(*?section:notAccessibleFor?user*)

The *personalised efficient evacuation routes* need to be ranked so as to select one route for each person in the building. We represent the evacuation network by a directed graph $G = (N, A)$, where N is a set of n nodes representing sections, and A is the set of m arcs $a = (i, j)$, $i, j \in N$ and $i \neq j$, representing walkways, doors, gateways, and passages connecting sections i and j. Let $O \subseteq N$ and $D \subseteq N$ be a set of all evacuees' origins and safe exit destinations, respectively. We modelled the evacuation as a unified crowd flow, where each individual is seen as a unit element (particle) of that flow and the objective is to maximise the flow of demands (evacuation requests) with certain constraints. We considered travel time optimisation with path safety, envy-freeness (fairness), and agile paths.

Route safety. Our objective is to safely evacuate as many evacuees as possible from all origins o $\in$ O over the safest and the most efficient evacuation paths to any of the safe exits d $\in$ D. Let us assume that safety status S_a is given for each arc $a \in A$ as a function of safety conditions that can be jeopardised by a hazard. Safety can be calculated from sensor data (e.g., temperature, smoke, etc.), and using space propagation models and aggregation functions to combine different influences and variables measured. A thorough description of this field can be found in, for example [35,36]. We normalised it to the range [0, 1], such that 1 represents perfect conditions while 0 represents conditions impossible for survival, with a critical level for survival $0 < S^{cr} < 1$ depending on the combination of the previously mentioned parameters.

If each constituent arc $a \in k$ of a generic path k has safety $S_{a \in k} \geq S^{cr}$, then path k is considered to be safe. On the contrary, a path is considered unsafe and its harmful effects may threaten the evacuees' lives. The proposed evacuation paths should all satisfy safety conditions $S^k \geq S^{cr}$. However, when such a path is not available, a path with the maximal safety should be proposed where the travel time passed in the safety-jeopardised areas should be minimised. Since safety may vary throughout a path, here, we introduce a normalised path safety that balances the minimal and average arc safety values:

$$S^k = \sqrt[|a \in k|]{\prod_{a \in k} S_a}, \ \forall k \in P_o, \ o \in O$$

where P_o is the set of simple paths from origin $o \in O$ to an exit.

Fair route recommendation. An adequate notion of fairness of evacuation route recommendations is important to assure the trustworthiness of the system from the evacuees' viewpoint [33]: the guidance should not only achieve good overall performance of the evacuation process but must also generate evacuation routes for each of the evacuees that each of them perceives as efficient and fair. For example, if there are two close-by evacuees at some building location, they should be offered the same evacuation route or, if not possible, then routes with similar safety conditions and evacuation time.

We aim at proposing available safe simple paths with a maximised safety acceptable in terms of duration in free flow for each evacuation origin. By acceptable in terms of duration in free flow, we mean the paths whose traversal time in free flow is within an upper bound in respect to the minimum free flow duration among all the available evacuation paths for that origin.

The concept of envy-free paths was introduced in [37]. Basically, it defines a path allocation to be *α-envy-free* if there is no evacuee at origin o' that envies any other evacuee at origin o for getting assigned the path with a lower duration than the αth power of the path duration assigned to the evacuee on o'.

Agile routes. When an unpredicted hazard occurs on a part of the evacuation route, it becomes unsafe and impassable. If, in the computation of an evacuation route, we did not consider this fact and the related possibility to reroute to other efficient evacuation routes on its intermediate nodes, then, in case of contingency, rerouting towards safe areas might be impossible, causing imminent fatalities of evacuees. A similar case may occur if, for example, a high flow of evacuees saturates an evacuation path and causes panic. Therefore, we prefer routes where each intermediate node has a sufficient number of dissimilar efficient evacuation paths towards safe exits, if possible, within the maximum time of evacuation given for a specific emergency case. In that respect, evacuation centrality is defined in [38] as follows.

Evacuation centrality $C_\varepsilon(i)$ of node i is a parameter that represents the importance of node i for evacuation. The value of the evacuation centrality of the node is the number of available, sufficiently dissimilar, time-efficient evacuation paths from that node i towards safe exits.

Once we find the evacuation centrality measure for each node of the graph, the objective is to find an evacuation path that maximises the overall value of the intermediate nodes' centrality measures. We call every such path *agile evacuation path*: a path where an evacuee has higher chances to reroute in case of contingency in any of the intermediate nodes or arcs. *Path agility* $\Delta(k)$ is defined as:

$$\Delta(k) = \sqrt[|(i,j)\in k|]{\prod_{(i,j)\in k} C_\varepsilon(j)}$$

Since we are not concerned about the number of arcs in the path, we take the $|(i,j) \in k|$th root of the Nash product in this formula. We recommend the evacuation paths with the highest agility to evacuees and recompute this value every time the safety and/or congestion conditions change along the recommended path.

3.3. Emergency Medical Service Coordination

The domain of medical assistance includes many tasks that require flexible on-demand negotiation, initiation, coordination, information exchange, and supervision among different involved entities (e.g., ambulances, emergency centres, hospitals, patients, physicians, etc.). In addition, in the case of medical urgencies, the need for fast assistance is evident. It is of crucial importance for obtaining efficient results, improving care, and reducing mortality, especially in the case of severe injuries. Out-of-hospital assistance in medical urgencies is usually provided by emergency medical assistance (EMA) services, using vehicles (typically ambulances) of different types to assist patients appearing at any location in a given area. In such services, the coordination of the available resources is a key factor in order to assist patients as fast as possible. The main goal here is to improve one of the key performance indicators: the response time (the time between a patient starting to call an EMA service centre and the moment medical staff, e.g., an ambulance, arrives at his location and the patient can receive medical assistance).

One way to reduce response times consists of reducing the part that depends on the logistic aspects of an EMA service through an effective coordination of the assistance vehicle fleet (for simplicity, here we assume a fleet of ambulances). In this regard, there are two principal problems EMA managers are faced with: the assignment or allocation of ambulances to patients and the location and redeployment of the ambulance fleet. The assignment or allocation problem consists of determining, at each moment, which ambulance should be sent to assist a given patient. Moreover, the location and redeployment consist of locating and possibly relocating available ambulances in the region of influence in a way that new patients can be assisted in the shortest time possible.

Most of recent works for coordinating ambulance fleets for EMA have been dedicated to the redeployment problem. A lot of work has concentrated on the dynamic location of ambulances, where methods are proposed to redeploy ambulances during the operation of a service in order to take into account the intrinsic dynamism of EMA services (e.g., [39–41]). Most proposals on dynamic redeployment of ambulances only consider the possibility to relocate ambulances among different, predefined sites (stations). This requirement is relaxed in the work proposed in [42], where a number of ambulances can be relocated to any place in the region. Regarding dispatching strategies (the patient allocation problem), most works use the "nearest available ambulance" rule for assigning ambulances to patients in a first-came first-served manner. Some works analyse priority dispatching strategies to account for severity level of patients [42,43].

In our previous work [44], we proposed a system that re-allocates ambulances to patients and redeploys available ambulances in a dynamic manner in order to reduce the average response times. Our redeployment approach differs from others in the sense that it does not try to maximise the zones that are covered in a region, with respect to some time limits. Instead, the approach is based on geometric optimisation that tends to optimise, in each moment, the positions of all ambulances that are still available such that the expected arrival time to potential new emergency patients is minimised. Here, with regard to the allocation of patients to ambulances, we propose a dynamic approach similar to [45] but, instead of optimising the global travel times of all ambulances, we concentrate only on the sum of the arrival times of ambulances to the pending emergency patients. This system is summarised in this section.

We use the following notation to describe the problem and to present our solution. The set of ambulances of an EMA service is denoted by $A = \{a_1, \ldots, a_n\}$, where n is the cardinality of A. Even though most EMA services employ different types of ambulances, for reasons of simplicity, we just consider a single type. Each ambulance has a position and an operational state which vary during time. $p(a_i)$ and $s(a_i)$ denote the current position and the current state of ambulance a_i, respectively. The position refers to a geographical location, and the state can be one of the following:

- *assigned*: An ambulance that has been assigned to a patient and is moving to the patient's location.
- *occupied*: An ambulance that is occupied, either attending a patient "in situ" or transferring him/her to a hospital.
- *idle*: An ambulance that has no mission in this moment.

We denote the sets of available, occupied, and idle ambulances at a given moment by A^A, A^O, and A^I.

Regarding the patients, $P = \{p_1, \ldots, p_m\}$ denotes the current set of unattended patients in a given moment (e.g., patients that are waiting for an ambulance), where m is the cardinality of P. Each patient $p_j \in P$ has a location (denoted by $p(p_j)$). We assume that patients do not move while they are waiting for an ambulance, thus $p(p_j)$ is constant. Furthermore, once an ambulance has reached a patient's location in order to provide assistance, this patient is removed from P.

3.3.1. Dynamic Reassignment

The ambulance allocation problem consists of finding an assignment of (available) ambulances to the emergency patients that have to be attended to. In current EMA services, mostly a priority dispatching strategy is used, where patients are assigned in a sequential order of appearance and patients with a higher severity level are assigned first. In each case, usually the nearest idle ambulance $a_i \in A^I$ is assigned. This can be seen as a first-call first-served (FCFS) rule, where patients with the same security level that called first are also assigned first to an ambulance. After an ambulance has been assigned to a patient, this assignment is usually fixed.

The FCFS approach is not always optimal from a global perspective. First, if more than one patient has to be attended, it is not optimal in the sense that is does not minimise the response times to all patients. Furthermore, the dynamic nature of an EMA system implies that a given assignment of ambulances to patients at one point in time might not be optimal at a later point, for example, if new patients appear or if an ambulance that was occupied before is becoming available again.

In order to reduce the average arrival time in the dynamic environment of an EMA service, the assignments of ambulances to patients could be optimised globally and the assignments should be recalculated whenever relevant events take place and a better solution may exist. Based on this idea, we propose a dynamic assignment mechanism of ambulances to patients, which optimises the assignments at a given point in time and recalculates optimal assignments when the situation changes.

Given a set of patients to be attended P and a set of ambulances that are not occupied $A^A \cup A^I$ at a specific moment, the optimal assignment of ambulances to patients is a one-to-one relation between the sets $A^A \cup A^I$ and P, that is, a set of pairs $AS = \{<a_k,p_l>,<a_s,p_q>, \dots \}$ such that the ambulances and the patients are all distinct, and that fulfils the following conditions:

- The maximum possible number of patients is assigned to ambulances, that is:

    $\forall p_j \in P: \exists <a_i,p_j> \in AS$ if $n \geq m$ and $\forall a_i \in A^A \cup A^I: \exists <a_i,p_j> \in AS$ if $n < m$
- The total expected travel time of the ambulances to their assigned patients:

    $$\sum_{<a_i,p_j> \in AS} ETT(p(a_i), p(p_j)) \text{ is minimised}$$

$ETT(x,y)$ denotes the expected travel time for the fastest route from one geographical location x to another location y.

Calculating such an optimal assignment is a well-known problem which can be solved in cubic time, such as with the Hungarian method [46] or with Bertsekas' auction algorithm [47]. We propose to use the second approach because it has a naturally decentralised character and can be optimised in settings, as the one analysed here.

An optimal assignment AS at a moment t, due to the dynamic nature of an EMA service, might become suboptimal at a time t' ($t' > t$). The following cases need to be considered:

1. One or more new patients require assistance: in this case, the set of patients that have to be attended changes and the current assignment AS may not be optimal any more.
2. Some ambulances that have been occupied at time t have finished their mission and are idle at time t'. These ambulances could eventually improve the current assignment.

Based on this analysis, we propose a dynamic system based on an event-driven architecture that recalculates the global assignment whenever one of the following events occur: $newPatient(p_j)$ (a new patient has entered the set P) or $ambFinishedEvent(a_i)$ (an ambulance that was occupied before is becoming idle again). In the recalculation of an existing assignment, ambulances that have been already dispatched to a patient, but did not reach the patient yet, may be deassigned from their patients or might be reassigned to other patients. This approach assures that the assignment AS is optimal, with regard to the average travel time to the existing patients, at any point in time.

3.3.2. Dynamic Redeployment

The second part of the proposed coordination approach for emergency medical services (EMS) consists of locating and redeploying ambulances in an appropriate manner. Here, the objective is to place ambulances in such a way that the expected travel time to future emergency patients is minimised.

We addressed this problem by using Voronoi tessellations [48]. A Voronoi tessellation (or Voronoi diagram) is a partition of a space into a number of regions based on a set of generation points such that for each generation point there will be a corresponding region. Each region consists of the points in the space that are closer to the corresponding generation point than to any other. Formally, let $\Omega \in R^2$ denote a bounded, two-dimensional space and let $S = \{s_1, \dots, s_g\}$ denote a set of generation points in Ω. For simplicity, let Ω be a discrete space. The Voronoi region V_i corresponding to point s_i is defined by:

$$V_i = \{y \in \Omega : |y - s_i| < |y - s_j| \text{ for } j = 1, \dots, k \text{ and } j \neq i\}$$

where $|\cdot|$ denotes the Euclidean norm. The set $V(S) = \{V_1, \dots, V_k\}$ with $\cup_{i=1}^{k} V_i = \Omega$ is called a Voronoi tessellation of S in Ω. A particular type of tessellation is centroidal Voronoi tessellation (CVT). A centroidal Voronoi tessellation is one where each generation point s_i is located in the mass centroid of its Voronoi region with respect to some positive density function ρ on Ω. A CVT is a necessary condition for minimising the cost function and, thus, provides a local minimum for the following cost function:

$$F(S) = \sum_{V_i \in V(S)} \sum_{y \in V_i} \rho(y)|y - s_i|^2 \tag{1}$$

A common approach to calculate CVTs and, thus, to minimise (1) is the algorithm proposed by Lloyd [49]. The algorithm is an iterative gradient descent method that finds a set of points S that minimises $F(S)$ in each iteration and converges to a local optimum. Lloyd's algorithm can be summarised in the following steps:

1. Select an initial set S of k sites in Ω.
2. Calculate the Voronoi regions V_i for all generation points $s_i \in S$.
3. Compute the mass centroid of each region V_i with respect to the density function q. These centroids compose the new set of points S.
4. If some termination criteria are fulfilled, finish; otherwise return to Step 2.

Lloyd's algorithm is not assured to find a global minimum, but it finds solutions of reasonable quality very fast—after a few iterations. Therefore, we applied Lloyd's algorithm to find suboptimal positions of ambulances. The application is straightforward: Ω represents the region of interest, and the set of generation points S represents the set of all idle ambulances and their positions. Each Voronoi region V_i corresponds to the area that is covered by ambulance s_i (e.g., the area for which s_i is the closest ambulance in case any patient requires help). Furthermore, as the density function ρ, we use an estimation of the probability distribution of the appearance of a patient in any point in the region of interest. In particular, to define ρ, we divide the region of interest in a set of equally sized cells $C = \{c_1, \dots, c_u\}$, where u is the cardinality of C. That means we discretise the region Ω into u points, each of which represents the centre of one of the cells c_i. Then, we estimate, for each cell c_i, the conditional probability p_{ci} that a new emergency patient will appear in this cell. With this setting, Lloyd's algorithm finds a distribution of ambulances that minimises (1). In particular, in our case, this is a reasonable approximation for minimising the expected distance and, thus, the arrival time, to future emergency patients.

The probabilities p_{ci} can be obtained by tracking historical data on emergency cases. Furthermore, it is possible to model different situations (like seasons, day of the week, time interval, etc.) through different probability distributions.

In the application of Lloyd's algorithm, at a given point in time, we used the positions of all idle ambulances as the set of initial generation points S. We applied a fixed number of iterations (in the experiments we used 50) and the resulting set S represents the "recommended" distribution of ambulances at this particular moment. The new positions are sent to the ambulances, and the ambulances are situated such that they can move towards that positions. Because of road and parking conditions, the obtained positions are considered as indications of an area. That is, once given such an area, the ambulance driver will decide which is the most appropriate waiting location in that area.

We used the Euclidean norm as a distance measure to generate the Voronoi regions for the ambulances. While in a real traffic scenario, as is our case, the Euclidean distance is a rather imprecise approximation of real distances on the road network, from a global perspective, and assuming a rather homogeneous connection between different points of the region of interest (as is usually the case in many urban areas), the Euclidean norm seems to work reasonably well for our purposes. Furthermore, using "road-network distances" when calculating the Voronoi regions is a rather complicated task that would increase the computation complexity considerably.

Similar to the ambulance assignment problem, the dynamic nature of an EMA service implies that the optimal positions of the idle ambulances will change when changes in the environment occur. In order to cope with such changes, the ambulance positions are recalculated dynamically whenever any of the following events occur:

- An ambulance that was assigned to a patient has been deassigned.
- An ambulance has finished a patient assistance mission and has changed its state from occupied to idle.
- An ambulance that was idle has been requested to assist a patient. It changes its state from idle to assigned.
- The situation that determines the underlying probability distribution changes (e.g., a distribution for a new time interval, day of the week, or season should be used).

3.3.3. Experimental Results

We tested the effectiveness of the dynamic reassignment and redeployment approaches in different experiments simulating the operation of SUMMA112, the EMA service provider organisation in the Autonomous Region of Madrid in Spain. We used a simulation tool that allows for a semi-realistic simulation of intervals of times of the normal operation of an EMA service. The tool reproduces the whole process of attending emergency patients, beginning with their appearance and communication with the emergency centre, the schedule of an ambulance, the "in situ" attendance, and, finally, the transfer of the patients to hospitals. The simulator operates in a synchronised manner based on a stepwise execution, with a step frequency of 5 s. That is, every 5 s, the activities of all agents are reproduced, leading to a new global state of the system. In the simulations, we are mainly interested in analysing the movements of ambulances and the subsequent arrival times to the patients. The movements are simulated using an external route service to reproduce semi-realistic movements on the actual road network with a velocity adapted to the type of road. External factors, like traffic conditions or others, are ignored. The duration of the phone call between a patient and the emergency centre and the attendance time "in situ" is set to 2 and 15 min, respectively. As the area of consideration, we used a rectangle of 125 × 133 km that covers the whole region of Madrid. For calculating the probability distribution of upcoming patients, we divided the region into cells of size 1300 × 1300 m. A different probability distribution is estimated for each day of the week and each hour from statistical data (patient data from the whole year of 2009). We used 29 hospitals (all located at their real positions) and 29 ambulances with advanced life support (as it was used by SUMMA112 in 2009) and we simulated the operation of the service for 10 different days (24 h periods) with patient data from 2009 (1609 patients in total). The days were chosen to have a representation of high, medium, and low workloads. We only accounted for so-called level 0 patients, (e.g., patients with a live threatening situation).

We compared two approaches:

- SUMMA112: The classical approach (used by SUMMA112). Patients are assigned to the closest ambulances using a fixed FCFS strategy. Furthermore, ambulances are positioned on fixed stations (at the hospitals), waiting for missions. After finishing any mission, the ambulances return to their station.
- DRARD: In this case, the dynamic reassignment and redeployment methods are employed. With regard to dynamic re-employment, idle ambulances only move to a new recommended position if it is further away than 500 m. This is to avoid short, continuous movements.

Table 1 presents the average arrival times (in minutes) obtained with the two models in simulations for the 10 selected days. As the results show, the use of the DRARD approach provides a considerable improvement (between around 10% and 20%). If we look at all 1609 patients, the average times are 11:45 and 9:54 min, respectively, which implies an improvement of 15.8%.

Table 1. Average arrival times in minutes for 10 different days.

Day/#Patients	1/221	2/152	3/199	4/124	5/137
SUMMA	11:42	11:52	11:03	11:23	11:50
DRARD	9:46	9:50	9:29	9:39	10:09
Improvement %	16.6	15.7	12.8	9.3	14.0
Day/#Patients	6/175	7/96	8/160	9/144	10/201
SUMMA	12:30	12:51	12:42	10:11	11:49
DRARD	10.51	9:48	10:05	9:05	10:08
Improvement %	13.1	23.7	20.7	10.8	14.3

In Figure 9 we present the results of the distribution of arrival times for the different approaches for all 1609 patients of the 10 selected days. The patients in each curve are ordered by increasing arrival time. A clear difference can be observed between the DRARD method, with respect to the current operation model of SUMMA112. The results are clearly better for almost all arrival time ranges. Furthermore, the most important improvements can be observed in the range of higher times. This is a very positive effect because it assures that more patients can be attended to within given response time objectives. For example, out of the 1609 patients, 1163 (72.3%) are reached within 14 min with the SUMMA model, whereas this number increases to 1356 patients (84.3%) with DRARD.

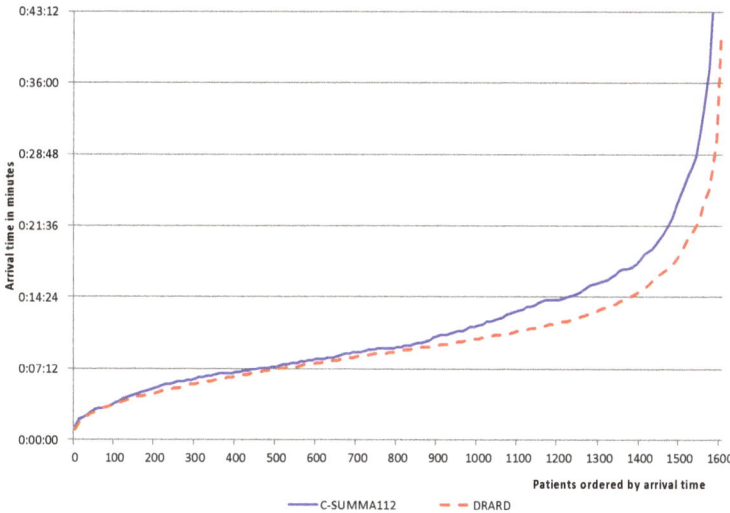

Figure 9. Comparison of arrival times to patients. Here, the 1609 patients from the 10 analysed days are ordered with respect to the arrival time in each curve.

As shown in the results, the proposed dynamic reassignment and redeployment methods clearly improve the efficiency of an EMA service in terms of reducing response time. However, the approaches, particularly dynamic redeployment, introduce an extra cost. Since the mechanism is based on an almost continuous repositioning of idle ambulances, the travel distances of the ambulances increase. Considering the 10 days, the average distance each ambulance has to cover each day in the SUMMA model is 95.48 km, whereas it is 299.97 km for the DRARD approach. That is, ambulances have to travel about three times the distance because of frequent location changes. It is a political decision whether this extra effort is acceptable in order to improve the quality of service. In any case, compared to augmenting the number of ambulances in order to reduce response times, the DRARD approach appears to be a less costly alternative. In this sense, we have executed the DRARD method. Further, fewer ambulances achieve roughly the same average arrival time as that obtained in the SUMMA model. That is, with 29 ambulances in the SUMMA model, the same average arrival time can be obtained with 21 ambulances in the DRARD approach.

3.4. Distributed Coordination of Emergency Medical Service for Angioplasty Patients

Based on the World Health Organization data, ischemic heart disease (IHD) is the single most frequent cause of death, killing 8.76 million people in 2015, and one of the leading causes of death globally in the last 15 years [50]. It is a disease characterised by ischaemia (reduced blood supply) of the heart muscle, usually due to coronary artery disease. At any stage of coronary artery disease, the acute rupture of an atheromatous plaque may lead to an acute myocardial infarction (AMI), also called a heart attack. AMI can be classified into acute myocardial infarction with ST-segment elevations (STEMI) and without ST elevation (NSTEMI). Effective and rapid coronary reperfusion is the most important goal in the treatment of patients with STEMI.

One of the reperfusion methods is angioplasty or primary percutaneous coronary intervention (PCI). It is the preferred treatment when feasible and when performed within 90 min after the first medical contact [51,52]. Due to insufficient EMS coordination and organisational issues, elevated patient delay time—defined as the period from the onset of STEMI symptoms to the provision of reperfusion therapy—remains a major reason why angioplasty has not become the definitive treatment in many hospitals.

Conventional EMS procedure in assisting AMI emergencies is the following. Patients are diagnosed in the place where they suffer chest pain: at their momentary out-of-hospital location or at a health centre without angioplasty. In both cases, the emergency coordination centre (ECC) applies first-come first-served (FCFS) strategy and locates the nearest available (idle) ambulance with advanced life support (ALS) and dispatches it to pick up the patient. After the ambulance arrives to the scene and diagnoses AMI by an electrocardiogram, the ambulance confirms the diagnosis to the ECC, which has real-time information of the states of ambulances. The ECC applies FCFS strategy for hospital and cardiology team assignment by locating the nearest available hospital with a catheterisation laboratory and alerting the closest hospital cardiology team of the same hospital.

The improvements of the EMS coordination in the literature were achieved both by novel fleet real-time optimisation and communication methods, using new multiagent models (see, e.g., [44,52–55]). Despite an exhaustive quantity of work on the optimisation of EMA, to the best of our knowledge, there is little work on optimisation models for the coordination of EMS when the arrival of multiple EMS actors needs to be coordinated for the beginning of the patients' treatment. This is the case with STEMI patients assigned for angioplasty treatment where, in the case of multiple angioplasty patients, the FCFS strategy discriminates the patients appearing later.

EMA coordination for STEMI patients includes the assignment of three groups of actors: assignment of idle ambulances to patients, assignment of catheterisation laboratories in available hospitals to patients receiving assistance in situ, and assignment of available cardiology teams to hospitals for the angioplasty procedure performance. All of the three assignments need to be combined in a region of interest such that the shortest arrival times are guaranteed to all patients awaiting angioplasty at the same time. In continuation, we present the solution approach from [52], which presents a coordination model for EMS participants for the assistance of angioplasty patients. The proposed approach is also applicable to emergency patients of any pathology needing prehospital acute medical care and urgent hospital treatment.

We concentrated on the minimisation of the patient delay, defined as the time passed from the moment the patient contacts the medical ECC to the moment the patient starts reperfusion therapy in the hospital. The patient delay defined in this way is made of the following parts, and is shown in Figure 10:

T1   Emergency call response and decision making for the assignment of EMS resources;
T2   Mobilisation of an idle ambulance and its transit from its current position to the patient;
T3   Patient assistance in situ by ambulance staff;
T4   Patient transport in the ambulance to assigned hospital;
T5   Cardiology team transport from its momentary out-of-hospital position to the hospital;
T6   Expected waiting time due to previous patients in the catheterisation laboratory (if any).

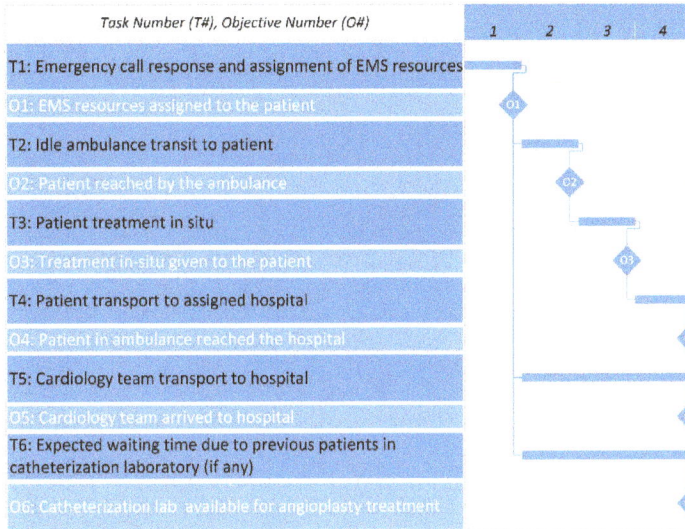

Figure 10. Gantt diagram of the coordination of emergency medical services (EMS) for angioplasty treatment.

The optimal patient delay time for a single patient is the lowest among the highest values of the following three times for all available ambulances and angioplasty-enabled hospitals (Figure 10):

- The expected patient delay time to hospital (the sum of times T2, T3, and T4, in continuation represented by parameters $t(a,p)$, $t(p)$, and $t(p,h)$, respectively);
- The expected minimal arrival time among cardiology teams to the same hospital (T5), represented by $min_{c \in C_{av}} t(c,h)$;
- The expected shortest waiting time until hospital h gets free for patient p, $min\ \rho_{h,p}$ (T6).

For simplicity, we let $t_{php} = max_{h \in H_{av}} (t(p,h),\ min\ \rho_{h,p})$ for all patients $p \in P$. Then, from the global point of view, considering all pending out-of-hospital patients, the problem transforms into:

$$min \Delta t_P = \sum_{p \in P} \Delta t_P = \sum_{p \in P} t(a,p) + \sum_{p \in P} t(p) + \sum_{p \in P} \left(max_{h \in H_{av}} \left(t_{php},\ min_{c \in C_{av}} t(c,h) \right) \right) \qquad (2)$$

subject to

$$\Delta t_p \le t_p^{max},\ \forall p \in P. \qquad (3)$$

The overall patient delay time Δt_P in Figure 10 is an additive function. Since the minimum arrival times cannot always be guaranteed for all patients due to the limited number of EMS resources, a sum of the EMA tasks' durations should be minimised for each patient individually and for the system globally considering individual constraints. This gives an underlying linear programming structure to the EMS coordination problem. Therefore, it is possible to guarantee optimal outcomes even when the optimisation is performed separately on individual sum components (i.e., when ambulance assignments are negotiated separately from the hospital and cardiology team assignment, e.g., [56,57]). This fact significantly facilitates the multiagent system's distribution and enables a multilevel optimisation. Hence, we decomposed optimisation Problems (1)–(2) as follows. On the first level, we assign ambulances to patients such that the expected arrival time of ambulances to patients $t(a,p)$ is minimised. Note that since $t(p)$ in (1) is a constant for every patient p, depending only on the patient's pathology and not on the assigned ambulance, we can exclude it from the optimisation.

Then, on the second optimisation level, we approached the second part of (1) $\sum_{p \in P}\left(max_{h \in H_{av}}\left(t_{php}, min_{c \in C_{av}} t(c, h)\right)\right)$, which is an NP-hard combinatorial problem. However, by approximating (1) with a sequence of problems where we first decide on the assignment of hospitals to pending patients and then assign cardiologists to patients already assigned to hospitals, we obtain two linear programs to which we can apply tractable optimal solution approaches, such as the auction algorithm [47]. By decomposing (1), as can be seen in Figure 11, and allowing for reassignment of resources based on the adaptation to contingencies in real time, we obtain a flexible EMS coordination solution.

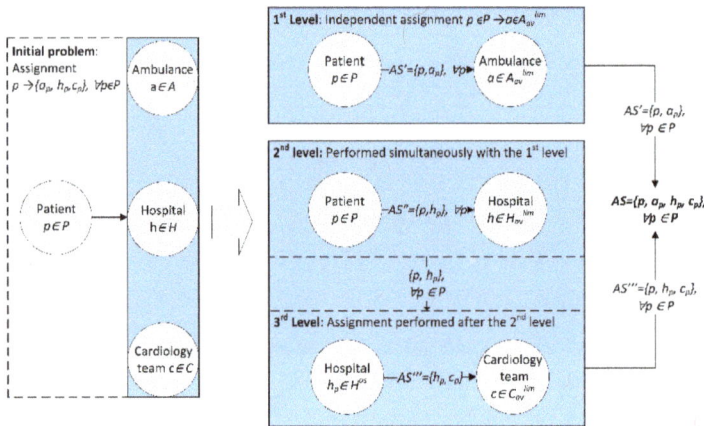

Figure 11. Proposed three-level decomposition of the problem of EMS coordination for with ST-segment elevation acute myocardial infarction (STEMI) patients awaiting angioplasty.

In the following, we propose a change in the centralised hierarchy-oriented organisational structure to a patient-oriented distributed organisational structure of EMS that increases the flexibility, scalability, and the responsiveness of the EMS system. The proposed decision-support system is based on the integration and coordination of all the phases EMS participants go through in the process of emergency medical assistance (EMA). The model takes into consideration the positions of ambulances, patients, hospitals, and cardiology teams for real-time assignment of patients to the EMS resources.

3.4.1. EMA for Angioplasty Patients

The emergency medical system for the assistance of patients with STEMI is made of the following participants: patients, hospitals with angioplasty facilities, medical emergency coordination centre (ECC), ambulance staff, and cardiology teams, each one being a compound of a cardiologist and one or more nurses.

Usually, each hospital with angioplasty has assigned to it its own cardiology team(s) positioned at alert outside the hospital and obliged to come to the hospital in the case of emergency. This is because the cardiology teams' costs make a large portion of the overall costs in surgical services [55].

The objective of the proposed system is the reduction of patient delay times by distributed real-time optimisation of decision-making processes. In more detail, we modelled patient delay time, and present a three-level problem decomposition for the minimisation of combined arrival times of multiple EMS actors necessary for angioplasty. For the three decomposition levels, we propose a distributed EMS coordination approach and modify the auction algorithm proposed by Bertsekas in [47] for the specific case. The latter is a distributed relaxation method that finds an optimal solution to the assignment problem.

On the first level, agents representing ambulances find, in a distributed way, the patient assignment that minimises arrival times of available ambulances to patients. After the treatment in situ, on the second optimisation level, ambulances carrying patients are assigned to available hospitals. On the third level, arrival times of cardiology teams to hospitals are coordinated with the arrival times of patients. The proposed approach is based on a global view, not concentrating only on minimising single patient delay time but obtaining the EMS system's best solution with respect to the (temporal and spatial) distribution of patients in a region of interest.

3.4.2. Simulation Experiments

In this section, we describe settings, experiments, and results of the simulated emergency scenarios that demonstrate the efficiency of the coordination procedure and a significant reduction in the average patient delay. We tested the proposed approach for the coordination of EMS resources in angioplasty patients' assistance, focusing on the average patient delay time in the case of multiple pending patients. We compared the performance of our approach with the FCFS method, since it is applied by most of the medical emergency-coordination centres worldwide.

To demonstrate the scalability of our solution and its potential application to small, medium, and large cities and regions, in the experiments, we varied the number of EMA ambulances from 5 to 100 (with increments of 5) and the number of angioplasty-capable hospitals from 2 to 50 (with increments of 2). The number of cardiology teams $|C|$ in each experiment equals the number of hospitals $|H|$. Thus, the number of setup configurations used, combining different numbers of ambulances and hospitals with cardiology teams, sums up to 500.

For each configuration, we performed the simulation on three different instances of random EMS participants' positions, since we wanted to simulate a sufficiently general setting applicable to any urban area that does not represent any region in particular. The EMS participants were distributed across the environment, whose dimensions are 50 × 50 km. In each instance, we modelled hospital positions and the initial positions of ambulances, out-of-hospital cardiology teams, and patients based on a continuous uniform distribution. Therefore, each configuration can be considered as a unique virtual city with its EMS system. Assuming that the EMS system is placed in a highly dense urban area, this kind of modelling of the positions of EMS participants represents a general enough real case, since the election of the hospital positions in urban areas is usually the result of a series of decisions developing over time with certain stochasticity, influenced by multiple political and demographical factors.

In the simulations, ambulances were initially assigned to the base stations in the hospitals of the region of interest. Additionally, we assumed that after transferring a patient to the hospital, an ambulance is redirected to the base station, where it waits for the next patient assignment. Furthermore, we assumed that the hospitals have at their disposal a sufficient number of catheterisation laboratories, so that the only optimisation factor from the hospital point of view is the number of available cardiology teams. If there were more patients with the same urgency already assigned and waiting for treatment in the same hospital, they were put in a queue.

The simulation of each instance was run over a temporal horizon, in which new patients are generated based on a certain appearance frequency. The EMS resources were dynamically coordinated from the appearance of a patient until the time s/he is assisted in hospital by a cardiology team. Each instance simulation was run over a total of 300 patients, whose appearance was distributed equally along the overall time horizon based on the following two predetermined frequency scenarios: low (1 new patient every 10 time periods) and high (1 new patient every 2 time periods).

The period between two consecutive executions of the EMS coordination algorithm is considered here as the minimum time interval in which the assignment decisions are made; usually, this ranges from 1 to 15 min. In each period, the actual state of EMS resources and pending patients is detected, and the EMS coordination is performed such that the EMS resources are (re)assigned for all patients. To achieve an efficient dynamic reassignment of ambulances, the execution of the EMS coordination algorithm is furthermore performed with every new significant event (i.e., any time there is a significant change in the system due to new patients) or with a significant change in travel time or state of any of the EMS participants.

In the experiments, we tested the performance of the proposed EMS coordination method with respect to the FCFS benchmark approach. The comparison is based on the relative performance function $P = (t_{FCFS} - t_{OR})/t_{FCFS} \times 100$ (%), where t_{FCFS} and t_{OR} are average patient delay times of the benchmark FCFS approach and the proposed model, respectively.

The simulation results of performance function P for the two simulated cases of patient frequency appearance of 1 and 5 new patients every 10 time periods are presented in the following. The performance of the proposed approach increases as the number of angioplasty-enabled hospitals increases from almost identical average patient delay in the configuration setup with 2 hospitals up to 87.14% with 50 hospitals, as can be seen in Figure 12. Observing the performance dynamics with respect to the varying number of hospitals, it is evident that the performance of the proposed EMS coordination method increases, on average, proportionally to the increase of the number of hospitals.

With a relatively low number of angioplasty-enabled hospitals (less than 15), our proposed EMS coordination approach performs, on average, better than FCFS by up to 15%. As the number of hospitals increases, the performance improves, on average, up to a maximum of 39.98% for the first case (Figure 13) and up to 87.14%, for the second case (Figure 12). However, mean patient delay improvement for the two cases is 35% and 45.5%, respectively.

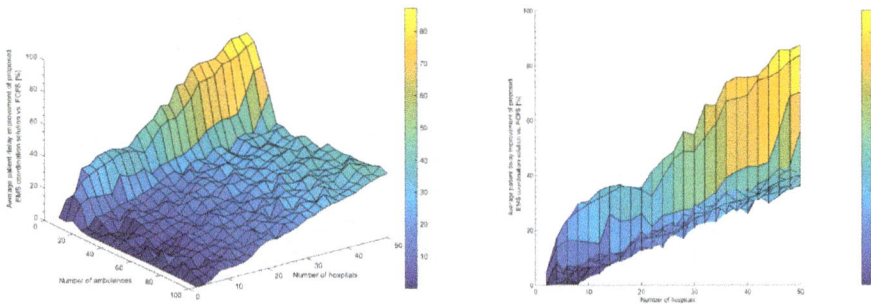

Figure 12. Average patient delay time of the proposed EMS coordination approach vs. the FCFS strategy (%) for the frequency of appearance of 1 patient each 2 time periods.

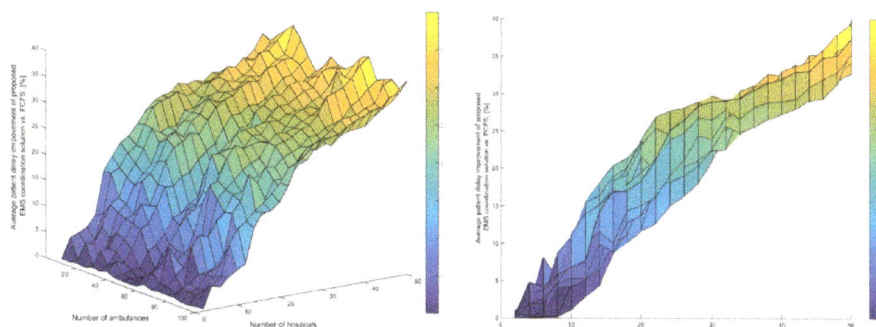

Figure 13. Average patient delay time of the proposed EMS coordination approach vs. the first-come first-served (FCFS) strategy (%) for the frequency of appearance of 1 patient each 10 time periods.

The static assignment of the FCFS principle discriminates against patients appearing later. Since ambulances are not equally distributed in the area, the proposed EMS coordination method compensates for the lack of EMS resources and their unequal distribution by reassigning them dynamically to pending patients. Dynamically optimised reassignment of EMS resources in real time is the main key to the improvement of the system's performance.

Thus, proportional to the increase of the number of hospitals, there is a constant improvement of performance. Even though the velocity of the EMS actors is not a relevant factor in the comparison of the performance of our proposed EMS coordination solution nor of the FCFS method, looking individually at the performance of each one of these methods, it is evident that the assignment cost accumulated through the time will be lower when the velocity of the EMS actors is higher.

Our simulation results show the efficiency of the proposed solution approach, resulting in significantly lower delay times for angioplasty, on average. Of course, the effectiveness of the proposed model depends on the initial classification of patients, and the related determination of the urgency of their cases, as well as on the timely availability of cardiology teams and hospitals. Still, as the current experience shows, good quality patient assessments and the EMS resource availabilities can be assured in practice.

To implement our approach in practice, a patient's location needs to be known to the system. Ideally, patients should contact the ECC through a mobile phone with GPS for easier location. In addition, ambulances should have a GPS and a navigator for localising the patient and navigating the way to him/her, as well as a means of communication with the rest of the EMS participants, and a digitalised map showing ambulances, patients, and hospitals.

Moreover, hospitals should have a digitalised receptionist service to receive and process relevant data of a patient before his/her arrival. None of these requirements go significantly beyond the current state of affairs in major cities (such as Madrid). Moreover, there are intrinsic uncertainties present in the EMS coordination. In our experiments, we assumed that travel times can be accurately forecasted, which, of course, is an important factor for the performance of the proposed system. In reality, this may not always be the case, as real-world traffic conditions are notoriously hard to predict. However, there is abundant literature on traffic-aware vehicle route guidance systems tackling this problem, and we believe that such systems can be easily integrated into our approach. Still, an effective proof of this conjecture is left to future work.

3.5. Coordination of Transportation Fleets of Autonomous Drivers

A similar problem to the emergency medical service coordination consists of coordinating fleets for transportation in an urban area (e.g., messaging services or taxi fleets). However, in contrast to medical emergency services, here the objective is not only focused on response time, but also on cost efficiency. Furthermore, a primary characteristic of such systems, at least with the boom of collaborative economy, is that such types of fleets may be open [58] in the sense that private persons may participate in the fleet as autonomous workers, with their own vehicle and at different time intervals. With regard to the coordination of such fleets, the autonomy of the drivers is a crucial characteristic. It implies that the drivers get their income on a per-service basis instead of a monthly salary. This means that, besides accepting a set of basic rules, drivers are more concerned with the actual service they have been assigned from the system and may have more freedom to accept or decline assignment decisions.

Maciejewski et al. [59] presented a real-time dispatching strategy based on solving the optimal taxi assignment among idle taxis and pending requests at certain intervals or whenever new events (new customer/available taxi) take place. Zhu and Prabhakar [60] analysed how suboptimal individual decisions lead to global inefficiencies. While most existing approaches try to minimise the average waiting time of customers, other works have a different focus. BAMOTR [61] provided a mechanism for fair assignment of drivers, that is, to minimise the differences in income among the taxi drivers. For that, they minimised a combination of taxi income and extra waiting time. Gao et al. [62] proposed an optimal multi-taxi dispatching method with a utility function that combines the total net profits of taxis and waiting time of passengers. Meghjani, and Marczuk [63] proposed a hybrid path search for fast, efficient, and reliable assignment to minimise the total travel cost with a limited knowledge of the network. In contrast to the previous works, the main characteristic of our approach is the possibility of modifying the assignment when a taxi has been dispatched but has not yet picked up a customer. In this sense, we followed a similar approach to the emergency medical service coordination presented in Section 3.3. One of the few other works in this line is [64], which presented an adaptive scheduling, where reassignment is possible during a time interval until a pick-up order is sent to the taxi and customer. In our case, we do not restrict reassignment to a specific interval. Furthermore, we propose a method that economically compensates the negatively affected taxis in the new schedule such that they do not have a loss in their income.

We considered a system that uses some mediator in charge of matching the transportation requests with vehicles of the fleet. On one hand, customers contact the mediator via some telematic means, requesting a transportation service. On the other hand, drivers subscribe to the system, offering their services during specific time intervals where they are available.

We assumed a payment structure for transportation services where clients pay a fixed price plus a *fare* per distance, as described in the following:

- From the client side, the price of a transportation service s is determined by the distance of the requested service plus a fixed cost:

$$Price(s) = fcost + fare \cdot d(s),$$

    where $d(s)$ denotes the distance from the service origin to its destination, *fcost* is a fixed cost, *fare* is a rate a client has to pay per distance unit.

- From the driver or vehicle side, for a driver v, the earnings depend on the price the client pays minus the cost of the vehicle for travelling the distance from the current position of v towards the position $p_o(s)$ and later to the destination $d(s)$:

$$Earn(v,s) = Price(s) - vcost \cdot (d(v,s) + d(s))$$

    where $d(v,s)$ denotes the distance from the current position of the vehicle in the moment of assigning a service s to the origin point of that service, and *vcost* is the actual cost rate of moving the vehicle on a per distance basis. *vcost* will implicitly include petrol, maintenance, depreciation of the vehicle, etc.

For simplicity, here we assumed that *fcost*, *fare*, and *vcost* are the same for all services and vehicles, that is, in the price structure we do not distinguish between different vehicle costs nor between different requests. If the use of different cost factors is important, the proposed model could be adapted accordingly. Furthermore, part or all of the amount of *fcost* could be retained by the mediator service as income. In this case, the earnings of the driver would be reduced by this amount.

Like in the emergency medical case, a typical approach for assigning services to vehicles in such a system is the first-call/first-served (FCFS) rule, where each incoming request is assigned to the closest available driver in that moment and no reassignments are done. As shown in [59], if there are more unassigned requests than available vehicles, it is better to assign vehicles to service requests. We call this strategy nearest-vehicle/nearest-request (NVNR).

Dynamic assignment strategies can improve the overall efficiency of a fleet. We assumed that drivers, once they are available, are obliged to accept a transportation service assigned to them. However, in contrast to the emergency medical scenario presented in the previous section, drivers do not have to accept changes in the assignments. That means they are free to accept or to decline proposed reassignments from the system, considering their own objectives and benefit. So, the dynamic reassignment approach, as proposed in Section 3.3.1, cannot be applied directly, as a driver would not be willing to accept a reassignment that reduces his/her net income. In order to still take advantage of cost reduction through reassignment, we developed an incentive schema so as to convince drivers to accept reassignments that are economically efficient from the global perspective of the system. The approach is detailed in the next subsection.

3.5.1. Dynamic Reassignment with Compensations

The idea of coordinating the assignments of transportation tasks to (autonomous) drivers is similar to the one proposed in Section 3.3.1: we want to reduce globally the total travel distance (and proportionally time) towards the origin points of the requested services. However, due to the rules of the system, a driver will usually not accept a "worse" task than the one he is already assigned to. We assumed drivers to be economically rational, that is, they want to maximise their net income and minimise the time spent on their trips. In particular, we made the following assumptions:

- A driver would always prefer a task with the same net income but that requires less time (e.g., less travel distance).
- A driver would always accept extra distance d if he would get extra net earnings of $d \cdot (fare - vcost)$. This is actually the current rate a driver earns when accomplishing a service and, thus, he would always be willing to provide his service for this rate.

Let us consider that a driver v is currently assigned to a service s_k and the mediator wants the driver to do other services s_j instead of s_k. Furthermore, let $td(v,s) = d(s_k) + d(v,s_k)$ denote the total distance driver v has to accomplish in order to provide service s. In order to convince the driver, we define a compensation c that is applied if a driver accepts the reassignment. This compensation is calculated as follows:

Case 1: If $td(v,s_k) < td(v,s_j)$: $c = [Earn(v,s_k) + (td(v,s_j) - td(v,s_k)) \cdot (fare - vcost)] - Earn(v,s_j)$

In this case, the effective income of the driver when accepting the reassignment and receiving the compensation c would be $Earn(v,s_k) + (td(v,s_j) - td(v,s_k)) \cdot (fare - vcost)$. That is, the driver receives the same income as before, plus the normal fare for the extra distance.

Case 2: If $td(v,s_k) \geq td(v,s_j)$: $c = Earn(v,s_k) - Earn(v,s_j)$

In this case, the effective new income of the driver with the compensation is $Earn(v,s_k)$. Thus, the driver would have the same earnings as before, but for less distance (and less time).

It is clear that, with the assumptions mentioned above, an economically rational driver would accept any reassignment with the defined compensations.

It should be noted that compensations may be positive or negative (e.g., a driver may get extra money for accepting a new service or s/he may have to pay some amount to the mediator).

For instance, if a driver is offered a reassignment from a service s_k to another service s_j with $d(s_k) = d(s_j)$ and $d(v,s_k) > d(v,s_j)$ (Case 2), the situation is a priori positive for the driver. S/he would have less distance to the starting point of the transportation service request but would earn the same money for the service itself. Thus, his net income would be higher. In this case, the compensation would be negative, with the amount $c = vcost(d(v,s_j) - d(v,s_k))$. That is, the driver would have to pay the cost of the difference in distances towards s_j with respect to the previous service s_k.

The idea of the mediator is to dynamically find global reassignments with compensations, such that the overall outcome of the mediator is zero or positive, for example, there would be no extra mediation cost. Given an existing assignment A_c at a given time, the algorithm we propose for calculating a new assignment A_n with compensations is summarised as follows:

1. Assign all pending transportation requests to vehicles using the NVNS rule and add the assignments to A_c
2. Calculate an optimal assignment A_n between all vehicles and requests assigned in A_c
3. Calculate the overall compensation C_o to be paid/received to/from drivers for the change from A_c to A_n
4. If $mediatorEarning - C_o > 0$ then
5. $mediatorEarning := mediatorEarning - C_o$
6. return A_n
7. else return A_c

The algorithm is executed by the mediator whenever either a new transportation request (service) is registered, or a driver becomes available (either after terminating a previous mission or because he starts working). In the first step, the system tries to assign pending requests in a rather standard fashion. Then, in Steps 2 and 3, a more efficient global assignment is searched for and the compensation cost of this new assignment is estimated. The new assignment is applied if the accumulated overall mediator earnings, together with the compensation cost, remains positive. This last part assures that the mediator has no extra mediation cost.

Regarding Step 2, we use Bertsekas' auction algorithm [47] to calculate an optimal assignment. In particular, we calculate the assignment A_n that minimises $D(A_n) + \gamma \cdot C_o$, where $D(A_n)$ is the sum of the distances of all vehicles in A_n to the corresponding original positions of the assigned service requests. This means we look for assignments that minimise the sum of the distances and also the potential cost of the compensations. γ is a factor for scaling monetary earnings into distance values (meters).

3.5.2. Evaluation

We tested the proposed approach in different experiments simulating the operation of a taxi fleet which basically has the characteristics of the type of fleets we want to address here. We used an operation area of about 9×9 km, an area that roughly corresponds to the city centre of Madrid, Spain. In the simulations, we randomly generated service requests (customers) who are assigned to available taxis, and we simulated the movement of taxis to pick up a customer, to drive him/her to his/her destination, and then waits for the assignment of a new customer. The simulations are not aimed at reproducing all relevant aspects of the real-world operation of a taxi fleet, but to analyse and compare the proposed coordination strategy (here called DYNRA) to the standard strategies FCFS and NVNR. Thus, we simplified the movements of taxis to straight-line movements with a constant velocity of 17 km/h. This velocity is within the range of the average velocity in the city centre of Madrid. Hence, we do not take into account either the real road network, nor the possibility of different traffic conditions.

The general parameters used in the simulation are as follows. We use 1000 taxis (initially distributed randomly in the area with a uniform distribution) and a simulation interval of 5 h. The taxis do not cruise, that is, they only move if they are assigned to a customer. We accomplish different simulation runs with different numbers of customers in order to represent different supply/demand ratios. We generate a fixed number of customers every 15 min (ranging from 250 to 1000 in steps of 125). For each customer, his/her origin (point of appearance) and destination location are randomly chosen such that each trip goes either from the outside of the area to the centre, or vice versa. The origin and destination points are generated using a normal distribution (for centre and outside points). When a taxi arrives at a customer's location, a pick-up time of 30 s is used where the taxi does not move. In the same sense, the simulated drop-off time is 90 s. The system assignment process is accomplished every 5 s, and only if a new client appears or a taxi becomes available again after a previous trip.

The payment scheme we used in the experiments is the one that has been used in the city of Madrid in the last years. A taxi trip has a fixed cost *fcost* = 2.4 euros and *fare* = 1.05 euros/km. Furthermore, the cost factor is *vcost* = 0.2 euros/km. This factor roughly corresponds to the actual cost of a vehicle, including petrol, maintenance, as well as other fixed costs. Finally, we applied a factor of $\gamma = 1/0.00085$, which corresponds to the net benefit a taxi receives per metre when transporting a client in the used payment scheme.

Each experiment was repeated 10 times with a different random seed, in order to avoid biased results due to a particular distribution of clients. The presented results are averages over those 10 runs.

Table 2 presents the average waiting times of the customers for the three methods and the different numbers of generated customers per hour. As it can be observed, for between 2000 and 2500 customers per hour, the FCFS approach starts to be perform really badly. Basically, the system becomes saturated and the rate of serving customers is lower than the rate of appearance of new customers. The other two methods, NVNR and DYNRA, can deal much better with this situation and their saturation point is higher (between 2500 and 3000 customers per hour). There is a clear improvement in the waiting times of these two methods, with respect to FCFS, if there are more than 2000 customers per hour. The dynamic reassignment approach with compensations performs better than the other two methods in all cases. The improvement is rather low if there are fewer customers, but increases with the number of customers up to 101.4 and 2.4 min with respect to FCFS and NVNR, respectively, for 4000 customers a hour. In terms of relative improvement, the highest peak is reached at 2500 customers, with an improvement of 94.6% with respect to FCFS and 44.9% with respect to NVNR.

It should be noted that the DYNRA approach is based on the compensation scheme presented above, that is, reassignments include compensations and (economically rational) taxi drivers will accept such reassignments. In Figure 14, we analyse the net income of the system, composed of the income of the taxi drivers plus the income of the mediator, in the case of the DYNRA approach. The presented results are normalised to the income of 1000 drivers and 1000 customers. The overall system income is highest for the DYNRA approach for all numbers of customers. The difference to the FCFS approach is considerable, between 473 and 565 euros, when above 2500 customers per hour (where FCFS is saturated). The difference of DYNRA with respect to NTNR is highest at 2500 customers (79 euros) and about 6–7 euros above that point. The taxi drivers always earn more money with the DYNRA approach, up to 2500 customers per hour. However, their net income is slightly lower than in the NTNR approach for more customers. Nevertheless, it should be noted that the mediator could redistribute its income among all drivers and, thus, the drivers would have a higher income in all cases.

Table 2. Average waiting times for customers (in minutes).

Method	# Customers per Hour						
	1000	1500	2000	2500	3000	3500	4000
FCFS	1.2	1.56	1.92	39.02	75.29	112.01	148.01
NVNR	1.2	1.56	1.92	3.83	8.29	27.38	49.01
DYNRA	1.19	1.48	1.73	2.11	6.83	25.24	46.62

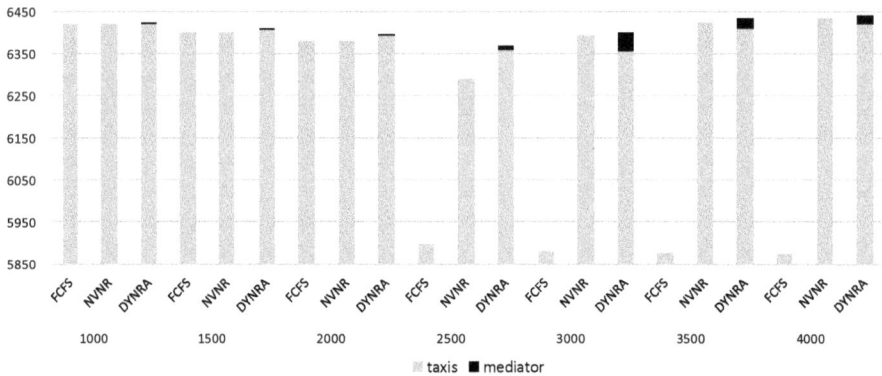

Figure 14. Average net income of taxi drivers and mediator in euros. The data are normalised to 1000 taxis serving 1000 customers.

Summarising, the proposed dynamic reassignment strategy can improve the performance of a transportation fleet of autonomous, self-interested drivers in terms of higher income and fewer movements (thus, also more environmentally friendly). The improvements are, in general, rather small if there are few movements (few service requests) and higher if the demand of transportation services is increasing.

The approach relies on a mediator service that manages the assignments of transportation tasks to drivers and pays compensations if necessary. Besides this fact, as proposed here, the mediator does not incur extra costs. Instead, it may have some positive income itself. The overall travel times and distances may still be reduced if the compensation system allowed for a negative balance. This could be of interest if, for example, a municipality would be willing to invest money in order to reduce CO_2 emissions.

4. Discussion

In this paper, we have argued that recent technological advances open up new possibilities for computers to support people's interactions in a variety of domains with high socioeconomic potential. In these domains, the choices and actions of a large number autonomous stakeholders need to be coordinated, and interactions can be regulated, by some sort of intelligent computing infrastructure, through institutions and institutional agents, or simply by providing information in an environment with a significant level of uncertainty. Many problems related to the vision of smart cities fall under this umbrella.

While centrally designed systems may be a suitable choice to address certain challenges related to smart cities, others are unlikely to be dealt with satisfactorily, either because stakeholders are *unwilling* to implement system recommendations that they do not understand and that they may not trust, or because it is *impossible* to compute good global solutions based on the information provided by stakeholders, which can be insufficient or biased by their personal interests. While the former problem can be addressed by providing stakeholders with their own trusted software agent, which represents them and acts on their behalf, the latter requires coordination mechanisms that take into account the *autonomy* of the stakeholders (and their software agents). Still, designing and implementing such coordination mechanisms in open systems is challenging, especially if the systems are large in scale, as in the case of most smart city applications.

We argued that technologies from the AT sandbox are suitable for fostering coordination in such scenarios. To back this claim, we reported on a variety of real-world applications, ranging from truly open systems, where coordination among agents is achieved either though (economic) incentives or by offering relevant information, to more closed domains, where AT techniques are used to "simulate" interactions among autonomous agents, which may, for instance, take the shape of auctions or market equilibria. While in some of the applications the use of AT (market-based approaches, in particular) was a mere design choice, in other more open domains, their use enables the provision of new functionalities and services.

In fact, a key lesson learnt from our work is that market-based coordination schemes can be successfully applied to quite different problems and domains, even though their particular shape needs to be carefully tailored and adapted based on the degree of autonomy of the stakeholders. In the applications outlined in Sections 3.3 and 3.4, the degree of autonomy of the different agents is low, as ambulance drivers, for instance, have to follow the assignments that they are given by the coordination mechanism. Evacuees in Section 3.2 do have a choice but, due to the specific characteristics of the emergency situation and the scarceness of adequate information, the suggestions of the system are likely to be followed. A similar "take it or leave it" situation is present in the taxi fleet coordination example of Section 3.5, but the stakeholders can make more informed decisions, so it is important that the incentives offered as part of the system proposal are such that taxi drivers can conclude that they are better off following the recommendations than not following them. Finally, within the case study related to networks of reservation-based intersections in Section 3.1, there are no explicit proposals for drivers to follow a particular route, but traffic assignment is achieved implicitly by coordinating the intersections' reserve prices. In addition, the auction protocol used at each reservation-based intersection needs to be such that the mechanism is resilient to attempts of strategic manipulation.

A limitation of our approach is related to the level of scalability needed for a particular domain. The applications outlined in this article have been evaluated in simulations with hundreds or thousands of agents, but going beyond these numbers may require the use of different approximations algorithms (e.g., to determine winner determination in auctions). Also, it should be noticed that take-it-or-leave-it recommendations may not work well when users can try to go for "outside options", that is, when competing service providers exist and users are allowed to use them. Finally, making mechanisms stable against strategic manipulation attempts often relies on assumption regarding the underlying communication infrastructure, which may hold in simulation experiments but are not achievable in all real-world situations.

We intend to continue developing applications for the aforementioned type of domains making use of the AT sandbox. We will particularly be looking into applications for which the semantic as well as the trust and reputation layers of the AT tower are of foremost importance. This will help us broaden the set of models and tools based on AT. In much the same way, we plan to extract further guidelines for designing these type of systems, based on descriptions of problem characteristics and requirements.

Author Contributions: All authors participated in the idea conception, organization and discussions. Each application presented in Section 3 had main contributors, namely: Intelligent intersections (S.O. and A.F.), smart building evacuation (A.F. and M.L.), emergency medical service (H.B. and M.L.), EMS for angioplasty patients (M.L. and H.B.), transportation fleets of autonomous drivers (H.B., A.F. and S.O).

Acknowledgments: This work has been partially supported by the Spanish Ministry of Economy and Competitiveness through grant TIN2015-65515-C4-X-R ("SURF"), by the Autonomous Region of Madrid through grant S2013/ICE-3019 ("MOSI-AGIL-CM", co-funded by EU Structural Funds FSE and FEDER), and by URJC-Santander grant 30VCPIGI15. The authors would like to thank Matteo Vasirani for his contribution to the application for coordinating of traffic flows through intelligent intersections outlined in Section 3.1.

Conflicts of Interest: The authors declare no conflict of interest.

References

1. Ossowski, S.; Andrighetto, G.; Polleres, A.; Argente, E.; Rodríguez-Aguilar, J.A.; Billhardt, H.; Sabater-Mir, J.; Botti, V.J.; Sierra, C.; Castelfranchi, C.; et al. *Agreement Technologies*; Law, Governance and Technology Series (LGTS) No. 8; Springer: Dordrecht, The Netherlands, 2013; ISBN 978-94-007-5582-6.
2. Gelernter, D.; Carriero, N. Coordination Languages and their Significance. *Commun. ACM* **1992**, *35*, 96–107. [CrossRef]

3. Ossowski, S.; Menezes, R. On coordination and its significance to distributed and multi-agent systems. *Concurr. Comput. Pract. Exp.* **2006**, *18*, 359–370. [CrossRef]
4. Vito, A.; Berardi, U.; Dangelico, R. Smart cities: Definitions, dimensions, performance, and initiatives. *J. Urban Technol.* **2015**, *22*, 3–21.
5. Wooldridge, M.; Jennings, N. Intelligent Agents—Theory and Practice. *Knowl. Eng. Rev.* **1995**, *10*, 115–152. [CrossRef]
6. Sierra, C.; Botti, V.; Ossowski, S. Agreement Computing. *Künstliche Intell.* **2011**, *25*, 115–152. [CrossRef]
7. Ossowski, S.; Sierra, C.; Botti, V. Agreement Technologies—A Computing perspective. *Agreem. Technol.* **2012**, *8*, 5–18.
8. Schumacher, M.; Ossowski, S. The Governing Environment. In *Environments for Multi-Agent Systems II (E4MAS)*; Springer: Basel, Switzerland, 2005; pp. 88–104.
9. Omicini, A.; Ossowski, S.; Ricci, A. Coordination Infrastructures in the Engineering of Multiagent Systems. In *Methodologies and Software Engineering for Agent Systems—The Agent-Oriented Software Engineering Handbook*; Bergenti, F., Gleizes, M.P., Zambonelli, F., Eds.; Springer: New York, NY, USA, 2004; pp. 273–296. ISBN 1-4020-8057-3.
10. Ossowski, S. Coordination in Multi-Agent Systems—Towards a Technology of Agreement. In *Multiagent System Technologies (MATES-2008)*; LNCS 5244; Springer: Kaiserslautern, Germany, 2008; pp. 2–12.
11. Ossowski, S. Constraint Based Coordination of Autonomous Agents. *Electron. Notes Theor. Comput. Sci.* **2001**, *48*, 211–226. [CrossRef]
12. Von Martial, F. *Co-Ordinating Plans of Autonomous Agents*; LNAI 610; Springer: Berlin/Heidelberg, Germany, 1992.
13. Shoham, Y.; Leyton-Brown, L. *Multiagent Systems—Algorithmic, Game-Theoretic, and Logical Foundations*; Cambridge University Press: Cambridge, UK, 2009.
14. Fornara, N.; Lopes Cardoso, H.; Noriega, P.; Oliveira, E.; Tampitsikas, C. Modelling Agent Institutions. *Agreem. Technol.* **2012**, *8*, 277–308.
15. Centeno, R.; Billhardt, H.; Hermoso, R.; Ossowski, S. Organising MAS—A formal model based on organisational mechanisms. In Proceedings of the 2009 ACM Symposium on Applied Computing (SAC), Honolulu, HI, USA, 8–12 March 2009; pp. 740–746.
16. Nam, T.; Pardo, T.A. Conceptualizing smart city with dimensions of technology, people, and institutions. In Proceedings of the 12th Annual International Digital Government Research Conference: Digital Government Innovation in Challenging Times, College Park, MD, USA, 12–15 June 2011; ACM: New York, NY, USA, 2011; pp. 282–291.
17. Giffinger, R.; Fertner, C.; Kramar, H.; Kalasek, R.; Pichler-Milanovic, N.; Meijers, E. *Smart Cities-Ranking of European Medium-Sized Cities*; Vienna University of Technology: Vienna, Austria, 2007.
18. O'Grady, M.; O'Hare, G. How smart is your city? *Science* **2012**, *335*, 1581–1582. [CrossRef] [PubMed]
19. Ahvenniemi, H.; Huovila, A.; Ointo-Seppä, I.; Airaksinen, M. What are the differences between sustainable and smart cities? *Cities* **2016**, *60*, 234–245. [CrossRef]
20. Bibri, S.; Krogstie, J. Smart sustainable cities of the future: An extensive interdisciplinary literature review. *Sustain. Cities Soc.* **2017**, *31*, 183–212. [CrossRef]
21. Kondepudi, S.N.; Ramanarayanan, V.; Jain, A.; Singh, G.N.; Nitin Agarwal, N.K.; Kumar, R.; Gemma, P. *Smart Sustainable Cities: An Analysis of Definitions*; The ITU-T Focus Group for Smart Sustainable Cities; International Telecommunication Union (ITU): Geneva, Switzerland, 2014.
22. Mohanty, S.; Choppali, U.; Kougianos, E. Everything you wanted to know about smart cities: The internet of things is the backbone. *IEEE Consum. Electron. Mag.* **2016**, *5*, 60–70. [CrossRef]
23. Petrolo, R.; Loscri, V.; Mitton, N. Towards a smart city based on cloud of things, a survey on the smart city vision and paradigms. *Trans. Emerg. Telecommun. Technol.* **2015**, *28*, e2931. [CrossRef]
24. Dresner, K.; Stone, P. A Multiagent Approach to Autonomous Intersection Management. *J. Artif. Intell. Res.* **2008**, *31*, 591–656.
25. Vasirani, M.; Ossowski, S. Evaluating Policies for Reservation-Based Intersection Control. In Proceedings of the 14th Portuguese Conference on Artificial Intelligence, Aveiro, Portugal, 12–15 October 2009; pp. 39–50.
26. Schepperle, H.; Bohm, K. Agent-based traffic control using auctions. In *Cooperative Information Agents XI (CIA-2007)*; LNCS 4676; Springer: Berlin/Heidelberg, Germany, 2007; pp. 119–133.
27. Vasirani, M.; Ossowski, S. A market-inspired approach for intersection management in urban road traffic networks. *J. Artif. Intell. Res.* **2012**, *43*, 621–659.

28. Vasirani, M.; Ossowski, S. Learning and Coordination for Autonomous Intersection Control. *Appl. Artif. Intell.* **2011**, *25*, 193–216. [CrossRef]

29. Codenotti, B.; Pemmaraju, S.; Varadarajan, K. The computation of market equilibria. *SIGACT News* **2004**, *35*, 23–37. [CrossRef]

30. Avery, W.H.; Soo, J. *Emergency/Disaster Guidelines and Procedures for Employees*; CCH Canadian Limited: Toronto, ON, Canada, 2003.

31. Chen, P.H.; Feng, F. A fast flow control algorithm for real-time emergency evacuation in large indoor areas. *Fire Saf. J.* **2009**, *44*, 732–740. [CrossRef]

32. Filippoupolitis, A.; Gelenbe, E. A distributed decision support system for building evacuation. In Proceedings of the 2nd IEEE Conference on Human System Interactions, Catania, Italy, 21–23 May 2009; pp. 323–330.

33. Lujak, M.; Ossowski, S. Intelligent People Flow Coordination in Smart Spaces. In *Multi-Agent Systems and Agreement Technologies*; Rovatsos, M., Vouros, G., Julian, V., Eds.; EUMAS 2015, AT 2015; Lecture Notes in Computer Science, Volume 9571; Springer: Cham, Switzerland, 2016; pp. 34–49.

34. Etzion, O.; Niblett, P. *Event Processing in Action*; Manning Publications: Shelter Island, NY, USA, 2010; ISBN 1935182218.

35. Khaleghi, B.; Khamis, A.; Karray, F.O.; Razavi, S.N. Multi-sensor data fusion—A review of the state-of-the-art. *Inf. Fusion* **2013**, *14*, 28–44. [CrossRef]

36. Zervas, E.; Mpimpoudis, A.; Anagnostopoulos, C.; Sekkas, O.; Hadjiefthymiades, S. Multi-sensor data fusion for fire detection. *Inf. Fusion* **2011**, *12*, 150–159. [CrossRef]

37. Lujak, M.; Giordani, S.; Ossowski, S. Route guidance: Bridging system and user optimization in traffic assignment. *Neurocomputing* **2015**, *151*, 449–460. [CrossRef]

38. Lujak, M.; Giordani, S. Centrality measures for evacuation: Finding agile evacuation routes. *Future Gener. Comput. Syst.* **2017**, *83*, 401–412. [CrossRef]

39. Rajagopalan, H.K.; Saydam, C.; Xiao, J. A multiperiod set covering location model for dynamic redeployment of ambulances. *Comput. Oper. Res.* **2008**, *35*, 814–826. [CrossRef]

40. Maxwell, M.S.; Restrepo, M.; Henderson, S.G.; Topaloglu, H. Approximate dynamic programming for ambulance redeployment. *INFORMS J. Comput.* **2010**, *22*, 266–281. [CrossRef]

41. Naoum-Sawaya, J.; Elhedhli, S. A stochastic optimization model for real-time ambulance redeployment. *Comput. Oper. Res.* **2013**, *40*, 1972–1978. [CrossRef]

42. Andersson, T.; Varbrand, P. Decision support tools for ambulance dispatch and relocation. *J. Oper. Res. Soc.* **2007**, *58*, 195–201. [CrossRef]

43. Bandara, D.; Mayorga, M.E.; McLay, L.A. Priority dispatching strategies for EMS systems. *J. Oper. Res. Soc.* **2013**, *65*, 572–587. [CrossRef]

44. Billhardt, H.; Lujak, M.; Sánchez-Brunete, V.; Fernández, A.; Ossowski, S. Dynamic coordination of ambulances for emergency medical assistance services. *Knowl. Based Syst.* **2014**, *70*, 268–280. [CrossRef]

45. Haghani, A.; Hu, H.; Tian, Q. An optimization model for real-time emergency vehicle dispatching and routing. In Proceedings of the 82nd Annual Meeting of the Transportation Research Board, Washington, DC, USA, 12–16 January 2003.

46. Munkres, J. Algorithms for the assignment and transportation problems. *J. Soc. Ind. Appl. Math.* **1957**, *5*, 32–38. [CrossRef]

47. Bertsekas, D. The auction algorithm: A distributed relaxation method for the assignment problem. *Ann. Oper. Res.* **1988**, *14*, 105–123. [CrossRef]

48. Du, Q.; Faber, V.; Gunzburger, M. Centroidal voronoi tessellations: Applications and algorithms. *SIAM Rev.* **1999**, *41*, 637–676. [CrossRef]

49. Lloyd, S. Least squares quantization in PCM. *IEEE Trans. Inf. Theory* **1982**, *28*, 129–137. [CrossRef]

50. World Health Organization. *WHO Fact Sheet No. 310*; World Health Organization: Geneva, Switzerland, 2015.

51. Van De Werf, F.; Bax, J.; Betriu, A.; Blomstrom-Lundqvist, C.; Crea, F.; Falk, V.; Filippatos, G.; Fox, K.; Huber, K.; Kastrati, A.; et al. Management of acute myocardial infarction in patients presenting with persistent st-segment elevation. *Eur. Heart J.* **2008**, *29*, 2909–2945. [CrossRef] [PubMed]

52. Wilde, E.T. Do emergency medical system response times matter for health Outcomes? *Health Econ.* **2013**, *22*, 790–806. [CrossRef] [PubMed]

53. Lujak, M.; Billhardt, H.; Ossowski, S. Distributed Coordination of Emergency Medical Assistance for Angioplasty Patients. *Ann. Math. Artif. Intell.* **2016**, *78*, 73–100. [CrossRef]

54. Lujak, M.; Billhardt, H. Coordinating Emergency Medical Assistance. *Agreem. Technol.* **2012**, *8*, 507–509.

55. Guerriero, F.; Guido, R. Operational research in the management of the operating theatre: A survey. *Health Care Manag. Sci.* **2011**, *14*, 89–114. [CrossRef] [PubMed]

56. Chevaleyre, Y.; Endriss, U.; Estivie, S.; Maudet, N. Multiagent resource allocation in k-additive domains: Preference representation and complexity. *Ann. Oper. Res.* **2008**, *163*, 49–62. [CrossRef]

57. Endriss, U.; Maudet, N.; Sadri, F.; Toni, F. On optimal outcomes of negotiations over resources. In Proceedings of the Second International Joint Conference on Autonomous Agents and Multiagent Systems (AAMAS), Melbourne, Australia, 14–18 July 2003; pp. 177–184.

58. Billhardt, H.; Fernández, A.; Lujak, M.; Ossowski, S.; Julián, V.; De Paz, J.F.; Hernandez, J.Z. Coordinating open fleets. A taxi assignment example. *AI Commun.* **2017**, *30*, 37–52. [CrossRef]

59. Maciejewski, M.; Bischoff, J.; Nagel, K. An Assignment-Based Approach to Efficient Real-Time City-Scale Taxi Dispatching. *IEEE Intell. Syst.* **2016**, *31*, 68–77. [CrossRef]

60. Zhu, C.; Prabhakar, B. Reducing Inefficiencies in Taxi Systems. In Proceedings of the IEEE 56th Annual Conference on Decision and Control (CDC), Melbourne, Australia, 12–15 December 2017.

61. Dai, G.; Huang, J.; Wambura, S.M.; Sun, H.A. Balanced Assignment Mechanism for Online Taxi Recommendation. In Proceedings of the of the 18th IEEE International Conference on Mobile Data Management (MDM), Daejeon, Korea, 29 May–1 June 2017; pp. 102–111.

62. Gao, G.; Xiao, M.; Zhao, Z. Optimal Multi-taxi Dispatch for Mobile Taxi-Hailing Systems. In Proceedings of the 45th International Conference on Parallel Processing (ICPP), Philadelphia, PA, USA, 16–19 August 2016; pp. 294–303.

63. Meghjani, M.; Marczuk, K. A hybrid approach to matching taxis and customers. In Proceedings of the Region 10 Conference (TENCON), Singapore, 22–25 November 2016; pp. 167–169.

64. Glaschenko, A.; Ivaschenko, A.; Rzevski, G.; Skobelev, P. Multi-Agent real time scheduling system for taxi companies. In Proceedings of the 8th International Conference on Autonomous Agents and Multiagent Systems (AAMAS), Budapest, Hungary, 10–15 May 2009; pp. 29–36.

applied
sciences

MDPI

Article

Cognitive Assistants—An Analysis and Future Trends Based on Speculative Default Reasoning

João Ramos [1,*], Tiago Oliveira [2], Ken Satoh [2], José Neves [1] and Paulo Novais [1]

[1] Algoritmi Centre, Department of Informatics, University of Minho, 4710 Braga, Portugal;
     jneves@di.uminho.pt (J.N.); pjon@di.uminho.pt (P.N.)
[2] National Institute of Informatics, Hitotsubashi, Chyoda-ku, Tokyo 101-8430, Japan; toliveira@nii.ac.jp (T.O.);
     ksatoh@nii.ac.jp (K.S.)
*   Correspondence: jramos@di.uminho.pt; Tel.: +351-253604430

Received: 28 February 2018; Accepted: 4 May 2018; Published: 8 May 2018

check for
updates

Abstract: Once a person is diagnosed (or a caregiver suspects that the person may have) cognitive disabilities he may lose the state of being autonomous, which may range from partial to total loss of independence, according to the level of incidence. Smart houses may be used as a tentative solution to overcome this situation. However, when one goes outside their premises, this alternative may become unusable. Indeed, due to the decreased orientation ability, caregivers may prevent these people from going out, as they may get lost. Therefore, we are developing a system for people with mild or moderate cognitive disabilities that guides the user through an augmented reality interface and provides a localization tool for caregivers. The orientation method implements a speculative computation module, thus the system may calculate and anticipate possible user mistakes and issue alerts before he takes the wrong path. Through a trajectory mining module, the path is also adjusted to user preferences. These two modules enable the system to adapt to the user.

Keywords: ambient intelligence; cognitive disabilities; mobile communication; orientation; person tracking; trajectory mining

---

1. Introduction

Cognitive disability is, according to the Diagnostic and Statistical Manual for Mental Disorders (DSM-IV) [1], a medical condition associated to an individual who has more difficulties in one or more types of mental tasks. These tasks include self-care, communication, orientation, use of community resources, functional academic skills, work, leisure, health, and safety, among others. The diagnostic of a cognitive disability is not an easy task since it may be present in different forms, like stroke, Alzheimers, traumatic brain injury, and even with different levels of incidence (varying from mild to extreme). The main criteria or the most frequently observed is the intellectual function, which is compromised in the presence of cognitive disabilities. A person with this disability has more difficulties in accomplishing one or more types of mental tasks [2].

When a person is submitted for a diagnosis, physicians may provide two different types of diagnosis, the functional and/or the clinical. The clinical diagnosis is related to the technical name, e.g., Down's syndrome, cerebral palsy or autism, and usually coexists with the functional one. The functional diagnostic is more concerned with the deficits that the person may face, namely memorization deficit, attention deficit, and mathematical comprehension deficit [3]. The categorization of the disabilities into a functional or clinical perspective may create a distance from the medical point of view. Indeed, the functional diagnostic may ignore the causes of the disease and focus on the resulting challenges due to the capacity decreasing.

Cognitive disabilities are usually divided into four stages, namely mild, moderate, severe and extreme. The majority of diagnostics are for mild cognitive disabilities (about 80%), while about 14% are related to moderate to severe ones [4]. According to the Diagnostic and Statistical Manual of Mental Disorders 5 (DSM-V) [5] the degree of incidence is less dependent on the Intelligence Quotient (IQ) scores and has more emphasis on the amount and type of intervention needed. Despite considering the IQ scores for assessing the level of disability, mental health professionals must consider the ability or impairment of the person in three areas, namely conceptual, social, and practical life skill.

An adopted solution by physicians may consist of the use of drugs with the goal of slowing the progression of the disease, and of giving the person a better quality of life. Brain stimuli (e.g., in Alzheimers) attenuate the progression of the disease, which results in an improvement in the quality of life. Thus attaining a better interaction and communication with the society in which he is embedded. On the other hand, according to the trials, this intervention is not appropriate for people in the later stages of dementia [6]. When the degree of incidence is moderate or severe, the loss of autonomy may be inevitable. The presence of a caregiver may be needed or the person may be reallocated to a nursing home or to a relative's house.

Despite the possible disability that a person may face, it is of extreme importance to remember that each person must have equal recognition before the law, must not be submitted to inhumane treatment, has the right to respect and has the right to live independently and to be included in the community. These are some of the rights expressed in the United Nations Convention on the Rights of Person's with Disabilities, namely articles 12, 15, 17, and 19 [7]. Indeed, despite the possible existence of a disability, people must be treated equally and have the same rights. Thus, a disability should not in any case impose a barrier to ordinary living. However, this is sometimes forgotten by modern societies where the absence of, for example, ramps for wheelchairs in crosswalks can be found.

Another possibility is to apply the concept of smart house [8], in which several embedded devices in the environment may allow for the monitoring of the person with disabilities, enabling remote access to the collected data by health professionals or caregivers. There are projects under the Active and Assisted Living (AAL) initiatives in Europe like Mylife [9] designed for older persons with reduced cognitive disabilities where the goal is to provide software as a service, giving access to simple and intuitive services that are adapted to the individual needs and wishes of each person. The aim of this project is to increase the independence and wellbeing of the user in home environments, and to decrease his social isolation by letting the user easily contact other people, among others.

Previous identified alternatives do not guarantee safety when the monitored person goes outside. Indeed, the lack of orientation is one of the causes for the loss of autonomy by people with mild or moderate cognitive disabilities. To create applications taking into consideration the end-user, it is necessary to consider features in its design in order to maximize its usability and accessibility. The cognitive and/or physical limitations of the user can be minimized since the cognitive processing is reduced to a minimum. This reduction can be achieved through alternatives to written text, such as pictures, animations and sounds [10]. The development of such applications must also consider the ability of the user to understand the instructions and to follow them. Thus, despite the level of the disabilities, the user must be able to use the application in an independent way.

To minimize the loss of independence we are engaged in developing a system that through an interface based on augmented reality (in order to diminish the effort needed to understand the displayed information), guides the user when traveling outside its premises. This system has a Speculative Computation unit which may anticipate possible user mistakes and alert him before taking the wrong turn. The path is also adapted to user preferences through a trajectory mining module. Our system also has a localization feature in order to keep caregivers aware of where the person with disabilities is. The developed system is aimed for people with mild or moderate cognitive disabilities with orientation deficit considering acquired disabilities such as the ones resulting from a traumatic brain injury, or those that are life long, such as Down's Syndrome, which do not evolve to higher degrees of incidence over time. However, it is assumed that the user is able to use and remembers how to use the mobile device during the guidance process.

The main contributions of this work are: (1) a brief description of orientation systems for people with cognitive disabilities and the identification of their main features; (2) the development of a system that guides the user, taking into consideration his preferences; (3) the development of a system that enables caregivers to remotely access the location of the person with disabilities; (4) the description of the module that adapts the path to the user preferences; (5) the description of the module that anticipates possible user mistakes and alerts him before he takes the wrong turn. The developed system has a double goal: (1) guide the user through an adapted trajectory, considering his preferences; (2) provide a localization feature for caregivers to enable them to know, in real-time, the current location of the person with disabilities. Using CogHelper, both users may see his independency increased since the person with disabilities may travel alone without the feeling of getting lost, and caregivers may develop another activity without neglecting the provided care due to the ability of remotely knowing the location of the user. Another topic that should be highlighted is the previously cited articles of the United Nations Convention on the Rights of Person's with Disabilities [7]. With this project there is the social goal of providing simple tools to enhance normal living.

This document is organized as follows: Section 1 introduces the scope of the document. Section 2 presents the main features on which orientation systems are grounded and the current state of the art with respect of orientation methods. Section 3 describes the proposed system. Section 4 explains the use of Speculative Computation in the system. Section 5 details the process of adapting the user path to his preferences through a trajectory mining process. Finally, Section 6 presents conclusions and future work directions.

2. Orientation Systems

The reduction in the quality of life caused by the lack of spatial orientation was studied in [11,12], where it is stated that there is a low applicability of orientation systems from a user perspective, i.e., people with diminished cognitive or physical capacities may not use these systems.

In traditional methods there is the presence of an instructor dedicated to the person which, through a lengthy and labor intensive training, was able to teach the person with disabilities to use a public transportation system, the bus, in an independent way [13]. This process that, on average, took one year considers the ability of the person to learn one route. Newer routes could take between one to eight weeks. However, if the person stopped performing the trained route, it could easily be forgotten and all intensive and exclusive work would be rapidly wasted [13].

Orientation methods may epitomize an important progress in assistive technology. These are very rigorous methods that may not tolerate faults and correctly respond to user stimuli. As such, existing systems place an extreme importance on the interfaces. They must be, simultaneously, simple and complete, i.e., they have to show/provide available features/functionalities in an easy way, so the person with disabilities may use the application without feeling confused or losing his confidence in these systems.

One major factor for the acceptance and use of an application is, according to the study of Dawe [4,14], the user interface. In her work, both caregivers and people with disabilities were integrated in the development of the application and, according to the results, this integration led to a better adoption of the system. It was also confirmed that the system was developed according to the user preferences and that they were not assumed by the programmer.

In Table 1, a perspective is given of the system features considered in different works. Here, the orientation method is related to the user interface, while prompts consider the type of alerts given to the user. The localization feature (useful for caregivers) is expressed as real time monitoring, i.e., on the fly. As depicted in the table, there are few proposals for guiding the user with cognitive disabilities, and, even in those, the information is only given through audio and visual (textual) prompts. In Table 1, with a tick (✓), one has the features that were implemented in each project. The last two columns of the table present works in which the main goal is not the user orientation, but his remote monitoring and next movement detection. These tasks are accomplished under a context aware setting.

Table 1. System features covered by exintings projects.

Features \ Projects		Carmien et al. [13]	Liu et al. [15]	Ramos et al. [16,17]	Fraunhofer Portugal [18]	Patterson et al. [19]	Sadilek & Kautz [20]
Orientation Method	Images		✓				
	Compass				✓		
	Augmented Reality			✓			
Prompts	Audio	✓	✓	✓	✓		
	Visual	✓	✓	✓	✓		
Context Aware						✓	✓
User Frequent Behaviour				✓		✓	
Real Time Monitoring				✓	✓	✓	✓

Although these orientation methods have a significant impact on the quality of life, and use up-to-date technologies and different types of hardware (like accelerometers, magnetometers, and GPS), they lack reasoning methods. These methods allied to the most recent technology create extremely promising systems in assistive technology, since they conciliate the fast technological progress with embedded artificial intelligence features. These systems allow for a "humanization" of the technology, enabling it to take into account user necessities. In a robotic scenario, the goal is to let an agent (or robot) to serve the purpose of moving an object from one place to another, taking the context into consideration. On the one hand, these decisions must be taken on-the-fly, and also in situations where the sensory data may not always be available. Indeed, the robot moves under an incomplete (data/information/knowledge) scenario. On the other hand, new data obtained from the sensors is continuously streamed to update the belief state of the robot (or agent) [21,22]. Under these frameworks and in order to develop an autonomous agent-based system, there are multiple ways to treat default data, information or knowledge, which are key factors in the process of enabling an agent to learn and to plan its actions. Table 2 provides an overview of the most prominent reasoning methods used in localization systems. It is outside of the scope of the paper to provide a detailed description of each one. The goal is to highlight the ones considered more suitable to be applied to the problem under analysis.

Table 2. Comparison between different reasoning methods.

Reasoning Methods \ Features	Complexity	Fault Tolerance	Learning Mechanism	Reward/Feedback	Input
Decision Tree Learning	NP-hard	✗	Supervised		✓
Artificial Neural Networks	High processing	✓	Unsupervised		✓
Support Vector Machines		✓	Supervised		✓
Bayesian Networks		✓			✓
Markov Decision Process		✓	Unsupervised	Reward function	✗
Case Based Reasoning		✓	Supervised	User Feedback	Knowledge Base
Speculative Computation		✓	Unsupervised	✗	Default Values

Decision Tree Learning is a method with predictive abilities through knowledge extraction and learning [23]. However, the Decision Tree has to be fed with a set of input values [24]. Thus, an initial supervised learning phase is necessary in order to enable this reasoning method to autonomously classify future inputs.

Inspired on biological systems, Artificial Neural Networks (ANNs) are able to model complex problems [25,26]. Through massive parallelism, these networks are able to learn and adapt to future cases. ANNs are also fault tolerant, i.e., when there is some missing information the ANNs are able to keep the computation.

Data classification may also be automated by Support Vector Machines [27,28]. After a training phase, the Support Vector Machine may handle new data and label it. Although this method has a predictive feature, it has to be previously trained.

Based on the Bayes Theorem, a Bayesian Network is able to reason under incomplete information scenarios [29]. A Bayesian Network establishes causal relations between its nodes through a probability distribution. Thus, given an input value this reasoning method calculates possible outcomes with the respective probability. Applying this method to the guiding context has its advantages since it could predict the next movement of the user. However, there is still the need for previous user information to be fed into the network in order to train it.

The Partially Observed Markov Decision Process (POMDP) has proven its advantages in fields like robotic navigation and planning under uncertainty. This reasoning method has been successfully applied in orientation systems for people with cognitive disabilities [30]. A POMDP is able to select the next action and, based on a reward function, evaluate if that action produced a positive or negative impact on the user. However, in an initial phase this method starts with a naive approach, in which it is assumed that the options will be correctly followed every time. With the usage, the POMDP evolves and better adapts to future cases.

Case-based Reasoning is a method that resembles the human thinking method, i.e., when facing a problem the user recalls similar situations and adapts the previously used ones to the new one [31]. Past experiences may be used through some adaptation to the new situation or to know what should not be done in order to avoid negative outcomes. This reasoning method needs a knowledge base from which a previous case is retrieved and adapted to solve a new problem. Once the reuse phase is over, there is the need to evaluate the acquired solution before saving it back to the knowledge base. This evaluation may be dependent on user feedback.

In this work, it is assumed that Speculative Computation [32,33] has a more adjusted knowledge representation and reasoning method, as it may anticipate the next movement of the user and provide alerts. This feature may not be found in statistical learning, nor is a pattern recognition method, and may not be found in reinforcement learning (since in the reinforcement learning method the reward is returned after the action is taken). Speculative Computation has the ability to deal with incomplete information, i.e., when the missing data is received the information is processed, the computation is revised and the system keeps executing the orientation guidelines. Indeed, the system is previously loaded with a set of default values, with information about the travel paths (e.g., which directions should the user take, what are the turns that should be done), and information about possible missing turns (e.g., locations in which the user may take the wrong direction and get lost). One is faced with a scenario that uses default values and keeps the execution of the program. Thus, the system does not go under an idle state and when the information is received, the execution is revised. Another aspect that must be referred to is related with a process of movement anticipation, i.e., at any moment the system is able to predict how the user acts, and therefore prevents mistakes. This reasoning method has never been used, as far as we know, in a system intended to guide a user with cognitive disabilities. The speculative computation framework uses a set of default values which are obtained from a trajectory mining module (described in detail in Section 5). This method may not be considered a substitute to the previously described reasoning methods, but a complementary one. Indeed, the approach used to obtain the default values is independent from the execution of the speculative computation framework. In this work a trajectory data mining is applied, but it may be replaced by another machine learning method. The Speculative Computation method offers a way to deal with incomplete information and to update it in real-time.

3. CogHelper—An Orientation System for People with Cognitive Disabilities

Nowadays, it is possible to have small devices with huge processing power (e.g., smartphones). This kind of device brought more portability than existing solutions (like laptops), since the user may install and execute several applications according to his needs. Indeed, although the old solutions help the user in his tasks, some of these applications may be too complex to be used by people with disabilities, especially by those with cognitive ones.

Being able to set the user needs, the system may adapt itself. For instance, if the system detects that the user is confused or lost, it may put an emphasis on the orientation process, minimizing the risks of a wrong move. It uses data gathered by a context aware framework, which is fed to a system that works in a speculative way, i.e., the system may predict the user movements and adjust its alerts accordingly.

The name of our system is CogHelper and it helps people with cognitive disabilities and their caregivers [17]. This system is intended to work outdoors, and its structure (Figure 1) consists of internal services like the application for people with cognitive disabilities (*Cognitive Helper Mobile Solution*, Section 3.1), the application for caregivers (*Caregiver Applications*, Section 3.2), and the server, as well as external services that let the system use other facilities, like an external reminder. The framework of Speculative Computation and the trajectory mining module are described in detail in Sections 4 and 5, respectively. These features are considered internal services of the application *Cognitive Helper Mobile Solution*.

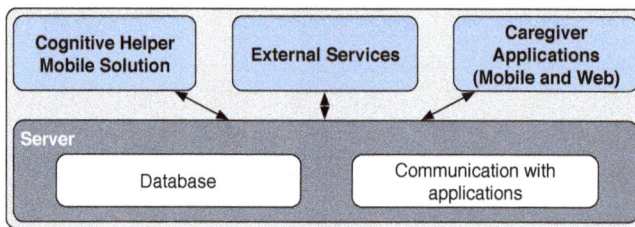

Figure 1. Simplified Framework of CogHelper.

The server is the core of the system, since all the other services are connected to it, namely in terms of a database (to store all the data needed to its correct operation, like usernames, locations, points of interest), and a communication module to connect to other internal services, as well as to external ones.

3.1. Application for People with Cognitive Disabilities

With the advent of high processing powered mobile devices, it was possible to overcome a major limitation of smart houses. Instead of having multiple fixed sensors at a home environment, it is possible to develop a portable monitoring bubble using mobile devices and sensors (e.g., over the body or clothes), thus leading to a new concept or abstraction, the Body Area Network (BAN) [34–36].

The system developed so far operates under the Android Operative System, targeting human beings with cognitive disabilities. It is based on the BAN concept. It fulfills two main objectives: to guide the user by adapting the user path and anticipating possible user mistakes, and to provide his localization to caregivers. Not only does it enable the user to travel between two locations without getting lost, but it also provides a localization system, so caregivers may know in real time the current location of the user.

A detailed framework of the *Cognitive Helper Mobile Solution* is presented in Figure 2 and consists of a Localization Layer, with methods to obtain the user location, necessary to let the Navigation system settle the route. The user location may be achieved trough the GPS module of the mobile device, or by a coarse location from the network. This layer also includes a Navigation Algorithm, which is based on a Speculative Computation Framework, described in detail in Section 4, and a Trajectory Mining Module (responsible for obtaining the default values for the speculative computation module), described in Section 5. With the information from the user location and the Speculative framework, the system may learn from the frequent paths taken by the user.

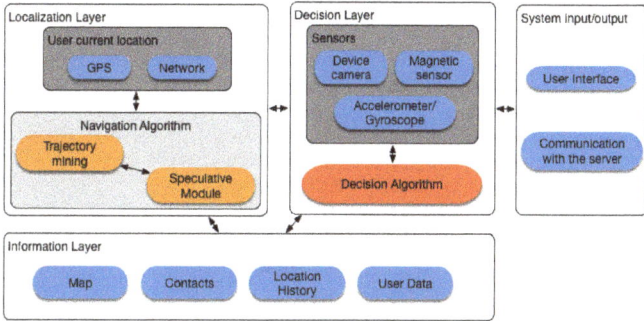

Figure 2. Detailed framework of the application for people with cognitive disabilities.

Once the destination is selected, using tools like cameras, accelerometers, or magnetic sensors, the route is presented to the user through augmented reality (Figure 3). In other words, one may say that the computational system is able not only to set the direction to where the user must point, but also to present it to him. If the user is pointing to the right direction, a green arrow will appear on the screen, signaling that the chosen track is the correct one. On the other hand, if the user is pointing to the wrong direction, a red cross is displayed. The decision process (at the center in Figure 2) ensures that the user is moving on the right track, and some warnings/prompts are brought to him, whenever necessary. These prompts may be seen as a stimulus to the user to keep himself on the right track, or to alert him to take the right decisions. To keep caregivers aware of user movements, an email or short message may also be sent to them, i.e., when the user reaches the destination point, passes a control one, or gets lost. Indeed, through an interface (right part in Figure 2) the user may communicate with the computational system and to set the destination points, namely in terms of starred destinations (e.g., home, school, or office), or generic ones (e.g., the mall). To communicate with the server in order to obtain and send the needed information, the computational system is provided with an Input/Output layer. The information exchanged between the system and the server may include the update of user destinations, the update of his current location, or information about the external services that are available.

Figure 3. Orientation system using augmented reality.

3.2. Applications for Caregivers

To look at and follow the user throughout his lifetime, caregivers have access to two different platforms, namely a mobile application for the Android Operative System (Figure 4a), and a Web application (Figure 4b). The former gives extra degrees of freedom to caregivers, since it may be used everywhere with an Internet connection. The detailed framework of these platforms is shown in Figure 5, and consists of three components, being the extra module exclusive for the Web application, which is pointed out as "Web". On the left it is possible to see the Input/Output layer, which enables the interaction between the caregiver and the computational system. There is also a Communication Module for the communication between the caregiver applications and the server, in order to send and receive requests, like getting the current position of the user.

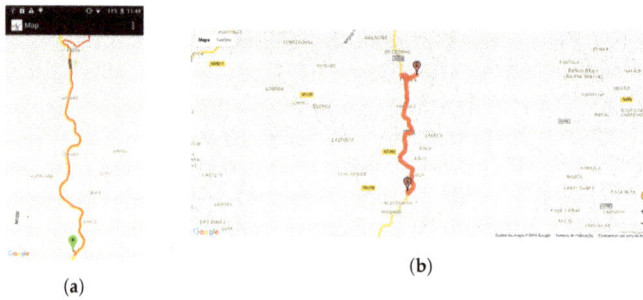

(a)
(b)

Figure 4. Localization system for caregivers. (**a**) Android application; (**b**) Web application.

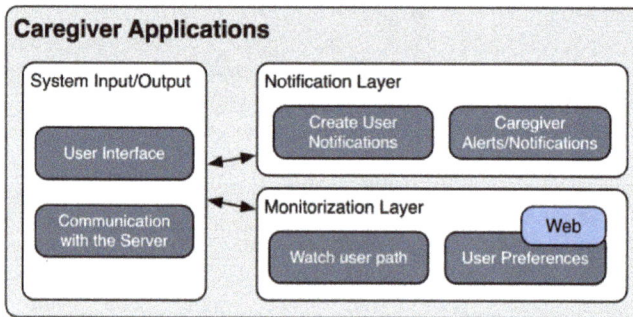

Figure 5. Detailed framework of the caregiver's application.

The Notification Layer is in charge of handling the alerts that come from the user application. These alerts may inform the caregiver that the user is lost or that they have reached their destination. The Create User Notification Module enables the caregiver to create notifications to the user, which are no more than small text messages that may include questions with a Yes/No answer (e.g., the user may be asked if he needs assistance). Finally, the framework has a Monitoring Layer, which enables the caregivers to be aware of user movements, like knowing his current location or the traveling path. The Web application is the only one that has a User Preferences Module, through which the caregiver may edit some preferences of the person with disabilities. This feature includes, for example, the creation of destination points, which may be done by a direct selection on the map or by searching an address. One is also able to state if the destination point is general, or favorite (i.e., more frequently used).

3.3. Ethical Issues

The development of an application which needs or collects sensitive information about the user must be handled with care. In CogHelper's particular case, the main users are considered to be specially vulnerable, people with disabilities, and sensitive information is stored. Indeed, despite storing all information in a ciphered form, regarding user position, user location, among other data, by using the system it is assumed that the user has given his consent to provide all necessary information for the correct operation of the system. The collected information is used to adapt the system to each individual user and, if requested, this is provided to the caregiver.

According to [37,38] and taking into consideration research purposes, privacy is a priority when dealing with vulnerable populations and the published results will not compromise or reveal sensitive information about the user. With this system there is the intention of maximizing the benefits, adapting the trajectory to the user preferences and anticipating possible user mistakes, and no information except what is absolutely necessary will be collected or demanded of the user.

4. Applying Speculative Computation to an Orientation System

Satoh [32] extended the procedure of Kakas et al. [33], giving rise to the theory of Speculative Computation and Abduction, which is a step forward in handling incomplete information. Instead of having the computational process in an idle state (i.e., waiting for incoming information to continue the computation), it moves ahead by replacing the unknown information by default information, therefore obtaining a tentative solution to the problem. Whenever the missing data is acquired, the computation process is re-examined. The execution of the speculative framework is based on two phases, *Process Reduction Phase* and *Fact Arrival Phase*. The former one stands for the normal execution of the computation process, while the latter denotes an interruption stage. Before starting the execution of the Speculative Computation framework, it is necessary to have the following information:

1. All the possible paths between the current user location and the intended destination as facts in the knowledge base;
2. The transitions between points usually performed by the user as default values;
3. Information regarding the inclusion of a point in the current recommended traveling path as default values; and
4. A set of rules that structure the execution of the computation regarding the most likely path that the user will follow and the issuing of alerts if a (potential) mistake happens.

4.1. The Speculative Computation Framework

The Speculative Computation Framework in the Orientation Method (SF_{OM}) for people with cognitive disabilities is defined in terms of the tuple $\langle \Sigma, \mathcal{E}, \Delta, \mathcal{A}, \mathcal{P}, \mathcal{I} \rangle$ [32], where:

-   Σ stands for a finite set of constants (an element of Σ is called a system module);
-   $\mathcal{E}$ denotes a set of functions called *external predicates*. When Q is a literal belonging to an external predicate and S is the identifier of the information source, $Q@S$ is called an *askable literal*. We define $\sim(Q@S)$ as $(\sim Q)@S$;
-   Δ is the *default answer* set, which is a set of ground askable literals that satisfy the condition: Δ does not contain both $p(t_1, \ldots, t_n)@S$ and $\sim p(t_1, \ldots, t_n)@S$ at once;
-   $\mathcal{A}$ is a set of predicates called *abducible predicates*. Q is called *abducible* when it is a literal with an abducible predicate;
-   $\mathcal{P}$ is a Logic Program (LP) and contains a set of rules in the form:

    ▷ $p \leftarrow p_1, p_2, \ldots, p_n$ where p is a positive ordinary literal, and each of $p_1, \ldots, p_n$ is an ordinary literal, an askable literal or an abducible; and
    ▷ p is the head of rule R and is named as $head(P)$ (always non-empty), in which R is a rule of the form $p \leftarrow p_1, \ldots, p_n$; where $p_1, \ldots, p_n$ is the body of the rule denoted as $body(P)$, that in some situations is replaced by the boolean value *true*.

- $\mathcal{I}$ is a set of integrity constraints in the form:

  ▷ $?(p_1, p_2, \ldots, p_n)$, where the symbol "?" denotes "falsity", the p_1, p_2, ..., p_n are ordinary literals or *askable literals* or *abducibles*. At least one of $p_1, p_2, \ldots, p_n$ is an *askable literal* or an *abducible*.

An *askable literal* may have different meanings, namely:

1. An *askable literal* $Q@S$ in a rule $\mathcal{P}$ stands for a question put to a system module S; and
2. An *askable literal* in Δ denotes a default truth value, either *true* or *false*, i.e., $p(t_1, \ldots, t_n)@S \in \Delta$, $p(t_1, \ldots, t_n)@S$ is usually *true* for a question to a system module S, and $\sim p(t_1, \ldots, t_n)@S \in \Delta$, $p(t_1, \ldots, t_n)@S$ is generally *false* for a question to a system module S.

In the logic program given below *path(a,b)* denotes that it is possible to travel between locations *a* and *b*; *show_next_point* states that the system must show the next location (which may be intermediate or final) to the user; *show_user_warning* indicates that the system must alert the user, given that he is going in the wrong direction; and the default values for the travel path of the user are defined in Δ. *user_travel(a,b)*, says that the user will travel from location *a* to location *b*; *included(a)* is evidence that location *a* is part the route.

To ensure that the user is traveling on the correct track and that he is alerted whenever he is off of it, a supporting structure was specified based on Speculative Computation, in terms of the logic programming suite (LPS):

▷ $\Sigma = \{gps\_sensor, recognizer\}$
▷ $\mathcal{E} = \{user\_travel, included\}$
▷ $\Delta = \{user\_travel(1, 2)@gps\_sensor,$
$\quad user\_travel(2, 3)@gps\_sensor,$
$\quad user\_travel(2, 4)@gps\_sensor,$
$\quad user\_travel(4, 5)@gps\_sensor,$
$\quad user\_travel(5, 6)@gps\_sensor,$
$\quad user\_travel(6, 3)@gps\_sensor,$
$\quad included(1)@recognizer,$
$\quad included(2)@recognizer,$
$\quad included(3)@recognizer,$
$\quad \sim included(4)@recognizer$
$\quad included(5)@recognizer,$
$\quad included(6)@recognizer\}$
▷ $\mathcal{A} = \{show\_next\_point, show\_user\_warning\}$
▷ $\mathcal{P}$ is the following set of rules:
$\quad guide(A, A) \leftarrow .$
$\quad guide(A, B) \leftarrow$
$\qquad\qquad path(A, F),$
$\qquad\qquad show\_next\_point(F),$
$\qquad\qquad user\_travel(A, F)@gps\_sensor,$
$\qquad\qquad guide(F, B).$
$\quad guide(A, B) \leftarrow$
$\qquad\qquad path(A, F),$
$\qquad\qquad user\_travel(A, F)@gps\_sensor,$
$\qquad\qquad show\_user\_warning(F),$
$\qquad\qquad guide(F, B).$
$\quad path(1, 2) \leftarrow .$
$\quad path(2, 3) \leftarrow .$
$\quad path(2, 4) \leftarrow .$

$$path(4,5) \leftarrow .$$
$$path(5,6) \leftarrow .$$
$$path(6,3) \leftarrow .$$

▷ $\mathcal{I}$ denotes the following set of integrity constraints or invariants:

$$?(show\_next\_point(F),$$
$$\sim included(F)@recognizer).$$
$$?(show\_user\_warning(F),$$
$$included(F)@recognizer).$$

To ensure program integrity, two invariants were added that state that the system may not show the next route point to the user if it is not part of it, or that the system may not alert the user if he is moving on the right track.

As an example, it is assumed that the user will travel between locations 1 and 3 through intermediate location 2. An elucidation of the possible paths that the user may use are presented in Figure 6.

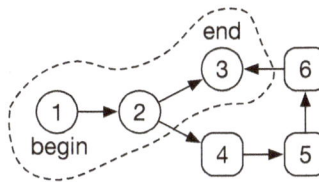

Figure 6. Possible ways to move among locations 1 and 3.

4.2. Preliminary Definitions

There are some aspects of the formal process set above that must be defined in order to translate it into a proof procedure.

Definition 1. *An extended literal is either a literal or an expression of the form $fail(\{l_1,\ldots,l_n\})$ where l_i is a literal. $fail(\{l_1,\ldots,l_n\})$ is used to prove that there is no proof for l_i [39].*

Definition 2. *A process is the tuple $\langle GS, OD, IA, ANS \rangle$ in which GS is a set of extended literals and called a Goal Set (GS), expressing the current status of an alternative computation; OD is a set of askable literals called Outside Defaults (OD), which denote a set of assumed information about the outside world during a computation; IA is a set of negative literals or abducibles called Inside Assumptions (IA) that stand for the values assumed during a computation; and ANS is a set of instantiations of variables in the initial inquiry (named Answers - ANS).*

Definition 3. *PS is a set of processes. A set of Already Asked Questions AAQ is a set of askable literals. A Current Belief State CBS is a set of askable literals.*

The set of processes *PS* expresses all the alternative computations that were considered. The *AAQ* set is used to avoid asking redundant questions to the sensors. *CBS* is the current belief state and expresses the current status of the outside world. It is also important to define an *active process* and a *suspended process*.

Definition 4. *Let $\langle GS, OD, IA, ANS \rangle$ be a process and CBS be a current belief state. A process is active with respect to CBS if $OD \subseteq CBS$. A process is suspended with respect to CBS otherwise.*

The definition of an active process emphasizes that it is a process whose outside defaults have to be consistent with the current belief state.

4.3. Process Reduction Phase

In this phase, changes may occur in the process set. In the following description, changed *PS*, *AAQ* and *CBS* are specified as *NewPS*, *NewAAQ* and *NewCBS*; otherwise they stay unchanged.

Initial Step: Let *GS* be an initial goal set. The tuple $\langle GS, \varnothing, \varnothing, ANS \rangle$ is given to the proof procedure where *ANS* is a set of variables in *GS*. That is, $PS = \{\langle GS, \varnothing, \varnothing, ANS \rangle\}$. Let $AAQ = \varnothing$ and $CBS = \Delta$.

Iteration Step: Do the following:

▷ **Case 1:** If there is an active process $\langle GS, \varnothing, \varnothing, ANS \rangle$ with respect to *CBS* in *PS*, terminate the process by returning outside defaults *OD*, inside assumptions *IA*, and instantiation for variables *ANS*. This case may only be applied on the first iteration step since *OD* and *IA* are empty sets;

▷ **Case 2:** If there is no active process, terminate the process by reporting a failure of the goal;

▷ **Case 3:** Select an active process $\langle GS, OD, IA, ANS \rangle$ with respect to *CBS* from *PS* and select an extended literal *L* in *GS*. Let $PS' = PS - \{\langle GS, OD, IA, ANS \rangle\}$ and $GS' = GS - \{L\}$. For the selected extended literal *L*, do the following:

— **Case 3.1:** If *L* is a positive ordinary literal, $NewPS = PS' \bigcup \{\langle (\{ \ body(R)\} \cup GS')\theta, OD, IA, ANS\theta \rangle | \exists R \in \mathcal{P}$ and $\exists$most general unifier θ so that $head(R)\theta = L\theta$.

— **Case 3.2:** If *L* is a ground negative ordinary literal or a ground abducible then:

  * **Case 3.2.1:** If $L \in IA$ then $NewPS = PS' \bigcup \{\langle GS', OD, IA, ANS \rangle\}$.
  * **Case 3.2.2:** If $\overline{L} \in IA$ then $NewPS = PS'$.
  * **Case 3.2.3:** If $L \notin IA$ then $NewPS = PS' \bigcup \{\langle NewGS, OD, \ IA, \ \cup\{L\}, ANS \rangle\}$ where $NewGS = \{fail(BS) | BS \in resolvent(\ L, \mathcal{P} \cup \mathcal{I})\} \bigcup GS'$ and $resolvent(L, T)$ is defined as follows:

    · If *L* is a ground negative ordinary literal, $resolvent(L, T) = \{\{L_1\theta, \ldots, L_k\theta\} | H \leftarrow L_1, \ldots, L_k \in T$ so that $\overline{L} = H\theta$ by a ground substitution $\theta\}$

    · If *L* is a ground abducible, $resolvent(L, T) = \{\{L_1\theta, \ldots, L_{i-1}\theta, L_{i+1}\theta, \ldots, L_k\theta\} | \perp \leftarrow L_1, \ldots, L_k \in T$ so that $L = L_i\theta$ by a ground substitution $\theta\}$.

— **Case 3.3:** If *L* is $fail(BS)$, then

  * If $BS = \varnothing$, $NewPS = PS'$;
  * IF $BS \neq \varnothing$, then do the following:
    (1) Select *B* from *BS* and let $BS' = BS - \{B\}$.
    (2) **Case 3.3.1:** If *B* is a positive ordinary literal, $NewPS = PS' \bigcup \{\langle NewGS \cup GS', OD, IA, ANS \rangle\}$ where $NewGS = \{fail((\{body(R)\} \cup BS')\theta | \exists R \in \mathcal{P}$ and $\exists$MGU θ so that $head(R)\theta = B\theta\}$

    Case 3.3.2: If *B* is a ground negative ordinary literal or a ground askable literal or an abducible, $NewPS = PS' \bigcup \{\langle \{ \ fail(BS')\} \cup GS', OD, IA, ANS \rangle\} \bigcup \{\langle \{\overline{B}\} \cup GS', OD, IA, ANS \rangle\}$.

— **Case 3.4:** If *L* is a ground askable literal, $Q@S$, then do the following:
    (1) If $L \notin AAQ$ and $\overline{L} \notin AAQ$, then send the question *Q* to the slave agent *S* and $NewAAQ = AAQ \cup \{L\}$.
    (2) If $\overline{L} \in OD$ then $NewPS = PS'$ else $NewPS = PS' \bigcup \{\langle GS', OD \cup \{L\}, IA, ANS \rangle\}$.

4.4. Fact Arrival Phase

In this phase the current belief state is revised according to the information received from the sensors. Supposing that an answer *Q* is returned from a sensor *S*. Let $L = Q@S$. After finishing a step of process reduction, let us do the following:

• If $\overline{L} \in CBS$, then $NewCBS = CBS - \{\overline{L}\} \cup \{L\}$

- Else if $L \notin CBS$, then $NewCBS = CBS \cup \{L\}$.

There might be some askable literals that are not included in the initial belief set. If this occurs, processes that are using such askable literals and those using their complements are suspended until the answers are returned.

4.5. Correctness of the Proof Procedure

The correctness of the procedure is guaranteed by stable model semantics [39]. Thus, the following definitions are given for the semantics of the previously described logical program.

Definition 5. *Let T be a set of rules and integrity constraints. The set of ground rules obtained by replacing all the variables in every rule or every integrity constraint T by every ground term is denoted as* $\prod_T$.

Definition 6. *Let T be a set of rules and integrity constraints. Let M be a set of ground atoms and* $\prod_T^M$ *be the following program:* $\prod_T^M = \{H \leftarrow B_1, \ldots, B_l | H \leftarrow B_1, \ldots, B_l, \sim A_1, \ldots, \sim A_h. \in \prod_P \text{ and } A_i \notin M \text{ for each } i = 1, \ldots, h.\}$. *Let* $min(\prod_T^M)$ *be the least model of* $\prod_T^M$. *A stable model for a logic program T is M iff* $M = min(\prod_M^T)$ *and* $\perp \notin M$.

Definition 7. *Let T be a set of rules and integrity constraints, and* Θ *be a set of ground abducibles.*

For any process evaluation strategy, when an answer is received with a set of outside defaults and a set of inside assumptions from the proof procedure, the answer is correct with respect to a generalized stable model with respect to inside assumptions and the program. This program is obtained from the original one and the current world belief.

Theorem 1. *Let* $SF_{OM} = \langle \Sigma, \mathcal{E}, \Delta, \mathcal{A}, \mathcal{P}, \mathcal{I} \rangle$ *be a speculative framework where* $\mathcal{P}$ *is a call-consistent logic program whose set of integrity constraints is satisfiable. Let GS be an initial goal set. Suppose that GS is reduced to* $\emptyset$ *when OD is outside defaults. IA are inside assumptions, ANS is the set of variables instantiations in GS, and CBS is the current belief set. Let GS' be a goal obtained from GS by replacing all the variables in GS by ANS. Then, there is a generalized stable model* $M(\Theta)$ *for* $\mathcal{P} \cup \mathcal{I} \cup \mathcal{F}(CBS)$, *such that* $M(\Theta) \vDash GS'$ *and* $OC \subseteq CBS$ *and* $IA \subseteq \Theta$.

Proposition 1. *Since* $\mathcal{P}$ *is a call consistent logic program, so is* $\mathcal{P} \cup \mathcal{I} \cup \mathcal{F}(CBS)$. *Then, an abducible derivation may be constructed* [39] *using a set of reduction steps applied to* $\langle GS, \emptyset, \emptyset, ANS' \rangle$ *to* $\langle \emptyset, OD, IA, ANS \rangle$ *where ANS' is a set of variables in GS. This derivation is correct for generalized stable model semantics for a call-consistent logic program* [39]. *Thus,* $M(\Theta) \vDash GS\theta$ *and* $OD \subseteq \Theta$.

An execution example of the program introduced in Section 4.1 is presented in Appendix A. For the reduction process, the following strategy is used:

▷ When a positive literal is reduced, new processes are created according to the rule order in the program, which are unifiable with the positive literal;
▷ A newly created or newly resumed process and the most left literal is always selected.

In the execution trace in Appendix A for $guide(1,3)$ the selected literal is underlined in the active process that was sorted out. *AAQ* and *CBS* are only shown when a change occurs. In order to reduce the execution example and to facilitate its reading and interpretation, the literals have been abbreviated. Thus, *gps_sensor* is represented by *g*, *recognizer* by *r*, *user_travel* by *u*, *included* by *i*, *path* by *p*, *show_next_point* by *snp*, and *show_user_warning* by *suw*.

The program starts with the objective of guiding the user from location 1 to location 3. According to Figure 6 there are two possibilities, i.e., making their way through location 2 or through locations 2, 4, 5 and 6. The system starts processing two branches for each next possible location: one branch lets the execution continue if the location is valid (i.e., it is included in the possible locations set) and the user is moving towards it, the other stands for the situation in which the location is not valid and the user is moving to it. In the former one, the system keeps guiding the user to the next point, whereas in the latter it alerts the user and guides him to the correct track. The execution of an alternative combination of what was previous described is suspended since that move is not valid (e.g., alert the user if the next location is valid).

During the execution of a particular move that may change positions, the information source may be queried. While this information is not returned the program continues its execution, so the moves which are queried $\sim user\_travel(1,2)$ @gps_sensor are paused (since the default value is negative). The only move that proceeds is the one that assumes the default value $user\_travel(1,2)$@gps_sensor. Whenever an answer is returned from the information source, the computation is revised. The current move may be paused and a previously paused move may be resumed. The system may re-initialize, pause, or resume moves until the destination point is reached.

Instead of suspending the computation due to not knowing the answer to set the next user's movement, Speculative Computation enables the continuity of the computational task, i.e., it sets a default value as the way to have an answer to set the next user's movement and therefore the computational process does not come to an end (e.g., Step 8).

At Step 3, $show\_next\_point(2)$ is assumed and the integrity constraints are checked. Thus, it is assured that there is no contradiction by checking if $\sim included(2)$@$recognizer$ is not derived. In this step an ordinary abduction operation is carried out.

When an answer is returned and it confirms the default value, nothing changes. Thus, in these situations Speculative Computation is in an advanced stage of the computation and it does not have to be revised (e.g., Step 11).

At different stages of the computational process, an ongoing procedure may have to be suspended once the askable literal is in opposition with the information contained in the CBS. For instance, this situation occurred at Step 33.

The execution trace referred to above denotes an example of the computational process. Its purpose is to check and to demonstrate that it is possible to set the orientation problem through speculative computation. Indeed, according to the proof procedure referred to above, the application of Speculative Computation is an expressive and adequate option.

5. Applying Trajectory Pattern Mining to Adapt the Path

The module of Speculative Computation described in Section 4 has the goal of guiding the user and predicting possible wrongdoings. Indeed, the user may be alerted before taking the wrong turn and possibly getting lost. However, this speculative module needs to be fed with the travelling route, including the points where an alert should be triggered.

The Trajectory Pattern Mining Module has a double goal: to adapt the route to the user and to load this information into the Speculative Computation Module. By adapting the path to the user preferences, it is possible to achieve better results in the guiding process, since it is the system that adapts to the user needs and not the opposite. On the other hand, if one considers the cognitive disabilities of the user, it is important to highlight that a known trajectory (or part of the entire trajectory), despite being longer, is preferred by the user. Thus, the usual rigid way of traditional guiding systems is overcome. Indeed, the default method for calculating the best route (e.g., shortest path) is only considered when the system does not have information about the user habits (already travelled paths).

A sequence of time-stamped Cartesian coordinates defines the trajectory (or route) done by a person with cognitive disabilities. This data is obtained from the GPS sensor of the mobile device. Indeed, the location is composed by latitude, longitude, and timestamp values. This process resorts to an active recording of the user location whenever he is receiving the necessary aid from the application, i.e., is travelling using the guiding information provided by the application. Figure 7 depicts the trajectory mining process. This may derive the trajectory by two different approaches based on the previous information that was stored. If there is no information about the user, since he may be using the application for the first time, or when the chosen destination does not have enough information to retrieve a pattern, the system calculates the path based on the default routing algorithm. This step is achieved through the external tool GraphHopper [40]. On the other hand, with the continuous use of the system, it is possible to retrieve previous information about the person with cognitive disabilities and to use this available spatial-temporal data to provide the best adapted trajectory to the user.

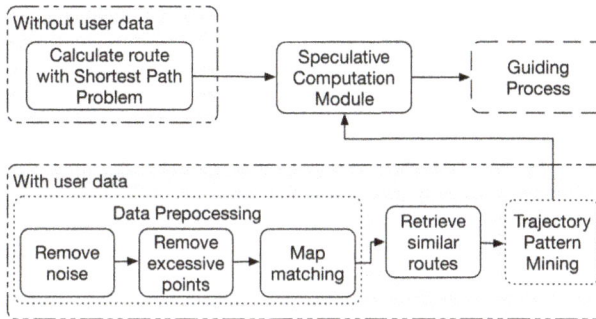

Figure 7. Process for generating adapted routes.

The raw data obtained from the device GPS module needs to be pre-processed in order to be used in the mining process to adapt the path to the user preferences. Indeed, this pre-processing phase (Figure 7) includes noise removal (from the raw data), excessive points removal (locations that do not increase the detail of the trajectory), and map matching. The noise removal phase is responsible for removing all locations where the GPS error is notorious, i.e., locations that, for some reason, like signal reflection in the surrounding buildings, have an error higher than the usual. In the second step, it is important to reduce the number of points by eliminating those that do not add value to the mining process, i.e., locations that are in the middle of a straight line. Finally, the last process is the map matching, where the goal is to get the remaining data and match it to the roads defined in the Open Street Map (Open Street Map, available at http://www.openstreetmap.org/) databases. Figure 8 illustrates the pre-processing stage. In Figure 8a, the raw data recorded from the mobile device GPS module is represented on the map, whereas in Figure 8b the final stage of the pre-processing phase is visible, which includes noise removal and map matching. For this pre-processing phase we use GraphHopper Directions Api [40], which has a map matching feature and is an open source tool.

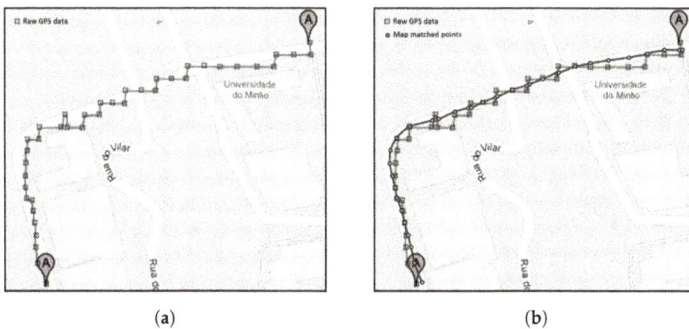

(a)   (b)

Figure 8. Trajectory data pre-processing (extracted from [41]). (a) Route obtained from database raw data; (b) Error removal from raw data.

When the pre-processing stage ends, the trajectory data mining method may start and extract the frequent trajectories (patterns). Without these previous steps, the results would be worse due to the noise that would be considered in the mining process. For the used trajectory data mining, two different methods are being used with good results: the PrefixSpan algorithm [42], which mines frequent sequential patterns; and OPTICS [43], which is a clustering algorithm that groups the locations in clusters. For this process we rely on an open source Application Programming Interface (API) named "SPMF - An Open-Source Data Mining Library" [44] that provides an implementation of both algorithms. Using the PrefixSpan algorithm, the extraction of frequent trajectories considers all travelling paths that have been done by the user. The results of the algorithm (frequent trajectories) consider the support of a sequential pattern, i.e., an input value that defines the minimum percentage of sequences in which the pattern occurs, considering the entire set of travelled paths. To adjust the path to the user preferences, in the mining process the system only considers paths travelled by that user, i.e., the user that is using the application and receiving the guiding information. The execution of the algorithm searches for patterns in the dataset according to the defined threshold. The obtained results (patterns) may represent an entire trajectory, or be part of the trajectory. In the former, the returned path represents the entire path that will be used to guide the user. There may be a need to adjust the starting point, but the remainder of the path (considering the selected destination) is the same. In the latter, a new travelling route is created based on the mined patterns, which is adapted to the user. When the recently created route is much longer than the default path, the system uses the shortest one, instead of the adapted one.

By applying the clustering method (OPTICS) the processed spatial-temporal data is grouped in clusters. For this process it is important to define some input parameters like the minimum number of points that are needed to define a core and its neighborhood, and the radius ϵ that defines the maximum distance that a point may be from the core of the cluster (defining its neighborhood). After executing the OPTICS algorithm, the retrieved locations are grouped in clusters. Thus, it is necessary to create the connections between them in order to obtain a trajectory and guide the user from his current location to the pre-selected destination.

In the case of using the shortest path or in the case of applying the trajectory data mining method, the obtained trajectory needs to be converted in order to be used in the execution of the speculative computation.

6. Conclusions

An active participation in society is vital for people with cognitive disabilities and, having this in mind, numerous scholars have studied and developed various methods to orientate them, both indoors and outdoors. The independent traveling allows the user to maintain an employment (within their capabilities). The degree of independence for the main user and for the caregiver is increased, since the person with disabilities can be remotely controlled, and the caregiver may be otherwise engaged without neglecting his role as a caregiver.

The present orientation systems have to be adapted to the characteristics of the users, i.e., these systems must be easy to operate when the orientation is being performed. If these system specifications are not matched, it may imply a more complex mental activity, leaving the user confused. In this situation the orientation system may become infeasible.

For people with reduced orientation capabilities there are several projects under consideration. Using these systems, people with cognitive disabilities may have a more active life, reducing the worry of getting lost both indoors and outdoors. Depending on the type of cognitive disability and its degree of incidence, the user can keep a job or perform other activities. However, the user has to learn two types of orientation systems, since a system adapted to the outside world cannot be used indoors or vice versa. The system presented in this work stands for an answer to all these shortcomings, making a move to providing a better quality of life to disabled people.

Through this work we present an orientation system for people with mild (or moderate) cognitive disabilities that has an anticipation feature that alerts the user before he takes the wrong turn. This is achieved by applying a Speculative Computation module, which needs to be loaded by the traveling path before its execution. This specificity enables the independent development of a pattern mining module, so that travel path is adapted to the user (instead of using the shortest one by default). Caregivers are not disregarded and a localization feature was considered. Thus, these users may remotely access the traveling path of the person with disabilities.

In the future we intend to test the entire system in real case scenarios in order to check and, if necessary, add new features or improve the ones that were developed. People with cognitive disabilities that are able to use a smartphone will be selected for these real tests, since the performed tests were done using non-impaired people who tried to resemble a cognitive disability issue.

Author Contributions: All the authors proposed ideas for developing and improving the system; J.R. was the main programmer of the system; J.R. and T.O. developed and programmed the Speculative Computation Module; all the authors contributed to the analysis and interpretation of the data; and all the authors wrote the paper in a collaborative way.

Acknowledgments: This work has been supported by COMPETE: POCI-01-0145-FEDER-007043 and FCT—Fundação para a Ciência e Tecnologia within the Project Scope: UID/CEC/00319/2013. The work of João Ramos is supported by a doctoral FCT grant SFRH/BD/89530/2012. The work of Tiago Oliveira is also supported by the FCT grant with the reference SFRH/BD/85291/2012.

Conflicts of Interest: The authors declare no conflict of interest.

Appendix A. Execution Example of the Speculative Computation Framework

An execution example is given below in which the user is guided from location 1 to location 3. In order to reduce the amount of text in the example, we used the abbreviated form of the elements of each tuple. Thus $u(1,2)@g$ is the askable literal $user\_travel(1,2)@gps\_sensor$; $i(1)@r$ is the literal $included(1)@recognizer$; $p(1,2)$ represents $path(1,2)$; $snp(2)$ is the predicate $show\_next\_point(2)$; and $suw(2)$ represents the predicate $show\_user\_warning$. The sets AAQ and CBS are shown when a change occurs.

In this execution example the turn instructions consider the locations depicted in Figure 6. Thus, the system calculates the adapted traveling path, which takes locations 1-2-3, and that the user will probably take the wrong turn towards location 4. In this case a warning is prompted to the user which, for this example, will make the user take the correct path.

1.
$$PS = \{< \{guide(1,3)\}, \varnothing, \varnothing >\}$$
$$AAQ = \varnothing$$
$$CBS = \{u(1,2)@g, u(2,3)@g, u(2,4)@g, u(4,5)@g,$$
$$u(5,6)@g, u(6,3)@g, i(1)@r, i(2)@r,$$
$$i(3)@r, \sim i(4)@r, i(5)@r, i(6)@r\}$$

2. By **Case 3.1**
$$PS = \{< \{p(1,Y), snp(Y), u(1,Y)@g, guide(Y,3)\},$$
$$\varnothing, \varnothing >, < \{p(1,Y), u(1,Y)@g,$$
$$suw(Y), guide(Y,3)\}, \varnothing, \varnothing >\}$$

3. By **Case 3.1**
$$PS = \{< \{snp(2), u(1,2)@g, guide(2,3)\}, \varnothing, \varnothing >,$$
$$< \{u(1,2)@g, suw(2), guide(2,3)\}, \varnothing, \varnothing >\}$$

4. By **Case 3.2.3**
$$PS = \{< \{fail(\{\sim i(2)@r\}), u(1,2)@g, guide(2,3)\},$$
$$\varnothing, \{snp(2)\} >, P_1 \ (P_1 = < \{u(1,2)@g, suw(2), guide(2,3)\}, \varnothing, \varnothing >)\}$$

5. By **Case 3.3.2**

$PS \quad = \quad \{< \{fail(\varnothing), u(1,2)@g, guide(2,3)\}, \varnothing, \{snp(2)\} >,$
$\qquad\qquad < \{i(2)@r, u(1,2)@g, guide(2,3)\}, \varnothing, \{snp(2)\} >,$
$\qquad\qquad P_1\}$

6. By **Case 3.3**

$PS \quad = \quad \{< \{i(2)@r, u(1,2)@g, guide(2,3)\}, \varnothing, \{snp(2)\} >,$
$\qquad\qquad P_1\}$

7. By **Case 3.4**

$i(2)$ **is asked to the sensor** r

$\qquad PS \quad = \quad \{< \{u(1,2)@g, guide(2,3)\}, \{i(2)@r\},$
$\qquad\qquad\qquad \{snp(2)\} >, P_1\}$
$\qquad AAQ \quad = \quad \{i(2)@r\}$

8. By **Case 3.4**

$u(1,2)$ **is asked to the sensor** g

$\qquad PS \quad = \quad \{< \{guide(2,3)\}, \{i(2)@r, u(1,2)@g\},$
$\qquad\qquad\qquad \{snp(2)\} >, P_1\}$
$\qquad AAQ \quad = \quad \{i(2)@r, u(1,2)@g\}$

From the beginning to this iteration the system shows location 2 as the next destination to the user and, since the real value has not been retrieved from the information source, the system assumed that the user will correctly travel towards location 2. Continuing the current execution branch, the system will derivate the possibility of correctly traveling to location 3 and indicate this to the user, or the user may turn to location 3 and this is considered as a user mistake.

9. By **Case 3.1**

$PS \quad = \quad \{< \{p(2,Y), snp(Y), u(2,Y)@g, guide(Y,3)\},$
$\qquad\qquad \{i(2)@r, u(1,2)@g\}, \{snp(2)\} >,$
$\qquad\qquad < \{p(2,Y), u(2,Y)@g, suw(Y), guide(Y,3)\},$
$\qquad\qquad \{i(2)@r, u(1,2)@g\}, \{snp(2)\} >, P_1\}$

10. By **Case 3.1**

$PS \quad = \quad \{< \{snp(3), u(2,3)@g, guide(3,3)\},$
$\qquad\qquad \{i(2)@r, u(1,2)@g\}, \{snp(2)\} >,$
$\qquad\qquad < \{snp(4), u(2,4)@g, guide(4,3)\},$
$\qquad\qquad \{i(2)@r, u(1,2)@g\}, \{snp(2)\} >,$
$\qquad\qquad < \{u(2,3)@g, suw(3), guide(3,3)\},$
$\qquad\qquad \{i(2)@r, u(1,2)@g\}, \{snp(2)\} >,$
$\qquad\qquad < \{u(2,4)@g, suw(4), guide(4,3)\},$
$\qquad\qquad \{i(2)@r, u(1,2)@g\}, \{snp(2)\} >,$
$\qquad\qquad P_1\}$

11. $u(1,2)$ is returned from g

Nothing changes.

12. By **Case 3.2.3**

$PS \quad = \quad \{< \{fail(\{\sim i(3)@r\}), u(2,3)@g, guide(3,3)\},$
$\qquad\qquad \{i(2)@r, u(1,2)@g\}, \{snp(2), snp(3)\} >,$
$\qquad\qquad P_2 \ (P_2 =< \{snp(4), u(2,4), guide(4,3)\}, \{i(2)@r, u(1,2)@g\}, \{snp(2)\} >),$
$\qquad\qquad P_3 \ (P_3 =< \{u(2,3)@g, suw(3), guide(3,3)\}, \{i(2)@r, u(1,2)@g\}, \{snp(2)\} >),$
$\qquad\qquad P_4 \ (P_3 =< \{u(2,4)@g, suw(4), guide(4,3)\}, \{i(2)@r, u(1,2)@g\}, \{snp(2)\} >),$
$\qquad\qquad P_1 \}$

13. By **Case 3.3.2**

$$PS = \{< \{fail(\varnothing), u(2,3)@g, guide(3,3)\},$$
$$\{i(2)@r, u(1,2)@g\}, \{snp(2), snp(3)\} >,$$
$$< \{i(3)@r, u(2,3)@g, guide(3,3)\}, \{i(2)@r,$$
$$u(1,2)@g\}, \{snp(2), snp(3)\} >,$$
$$P_2, P_3, P_4, P_1\}$$

14. By **Case 3.3**

$$PS = \{< \{i(3)@r, u(2,3)@g, guide(3,3)\},$$
$$\{i(2)@r, u(1,2)@g\}, \{snp(2), snp(3)\} >,$$
$$P_2, P_3, P_4, P_1\}$$

15. By **Case 3.4**

$i(3)$ **is asked to the sensor** r

$$PS = \{< \{u(2,3)@g, guide(3,3)\},$$
$$\{i(2)@r, u(1,2)@g, i(3)@r\}, \{snp(2), snp(3)\} >,$$
$$P_2, P_3, P_4, P_1\}$$
$$AAQ = \{i(2)@r, u(1,2)@g, i(3)@r\}$$

16. By **Case 3.4**

$u(2,3)$ **is asked to the sensor** g

$$PS = \{< \{guide(3,3)\}, \{i(2)@r, u(1,2)@g,$$
$$i(3)@r, u(2,3)@g\}, \{snp(2), snp(3)\} >,$$
$$P_2, P_3, P_4, P_1\}$$
$$AAQ = \{i(2)@r, u(1,2)@g, i(3)@r, u(2,3)@g\}$$

17. By **Case 3.1**

$$PS = \{< \varnothing, \{i(2)@r, u(1,2)@g, i(3)@r, u(2,3)@g\},$$
$$\{snp(2), snp(3)\} >,$$
$$< \{snp(4), u(2,4), guide(4,3)\},$$
$$\{i(2)@r, u(1,2)@g\}, \{snp(2)\} >,$$
$$P_3, P_4, P_1\}$$

At this iteration the current branch has reached the end, i.e., following this execution branch the user will correctly travel between location 1 and 3. Thus, after considering this as a possible solution to the guidance process, the system starts derivating a suspended branch.

18. By **Case 3.2.3**

$$PS = \{< \{fail(\sim i(4)@r), u(2,4), guide(4,3)\},$$
$$\{i(2)@r, u(1,2)@g\}, \{snp(2), snp(4)\} >,$$
$$P_3, P_4, P_1, P_5 \ (P_5 = < \varnothing, \{i(2)@r, u(1,2)@g, i(3)@r, u(2,3)@g\}, \{snp(2), snp(3)\} >) \}$$

19. By **Case 3.3.2**

$$PS = \{< \{fail(\varnothing), u(2,4), guide(4,3)\},$$
$$\{i(2)@r, u(1,2)@g\}, \{snp(2), snp(4)\} >,$$
$$< \{i(4)@r, u(2,4), guide(4,3)\},$$
$$\{i(2)@r, u(1,2)@g\}, \{snp(2), snp(4)\} >,$$
$$P_3, P_4, P_1, P_5\}$$

20. By **Case 3.3**

$$PS = \{< \{i(4)@r, u(2,4), guide(4,3)\},$$
$$\{i(2)@r, u(1,2)@g\}, \{snp(2), snp(4)\} >,$$
$$P_3, P_4, P_1, P_5\}$$

While waiting for the answer about the inclusion of location 4 as a valid travel point, the system assumes that it is not part of the included set and suspends this execution branch.

21. By **Case 3.4**

$i(4)$ **is asked to the sensor** r

$$PS \quad = \quad \{< \{u(2,4), guide(4,3)\}, \{i(2)@r,$$
$$u(1,2)@g, i(4)@r\}, \{snp(2), snp(4)\} >,$$
$$< \{u(2,3)@g, suw(3), guide(3,3)\},$$
$$\{i(2)@r, u(1,2)@g\}, \{snp(2)\} >,$$
$$P_4, P_1, P_5\}$$

$$AAQ \quad = \quad \{i(2)@r, u(1,2)@g, i(3)@r, u(2,3)@g, i(4)@r\}$$

22.

$u(2,3)$ **has already been asked to the sensor** g

$$PS \quad = \quad \{< \{suw(3), guide(3,3)\}, \{i(2)@r,$$
$$u(1,2)@g, u(2,3)@g\}, \{snp(2)\} >,$$
$$P_4, P_1, P_5, P_6 \ (P_6 =< \{u(2,4), guide(4,3)\}, \{i(2)@r, u(1,2)@g, i(4)@r\}, \{snp(2), snp(4)\} >) \}$$

23. By **Case 3.2.3**

$$PS \quad = \quad \{< \{fail(i(3)@r), guide(3,3)\}, \{i(2)@r,$$
$$u(1,2)@g, u(2,3)@g\}, \{snp(2), suw(3)\} >,$$
$$P_4, P_1, P_5, P_6\}$$

24. By **Case 3.3.2**

$$PS \quad = \quad \{< \{fail(\varnothing), guide(3,3)\}, \{i(2)@r,$$
$$u(1,2)@g, u(2,3)@g\}, \{snp(2), suw(3)\} >,$$
$$< \{\sim i(3)@r, guide(3,3)\}, \{i(2)@r,$$
$$u(1,2)@g, u(2,3)@g\}, \{snp(2), suw(3)\} >,$$
$$P_4, P_1, P_5, P_6\}$$

25. By **Case 3.3**

$$PS \quad = \quad \{< \{\sim i(3)@r, guide(3,3)\}, \{i(2)@r,$$
$$u(1,2)@g, u(2,3)@g\}, \{snp(2), suw(3)\} >,$$
$$P_4, P_1, P_5, P_6\}$$

Since location 3 is in the set of the valid points to travel, the current execution branch is suspended since it is contrary to the real value.

26.

$i(3)$ **has already been asked to the sensor** r

$$PS \quad = \quad \{< \{guide(3,3)\}, \{i(2)@r, u(1,2)@g,$$
$$u(2,3)@g, \sim i(3)@r\}, \{snp(2), suw(3)\} >,$$
$$< \{u(2,4)@g, suw(4), guide(4,3)\},$$
$$\{i(2)@r, u(1,2)@g\}, \{snp(2)\} >, P_1, P_5, P_6\}$$

27. By **Case 3.4**

$u(2,4)$ **is asked to the sensor** g

$$PS \quad = \quad \{< \{suw(4), guide(4,3)\}, \{i(2)@r,$$
$$u(1,2)@g, u(2,4)@g\}, \{snp(2)\} >,$$
$$P_1, P_5, P_6, P_7 \ (P_7 =< \{guide(3,3)\}, \{i(2)@r, u(1,2)@g, u(2,3)@g, \sim i(3)@r\}, \{snp(2), suw(3)\} >) \}$$

$$AAQ \quad = \quad \{i(2)@r, u(1,2)@g, i(3)@r, u(2,3)@g$$
$$i(4)@r, u(2,4)@g\}$$

$\cdots$

28. By **Case 3.2.3**

$$PS \quad = \quad \{< \{fail(i(4)@r, guide(4,3)\}, \{i(2)@r,$$
$$u(1,2)@g, u(2,4)@g\}, \{snp(2), suw(4)\} >,$$
$$P_1, P_5, P_6, P_7\}$$

29. By **Case 3.3.2**

$$PS \quad = \quad \{< \{fail(\varnothing), guide(4,3)\}, \{i(2)@r,$$
$$u(1,2)@g, u(2,4)@g\}, \{snp(2), suw(4)\} >,$$
$$< \{\sim i(4)@r, guide(4,3)\}, \{i(2)@r,$$
$$u(1,2)@g, u(2,4)@g\}, \{snp(2), suw(4)\} >,$$
$$P_1, P_5, P_6, P_7\}$$

30. By **Case 3.3**

$$PS \quad = \quad \{< \{\sim i(4)@r, guide(4,3)\}, \{i(2)@r,$$
$$u(1,2)@g, u(2,4)@g\}, \{snp(2), suw(4)\} >,$$
$$P_1, P_5, P_6, P_7\}$$

31.

$i(4)$ **has already been asked to the sensor** r

$$PS \quad = \quad \{< \{guide(4,3)\}, \{i(2)@r, u(1,2)@g,$$
$$u(2,4)@g, \sim i(4)@r\}, \{snp(2), suw(4)\} >,$$
$$P_1, P_5, P_6, P_7\}$$

32. $u(2,3)$ is returned from g
Nothing changes.

33. $\sim u(2,4)$ is returned from g
By **Fact Arrival Phase**:

$$CBS \quad = \quad \{u(1,2)@g, u(2,3)@g, \sim u(2,4)@g, u(4,5)@g,$$
$$u(5,6)@g, u(6,3)@g, i(1)@r, i(2)@r,$$
$$i(3)@r, \sim i(4)@r, i(5)@r, i(6)@r\}$$

At this iteration all branches have been derived to their maximum considering the real values returned from the information sources or the ones in the initial default set. Thus, the system is able to return the outside defaults and inside assumptions which, for this execution example, is to indicate the user to travel to location 2 and then to location 3.

34.
$$\{i(2)@r, u(1,2)@g, i(3)@r,$$
$$u(2,3)@g\} \text{ and } \{snp(2), snp(3)\}$$
are returned as outside defaults and inside assumptions.

References

1. American Psychiatric Association. *Diagnostic and Statistical Manual of Mental Disorders: DSM-IV*, 4th ed.; American Psychiatric Association: Washington, DC, USA, 1994; p. 915.
2. Schalock, R.L.; Borthwick-Duffy, S.A.; Bradley, V.J.; Buntinx, W.H.E.; Coulter, D.L.; Craig, E.M.; Gomez, S.C.; Lachapelle, Y.; Luckasson, R.; Reeve, A.; et al. *Intellectual Disability: Definition, Classification, and Systems of Supports*, 11th ed.; American Association on Intellectual and Developmental Disabilities: Washington, DC, USA, 2010; p. 259.
3. WebAIM—Web Accessibility In Mind. *Cognitive Disabilities*; WebAIM: Logan, UT, USA, 2007. Available online: http://www.webaim.org/articles/cognitive/ (accessed on 15 June 2016).
4. Dawe, M. Desperately seeking simplicity: How young adults with cognitive disabilities and their families adopt assistive technologies. In Proceedings of the SIGCHI Conference on Human Factors in Computing Systems, Montréal, QC, Canada, 22–27 April 2006; ACM: New York, NY, USA, 2006; pp. 1143–1152.
5. American Psychiatric Association. *Diagnostic and Statistical Manual of Mental Disorders (DSM-5)*, 5th ed.; American Psychiatric Publishing: Washington, DC, USA, 2013; p. 991.
6. Woods, B.; Aguirre, E.; Spector, A.E.; Orrell, M. Cognitive stimulation to improve cognitive functioning in people with dementia. *Cochrane Database Syst. Rev.* **2012**, *2*. [CrossRef] [PubMed]

7.   United Nations—Division for Social Policy and Development Disability. *United Nations Convention on the Rights of Person's with Disabilities*; United Nations: New York, USA, 2006.

8.   Sadri, F. Multi-Agent Ambient Intelligence for Elderly Care and Assistance. In Proceedings of the AIP Conference, Vancouver, BC, Canada, 25–29 June 2007; Volume 2007, pp. 117–120.

9.   *AAL Joint Programme Call AAL-2010-3 Project no. AAL 2010-3-012*; Mylife: Losangeles, CA, USA, 2012.

10.   Lanyi, C.S.; Brown, D.J. Design of Serious Games for Students with Intellectual Disability. In Proceedings of the International Conference on Interaction Design International Development, Mumbai, India, 20–24 March 2010; Joshi, A., Dearden, A., Eds.; British Computer Society: Swinton, UK, 2010; pp. 44–54.

11.   Liu, A.L.; Hile, H.; Kautz, H.; Borriello, G.; Brown, P.A.; Harniss, M.; Johnson, K. Indoor Wayfinding: Developing a Functional Interface for Individuals with Cognitive Impairments. In Proceedings of the 8th International ACM SIGACCESS Conference on Computers and Accessibility (ASSETS), Portland, OR, USA, 23–25 October 2006.

12.   Liu, A.L.; Hile, H.; Kautz, H.; Borriello, G.; Brown, P.A.; Harniss, M.; Johnson, K. Indoor wayfinding: Developing a functional interface for individuals with cognitive impairments. *Disabil. Rehabil. Assist. Technol.* **2008**, *3*, 69–81. [CrossRef] [PubMed]

13.   Carmien, S.; Dawe, M.; Fischer, G.; Gorman, A.; Kintsch, A.; Sullivan, J.F. Socio-technical environments supporting people with cognitive disabilities using public transportation. *ACM Trans. Comput.-Hum. Interact.* **2005**, *12*, 233–262. [CrossRef]

14.   Dawe, M. Let Me Show You What I Want: Engaging Individuals with Cognitive Disabilities and their Families in Design. In Proceedings of the Extended Abstracts on Human Factors in Computing Systems, San Jose, CA, USA, 28 April–3 May 2007; pp. 2177–2182.

15.   Liu, A.L.; Hile, H.; Borriello, G.; Kautz, H.; Brown, P.A.; Harniss, M.; Johnson, K. Informing the Design of an Automated Wayfinding System for Individuals with Cognitive Impairments. In Proceedings of the 3rd International Conference on Pervasive Computing Technologies for Healthcare, London, UK, 1–3 April 2009; Volume 9, p. 8.

16.   Ramos, J.; Anacleto, R.; Costa, A.; Novais, P.; Figueiredo, L.; Almeida, A. Orientation System for People with Cognitive Disabilities. In *Ambient Intelligence—Software and Applications*; Novais, P., Hallenborg, K., Tapia, D.I., Rodríguez, J.M.C., Eds.; Springer: Berlin/Heidelberg, Germany, 2012; Volume 153, pp. 43–50.

17.   Ramos, J.; Anacleto, R.; Novais, P.; Figueiredo, L.; Almeida, A.; Neves, J. Geo-localization System for People with Cognitive Disabilities. In *Trends in Practical Applications of Agents and Multiagent Systems*; Pérez, J.B., Hermoso, R., Moreno, M.N., Rodríguez, J.M.C., Hirsch, B., Mathieu, P., Campbell, A., Suarez-Figueroa, M.C., Ortega, A., Adam, E., et al., Eds.; Springer: Cham, Switzerland, 2013; Volume 221, pp. 59–66.

18.   Fraunnhover Portugal. AlzNav. Available online: http://www.fraunhofer.pt/en/fraunhoferaicos/-projects/internalresearch/alznav.html (accessed on 10 December 2012).

19.   Patterson, D.J.; Liao, L.; Fox, D.; Kautz, H. Inferring High-Level Behavior from Low-Level Sensors. In *International Conference on Ubiquitous Computing*; Springer: Berlin/Heidelberg, Germany, 2003; Volume 2864, pp. 73–89.

20.   Sadilek, A.; Kautz, H. Recognizing Multi-Agent Activities from GPS Data. *Artif. Intell.* **2010**, *39*, 1134–1139.

21.   Russell, S.; Norvig, P. *Artificial Intelligence: A Modern Approach*; Prentice Hall Series In Artificial Intelligence; Chapter Probabilis; Prentice Hall: Upper NJ River, NJ, USA, 2003; Volume 60, pp. 269–272.

22.   Hertzberg, J.; Chatila, R. AI Reasoning Methods for Robotics. In *Springer Handbook of Robotics*; Siciliano, B., Khatib, O., Eds.; Springer: Berlin/Heidelberg, Germany, 2008; Chapter 10, pp. 207–223.

23.   Dahan, H.; Cohen, S.; Rokach, L.; Maimon, O. *Proactive Data Mining with Decision Trees*; Springer Science & Business Media: Berlin, Germany, 2014; p. 94.

24.   Rokach, L.; Maimon, O. Decision Trees. In *Data Mining and Knowledge Discovery Handbook*; Springer US: Boston, MA, USA, 2005; pp. 165–192.

25.   Basheer, I.A.; Hajmeer, M. Artificial neural networks: Fundamentals, computing, design, and application. *J. Microbiol. Methods* **2000**, *43*, 3–31. [CrossRef]

26.   Jain, A.K.; Mao, J.C.; Mohiuddin, K.M. Artificial neural networks: A tutorial. *Computer* **1996**, *29*, 31–44. [CrossRef]

27.   Noble, W.S. What is a support vector machine? *Nat. Biotechnol.* **2006**, *24*, 1565–1567. [CrossRef] [PubMed]

28.   Bennett, K.P.; Campbell, C. Support vector machines: Hype or hallelujah? *ACM SIGKDD Explor. Newslett.* **2000**, *2*, 1–13. [CrossRef]

29. Uusitalo, L. Advantages and challenges of Bayesian networks in environmental modelling. *Ecol. Model.* **2007**, *203*, 312–318. [CrossRef]

30. Liu, A.L.; Hile, H.; Borriello, G.; Brown, P.A.; Harniss, M.; Kautz, H.; Johnson, K. Customizing directions in an automated wayfinding system for individuals with cognitive impairment. In Proceedings of the Eleventh International ACM SIGACCESS Conference on Computers and Accessibility (ASSETS), Pittsburgh, PA, USA, 25–28 October 2009; p. 27.

31. Kolodner, J.L. An introduction to case-based reasoning. *Artif. Intell. Rev.* **1992**, *6*, 3–34. [CrossRef]

32. Satoh, K. Speculative Computation and Abduction for an Autonomous Agent. *IEICE Trans. Inf. Syst.* **2005**, *88*, 2031–2038. [CrossRef]

33. Kakas, A.C.; Kowalski, R.A.; Toni, F. The Role of Abduction in Logic Programming. *Handb. Log. Artif. Intell. Log. Progr.* **1998**, *5*, 235–324.

34. Jain, P. Wireless Body Area Network for Medical Healthcare. *IETE Tech. Rev.* **2011**, *28*, 362. [CrossRef]

35. Montón, E.; Hernandez, J.; Blasco, J.; Hervé, T.; Micallef, J.; Grech, I.; Brincat, A.; Traver, V. Body area network for wireless patient monitoring. *IET Commun.* **2008**, *2*, 215–222. [CrossRef]

36. Wolf, L.; Saadaoui, S. Architecture Concept of a Wireless Body Area Sensor Network for Health Monitoring of Elderly People. In Proceedings of the 4th IEEE Consumer Communications and Networking Conference, Las Vegas, NV, USA, 11–13 Janurary 2007; pp. 722–726.

37. Shamoo, A.E.; Resnik, D.B. *Responsible Conduct of Research*, 3rd ed.; Oxford University Press: Oxford, UK, 2015; p. 360.

38. Yip, C.; Han, N.L.; Sng, B. Legal and ethical issues in research. *Indian J. Anaesth.* **2016**, *60*, 684–688. [PubMed]

39. Kakas, A.C.; Mancarella, P. On the relation between Truth Maintenance and Abduction. In Proceedings of the First Pacific Rim International Conference on Artificial Intelligence (PRICAI), Nagoya, Japan, 1990; pp. 438–443.

40. Karich, P.; Schroder, S. GraphHopper Directions API with Route Optimization. Available online: https://graphhopper.com (accessed on 20 June 2016).

41. Ramos, J.; Oliveira, T.; Satoh, K.; Neves, J.; Novais, P. An Orientation Method with Prediction and Anticipation Features. *Iberoam. J. Artif. Intell.* **2017**, *20*, 82–95. [CrossRef]

42. Pei, J.; Han, J.; Mortazavi-Asl, B.; Wang, J.; Pinto, H.; Chen, Q.; Dayal, U.; Hsu, M.C. Mining sequential patterns by pattern-growth: The PrefixSpan approach. *IEEE Trans. Knowl. Data Eng.* **2004**, *16*, 1424–1440.

43. Ankerst, M.; Breunig, M.M.; Kriegel, H.P.; Sander, J. Optics: Ordering points to identify the clustering structure. *ACM Sigmod Rec.* **1999**, *28*, 49–60. [CrossRef]

44. Fournier-Viger, P. SPMF—Open-Source Data Mining Library. Available online: http://www.philippe-fournier-viger.com/spmf/index.php (accessed on 30 June 2016).

![applied sciences logo] *applied sciences*

MDPI

Article

Computational Accountability in MAS Organizations with ADOPT

Matteo Baldoni *,†, Cristina Baroglio †, Katherine M. May †, Roberto Micalizio †
and Stefano Tedeschi †

Dipartimento di Informatica, Università degli Studi di Torino, via Pessinetto 12, I10149 Torino, Italy;
baroglio@di.unito.it (C.B.); katherine.may@edu.unito.it (K.M.M.); roberto.micalizio@unito.it (R.M.);
tedeschi@di.unito.it (S.T.)
* Correspondence: baldoni@di.unito.it; Tel.: +39-011-670-6756
† These authors contributed equally to this work.

Received: 28 February 2018; Accepted: 20 March 2018; Published: 23 March 2018

check for
updates

Abstract: This work studies how the notion of accountability can play a key role in the design and realization of distributed systems that are open and that involve autonomous agents that should harmonize their own goals with the organizational goals. The socio–technical systems that support the work inside human companies and organizations are examples of such systems. The approach that is proposed in order to pursue this purpose is set in the context of multiagent systems organizations, and relies on an explicit specification of relationships among the involved agents for capturing who is accountable to whom and for what. Such accountability relationships are created along with the agents' operations and interactions in a shared environment. In order to guarantee accountability as a design property of the system, a specific interaction protocol is suggested. Properties of this protocol are verified, and a case study is provided consisting of an actual implementation. Finally, we discuss the impact on real-world application domains and trace possible evolutions of the proposal.

Keywords: methodologies for agent-based systems; organizations and institutions; socio–technical systems; computational accountability; social commitments; agent-based programming

1. Introduction

The design of complex distributed systems, such as socio–technical systems (STSs), requires the coordination of activities that are carried out simultaneously by different software components, that is, the interfaces through which humans interact. Multiagent systems (MASs) allow tackling this problem by providing modeling, like [1–5], development approaches, like [6,7], and frameworks, like [8–11]. Broadly speaking, such solutions represent software components as goal-oriented, autonomous agents, which act in a shared environment and need to coordinate so as to achieve their goals [12].

This goal-oriented approach, in which modularity is realized through the assignment of subgoals to the agents, is criticized in [13–15] because it does not fit the realization of open systems, which comprise autonomous agents, each with their own goals to accommodate in a greater picture. For instance, an agent could be assigned a goal it has no capability to achieve, making the whole system vulnerable. Interaction and a representation of the responsibilities that are explicitly taken by the agents are suggested by those works to play a central role. The proposal that we present in this paper develops this perspective (i) by supplying a definition of computational accountability [14,16] and (ii) by providing both modeling and computational tools that can be used in actual implementations. Accountability is a well-known key resource inside human organizations: it fosters the creation of commitments that will help the organization to meet expectations, develop and align strategies, assign resources, analyze and react to failures, and adapt processes to evolving environmental conditions. In decision making, accountability constrains the set of acceptable alternatives, and thus it allows organization members to take decisions in a way that goes beyond the private beliefs, goals and psychological dispositions of the decision maker [17]. At the core of the proposal lies an interpretation of accountability as a design property. We will explain how it is possible to design systems where accountability is a property that is guaranteed by design, and where the monitoring of ongoing interactions will allow identifying behaviors that diverge from the expected. The proposed solution builds upon the equally important notions of control and of accountability relationship. In particular, the realization of computational accountability follows principles, explained in Section 3, that find a realization in the ADOPT (accountability-driven organization programming technique) protocol for creating and manipulating accountability relationships. Technically, the core of the proposal builds upon the notion of role and in the action of role adoption (or enactment), on one side, and on the concept of social commitment [18,19] on the other side. An early version of ADOPT was presented in [16]. In this paper we enhance the protocol by clearly separating the role adoption phase from the goal agreement phase, and by relying on well-known FIPA protocols to capture the message exchanges which create the accountability relationships (represented by social commitments), and make them evolve. Moreover, we present, as a case study, an extension of the JaCaMo framework [11] that implements the proposal.

From a practical perspective, the proposal can find a natural application in supporting self-regulatory initiatives in human organizations. For instance, many organizations and companies voluntarily adopt monitoring and accountability frameworks, for example [20,21], but such frameworks are currently very-little supported by software and information systems. The commitments that involve the parties are basically hand-written by filling in forms [22,23], and the assessment of satisfaction or violation of the involved liabilities, as well as the actual accounting process, are totally handled by authorized human parties [24]. The problem is that when accountability channels are informal or ambiguous, in the long run, all these processes that are at the heart of a healthy organization will be thwarted, leading to little effectiveness and poor performance. The ADOPT protocol and, more in general, MASs that provide accountability as a design property, would find immediate application in such contexts.

The paper is organized as follows. Section 2 explains the lack, due to relying mostly on goal-orientation, of current MAS approaches (in particular those concerning MAS organizations). It motivates the need of a change of perspective in order to enable the realization of accountability frameworks. Section 3 explains accountability, getting into the depths of computational accountability in organizational settings. Section 4 explains the ADOPT accountability protocol, including also the verification of properties that are granted by the protocol. Section 5 reports, as a case study, the implementation of ADOPT in the JaCaMo framework, which is one of the best-known, and widely used, MAS programming frameworks. Section 6 discusses the impact and related works, and Section 7 ends the paper with some conclusions.

2. Multiagent Systems Need Accountability

A common way to tackle coordination in MASs is to define an organization, that is, a "structure", within which interaction occurs. Inspired by the human organizations, a MAS organizational model includes a specification of organizational goals, and of a functional decomposition for achieving these goals through subgoals distributed among the agents in the organization. One key notion that is used in defining an organization is that of role. Organizational roles are typically understood as a way to delegate a complex task to different principals, abstracting from the actual individuals (the agents) who will play the roles. Following [25], when adopting a role, an agent, who satisfies some given requirements, acquires some powers inside that organizational context. A power extends the agent's capabilities, allowing it to operate inside the organization and to change the organizational state. Requirements, in turn, are needed capabilities that a candidate player must exhibit to play a particular role. This definition of role follows the one given in [26] in which roles are definitionally dependent from the organization they belong to: Roles do not exist as independent entities, but rather, they are linked by definition to the organization they belong to.

Typically, in organization-oriented approaches, the MAS designer focuses on the organization of the system, taking into account its main objectives, structure and social norms. OperA [2], for instance, defines an organization along three axes: An organizational model describes the desired overall behavior of the organization, and specifies the global objectives and the means to achieve them; a social model defines the agent population, maps agents to organizational roles, and defines social contracts (agreements) on role enactment; and the interaction model correlates role-enacting agents through specified interactions. OMNI [3], based on OperA, provides a top-down, layered model of organization. An abstract level describes the general aims of the organization through a list of externally observable objectives. A concrete level specifies the means to achieve the objectives identified in the abstract level. Finally, an implementation level describes the activity of the system as realized by the individual agents. In this perspective, individual agents will enact organizational roles as a means to realize their own goals. E-institutions [1] model organizations as social groups of individuals built to achieve common (shared or antagonistic) goals. Here, organizations usually provide a set of norms, which agents in the system need to follow, depending on the role they are playing. If an agent violates a norm, the system must detect the violation and act consequently, for example, applying a sanction or warning the other participants. OCeAN [4] is a conceptual framework that extends the concept of the E-institution. The authors state that a suitable meta-model for open interaction systems must comprise, besides norms, an ontology, a set of roles associated to institutional actions, and a set of linguistic conventions. Roles provide agents the capability to perform institutional actions, which are actions whose meaning and effects require a convention among participants of the system. OCeAN foresees that institutional actions are messages, whose sending is bound to an institutional action by means of a count-as relation. 2OPL [10] is a rule-based language to define norms within a MAS, in order to achieve global goals of the system. The state of the world is represented by means of brute facts, first-order atoms, and of institutional facts which model the normative state of the system, tracing occurred violations. In Tropos [7], a MAS is seen as an organization of coordinated autonomous agents that interact in order to achieve common goals. Considering real-world organizations as a metaphor, the authors propose some architectural styles which borrow concepts from organizational theories, such as joint ventures. The styles are modeled using the notions of actor, goal and actor dependency, and are intended to capture concepts like needs/wants, delegations and obligations. Finally, the organizational model adopted in JaCaMo [11] decomposes the specification of an organization into three dimensions. The structural dimension specifies roles, groups and links between roles in the organization. The functional dimension is composed of one (or more) scheme(s) that elicits how the global organizational goal(s) is (are) decomposed into subgoals and how these subgoals are grouped in coherent sets, called missions, to be distributed to the agents. Finally, the normative dimension binds the two previous dimensions by specifying the roles' permissions and obligations for missions.

Recently, however, such a goal-oriented perspective on organizations has been criticized [13,27]. The point is that this approach provides a viable solution for modularizing complex tasks in terms of subtasks, each of which is assigned to a specific agent, but it is not a good solution for handling open distributed systems where agents are autonomous. In such scenarios, in fact, agents' goals might be in contrast to each other. Indeed, the organizational goals may not overlap completely with agents' goals. It follows that agents' obedience to the system norms cannot be taken for granted. Agent autonomy demands a different way of conceptualizing software modularity: not in terms of subgoals that are assigned to the agents, but rather in terms of responsibilities that are explicitly taken on by the agents. In [13], the authors see interaction as a central element of their modeling approach, and model interactions in terms of mutual expectations among the agents, bringing the conception of accountability as a way of characterizing the "good behavior" of each of the involved agents. Here, accountability is a directed relationship from one agent to another, and reflects the legitimate expectations the second principal has of the first. The resulting approach, dubbed interaction-oriented software engineering (IOSE), focuses on social protocols, which specify how accountability relationships among the concerned principals progress through their interactions. In particular, as the authors themselves state, "in IOSE, it makes little sense to ask what functionality a role provides [...]; it makes sense though to ask to whom and for what is a role accountable [...]. A social protocol essentially describes how a principal playing a role would be embedded in the social world by way of accountability".

The essential message we want to stress, thus, is that the lack of an explicit treatment of the accountability relationship, within the existing organizational models, makes those models vulnerable. In this paper we have demonstrated this vulnerability in practice in the context of the JaCaMo platform. In particular, we show how the lack of an explicit take-on of responsibilities, for the distributed goals, thwarts goal achievement as well as the overall system capability of answering when some exceptional event occurs. We show, in fact, that when an agent joins an organization by playing a role, it is not aware of all the possible goals it will be asked to achieve as a role player. Thus, an agent can be assigned a goal for which it has no plan. The failure of the goal, though, cannot be attributed to the agent, nor to the organization, in an easy way. In fact, on the agent's side, the lack of a proper plan for the goal does not make the agent responsible. On the organization's side, the organization has no means to know what an agent can actually do, thus, the system cannot be considered as responsible either. In other terms, if an agent does not have the capabilities for achieving a goal, assigning that goal to it, even through an obligation, does not bring any closer the achievement of the goal of interest. In JaCaMo, the possibility to create goals along with the execution is a desired feature whose rationale is that goals pop up dynamically and cannot be foreseen; the problem is that agents, who are aware of their own capabilities, do not possess instruments for accepting or negotiating their goals.

Notably, these are not just features of JaCaMo, but of all the organizational models based on functional decomposition of goals. Those approaches substantially assume that the agents playing roles have been specifically designed for that purpose. This limits the openness of the system and the reuse of code. Instead, the design of an organization should rely on explicit relationships between agents, and also between agents and their organizations, capturing assumptions of responsibility by the agents. This would make the system accountable.

The organizational model, thus, should no longer be a structure that distributes goals to its agents, but it should become a way for coordinating responsibility assumption by the agents. More precisely, since agents are opaque (i.e., not inspectable even by the organization), and since no assumption on their capabilities can be done, an organization cannot assign a goal to an agent without putting the system in danger (as we have seen). The organization, however, can safely undertake an interaction protocol through which it negotiates with agents the attribution of goals. At the end of such a protocol, the agent itself takes on the responsibility of achieving a specific goal. The rationale here is that only the agent knows whether it has the control over a goal. Hence, if the agent accepts to bring about that goal, it also takes on the responsibility for the very same goal. The organization has therefore a legitimate expectation that the goal will be obtained. If the final outcome is not satisfactory, then, the agent is held to account for its conduct. Before explaining the accountability protocol, we characterize in the next section the key features of computational accountability.

3. Computational Accountability in Organizational Settings

Different research communities have dealt with the topic of accountability, such as [28–36]. While its main features remain relatively static, definitions vary in approach, scope and understanding in different communities. The cause of such variability lies with its socio–cultural nature and is one of the main reasons for the lack of a comprehensive support for the realization of accountability frameworks in current socio–technical systems. Accountability's most general definition refers to the assumption of responsibility for decisions and actions that a principal, individual or organization has towards another party. In other words, principals must account for their behavior to another when put under examination. The concept is inherently social, and provides a mechanism by which entities constrain one another's behavior [17]. The previously cited examination is usually carried out by an investigative entity, a forum of auditors [37,38]. The process can be divided into three main phases: (1) The forum receives all information regarding the principals' actions, effects, permissions, obligations and so on that led to the situation under scrutiny; (2) the forum contextualizes actions to understand their adequacy and legitimacy; and finally, (3) the forum passes judgment on agents with sanctions or rewards. Our goal consists of automating the entire process for use in multiagent organizations, although we will presently leave out the sanctioning piece of the third phase due to its domain-specific application.

With the term *computational accountability*, we mean the abilities, realized via software, to trace, evaluate and communicate accountability, in order to support the interacting parties and to help solve disputes. In modern organizations, in particular, accountability determination can be a way to obtain feedback useful to evaluate and possibly improve the processes put in place and, eventually, the overall structure, too. Accountability determination is an extremely complex task, as the following example highlights.

Example 1. *Alice and Bob are a painter and a bricklayer, called by Carol for estimating the cost of renovating a room. The walls in the room are very old and for this reason, before being painted, they should be spackled. Since the prices seem reasonable, Carol decides to hire both Alice and Bob with the following work plan. Firstly, Bob should spackle the walls and Alice should paint them white afterwards. Come execution time, Bob decides to use a new variety of dark-colored spackle, since he has an open tin of it and he had not received any precise instruction from Carol. The following day, when Alice finds the dark colored walls she realizes she will not be able to satisfy the commitment she made with Carol because, in order to do a nice job, she will have to use twice as much paint as expected. This simple example shows many challenges brought about when trying to tackle accountability in a computational way. Let us suppose the agreement between Alice and Carol was somehow formalized (for example, with a contract). Alice is unable to fulfill her contract with Carol. Should the simple fact that she made a commitment, which is now impossible for her to fulfill, cause a computational system to conclude she is accountable for the failure? Clearly, conditions have changed since the original room inspection. One may argue that contextual conditions, constituting the prerequisites for Alice for the execution of the work, should have been formalized. In the real world, however, contextual conditions that hardly change over time are presumed implicitly stipulated even when they are not formalized. How could Alice foresee that Bob would have used a particular kind of spackle while the great majority of the bricklayers use white ones? Is, then, Carol accountable? Perhaps she should have checked all the involved parties to be in condition to fulfill their tasks properly when she organized the work. On the other hand, we know that Bob is the one who arbitrarily changed the color of the spackle. However, he received no instruction about the desired color of the spackle from Alice nor from Carol. In other words, there was no reason to assume that a similar decision would have caused problems. Here the problem arises from the fact that the involved parties, despite being in a collaborative environment, have different expectations and rely on (conflicting) assumptions about some contextual conditions. An alternative ending of the story is that Alice, feeling responsible, will paint the room at the agreed price because she values the satisfaction of the contract more than earning money.*

In our society, accountability becomes possible because of shared meanings culturally accepted as interpretations of events. Without such attributed meanings, mechanisms of accountability become difficult to be realized for lack of consensus concerning the interpretation of the events in question. This means that accountability becomes impossible in the absence of collective interpretations and meaning attributions. Another difficulty lies in value attribution. How does one decide whether a given outcome is "good" or "bad"? Who decides? This depends on the value attribution mechanism adopted in the system. In different systems the same outcome could be judged in different ways. In other words, accountability requires the presence of a social-meaning-defining structure in which actions and outcomes can be interpreted and evaluated in a uniform manner. Software systems can provide such a structure in the form of an organization, which provides the tools and infrastructure needed to communicate and collaborate with the reassurance of shared meanings.

Our interest in applying accountability to the software world lies in two of its natural consequences: its diagnostic ability, that is, the ability to understand what went wrong, and its society-building aspect, that is, a system in which entities can know what most helps the encompassing organization and find encouragement to act on that knowledge to better it. With our concept of computational accountability, we strive towards a general goal of teaching agents correct behavior in the context of a particular organization. The mechanism of accountability contains two sides that we call a *positive* and a *negative* approach. Positive accountability means that an entity is socially expected to act in a certain way and will be held to account for that expectation's fulfillment. Negative accountability means that an entity is expected to not impede organizational progress and negatively impact others. In this work we focus on positive accountability, leaving the discussion related to negative accountability for future studies.

Accountability implies agency because if an agent does not possess the qualities to act "autonomously, interactively and adaptively", that is, with agency, there is no reason to speak of accountability, for the agent would be but a tool, and a tool cannot be held accountable [39]. As discussed in [40], entities are free to act as they wish even against behavioral expectations. Because accountability's domain lies in the past, it is only concerned with how entities acted, and places no restrictions on future actions. So, accountability does not hinder autonomy and allows entities the freedom to choose. On the other hand, neither does accountability unjustly single out those who have no control and are unable to act as autonomous beings. For example, during a robbery, a bank teller who hands over money would not be held accountable even though that person is an autonomous being who directly caused a financial loss, because that person had no "avoidance potential" [41], that is, no control over the situation. In an analogous fashion for positive accountability, entities must exhibit the possibility of action for accountable expectations: they must have control. Control is an extremely complex concept, related to the philosophical notion of free will. Restricting our attention to the scope of software agents, we cannot say, broadly speaking, that they have free will, but that they can exhibit some kind of control. Referring to [42], control can be defined as the the capability, possibly distributed among agents, of bringing about events. Ref. [43] gives a slightly different definition of control as the ability of an agent to maintain the truth value of a given state of affairs.

Accountability requires but is not limited to causal significance. Intuitively, for an entity to be held accountable for an outcome, that entity must be causally significant to that outcome [38]. However, accountability cannot be reduced only to a series of causes, that is, be reduced to the concept of traceability. Some authors, namely [40], even suggest that traceability is neither necessary nor sufficient for accountability. Accountability comes from social expectations that arise between principals in a given context. However, as implied by the word social, both principals must be aware of and approve the stipulated expectation in order to hold one another accountable. Accountability cannot be solely a result of principals' internal expectations. Expectations, here, follow the definition in [13], of ways to represent a given social state.

In order to make correct judgments, a forum must be able to observe the necessary relevant information. However, in order to maintain modularity, a forum should not observe beyond its scope. Organizations can be made up of other organizations. Societies, for example, contain micro-societies in which actions are encoded differently. In each micro-context the forum must be able to observe events and actions strictly contained in that context and decipher accountability accordingly. In each context, the forum must be able to observe events and actions strictly contained in its scope and decipher accountability accordingly. As context changes, accountability will change accordingly. Observability, thus, becomes integral for the forum to exercise its ability to process information. For this reason, a mechanism to compose different contexts and decide accountability comprehensively is essential.

The object under a forum's scrutiny can take the form of either an action or an outcome. The evaluation of the former implicates the recognition of social significance inherent in that action, relatively independently of where it leads. On the other hand, an evaluation of the latter denotes an importance in a state. When speaking of evaluating actions rather than outcomes in accountability, the examination implicitly involves a mapping between individual and social action, since the same individual action performed in different social contexts can take on different social significance. Therefore, a fundamental part of holding an individual accountable consists of identifying the social significance mapped from that individual's actions. In a social context an agent is accountable for an action to others, because of the realization that the others' goals depend on the outcome of the given actions.

When Does a MAS Support Accountability?

A MAS can be said to support accountability when it is built in such a way that accountability can be determined from any future institutional state. Consequently, the MAS must necessarily provide a structure that creates and collects contextualized information, so that accountability can actually be determined from any future institutional state. We consider integral to this process the following steps. A forum must receive all information (including all causal actions) regarding a given situation under scrutiny. The forum must be able to contextualize actions to understand their adequacy and legitimacy. Finally, the forum must be able to pass judgment on agents.

We identify the following necessary-but-not-sufficient principles a MAS must exhibit in order to support the determination of accountability.

Principle 1 All collaborations and communications subject to considerations of accountability among the agents occur within a single scope that we call organization.

Principle 2 An agent can enroll in an organization only by playing a role that is defined inside the organization.

Principle 3 An agent willing to play a role in an organization must be aware of all the powers associated with such a role before adopting it.

Principle 4 An agent is only accountable, towards the organization or another agent, for those goals it has explicitly accepted to bring about.

Principle 5 An agent must have the leeway for putting before the organization the provisions it needs for achieving the goal to which it is committing. The organization has the capability of reasoning about the requested provisions and can accept or reject them.

Principle 1 calls for situatedness. Accountability must operate in a specific context because individual actions take on their significance only in the presence of the larger whole. What constitutes a highly objectionable action in one context could instead be worthy of praise in another. Correspondingly, a forum can only operate in context, and an agent's actions must always be contextualized. The same role in different contexts can have radically diverse impacts on the organization and consequently on accountability attribution. When determining attribution, thus, an organization will only take into account interactions that took place inside its boundaries.

Placing an organizational limit on accountability determination serves multiple purposes. It isolates events and actors so that when searching for causes/effects, one need not consider all actions from the beginning of time nor actions from other organizations. Agents are reassured that only for actions within an organization will they potentially be held accountable. Actions, thanks to agent roles (Principle 2), also always happen in context.

To adequately tackle accountability by categorizing action, we must deal with two properties within a given organization: (1) An agent properly completes its tasks, and (2) an agent does not interfere with the tasks of others. The principles 2–5 deal more explicitly with the first property, that is, how to ensure that agents complete their tasks in a manner fair for both the agents and the organization. The second property is also partially satisfied by ensuring that, in the presence of goal dependencies, the first agent in sequence not to complete its goal will bear accountability, not only for its incomplete goal, but for all dependent goals that will consequently remain incomplete. That is, should an agent be responsible for a goal on whose completion other agents wait, and should that agent not complete its goal, then it will be accountable for its incomplete goal and for that goal's dependents as well.

As an organizational and contextual aid to accountability, roles attribute social significance to an agent's actions. Following the tradition initiated by Hohfeld [44], a power is "one's affirmative 'control' over a given legal relation as against another." The relationship between powers and roles has long been studied in fields like social theory, artificial intelligence and law. By Principle 3 we stipulate that an agent can only be accountable for exercising the powers that are publicly given to it by the roles it plays. Such powers are, indeed, the means through which agents affect their organizational setting. An agent cannot be held accountable for unknown effects of its actions but, rather, only for consequences related to an agent's known place in sequences of goals. On the other hand, an agent cannot be held accountable for an unknown goal that the organization attaches to its role, and this leads us to Principle 4. An organization may not obligate agents to complete goals without prior agreement. In other words, an organization must always communicate to each agent the goals it would like the agent to pursue, and accountability will not be attributable in the presence of impossibilities, that is, when the agent does not have control of the condition or action to perform. Correspondingly, agents must be able to stipulate the conditions under which a given goal's achievement becomes possible, that is, the agent's requested provisions. The burden of discovery for impossibilities, therefore, rests upon an agent collective: A goal becomes effectively impossible for a group of agents should no agent stipulate a method of achievement. Conversely, an agent declares a goal possible the moment it provides provisions to that goal. Should a uniformed agent stipulate insufficient provisions for an impossible goal that is then accepted by an organization, that agent will be held accountable because by voicing its provisions, it declared an impossible goal possible. The opportunity to specify provisions, therefore, is fundamental in differentiating between impossibilities and possibilities.

The next section introduces a high-level protocol that enables the creation and collection of that contextualized information, which is necessary for accountability to be determined from any future institutional state. The adoption of this protocol allows an organization to support accountability.

4. The ADOPT Accountability Protocol

ADOPT is a protocol that allows the realization of accountable MAS organizations. Agents and organization will, thus, share the relevant information by exchanging messages, whose structure follows the FIPA ACL specification [45]. In the protocol, the organization is considered as a persona juris, a principal as any other principal, on which mutual expectations can be put. The protocol is divided into two main phases, a role adoption phase and a goal agreement one (explained in Sections 4.1 and 4.2, respectively), which are shown in Figures 1 and 2 as UML sequence diagrams. Such diagrams provide a sequencing of the messages, but what produces the accountability is the set of commitments that is created and that evolves with the messages, of both the role-adoption and the goal-agreement phases. Along with the message exchanges, in fact, the protocol records and tracks the evolution of accountability relationships between the parties, that we represent by way of social commitments [19,46]. In principle, other sequencings can be allowed as long as the the commitments are satisfied.

A social commitment is formally specified as $C(x, y, p, q)$, where x is the debtor, who commits to the creditor y to bring about the consequent condition q should the antecedent condition p hold. Social commitments embody the capacity of an agent to take responsibilities autonomously towards bringing about some conditions. They can be manipulated by the agents through the standard operations create, cancel, release, discharge, assign, delegate [19]. Commitment evolution follows the life-cycle formalized in [47]. A commitment is *Violated* either when its antecedent is true but its consequent is false, or when it is canceled when detached. It is *Satisfied* when the engagement is accomplished. It is *Expired* when it is no longer in effect. A commitment should be *Active* when it is initially created. Active has two sub-states: *Conditional* as long as the antecedent does not occur, and *Detached* when the antecedent has occurred. A commitment is autonomously taken by a debtor towards a creditor on its own initiative. This preserves the autonomy of the agents and is fundamental to harmonize deliberation with goal achievement. An agent will create engagements towards other agents while it is trying to achieve its goals or to the aim of achieving them. Commitments concern the observable behavior of the agents and have a normative value, meaning that debtors are expected to satisfy their engagements, otherwise a violation will occur. Commitment-based approaches assume that a (notional) social state is available and inspectable by all the involved agents. The social state traces which commitments currently exist and the states of these commitments according to the commitments life-cycle. By relying on the social state, an agent can deliberate to create further commitments, or to bring about a condition involved in some existing commitment.

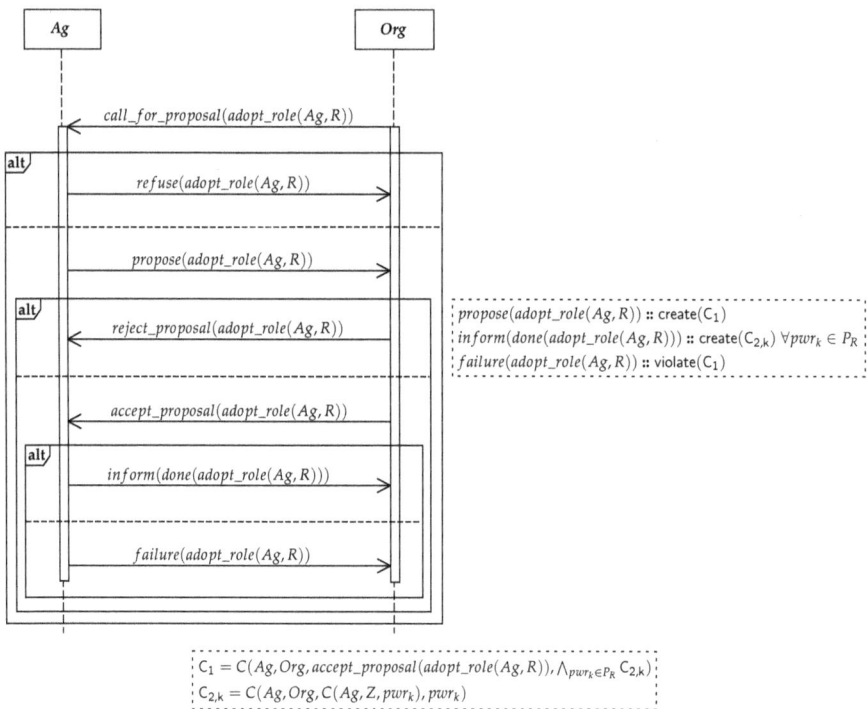

Figure 1. The first phase of the accountability protocol (role adoption).

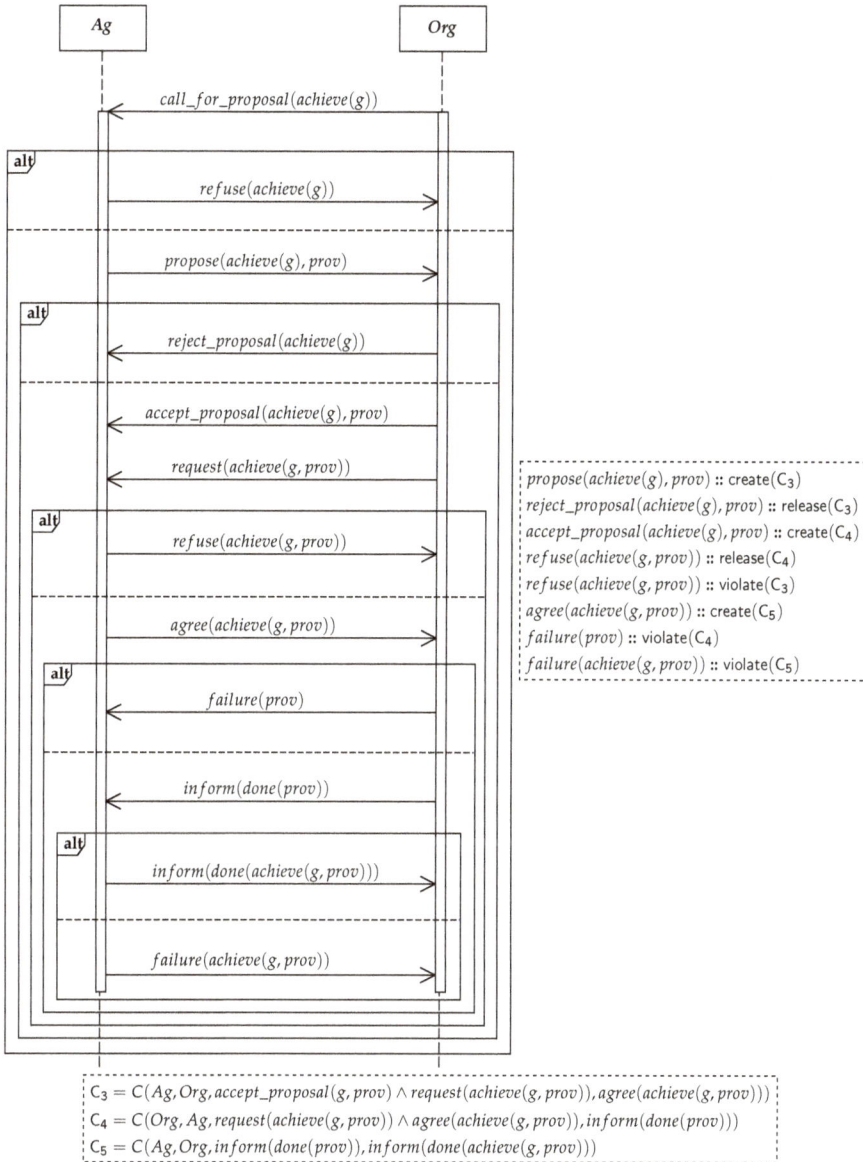

Figure 2. The second phase of the accountability protocol (goal agreement).

4.1. Role Adoption

The first phase of the protocol, reported in Figure 1, regulates the interaction that occurs when a new agent joins an organization by adopting an organizational role. The organization provides the context that gives significance to the actions that will be executed, thus satisfying Principle 1. Moreover, an agent will have an impact inside an organization only if the role adoption is successful, thus satisfying Principle 2. The interaction pattern follows the well-known Contract Net Protocol (CNP) [48,49]. (In the description of the messages we omit those arguments of the FIPA speech acts that are not strictly necessary to manage accountability.) There are two different types of agents, one initiator and one or more participants. CNP provides a means for contracting as well as subcontracting tasks, where the initiator is the agent willing to delegate the task and participants are contractors. In our case, the organization is the initiator, the agents in the system are participants and the task corresponds to role adoption. The messages that are exchanged in this phase are the following:

- *call_for_proposal(adopt_role(Ag, R))*: this message notifies an agent *Ag* that the organization *Org* is looking for someone to play role *R*. A specification of *R* is provided, including the set P_R of the powers a player of that role will be provided with. Such powers allow a role player to operate in the organizational context, and the role player will have to commit towards the organization for their use, thus becoming accountable to the organization (see *propose*). We invoke a knowledge condition, and stipulate that an agent can only be accountable for exercising the powers that are publicly given to it by the roles it plays, thus realizing Principle 3.
- *refuse(adopt_role(Ag, R))*: with this message the agent declares that it is not interested in playing role *R*.
- *propose(adopt_role(Ag, R))*: this message is, instead, used by an agent to candidate for playing role *R*. This amounts to creating a commitment $C_1 = C(Ag, Org, accept\_proposal (adopt\_role(Ag, R)), \bigwedge_{pwr_k \in P_R} C(Ag, Org, C(Ag, Z, pwr_k), pwr_k))$. In the following we denote by $C_{2,k}$ the inner commitment concerning power pwr_k. Note that this commitment amounts to a declaration of awareness by the agent to the organization of the powers it has, and also that it will exercise such powers when requested by the legal relationships it will create towards other principals. In this way, the agent's behavior becomes accountable towards the organization itself.
- *reject_proposal(adopt_role(Ag, R))*: with this message the organization rejects the agent's proposal.
- *accept_proposal(adopt_role(Ag, R))*: with this message the organization accepts the agent as a player of role *R*. The commitment C_1, having the organization as creditor and the agent as debtor, will be detached.
- *inform(done(adopt_role(Ag, R)))*: this message from the agent notifies the organization that the agent has successfully adopted role *R*, and creates all the commitments $C_{2,k}$ with $pwr_k \in P_R$. Commitment C_1 is satisfied.
- *failure(adopt_role(Ag, R))*: this message from the agent notifies the organization that the role adoption operation has not succeeded. This means that C_1 will not be satisfied.

We assume that *reject_proposal(adopt_role(Ag, R))* is mutually exclusive with *accept_proposal (adopt_role(Ag, R))*, and also that *failure(adopt_role(Ag, R))* is mutually exclusive with *inform (done(adopt_role(Ag, R)))*. This means that the occurrence of *reject_proposal(adopt_role(Ag, R))* will make C_1 expire. Instead, occurrence of *failure(adopt_role(Ag, R))* will cause the violation of C_1, neglecting what was previously stipulated.

After the role-adoption phase is successfully concluded, the organization can request the agents to pursue goals. This aspect is addressed in the second phase of the protocol.

4.2. Goal Agreement

This phase of the protocol regulates the agreement process between an agent and an organization for the achievement of a given organizational goal. Figure 2 shows the sequencing of the exchanged messages, which combines FIPA Request Interaction Protocol [50] with the already described CNP. When an organizational goal is to be pursued, the organization asks an agent to achieve it. The agent has, then, the possibility to accept or refuse the request made by the organization, or to make further requests of certain provisions that are considered by the agent as necessary to achieve the goal. This realizes Principles 4 and 5. The messages that are exchanged in this phase are the following:

- *call_for_proposal(achieve(g))*: by this message *Org* starts an interaction with *Ag* with the aim of assigning it a goal *g* to pursue.

- *refuse(achieve(g))*: the agent refuses the request made by the organization to achieve *g*.
- *propose(achieve(g), prov)*: with this message, an agent creates a commitment $C_3 = C(Ag, Org, accept\_proposal(g, prov) \wedge request(achieve(g, prov)), agree(achieve(g, prov)))$. This means that on receiving a request to achieve *g*, given that the provisions *prov* were accepted, the agent is expected to agree to pursue the goal. It is up to the organization to decide whether *prov* is acceptable or not. Only in case *prov* is accepted, the commitment will eventually be detached. When no provisions are needed, *prov* will amount to $\top$.

This is a necessary step to reach two ends: agent's awareness of assigned goals, and agent's acknowledgement to exert the necessary control to reach the goal. The rationale is that an agent, by being aware of its own capabilities, will not promise to pursue a goal it is not capable to pursue—even indirectly by enticing other agents to act—and that the relevant conditions that are necessary for goal achievement, but that the agent does not control (*prov*), are clearly stipulated.

- *reject_proposal(achieve(g), prov)*: with this message the organization rejects the request for provisions *prov* made by the agent for goal (*g*). It will release C_3.
- *accept_proposal(achieve(g), prov)*: with this message the organization accepts to provide the provisions *prov* that were requested by the agent for goal *g*. This creates a commitment $C_4 = C(Org, Ag, request(achieve(g, prov)) \wedge agree(achieve(g, prov)), inform(done(prov)))$. With this commitment the organization ties the request to the agent to pursue a goal with the supply of provisions *prov*. The commitment will be detached by the final word of the agent, and its possible agreement to the pursuit. The rationale is that, since provisions may come at a cost, before supplying them, the organization waits for a confirmation by the agent.
- *request(achieve(g, prov))*: with this message the organization *Org* asks agent *Ag* to achieve goal *g*, with provisions *prov*. Note that stipulation of what provisions should be supplied was done earlier through *propose* and *accept_proposal*, but while that was an interaction aimed at deciding whether to assign *g* to *Ag*, now *Org* wants the goal achieved. Generally, *request* may be uttered after *call_for_proposal*. The occurrence of a *request* contributes to detaching both C_3 and C_4.
- *refuse(achieve(g, prov))*: with this message the agent refuses to achieve *g* with provisions *prov*. It, thus, releases C_4, meaning that *Org* does not have to actually supply *prov*. We assume that this message is mutually exclusive with *agree(achieve(g, prov))*, thus, by this *refuse* the agent violates C_3, which at this point is already detached.
- *agree(achieve(g, prov))*: with this message the agent creates the commitment where $C_5 = C(Ag, Org, inform(done(prov)), inform(done(achieve(g, prov))))$, meaning that it will pursue the goal if *prov* is actually provided. *Org* is now expected to supply provisions.
- *failure(prov)*: the organization did not succeed in supplying provisions. We assume this message to be mutually exclusive with *inform(done(prov))*, thus, C_5 will expire and C_4 will be violated.
- *inform(done(prov))*: the organization succeeded in supplying the provisions, thus, C_5 will be detached and C_4 will be satisfied.
- *inform(done(achieve(g, prov)))*: the agent achieved *g* and C_5 is satisfied.
- *failure(achieve(g, prov))*: the agent failed in achieving *g* with *prov*. Commitment C_5 is violated.

4.3. Verifying ADOPT

We now verify the correctness of the ADOPT protocol discussed above. To this end, we have to verify two aspects: first, the adherence of ADOPT to the five principles we have identified as necessary conditions for accountability; and second, the satisfaction of some fundamental properties that any good protocol should possess, specifically *safety* and *liveness* conditions. According to the literature about protocol verification [51–53], safety means that a protocol never enters an unacceptable state, whereas liveness, strictly related to state reachability, means that the protocol always progresses towards its completion. As usual in protocol verification, we will turn to a temporal logic for formalizing and validating the protocol dynamics. In particular, the verification of ADOPT has to be focused on the treatment of the commitments that are created along the interaction, since these encode the relations upon which the accountability is established. Intuitively, we have to demonstrate that ADOPT allows the commitments to progress towards satisfaction. This demands for logics that are capable of handling commitments directly. El-Menshawy et al. [54] have proposed a temporal logic, named CTLC (i.e., computation-tree logics with commitments), which is an extension of CTL [55] with a modality operator for social commitments. In addition, the authors show how commitment-based protocols, specified in CTLC, can be checked by resorting to existing model-checking engines. Specifically, CTLC can be reduced to CTLK [56], an epistemic logic on branching time whose calculus has been implemented in the MCMAS model checker [57]. In a nutshell, the CTLC syntax is as follows:

$$\varphi := p \mid \neg\varphi \mid \varphi \vee \varphi \mid \mathtt{EX}\varphi \mid \mathtt{EG}\varphi \mid \mathtt{E}(\varphi\mathtt{U}\varphi) \mid \mathtt{C}(i,j,\varphi),$$

where p is an atomic proposition, and where the temporal modalities have the same semantics as in CTL. For example, $\mathtt{EX}\varphi$ means that "there is a path where φ holds at the next state in the path", whereas $\mathtt{EG}\varphi$ means that "there is a path where φ holds in every state along that path". Finally, $\mathtt{E}(\varphi_1\mathtt{U}\varphi_2)$ means that "there is a path along which φ_1 holds in every state until φ_2 holds". The same shortcuts defined for CTL are also valid in CTLC: $\neg\mathtt{EX}\neg\varphi \equiv \mathtt{AX}\varphi$ can be read as "φ holds in all the next states reachable from the current one"; $\neg\mathtt{EF}\neg\varphi \equiv \mathtt{AG}\varphi$: "$\varphi$ holds in every state of each path outgoing from the current state"; and $\neg\mathtt{EG}\neg\varphi \equiv \mathtt{AF}\varphi$: "$\varphi$ will eventually hold in every path outgoing from the current state".

The peculiar characteristic of CTLC is the modality $\mathtt{C}(i,j,\varphi)$, that is read: "agent i commits towards agent j to bring about proposition φ". Since the modality operator only tackles base commitments, a conditional commitment $\mathtt{C}(i,j,\psi,\varphi)$ must be reduced to the formula $\psi \to \mathtt{C}(i,j,\varphi)$, and in the following we will use $\mathtt{CC}(i,j,\psi,\varphi)$ as a shortcut of such a reduction. The interpretation of a CTLC formula is based on Kripke models where a specific accessibility relation R_{sc} is defined for the commitment modality $\mathtt{C}$. Providing a detailed reporting of the CTLC semantics is out of the scope of this paper. For our purposes, it is sufficient to say that a commitment modality $\mathtt{C}(i,j,\varphi)$ is satisfied in a model M at a state w iff φ is true in every accessible state from w using the accessibility relation R_{sc}.

For the sake of readability, we will just report some of the formulas that have actually been used for the verification of the ADOPT protocol. The usage of the CTLC language allows us to provide a synthetic yet formal account of how the protocol can be verified. Of course, in order to prove concretely the CTLC formulas above, we have to reduce them into equivalent CTLK formulas so as to use the MCMAS model checker. We leave these details out of this paper, as our intent, here, is to provide just some insights about the correctness of ADOPT, while the substantial contribution lies in the engineering of accountability in MAS enabled by ADOPT.

4.3.1. Principles 1, 2 and 3

Let us take into account the principles, and observe that the usage of commitments as a means for representing accountability relationships imposes some design choices. Commitments, in fact, are meaningful only when placed into a context, and the context is provided by the organization, the set of its roles, and its social state. Thus, when an agent is willing to play a role within an organization Org, it must be aware of the organization itself, the role R, and the powers P_R that come along with R. Only by having this knowledge, Ag can, during the role-adoption phase, create the commitment $C_{2,k}$ for each of the powers pwr_k in P_R. Summing up: (1) The organization must exist; (2) roles must be defined in the context of an organization; and (3) power associated with roles must be known at the time of role enactment. When these elements are all known to an agent before joining an organization, the system implicitly satisfies the Principles 1, 2 and 3, which are structurally satisfied by the adoption of commitments as a means to represent accountability relations. Indeed, in the role-adoption phase of ADOPT, the only elements that come into play are the structural (i.e., static) properties of the organization.

4.3.2. Reachability Property and Principles 4 and 5

Principles 4 and 5, instead, concern goals, which are dynamic by nature. Specifically, Principle 4 states that an agent is only accountable for those goals for which it has taken an explicit commitment, whereas Principle 5 allows an agent to negotiate the provisions necessary for achieving a specific goal. The verification of these principles requires one to consider the dynamics of the accountability protocol, and overlaps with the *reachability* property we are interested in verifying in ADOPT. Relying on CTLC, it is possible to express useful properties about commitment satisfaction, and hence, about the correctness of the ADOPT protocol. For instance, for each commitment $C(i, j, \psi, \varphi)$ that is foreseen by the protocol, one can verify that when the commitment is created it can also be satisfied. This corresponds to expression of the following reachability property in CTLC: $\texttt{AF EF}(\texttt{createdC} \rightarrow \texttt{CC}(i, j, \psi, \varphi))$, where $\texttt{createdC}$ is an atomic proposition that becomes true when the given commitment $C(i, j, \psi, \varphi)$ is created (i.e., when a specific message is sent by the debtor agent). This formula is valid if and only if in all the possible paths (i.e., runs of ADOPT), a state will be reached from which there exists at least one path along which either $\texttt{createdC}$ is false (the commitment is not created), or $\texttt{createdC}$ is true and the modality $\texttt{CC}(i, j, \psi, \varphi)$ will eventually be satisfied. Please recall that $\texttt{CC}(i, j, \psi, \varphi)$ is just a shortcut for $\psi \rightarrow C(i, j, \varphi)$. This formula is satisfied either when ψ is false, corresponding to the case in which the commitment expires (for example, released by the creditor); or when ψ is true and hence $C(i, j, \varphi)$ must hold. The latter corresponds to the case in which the commitment is, firstly, detached by the creditor and, then, satisfied by the debtor. Intuitively, this can be verified simply by looking at the sequence diagrams in Figures 1 and 2, and observing that whenever a commitment is created by a message of the debtor, the creditor has always the chance to release that commitment or detach it. Similarly, whenever a commitment is detached, the protocol encompasses at least a run along which the debtor can send a message whose meaning consists of the progression of that commitment to satisfaction.

In short, it is possible to show that the above formula holds for every commitment that may arise in ADOPT. This means that in every possible run of ADOPT, there is always an execution path along which the conditional commitment $C(i, j, \psi, \varphi)$, once created, can always be released by the creditor, or satisfied by the debtor. As a consequence, we can conclude that Principles 4 and 5 of the accountability requirements are actually satisfied, since the agent will have also the chance to agree to achieve a goal by creating a commitment (specifically C_4 of the goal-agreement phase), and by negotiating its provisions again via a commitment (specifically C_3). An agent, thus, will only be accountable for the goals it has agreed upon, possibly having also established some necessary provisions.

4.3.3. Safety

To verify the safety condition, we have to show that "bad" conditions will never be reached. In the specific case of ADOPT, an unwanted condition is the impossibility to satisfy a commitment $C(i, j, \psi, \varphi)$ after the commitment has been created. This condition can be formulated in CTLC as the formula $\neg\texttt{AF EF}(\texttt{createdC} \rightarrow \texttt{CC}(i, j, \psi, \varphi))$. Also in this case, we can show intuitively that the above formula is false in every possible run of the protocol. In fact, the only situation in which the role-adoption phase completes without the satisfaction of any commitment is when the agent answers with a *refuse* message to the call for proposal from the organization, but in this case no commitment is created. In all the other possible runs of the first phase, for every commitment that is created there exists at least one run along which it will be eventually satisfied. A similar consideration holds for the goal-agreement phase, too. This means that ADOPT is safe.

4.3.4. Liveness and Nested Commitments

The nested commitments in ADOPT also deserve some special attention. In the role-adoption phase, the following nested commitments are created: $C_1 = C(Ag, Org, accept\_proposal(adopt\_role$

$(Ag, R))$, $\bigwedge_{pwr_k \in P_R} C_{2,k})$; where, as before, $C_{2,k} = C(Ag, Org, C(Ag, Z, pwr_k), pwr_k)$ for every power $pwr_k \in P_R$. In this case it is interesting to verify whether whenever C_1 is detached by Org, Ag has the chance to satisfy C_1 by creating the commitments $C_{2,k}$. That is, whether there exists at least one run of the protocol where these commitments will be actually created. This corresponds to verifying the *liveness* of the protocol by asking whether something "good" will eventually happen. The following CTLC formula serves this purpose: `AG(detachedC1` $\rightarrow$ `EF(createdC2k)))`, where `detachedC1` is an atomic proposition that becomes true when Org sends message *propose(adopt_role(Ag,R))*, whereas `createdC2k` is another proposition that becomes true when Ag sends the message *inform(done(adopt_role(Ag, R)))*. Indeed, looking at the sequence diagram in Figure 1, it is apparent that once the commitment C_1 is detached, there exists at least one run of the protocol along which the message *inform(done(adopt_role(Ag, R)))* is actually sent by the agent. The formula is thus satisfied, and the protocol enjoys the liveness property.

5. Case Study: Accountability in the JaCaMo Framework

JaCaMo [11] is a conceptual model and programming platform that integrates agents, environments and organizations. It is built on the top of three platforms, namely Jason [8] for programming agents, CArtAgO [9] for programming environments, and Moise [58] for programming organizations. More specifically, Jason is a platform for agent development based on the language AgentSpeak [59]. Here, an agent is specified by a set of beliefs, representing both the agent's current state and knowledge about the environment, a set of goals, and a set of plans which are courses of actions, triggered by events. CArtAgO, based on the Agents & Artifacts meta-model [60], is a framework for environment programming which conceives the environment as a layer encapsulating functionalities and services that agents can explore and use at runtime [61]. An environment is programmed as a dynamic set of artifacts, whose observable states can be perceived by the agents. Agents can act upon artifacts by executing the operations that are provided by the artifact's usage interface. Finally, Moise implements a programming model for the organizational dimension. It includes an organization modeling language, an organization management infrastructure [62] and a support for organization-based reasoning at the agent level. A JaCaMo multiagent system is, then, given by an agent organization, programmed in Moise, organizing autonomous agents, programmed in Jason, working in a shared, artifact-based environment, programmed in CArtAgO.

According to [62], the Moise organizational model, adopted in JaCaMo, decomposes the specification of an organization into three dimensions. The structural dimension specifies roles, groups and links between roles in the organization. The functional dimension is composed of one or more schemes that elicit how the global organizational goals are decomposed into subgoals and how these subgoals are grouped in coherent sets, called missions, to be distributed to the agents. Finally, the normative dimension binds the two previous dimensions by specifying the roles' permissions and obligations for missions. One important feature of Moise is to avoid a direct link between roles and goals. Roles are linked to missions by means of permissions and obligations, thereby keeping the structural and functional specifications independent. This independence, however, is source of problems from the viewpoint of accountability determination. The reason is that schemes, in principle, can be dynamically created during the execution (thus modifying the organizational specification), and assigned to groups within an organization when agents are already playing the associated roles. This means that agents, when entering into an organization by adopting an organizational role, in principle have no information about what they could be asked to do in the future. At the same time, the organizational infrastructure implemented in JaCaMo does not provide any mechanism for agents to explicitly accept a given organizational goal or to negotiate any provision for it. This is, indeed, in contrast with what was discussed in Section 3.

JaCaMo provides various kinds of organizational artifacts that altogether allow encoding the state and behavior of an organization, in terms of groups, schemes and normative states. These organizational artifacts provide both the actions the agents will use (to take part in an organization and act upon it) and the observable properties that allow the state of the organization (and its evolution) to be perceived by the agents. Artifacts' observable properties are automatically mapped to agents' beliefs. The main organizational artifacts are the following:

- **OrgBoard** artifacts, which keep track of the overall current state of the organizational entity, one instance for each organization (optional);
- **GroupBoard** artifacts, which manage the life-cycle of specific groups of agents, one for each group;
- **SchemeBoard** artifacts, each of which manages the execution of one scheme;
- **NormativeBoard** artifacts, used to maintain information concerning the agents' compliance to norms.

5.1. Accountability Issues and Their Reasons

In order to explain the lack of accountability of JaCaMo, let us now consider a scenario based on an excerpt of the *building-a-house* example presented in [11]:

Example 2. *An agent, called Giacomo, wants to build a house on a plot. In order to achieve the goal he will have to hire some specialized companies, and then ensure that the contractors coordinate and execute in the right order the various subgoals. Some tasks depend on other tasks while some other tasks can be performed in parallel. This temporal ordering is specified in the functional specification of the scheme describing the process.*

As soon as the building phase starts, Giacomo creates a GroupBoard artifact, called here `bh_group` (for sake of discussion, the example is slightly revised with respect to the original version, presented in [11]. This includes a change in the artifacts' names), following the organization specification. Roles are gathered in a group that will, then, be responsible for the house construction. After that, he adopts his role and asks the hired agents to adopt theirs. Roles are adopted by executing an operation that is provided by the GroupBoard artifact. Finally, a SchemeBoard artifact, called `bh_scheme`, is created. When all agents have adopted their roles (i.e., when the group is well-formed), this is added to the schemes the group is responsible for.

After this step, the involved agents could be asked to commit to some "missions"—in JaCaMo, the term "commit" does not refer to social commitments but it is used in a general sense. This is done by relying on obligations that are issued by the organization, according to the normative specification. Figure 3 shows the general interaction pattern which is used in JaCaMo for role adoption, mission distribution and goal assignment, as it would be instantiated for a `companyA` agent, a `plumber` role and an `install_plumbing` mission (which in turn contains a `plumbing_installed` goal). Agent `companyA` is asked to commit to `install_plumbing` with an obligation of the form `obligation(companyA, n8, committed(companyA, install_plumbing, ...), ...)`, where `n8` is the norm that binds the `plumber` role with the `install_plumbing` mission in the normative specification. This does not yet mean the agent has to pursue the goal; this other obligation, however, may now be issued by the organization. Indeed, the main purpose of the SchemeBoard artifact, `bh_scheme`, is to keep track of which goals are ready to be pursued and create obligations for the agents accordingly. For instance, when the `plumbing_installed` goal will be ready to be pursued, `companyA` will receive a new an obligation `obligation(companyA, ..., achieved(..., plumbing_installed, companyA),...)`, and so forth with other organizational goals as soon as they become ready. Such obligations are observed by the agents and the corresponding internal goals are automatically created.

Listing 1 shows an excerpt of the `companyA` agent. The above-mentioned obligation, through the plan at line 15, creates the (internal) goal that is then pursued by following the plan at line 21. After that, the organizational goal is set as achieved, too (line 19), using the `goalAchieved` operation provided by the SchemeBoard artifact.

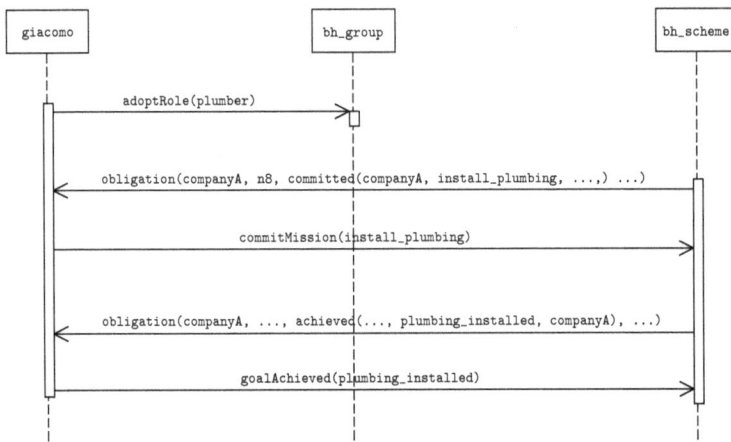

Figure 3. Interaction between the companyA agent and the organization in the *building-a-house* example.

Listing 1. Excerpt of the Jason code of the companyA agent.

```
1   task_roles("Plumbing", [plumber]).
2
3   +!contract(Task,GroupName)
4       :  task_roles(Task,Roles)
5       <- lookupArtifact(GroupName, GroupId);
6           for (.member(Role, Roles)) {
7               adoptRole(Role)[artifact_id(GroupId)];
8               focus(GroupId)
9           }.
10
11  +obligation(Ag,Norm,committed(Ag,Mission,Scheme),Deadline)
12      :  .my_name(Ag)
13      <- commitMission(Mission)[artifact_name(Scheme)].
14
15  +obligation(Ag,Norm,What,Deadline)[artifact_id(ArtId)]
16      :  .my_name(Ag) &
17          (satisfied(Scheme,Goal)=What | done(Scheme,Goal,Ag)=What)
18      <- !Goal[scheme(Scheme)];
19          goalAchieved(Goal)[artifact_id(ArtId)].
20
21  +!plumbing_installed
22      <- installPlumbing.
```

Now, let us suppose that Giacomo decides to have an air conditioning system installed, a thing he had initially not thought of for this house. Suppose also that he wants to exploit the contracted companies to achieve this purpose, which is related to the house construction, but was not discussed. Let us also suppose he decides to assign an air_conditioning_installed goal to the agent playing the plumber role. Giacomo's exploitive plan would work because when an agent adopts a role in a group, that agent has no information about the tasks that could be assigned to it. The bh_scheme SchemeBoard artifact could even not have been created yet. In fact, tasks could be created independently of roles, and only subsequently associated with them. In the example, however, the companyA agent, playing the plumber role reasonably will not have a plan to achieve the air_conditioning_installed goal (indeed, it has no such plan). Thus, when the corresponding obligation is created, this will not be fulfilled.

Given the above scenario, who could we consider accountable for the failure of the organizational goal air_conditioning_installed?

- Should the agent playing the `plumber` role be held accountable? The agent violated its obligation but it could not have reasonably anticipated the goal's introduction, which effectively made achievement impossible.
- Should Giacomo be held accountable since he introduced an unachievable goal, however licit?
- Perhaps the system itself ought to bear the brunt of accountability since it permits such kind of behavior? The system, however, does not know agent capabilities.

The inability to attribute accountability stems from the lack of adherence to Principles 4 and 5. Goal assignment is, in fact, performed through schemes, which can even be dynamically created and associated with an existing group. Moreover, the very independence between roles and goals violates Principle 4: When enacting a role in JaCaMo, agents do not have a say on the kind of goals they could be assigned. For this reason they cannot be held accountable later for some organizational goal they have not achieved. The problem, here, is that JaCaMo's organizational infrastructure agents do not have the possibility to discuss with the organization about the acceptability of organizational goals, which is, instead, encoded in the goal-agreement phase of the ADOPT protocol. In particular, it is impossible for agents to put before the organization the provisions needed to achieve a given organizational goal. This contradicts Principle 5.

Another critical issue concerns the powers the agents gain by taking part in an organization. Following JaCaMo's conceptual meta-model, the only ways for an agent to affect the organizational state are either to enter into a group, or to change the state of an organizational goal, for example, by achieving it. In order to join a group, an agent must have access to the GroupBoard artifact which manages that group—indeed, the `adoptRole` operation is provided by that artifact. Similarly, to set an organizational goal as achieved, the agent must have access to the SchemeBoard artifact, which manages the execution of the scheme to which the goal belongs. From an accountability standpoint, then, the role adoption does not end when a given agent enters into a group, but only when a scheme is associated with the group and, consequently, the agent gains access to it. However, should a new scheme be added to the group, the agents inside the group would also receive new powers (i.e., the powers to change the states of the goals in the new scheme). This clearly contradicts Principle 2, which requires a declaration of awareness of the powers, associated with a role, that should be done by each of its players. This awareness is, instead, guaranteed by the role-adoption phase of the ADOPT accountability protocol.

5.2. Achieving Accountability

We now show how we implemented the ADOPT accountability protocol in JaCaMo. A new kind of artifact, called *AccountabilityBoard* artifact, was added to the organizational infrastructure to preserve modularity. (The source code of the AccountabilityBoard artifact can be found here: http://di.unito.it/accountabilityboard.) In other words, this addition did not affect the implementation of the other organizational artifacts that were previously in JaCaMo. The new artifact supports both phases of ADOPT: the role-adoption phase and the goal-agreement phase. The artifact also provides some observable properties, informing the agents that focus on it about the overall state of the interaction. These observable properties represent the events occurring during the interaction, and allow encoding of the created commitments. More precisely, the artifact keeps track, in the first phase, of which calls for roles are pending, of the agents' replies, of which proposals have been accepted (and which have not), of which agents have successfully completed the role adoption, and which have failed. Later on, in the goal-agreement phase, the observable properties encode which provisions were accepted and which were not, which goals were agreed upon by which agents (and with what kinds of provisions), which ones were refused, and which provisions were confirmed to be holding. All such observable properties are created by executing the above described operations on the accountability artifact. In case of inspection, the observable properties help to identify the accountable parties.

Concerning the role-adoption phase, the artifact provides an organization with the operations to call for role players, and to accept or reject the agents' proposals. Agents, on the other hand, are provided with the operations to answer calls (either by refusing, or by proposing themselves), to declare the acquisition of organizational powers and to declare a failure in the role-adoption process. With respect to the power awareness declaration, we assume here that an agent acquires the powers to operate on a given scheme when the scheme is assigned to the group to which the agent belongs, and when the agent commits to the missions associated with its role in the scheme itself. To stipulate this awareness, the agent declares it has access both to the group and to the scheme artifacts, as player of the given role, which together define the scope of the agents' actions in the organizational context. By doing so, the agent ensures that it is conscious that it could be requested to achieve organizational goals defined in the given scheme as a role player in the given group (and it will have the possibility to refuse them or agree to their achievement). Should a new scheme be added to a given group, agents inside it should renegotiate the terms of role adoption (by executing again the first phase of the protocol) in order to acquire the organizational powers needed to act on the new scheme, too. The *AccountabilityBoard* artifact operations that allow all these things are:

- `callForRole(String addressee, String role)`: implements ADOPT's *call_for_proposal(adopt_role(Ag, R))*. It allows the organization's owner to open a call for a role, addressed to a given agent. An observable property `pendingRole(addressee, role)` is defined;
- `refuseRole(String role)`: implements *propose(adopt_role(Ag,R))*. It allows agents to answer to a call for role, refusing to adopt it. Its execution creates an observable property `refusedRole(agent, role)`;
- `proposeForRole(String role)`: implements *propose(adopt_role(Ag,R))*, and allows agents to propose themselves as role players. It creates an observable property `proposedForRole(agent, role)`;
- `rejectProposalForRole(String proposer, String role)`: implements *propose(adopt_role(Ag, R))*, and allows an organization's owner to reject a `proposer` agent's proposal. An observable property `roleProposalRejected(proposer, role)` is created;
- `acceptProposalForRole(String proposer, String role)`: implements *accept_proposal(adopt_role(Ag, R))*, and allows the organization's owner to accept an agent's proposal. It creates an observable property `roleProposalAccepted(proposer, role)`;
- `declareAdoptionSuccess(String role, String group, String scheme)`: implements *inform(done (adopt_role(Ag, R)))*, and allows a role player to declare awareness of the powers given by scheme `scheme`, belonging to group `group` as a player of role `R`. An observable property `powerAcquired(agent, role, group, scheme)` is defined. The execution of this operation completes the role-adoption phase of ADOPT;
- `declareAdoptionFailure(String role, String group, String scheme)`: implements *failure(done (adopt_role(Ag, R)))*, and allows an agent to inform the organization that its role-adoption process failed. An observable property `powerAcquisitionFailed(agent, role, group, scheme)` is defined.

With respect to the goal-agreement phase of ADOPT, instead, the artifact provides operations to the agents for refusing a goal, for proposing themselves as goal achievers, for proposing provisions for it, for agreeing to pursue a goal when the provisions are accepted by the organization, and for declaring goal achievement or failure. At the same time, it allows the organization (i.e., the agent that is the organization's owner) to open a call for goal achievement, to accept or reject provisions proposed by the agents, and to declare that a given provision holds or not. In particular, the operations provided by the artifact are:

- `callForGoal(String addressee, String goal)`: implements *call_for_proposal(achieve(g))* and allows the organization's owner to open a call for goal `goal` addressed to agent `addressee`, a `pendingGoal(addressee, goal)` observable property is defined;
- `refuseGoal(String goal)`: implements ADOPT's *refuse(achieve(g))* and *refuse(achieve(g,prov))*. With this operation, an agent that a goal is addressed to can refuse to achieve it. A `refusedGoal(agent, goal)` observable property is defined;

- `proposeProvision(String goal, String provision)`: realizes the *propose(achieve(g), prov)* message. An agent can propose a provision `provision` for goal `goal` assigned to it. A `pendingProvision (agent, goal, provision)` observable property is created;

- `acceptProvision(String proposer, String goal, String provision)`: implements *accept_proposal (achieve(g), prov)* and allows the organization's owner to accept a provision `provision` proposed by agent `proposer` for goal `goal`. An observable property `acceptedProvision(proposer, goal, provision)` is defined;

- `rejectProvision(String proposer, String goal, String provision)`: conversely, implements *reject_proposal(achieve(g), prov)*, and allows the organization's owner to reject the provision proposed by the agent for the goal. A `rejectedProvision(proposer, goal, provision)` property is defined;

- `requestGoal(String addressee, String goal)`: implements *request(achieve(g,prov))* in ADOPT. The organization's owner asks `addressee` to agree to achieve `goal` since the provisions proposed by it have been accepted. A `requestedGoal(addressee, goal)` property is created;

- `agreeGoal(String goal)`: implements *agree(achieve(g, prov))* and allows an agent to agree to achieve a given goal previously requested. An `agreedGoal(agent, goal)` property is defined;

- `informProvision(String goal, String provision)`: implements *inform(done(prov))*. The organization's owner agent confirms that a provision `provision` related to a goal `goal` is currently holding. This operation defines an observable property `holdingProvision (goal, provision)`;

- `failureProvision(String provision)`: implements *failure(prov)*. The organization's owner agent declares that a provision `provision` related to a goal `goal` cannot hold. This operation defines an observable property `failedProvision(provision)`.

- `informGoal(String goal)`: implements *inform(done(achieve(g,prov)))*. It allows an agent to confirm that a goal has been achieved. This operation defines an observable property `achievedGoal(goal)`;

- `failureGoal(String goal)`: implements *failure(achieve(g,prov))*. The agent declares that it has not been able to achieve goal `goal`. This operation defines an observable property `failedGoal(goal)`.

5.3. Building-a-House Revisited

Let us, now, see how the AccountabilityBoard artifact supports the execution of the ADOPT protocol with the help of a revised version of our example. (The full code of the revised example is available here: http://di.unito.it/buildingahouse.) In this case, before starting with the actual house construction, Giacomo will create an instance of the AccountabilityBoard artifact, thereby becoming the organization's owner. This will allow it to execute organization-reserved operations on the artifact, such as `acceptProposal`, `acceptProvision`, `confirmProvision`, and so on.

The original implementation of the example presented in [11] also includes a contracting phase before the building one, in which Giacomo hires the needed company agents and assigns roles to them through an auction mechanism. Thus, we mainly focus here on the agreement of goals. With respect to the first part of the protocol, however, agents, after having adopted their roles in the `bh_group`, still have to declare their awareness of the powers they are endowed with when the `bh_scheme` is assigned to the group. This is achieved by means of the plan reported in Listing 2. As soon as the scheme is added to the ones the `bh_group` is responsible for, agents inside it will focus on it (lines 7 and 11) and declare power awareness with respect to the `bh_group` and `bh_scheme` (line 8). At the same time, they will commit to their missions as requested by the organizational infrastructure.

Role players will now have the possibility to explicitly refuse or agree to pursue the organizational goals assigned to them (possibly asking for some provisions to hold) when asked by the organization's owner. This is achieved by means of the plans reported in Listing 3. For instance, the plan at line 2 is triggered when an acceptable goal (line 4) is proposed to the agent and there is no need to ask for provisions (line 5). In this case, the agent will simply propose a dummy *true_prov* provision to achieve the goal (line 6). The plan at line 9 deals with an unacceptable goal (line 11). In this case the agent simply refuses it (line 12). The plan at line 15 allows the agent to propose a provision for a pending goal (line 17). Should the provision be accepted (line 22) and the goal requested (line 20), the agent would then agree to pursue the goal given the provision (line 23). Finally, the plan at line 26 is triggered when a provision for a previously agreed-upon goal is declared to hold. In this case the agent works in order to achieve the goal. An internal goal corresponding to the organizational one will be generated 30 and, if achieved, the organizational goal will be set to achieved (line 31), as well. As a final step, the goal must be declared as achieved on the AccountabilityBoard, too (line 32).

Listing 2. Jason plan needed by an agent to declare powers awareness.

```
1   //declare powers awareness
2   +schemes(L)[artifact_name(_,Group), workspace(_,_,W)]
3      : .my_name(Ag) &
4        play(Ag,Role,Group)
5     <- for (.member(Scheme,L)) {
6          lookupArtifact(Scheme,ArtId)[wid(W)];
7          focus(ArtId)[wid(W)];
8          declareAdoptionSuccess(Role,Group,Scheme);
9          .concat(Group,".",Scheme,NBName);
10         lookupArtifact(NBName,NBId)[wid(W)];
11         focus(NBId)[wid(W)];
12       }.
```

Listing 3. Excerpt of the Jason plans needed by an agent playing a role in an organization to be compliant with the goal-agreement phase of ADOPT.

```
1   //no provision needed
2   +pendingGoal(Ag,Goal)
3      : .my_name(Ag) &
4        acceptableGoal(Goal) &
5        not provision(Goal,Prov)
6     <- proposeProvision(Goal, true_prov).
7
8   // refuse goal
9   +pendingGoal(Ag,Goal)
10     : .my_name(Ag) &
11       not acceptableGoal(Goal)
12    <- refuseGoal(Goal).
13
14  //propose provision
15  +pendingGoal(Ag,Goal)
16     : .my_name(Ag) & acceptableGoal(Goal) & provision(Goal,Prov)
17    <- proposeProvision(Goal,Prov).
18
19  //agree goal, provision accepted
20  +requestedGoal(Ag,Goal)
21     : .my_name(Ag) &
22       acceptedProvision(Ag,Goal,_)
23    <- agreeGoal(Goal).
24
25  //provision confirmed, must satisfy goal
26  +holdingProvision(Goal,Prov)
27     : .my_name(Ag) &
28       agreedGoal(Ag,Goal) &
29       acceptedProvision(Ag,Goal,Prov)
30    <- !Goal[scheme(Scheme)];
31       goalAchieved(Goal)[artifact_id(ArtId)];
32       informGoal(Goal).
```

Listing 4, in turn, shows an excerpt of the plans needed by Giacomo, as organization owner, to be compliant with the second phase of the ADOPT protocol. First of all, the plan at line 1 allows it to open a call for the achievement of a given goal (line 3) when the corresponding obligation is issued by the organizational infrastructure. The plans at lines 5 and 10, in particular, allow it to accept or reject provisions for goals asked by the role players (lines 7 and 12) depending on whether they seem acceptable or not (lines 6 and 11). If a provision is accepted, then the goal is requested to the agent, too (8). The plans at lines 14 and 19, finally, allow the organization's owners to inform that a given provision which has been agreed by some agent for a goal is holding. To this end, the provision must be present in the agent's belief base (i.e., the agent must believe that the proposition representing the provision is true).

Going back to the *companyA* agent, playing the *plumber* role, should an *air_conditioning_installed* (*aci* for short) goal be requested via *call_for_proposal(achieve(aci))*, the agent would have the legitimate possibility to refuse the proposal or to conditionally accept by specifying a provision. It could therefore ask that the walls be built in order to successfully install the air conditioning by adding *provision(aci, walls_built)*. The plan in line 15 in Listing 3 would be consequently triggered. As a result of executing `proposeProvision`, the following commitment would be created: $C_1 = C(companyA, Giacomo, accept\_proposal(aci, walls\_built) \land request(achieve(aci, walls\_built)), agree(achieve(aci, walls\_built)))$. Giacomo in turn has the freedom to either accept or reject *companyA*'s proposal. Since in the scheme specification walls are built beforehand, Giacomo will reasonably accept the provision by executing the `acceptProvision` operation in line 7 of Listing 4, leading to the creation of the commitment $C_2 = C(Giacomo, companyA, request(achieve(aci, walls\_built)) \land agree(achieve(aci, walls\_built)), inform(done(walls\_built)))$. Moreover, it will subsequently execute `requestGoal`, which in turn detaches C_1 and partially satisfies the antecedent in C_2.

Listing 4. Jason plans needed by an organization's owner agent to be compliant with the second phase of the accountability protocol.

```
1   +obligation(Ag,_,What,_)[artifact_id(ArtId)]
2      : (satisfied(Scheme,Goal)=What | done(Scheme,Goal,Ag)=What)
3      <- callForGoal(Ag,Goal).
4
5   +pendingProvision(Agent,Goal,Provision)
6      : acceptableProvision(Goal,Provision)
7      <- acceptProvision(Agent,Goal,Provision);
8         requestGoal(Agent,Goal).
9
10  +pendingProvision(Agent,Goal,Provision)
11     : not acceptableProvision(Goal,Provision)
12     <- rejectProvision(Agent,Goal,Provision).
13
14  +agreedGoal(Ag,Goal)
15     : acceptedProvision(Ag,Goal,Prov) &
16       Prov
17     <- informProvision(Goal,Prov).
18
19  +Prov
20     : acceptedProvision(Ag,Goal,Prov) &
21       not achievedGoal(Goal)
22     <- informProvision(Goal,Prov).
```

At this point, *companyA* should agree to achieve the *aci* goal to satisfy C_1, given *walls_built*. The agreement will also detach C_2 and will create a commitment $C_3 = C(companyA, Giacomo, inform(done(walls\_built)), inform(done(achieve(aci, walls\_built))))$. Giacomo will, then, execute `informProvision`, thereby satisfying C_2 and detaching C_3. As a final step, *companyA* will achieve the goal by installing the air conditioning (see plan 26 in Listing 3), and will communicate success via `informGoal`. After this step, C_3 will be satisfied, and hopefully the house construction will be successfully completed as expected.

At the same time, in case of organizational goal failure, an investigative entity, possibly Giacomo himself, could inspect the observable properties of the AccountabilityBoard artifact defined during the interaction, which altogether allow one to reconstruct the previously described commitments, and, on this foundation, discern accountability comprehensively. For instance, should the walls not be built, Giacomo would not be able to inform *companyA* that the previously accepted provision *walls_built* holds. Consequently, *companyA* would not start to pursue the goal because C_3 would not be detached. At the same time, this would cause the violation of commitment C_2, which would correctly lead to considering Giacomo as accountable for the failure, since it was Giacomo who failed to provide an agreed-upon provision. Similarly, should the provision hold and *companyA* not pursue the goal and therefore be unable to preform informGoal, the violation of commitment C_3 would allow one to consider the company agent itself accountable. In any case, the investigation process is outside the scope of this work, but the infrastructure provided by the artifact keeps track of the necessary information.

6. Impact and Discussion

MASs have demonstrated to be a viable approach for developing complex, distributed systems. Indeed, several methodologies for designing MASs have been discussed in the literature [1–4,6,7] and, correspondingly, different agent platforms [5,8–11] have been proposed as technological means for practically implementing distributed applications. As discussed in Section 2, most of these design methodologies and platforms adopt the organizational metaphor as a way for distributing and coordinating agents. The organization provides a functional decomposition of organizational goals into subgoals that can be assigned to agents. In addition, norms and obligations arising within an organization specify the good conduct of agents. As we have already pointed out, this approach has been criticized by some recent work [13,16], since such organizations lack a fundamental characterization: the specification of accountability relationships among the agents. In [13], the need for accountability relationships is motivated by the necessity of robustness even in open and dynamic environments. Open systems, in fact, are particularly challenging from the point of view of software engineering, because agents see an organization as a complex resource to be used in order to get their own goals, which may not match with the organizational ones. As shown in [16], the functional decomposition of organizational goals, and their attribution to agents, is not a viable solution for assuring software robustness. In fact, when an agent fails an obligation and does not achieve a goal, it cannot be held accountable since the contextual information is not sufficient to find out the causes, and hence assign the blame. Did the agent have the control over the goal? In other words, did the agent have a plan for reaching the goal? Was it legitimate from the point of view of the organization to expect that the agent had and used such a plan? The impossibility to precisely determine the causes of a failure makes the overall system vulnerable and poorly debugged.

ADOPT provides a mechanism for answering the above questions. By imposing agents to follow a protocol, for enacting a role and for explicitly agreeing to accomplish a goal, our approach makes agents accountable for their activities within the organization. When an agent agrees to achieve a goal, it implicitly declares that is has control over that goal. This is consistent with the fact that agents are not inspectable, and hence only the agent itself knows the tasks it can perform. On the other hand, the agent's declaration allows an organization to expect that the agent will bring about the goal condition. When an agent fails to achieve a goal, thus, the fault can be easily isolated to the agent, if it agreed with the goal, or to the organization, if it assigned that goal to an agent despite the agent not previously accepting the goal. The ADOPT protocol is specified in terms of social commitments. Commitments are indeed a powerful software engineering tool, since they provide a standard for agent correctness. To verify the compliance of an agent to a protocol, in fact, it is sufficient to verify whether the agent is capable of fulfilling the commitments associated with a specific protocol role. This paves the way for methodologies implementing agents driven by the commitments (see for example, the CoSE methodology [63]), and also mechanisms of agent-type checking, in which agent behaviors are typed with the set of commitments they make progress on [64].

Concerning human organizations, business ethics and compliance programs (and, more in general, self-regulatory initiatives) are becoming more and more central [20,21,65,66], bringing to the forefront the importance of accountability. Many organizations and companies voluntarily adopt monitoring and accountability frameworks, for example, [20,21], or comply to social accountability standards, like [67], which aim at increasing awareness about values (like health and safety, sustainability, no use of child labor, gender equality, and so on.) and also behaviors that account for such values. Principals must be held accountable for their (mis)behavior and, therefore, provide feedback about the reasons of performance.

Accountability frameworks vary considerably, depending on the kind of actors that are involved, on the kind of commitments, and on the activities that may be put under scrutiny. Figure 4 summarizes a general cycle that can be found (declined in many ways) in many organizations. (The picture is inspired by the framework schemas described in [20,22].) An accountability framework relies on further frameworks, not in the picture for simplicity: for monitoring, for risk assessment, and for managing complaint responses. Of these, the monitoring framework is strictly necessary, as monitoring pairs with and completes the obligation to report that is inherent in accountability.

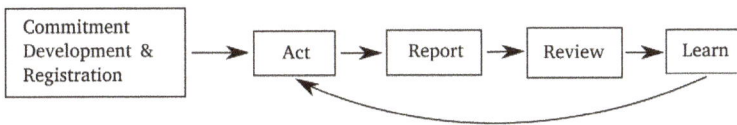

Figure 4. A general scheme for accountability frameworks.

Developing and implementing self-regulatory initiatives is a difficult task which requires human and financial resources. Stronger compliance mechanisms risk reducing participation to a small group of well-resourced organizations. Smaller organizations, often those most in need of support to improve quality and accountability, may be unable to participate [24]. Figure 5 shows the mechanisms through which self-regulation is realized: more often than not, realization of compliance (and its assessment) requires the intervention of external agencies and auditors. Informal structures have a limited impact. Accountability frameworks are currently very-little supported by software and information systems. The commitments that involve the parties are basically hand-written by filling in forms [22,23], and the assessment of satisfaction or violation of the involved liabilities, as well as the actual accounting process, is totally handled by authorized human parties [24]. The problem is that when accountability channels are informal or ambiguous in the long run, all these processes will be thwarted, leading to little effectiveness and poor performance, in particular when cross-organizational relationships must be established and maintained.

The proposal presented in this paper can help in supporting organizations, big and small, in realizing self-regulatory initiatives. Specifically, it allows creating software support to the commitment-creation phase, the monitoring-of-the-act phase, and the identification of the accountable parties who should report about circumstances under scrutiny. The subsequent revision, embodied by the review and learn phases, will instead be carried out by the involved principals. This can be done because the way in which accountability is modeled finds correspondence in the way accountability is understood in human accountability frameworks. Most of these (among which [20,22,23,65,66] share a common interpretation of accountability that is well described by [66] as a relationship between two or more parties that implies responsibilities and consequences) share obligation to take responsibility (belonging to the development of commitments and to the act phase), to demonstrate performance, to review. In particular, taking responsibility belongs to the development of commitments and to the act phase: it entails awareness to answer for what was (not) accomplished and also for the means used in the effort. Expectations about outcomes are intended to be agreed expectations (even in hierarchic contexts), and to stem from either a formal or informal agreement on what should be accomplished.

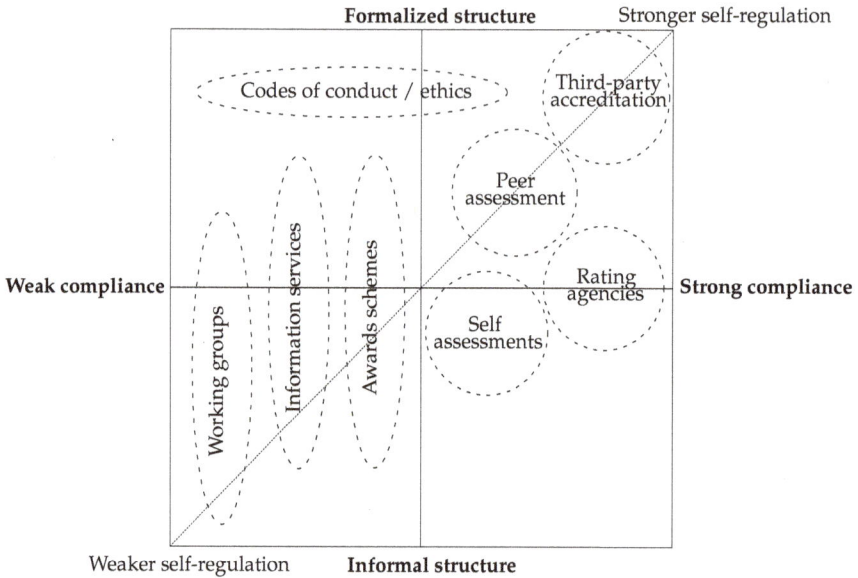

Figure 5. Types of self-regulatory initiatives within individual civil– social organizations [24].

7. Conclusions

Accountability is an important asset in human societies. In this paper we have shown how accountability plays a central role also in MAS engineering. Specifically, we have argued that while MASs rely on the metaphor of (human) organizations as a way for distributing goals and for establishing norms of good behavior, the organizational models in the state of the art are inadequate for developing robust open systems. In open systems, the agents see an organization as a means for reaching their own goals, which may not coincide with the organizational ones. Thus, when an organization assigns a goal to an agent, that organization cannot hold the expectation that the agent will pursue that goal. This poses a question about the robustness of an open MAS, which can only be obtained by relying on the commitments that each agent voluntarily takes towards the organization as a whole. This is the rationale that has driven us in proposing the ADOPT protocol, whose objective is to make agents, and the organization to which they belong, mutually accountable.

We plan to continue this study along two main research directions. The first concerns the specification of a conceptual model of accountability for use in MASs. We believe the notions of expectation and of control to be at the heart of accountability, and find a challenge in capturing the relational nature of this concept. Among the repercussions of such a model lies the possibility to provide organizations with a tool that will enable them to evaluate, and possibly improve, the processes put in place, as well as the whole organizational structure. The second is to account in the model, and also in the protocol, for a notion of negative accountability. This means not only accounting for what agents should have achieved and have not, but also for active interferences, by which agents may impede social progress. In addition, the ADOPT protocol presented in this paper could be complemented with some reasoning mechanism on the forum side. ADOPT, in fact, aims at tracing, during the interaction, the relevant information for the computation of accountability, but it does not address how to use such information to identify who is accountable. Indeed, this is the task of the forum [38]. A possible way to support the forum, thus, could be represented by the Delphi protocol, whose first formalization in terms of a MAS organization is given in [68] by means of the INGENIAS [69] methodology. In the Delphi protocol, a client submits to a group of experts a question, and one, or more, monitor agents set up an iterative interaction among the experts until a consensus is reached about the answer to the question. A similar mechanisms for consensus reaching can be adopted for automating, at least in part, the task of the forum.

Supplementary Materials: The source code of the AccountabilityBoard artifact and of the examples presented in Section 5 is available at http://di.unito.it/adoptjacamo.

Acknowledgments: This work was partially supported by the *Accountable Trustworthy Organizations and Systems (AThOS)* project, funded by Università degli Studi di Torino and Compagnia di San Paolo (CSP 2014).

Author Contributions: All authors equally contributed to this research.

Conflicts of Interest: The authors declare no conflict of interest.

Abbreviations

The following abbreviations are used in this manuscript:

CNP	Contract Net Protocol
CTL	Computation tree logic
CTLC	Computation tree logic with commitment modality
CTLK	Computation tree logic with epistemic modality
FIPA	Foundation for Intelligent Physical Agents
IOSE	Interaction-oriented software engineering
MAS	Multiagent system
STS	Socio–technical system

References

1. Esteva, M.; Rodríguez-Aguilar, J.A.; Sierra, C.; Garcia, P.; Arcos, J.L. On the Formal Specification of Electronic Institutions. In *Agent Mediated Electronic Commerce: The European AgentLink Perspective*; Dignum, F., Sierra, C., Eds.; Springer: Berlin/Heidelberg, Germany, 2001; pp. 126–147.

2. Dignum, V. *A Model for Organizational Interaction: Based on Agents, Founded in Logic*; SIKS: Lyon, France, 2004.

3. Dignum, V.; Vázquez-Salceda, J.; Dignum, F. *OMNI: Introducing Social Structure, Norms and Ontologies into Agent Organizations*; Lecture Notes in Artificial Intelligence; Springer: Berlin/Heidelberg, Germany, 2004; Volume 4, pp. 181–198.

4. Fornara, N.; Viganò, F.; Verdicchio, M.; Colombetti, M. Artificial institutions: A model of institutional reality for open multiagent systems. *Artif. Intell. Law* **2008**, *16*, 89–105.

5. Mariani, S.; Omicini, A. Coordinating activities and change: An event-driven architecture for situated MAS. *Eng. Appl. Artif. Intell.* **2015**, *41*, 298–309.

6. Zambonelli, F.; Jennings, N.R.; Wooldridge, M. Developing multiagent systems: The Gaia methodology. *ACM Trans. Softw. Eng. Methodol. (TOSEM)* **2003**, *12*, 317–370.

7. Kolp, M.; Giorgini, P.; Mylopoulos, J. Multi-agent architectures as organizational structures. *Auton. Agents Multi-Agent Syst.* **2006**, *13*, 3–25.

8. Bordini, R.H.; Hübner, J.F.; Wooldridge, M. *Programming Multi-Agent Systems in AgentSpeak Using Jason*; John Wiley & Sons: Hoboken, NJ, USA, 2007; Volume 8.

9. Ricci, A.; Piunti, M.; Viroli, M.; Omicini, A. Environment Programming in CArtAgO. In *Multi-Agent Programming: Languages, Tools and Applications*; Springer: Boston, MA, USA, 2009; pp. 259–288.

10. Dastani, M.; Tinnemeier, N.A.; Meyer, J.J.C. A programming language for normative multiagent systems. In *Handbook of Research on Multi-Agent Systems: Semantics and Dynamics of Organizational Models*; IGI Global: Hershey, PA, USA, 2009; pp. 397–417.

11. Boissier, O.; Bordini, R.H.; Hübner, J.F.; Ricci, A.; Santi, A. Multi-agent Oriented Programming with JaCaMo. *Sci. Comput. Program.* **2013**, *78*, 747–761.

12. Wooldridge, M.J. *Introduction to Multiagent Systems*; Wiley: Hoboken, NJ, USA, 2009.

13. Chopra, A.K.; Singh, M.P. From social machines to social protocols: Software engineering foundations for sociotechnical systems. In Proceedings of the 25th International Conference on World Wide Web, Montréal, QC, Canada, 11–15 April 2016.

14. Baldoni, M.; Baroglio, C.; May, K.M.; Micalizio, R.; Tedeschi, S. Computational Accountability. In Proceedings of the AI*IA Workshop on Deep Understanding and Reasoning: A challenge for Next-Generation Intelligent Agents, URANIA 2016, Genova, Italy, 28 November 2016; Chesani, F., Mello, P., Milano, M., Eds.; Volume 1802, pp. 56–62.

15. Baldoni, M.; Baroglio, C.; Capuzzimati, F.; Micalizio, R. Commitment-based Agent Interaction in JaCaMo+. *Fundam. Inform.* **2018**, *159*, 1–33.

16. Baldoni, M.; Baroglio, C.; May, K.M.; Micalizio, R.; Tedeschi, S. ADOPT JaCaMo: Accountability-Driven Organization Programming Technique for JaCaMo. In *PRIMA 2017: Principles and Practice of Multi-Agent Systems*; Bo, A., Bazzan, A., Leite, J., van der Torre, L., Villata, S., Eds.; Lecture Notes in Artificial Intelligence; Springer: Nice, France, 2017; Volume 10621, pp. 295–312.

17. Anderson, P.A. Justifications and Precedents as Constraints in Foreign Policy Decision-Making. *Am. J. Political Sci.* **1981**, *25*, 738–761.

18. Castelfranchi, C. Commitments: From Individual Intentions to Groups and Organizations. In Proceedings of the First International Conference on Multiagent Systems, ICMAS 1995, San Francisco, CA, USA, 12–14 June 1995; Lesser, V. R., Gasser, L., Eds.; pp. 41–48.

19. Singh, M.P. An ontology for commitments in multiagent systems. *Artif. Intell. Law* **1999**, *7*, 97–113.

20. Zahran, M. *Accountability Frameworks in the United Nations System*; UN Report A/66/710; United Nations: New York, NY, USA, 2012. Available online: http://repository.un.org/handle/11176/293914 (accessed on 22 March 2018).

21. United Nations Children's Fund. *Report on the Accountability System of UNICEF*; E/ICEF/2009/15; UNICEF: New York, NY, USA, 2009. Available online: https://www.unicef.org/about/execboard/files/09-15-accountability-ODS-English.pdf (accessed on 22 March 2018).

22. Sustainable Energy for All Initiative. *Accountability Framework*; United Nations: New York, NY, USA, 2014. Available online: https://sustainabledevelopment.un.org/content/documents/1644se4all.pdf (accessed on 22 March 2018).

23. Obrecht, A. *Effective Accountability? The Drivers, Benefits and Mechanisms of CSO Self-Regulation*; Technical Report Briefing No. 130; One World Trust: London, UK, 2012.

24. Warren, S.; Lloyd, R. *Civil Society Self-Regulation*; Technical Report Briefing Paper Number 119; One World Trust: London, UK, 2009.

25. Baldoni, M.; Boella, G.; Genovese, V.; Mugnaini, A.; Grenna, R.; van der Torre, L. A Middleware for Modelling Organizations and Roles in Jade. In Proceedings of the Programming Multi-Agent Systems (ProMAS 2009), Budapest, Hungary, 10–15 May 2009; Braubach, L., Briot, J.P., Thangarajah, J., Eds.; Lecture Notes in Artificial Intelligence; Springer: Berlin/Heidelberg, Germany, 2010; Volume 5919, pp. 100–117.

26. Boella, G.; van der Torre, L. The ontological properties of social roles in multiagent systems: Definitional dependence, powers and roles playing roles. *Artif. Intell. Law* **2007**, *15*, 201–221.

27. Chopra, A.K.; Dalpiaz, F.; Aydemir, F.B.; Giorgini, P.; Mylopoulos, J.; Singh, M.P. Protos: Foundations for engineering innovative sociotechnical systems. In Proceedings of the IEEE 22nd International Requirements Engineering Conference (RE 2014), Karlskrona, Sweden, 25–29 August 2014; pp. 53–62.

28. Bella, G.; Paulson, L.C. Accountability Protocols: Formalized and Verified. *ACM Trans. Inf. Syst. Secur.* **2006**, *9*, 138–161.

29. De Oliveira, A.S.; Charfi, A.; Schmeling, B.; Serme, G. A Model-Driven Approach for Accountability in Business Processes. In *Enterprise, Business-Process and Information Systems Modeling*; Bider, I., Gaaloul, K., Krogstie, J., Nurcan, S., Proper, H.A., Schmidt, R., Soffer, P., Eds.; Lecture Notes in Business Information Processing; Springer: Berlin/Heidelberg, Germany, 2014; Volume 175, pp. 184–199.

30. Yumerefendi, A.R.; Chase, J.S. The Role of Accountability in Dependable Distributed Systems. In Proceedings of the First Conference on Hot Topics in System Dependability, Yokohama, Japan, 28 June–1 July 2005; USENIX Association: Berkeley, CA, USA, 2005; p. 3.

31. Haeberlen, A.; Kouznetsov, P.; Druschel, P. PeerReview: Practical Accountability for Distributed Systems. *SIGOPS Oper. Syst. Rev.* **2007**, *41*, 175–188.

32. Kramer, S.; Rybalchenko, A. A Multi-Modal Framework for Achieving Accountability in Multi-Agent Systems. In Proceedins of the Workshop on Logics in Security, Copenhagen, Denmark, 9–13 August 2010; pp. 148–174.

33. Nissenbaum, H. Accountability in a computerized society. *Sci. Eng. Ethics* **1996**, *2*, 25–42.

34. Mao, W.; Gratch, J. Modeling Social Causality and Responsibility Judgment in Multi-agent Interactions. *J. Artif. Intell. Res.* **2012**, *44*, 223–273.

35. Feltus, C.; Petit, M. Building a Responsibility Model Including Accountability, Capability and Commitment. In Proceedings of the International Conference on Availability, Reliability and Security, Fukuoka, Japan, 16–19 March 2009; pp. 412–419.

36. Küsters, R.; Truderung, T.; Vogt, A. Accountability: Definition and Relationship to Verifiability. In Proceedings of the 17th ACM Conference on Computer and Communications Security, Chicago, IL, USA, 4–8 October 2010; ACM: New York, NY, USA, 2010; pp. 526–535.

37. Bovens, M.; Goodin, R.E.; Schillemans, T. *The Oxford Handbook of Public Accountability*; Oxford University Press: Oxford, UK, 2014.

38. Burgemeestre, B.; Hulstijn, J. Designing for Accountability and Transparency: A value-based argumentation approach. In *Handbook of Ethics, Values, and Technological Design: Sources, Theory, Values and Application Domains*; Springer: Basel, Switzerland, 2015.

39. Simon, J. *The Online Manifesto: Being human in a Hyperconnected Era*; Floridi, L., Ed.; Springer: Basel, Switzerland, 2015.

40. Chopra, A.K.; Singh, M.P. The thing itself speaks: Accountability as a foundation for requirements in sociotechnical systems. In Proceedings of the IEEE 7th International Workshop on Requirements Engineering and Law (RELAW), Karlskrona, Sweden, 26 August 2014; p. 22.

41. Braham, M.; van Hees, M. An Anatomy of Moral Responsibility. *Mind* **2012**, *121*, 601–634.

42. Marengo, E.; Baldoni, M.; Baroglio, C.; Chopra, A.K.; Patti, V.; Singh, M.P. Commitments with Regulations: Reasoning about Safety and Control in REGULA. In Proceedings of the 10th International Conference on Autonomous Agents and Multiagent Systems (AAMAS 2011), Taipei, Taiwan, 2–6 May 2011; Tumer, K., Yolum, P., Sonenberg, L., Stone, P., Eds.; Volume 2, pp. 467–474.

43. Dastani, M.; Lorini, E.; Meyer, J.C.; Pankov, A. Other-Condemning Anger = Blaming Accountable Agents for Unattainable Desires. In Proceedings of the 16th Conference on Autonomous Agents and MultiAgent Systems, São Paulo, Brazil, 8–12 May 2017; pp. 1520–1522.

44. Hohfeld, W.N. Some Fundamental Legal Conceptions as Applied in Judicial Reasoning. *Yale Law J.* **1913**, *23*, 16–59.

45. Foundation for Intelligent Physical Agents. *FIPA ACL Message Structure Specification*; Foundation for Intelligent Physical Agents: Geneva, Switzerland, 2002.

46. Castelfranchi, C. *Commitments: From Individual Intentions to Groups and Organizations*; The MIT Press: Cambridge, MA, USA, 1995; pp. 41–48.

47. Telang, P.R.; Singh, M.P.; Yorke-Smith, N. *Relating Goal and Commitment Semantics*; Lecture Notes in Artificial Intelligence; Springer: Berlin/Heidelberg, Germany, 2011; Volume 7217, pp. 22–37.

48. Smith, R.G. The Contract Net Protocol: High-Level Communication and Control in a Distributed Problem Solver. *IEEE Trans. Comput.* **1980**, *29*, 1104–1113.

49. Foundation for Intelligent Physical Agents. *FIPA Contract Net Interaction Protocol Specification*; Foundation for Intelligent Physical Agents: Geneva, Switzerland, 2002.

50. Foundation for Intelligent Physical Agents. *FIPA Request Interaction Protocol Specification*; Foundation for Intelligent Physical Agents: Geneva, Switzerland, 2002.

51. Lai, R.; Jirachiefpattana, A. Protocol Verification. In *Communication Protocol Specification and Verification*; Springer: Boston, MA, USA, 1998; pp. 143–163.

52. Alpern, B.; Schneider, F.B. Recognizing safety and liveness. *Distrib. Comput.* **1987**, *2*, 117–126.

53. Sajkowski, M. Protocol Verification Techniques: Status Quo and Perspectives. Protocol Specification, Testing and Verification IV. In Proceedings of the IFIP WG6.1 Fourth International Workshop on Protocol Specification, Testing and Verification, Skytop Lodge, PA, USA, 11–14 June 1984; pp. 697–720.

54. El-Menshawy, M.; Bentahar, J.; Dssouli, R. Symbolic model checking commitment protocols using reduction. In Proceedings of the 8th International Workshop on Declarative Agent Languages and Technologies, DALT 2010, Toronto, ON, Canada, 10 May 2010; Lecture Notes in Artificial Intelligence; Volume 6619, pp. 185–203.

55. Clarke, E.M.; Emerson, E.A.; Sistla, A.P. Automatic verification of finite-state concurrent systems using temporal logic specifications. *ACM Trans. Program. Lang. Syst.* **1986**, *8*, 244–263.

56. Penczek, W.; Lomuscio, A. Verifying Epistemic Properties of Multi-agent Systems via Bounded Model Checking. *Fundam. Inform.* **2003**, *55*, 167–185.

57. Lomuscio, A.; Qu, H.; Raimondi, F. MCMAS: An open-source model checker for the verification of multiagent systems. *Int. J. Softw. Tools Technol. Transf.* **2017**, *19*, 9–30.

58. Hubner, J.F.; Sichman, J.S.; Boissier, O. Developing Organised Multiagent Systems Using the MOISE+ Model: Programming Issues at the System and Agent Levels. *Int. J. Agent-Oriented Softw. Eng.* **2007**, *1*, 370–395.

59. Rao, A.S. AgentSpeak(L): BDI agents speak out in a logical computable language. In *Agents Breaking Away*; Lecture Notes in Artificial Intelligence; Springer: Berlin/Heidelberg, Germany, 1996; Volume 1038, pp. 42–55.

60. Omicini, A.; Ricci, A.; Viroli, M. Artifacts in the A&A meta-model for multiagent systems. *Auton. Agents Multi-Agent Syst.* **2008**, *17*, 432–456.

61. Weyns, D.; Omicini, A.; Odell, J. Environment as a first class abstraction in multiagent systems. *Auton. Agents Multi-Agent Syst.* **2007**, *14*, 5–30.

62. Hübner, J.F.; Boissier, O.; Kitio, R.; Ricci, A. Instrumenting multiagent organisations with organisational artifacts and agents. *Auton. Agents Multi-Agent Syst.* **2010**, *20*, 369–400.

63. Baldoni, M.; Baroglio, C.; Capuzzimati, F.; Micalizio, R. Empowering Agent Coordination with Social Engagement. In *AI*IA 2015, Advances in Artificial Intelligence*; Lecture Notes in Artificial Intelligence; Springer: Berlin/Heidelberg, Germany, 2015; Volume 9336, pp. 89–101.

64. Baldoni, M.; Baroglio, C.; Capuzzimati, F.; Micalizio, R. Type checking for protocol role enactments via commitments. *Auton. Agents Multi-Agent Syst.* **2018**, 1–38, doi:10.1007/s10458-018-9382-3.

65. World Health Organization. WHO Accountability Framework. 2015. Available online: http://www.who.int/about/who_reform/managerial/accountability-framework.pdf (accessed on 22 March 2018).

66. Office of the Auditor General of Canada. Modernizing Accountability in the Public Sector. In *Report of the Auditor General of Canada*; Minister of Public Works and Government Services Canada: Ottawa, ON, Canada, 2002; Chapter 9. Available online: http://www.oag-bvg.gc.ca/internet/English/parl_oag_200212_09_e_12403.html (accessed on 22 March 2018).

67. Social Accountability International. *Social Accountability 8000 International Standard*; Social Accountability International: New York, NY, USA, 2014.

68. García-Magariño, I.; Gómez-Sanz, J.J.; Pérez-Agüera, J.R. A multiagent based implementation of a Delphi process. In Proceedings of the 7th International Joint Conference on Autonomous Agents and Multiagent Systems (AAMAS 2008), Estoril, Portugal, 12–16 May 2008; Volume 3, pp. 1543–1546.

69. Gómez-Sanz, J.J.; Pavón, J. Implementing Multi-agent Systems Organizations with INGENIAS. In Proceedings of the Third International Workshop on Programming Multi-Agent Systems (ProMAS 2005), Utrecht, The Netherlands, 26 July 2005; Lecture Notes in Artificial Intelligence; Springer: Berlin/Heidelberg, Germany, 2005; Volume 3862, pp. 236–251.

applied
sciences

MDPI

Article

Modelling the Interaction Levels in HCI Using an Intelligent Hybrid System with Interactive Agents: A Case Study of an Interactive Museum Exhibition Module in Mexico

Ricardo Rosales [1,*], Manuel Castañón-Puga [2,*], Felipe Lara-Rosano [3], Josue Miguel Flores-Parra [2], Richard Evans [4], Nora Osuna-Millan [1] and Carelia Gaxiola-Pacheco [2]

[1] Accounting and Administration School, Autonomous University of Baja California, Tijuana 22390, Mexico; nora.osuna@uabc.edu.mx

[2] Chemistry and Engineering School, Autonomous University of Baja California, Tijuana 22390, Mexico; mflores31@uabc.edu.mx (J.M.F.-P.), cgaxiola@uabc.edu.mx (C.G.-P.)

[3] Complexity Science Center, National Autonomous University of Mexico, Mexico City 04510, Mexico; flararosano@gmail.com

[4] Business Information Management and Operations Department, University of Westminster, London NW1 5LS, UK; R.Evans@westminster.ac.uk

* Correspondence: ricardorosales@uabc.edu.mx (R.R.); puga@uabc.edu.mx (M.C.-P.); Tel.: +52-(664)-979-7500 (ext. 55000) (R.R.); +52(664)-979-7500 (ext. 54359) (M.C.-P.)

Received: 11 February 2018; Accepted: 13 March 2018; Published: 15 March 2018

Abstract: Technology has become a necessity in our everyday lives and essential for completing activities we typically take for granted; technologies can assist us by completing set tasks or achieving desired goals with optimal affect and in the most efficient way, thereby improving our interactive experiences. This paper presents research that explores the representation of user interaction levels using an intelligent hybrid system approach with agents. We evaluate interaction levels of Human-Computer Interaction (HCI) with the aim of enhancing user experiences. We consider the description of interaction levels using an intelligent hybrid system to provide a decision-making system to an agent that evaluates interaction levels when using interactive modules of a museum exhibition. The agents represent a high-level abstraction of the system, where communication takes place between the user, the exhibition and the environment. In this paper, we provide a means to measure the interaction levels and natural behaviour of users, based on museum user-exhibition interaction. We consider that, by analysing user interaction in a museum, we can help to design better ways to interact with exhibition modules according to the properties and behaviour of the users. An interaction-evaluator agent is proposed to achieve the most suitable representation of the interaction levels with the aim of improving user interactions to offer the most appropriate directions, services, content and information, thereby improving the quality of interaction experienced between the user-agent and exhibition-agent.

Keywords: human-machine interaction; ambient intelligence; user interaction levels; intelligent agents; intelligent hybrid systems; type-2 fuzzy inference system

1. Introduction

Since the dawn of the 21st century, technology has immersed itself in our everyday lives and become a necessary facilitator of daily activities; some technological devices assist or support us by completing set tasks or achieving desired goals with optimal affect and in the most efficient way, thereby improving our interactive experiences. However, what happens when a user interacts without technology? Is the interaction experienced better or worse? What are the interaction levels when using or not using technology and how does this change? Can we measure user interaction levels without metric variables, relying solely on body language, using linguistic variables? Which influencing factors increase or decrease our levels of interaction? What is the quality of the interaction? What is the interaction time? Which factors influence abandonment rates during interaction? This paper aims to address these questions by evaluating interaction levels in HCI and thereby improving user experiences. We consider the description of interaction levels using an intelligent hybrid approach to provide a decision-making system to an agent that self-evaluates interaction in interactive modules in a museum exhibition. The agents represent a high-level abstraction of the system, where communication takes place between the user, exhibition and the environment.

In our research, we analyse the evaluation made by an on-site observer from a sample of 500 users that visited "El Trompo" Museo Interactivo Tijuana in Mexico to set-up a Fuzzy Inference System (FIS) [1] using 3 hybrid techniques: (1) Empirical FIS (EF) [2,3], (2) a Fuzzy C-Means method of Data Mining named Data Mined Type-1 (DMT1F) [4,5] and (3) Neuro-Fuzzy System (NFS) [6,7]. The different user action inputs were represented to classify interaction levels using a FIS to improve the provision of content with the purpose of increasing interaction levels experienced in the Museum. The involved actors included the user and the exhibition module, which were represented by agents as a high-level abstraction of the system. We expressed the native user by User-Agent, the exhibition module by Exhibition-Agent (GUI) and interaction evaluation system by Interaction Evaluator-Agent (Interaction Evaluator).

1.1. Interaction Levels

In this research, based on Gayesky and Williams' Interaction Levels Theory [8], we set defined parameters for analysis, such as presence, interactivity, control, feedback, creativity, productivity, communication and adaptation, to identify the interaction level of users using a FIS. The interaction between user and exhibition is important to evaluation, including its related factors (interaction time, type of interaction, etc.). A secondary motivation for this research is that researchers typically evaluate user interaction levels using quantitative methods and not qualitative metrics. Moreover, it is important to understand each interaction (or lack of interaction) with the user to develop dependable user-exhibition interactions and inform him/her how much is truly valued; interactions are 'moments of truth' i.e., we can learn user preferences and guide them in their subsequent choices. This approach creates a new opportunity for developing improved interactive experiences. For museums, it is crucial that exhibitions can self-learn and adapt to user interactions at every stage of the interaction, based on user actions.

1.2. Museum User-Exhibition Interaction

This research provides a means to measure interaction levels based on the natural behaviour of users formed by their interactions with museum exhibitions. We consider that, by analysing user interactions in a museum, we can help to design better methods to interact with exhibition modules according to the preferences, characteristics and behaviours of its users. To evaluate this interaction requires the identification of objective criteria based on qualitative aspects of the users' behaviour. We consider this to be a complex task requiring specific considerations about not only the performance and/or interactions of users, but also the involved uncertainty in evaluating user perceptions, making it difficult to assess and draw conclusions.

2. Related Work

Emerging social phenomena are difficult to explain since traditional methods do not naturally identify them. Agent-based methodologies allow for the identification and explanation of the causes of agent interactions involved in the phenomena, providing a greater understanding of the context. Rosenfeld et al. [9], proposed a methodology for using automated agents in two scenarios: real-world and human-multi robot, for collaborative tasks. The agents in this research are able to learn from past interactions, creating policies that develop deeper planning capabilities. Rosenfeld and Sarit [10] proposed a methodology for developing agents to support people in argumentative discussions, including the ability to propose arguments. This research is based on Conversational Agents (CAs), where the CAs can converse with humans and provide information and assistance [11,12]. The CAs framework allows HCI because a human directly interacts with the CA starting a conversation and understanding user goals. The research analyzed the human argumentative behavior and demonstrated that effective predictions of argumentative behavior are considerably improved when merging four methods: Argumentative Theory, Relevance Heuristics, Machine Learning (ML) and Transfer Learning (TL), creating the ability to build intelligent argument agents. They concluded that ML techniques were the best option to allow predictions of human argumentative behavior and this behavior can be structured, semi-structured and free-form argumentation, as a training data on the wish topic is available, allowing argumentation in the real-world.

Garruzzo and Rosaci [13] argued that semantic negotiation is the key for agent clustering. They propose a form of 'groups of agents', based on similarities considering its ontologies. They also state that it is important to consider the context where the agent ontology is used and propose a novel clustering technique called HIerarchical SEmantic NEgotation (HISENE) which considers the structural semantic components. Their research also proposes an algorithm to compute ontology-based similarity. They build a 200 agent software with communication skills and use the semantic negotiation protocol. The research addresses the problem of developing MAS, when it is a is necessary form of the agents' group. We face several challenges relaing to cooperation and teamwork. Sometimes, it is necessary to collaborate among different groups of agents to achieve common goals or solve complex problems; if the agents are composed of the same group, it is easy to collaborate because they have the same ontologies, but on another hand, when the agent belongs to different groups, the communication is complicated because the ontologies are different, then this research responds to this issues exploiting the capabilities of HISENE.

Derive from complex problems, a novel proposal is required to find responses; sometimes, human capabilities are limited to do this. In this way, technology is needed on behalf of humans to find responses. One option is intelligent agents. Rosaci [14] proposes to build agents, considering the internal representation of behavior and interests of the owner creating ontologies; these ontologies are needed to help agents create inter-relationships of knowing-sharing. This paper proposes to construct semi-automated ontologies, based on observing behaviors. A MAS, named Connectionist Inductive Learning and Inter-Ontology Similarities (CICILOS), is proposed for recommending information agents. The agents are viewed as user models, applying the same process applied to ontologies checking similarities with humans. In this paper, the term ontology is used as Knowledge Kernel agent's. The CICILOS is composed of different levels of agents with different topologies. Level One is called 'Main'; the Main contains three agents: (1) Agent Management Systems (AMS), (2) Directory Facilitator (DF) and (3) Agent User Interface (GUI). This level is characterized by essential agent's based on JADE Platform. The second Level is called IACOM (connectionist) and contains inductive agents. These agents are related to humans, based on behavior and interests. The IACOM (symbolic) level is based on the neural-symbolic network. The third level is called IAOM. This level is connected to the IACOM level; this contains an ontology translator. The fourth level is called Ontologies Similarities Managers (OSMs). This is related to IAOM and computes similarities between agents based on IAOM ontology. The underlying enriches the capability of selected adequate agents for cooperation; the neural-symbolic network makes more efficient inductive mechanisms, improving the planning tasks of the agents based on learning run time.

On another hand, the use of agents can help us in our social lives by understanding our personality and creating relationships with other people with similarities. Cerekovic et al. [15] used Rapport by [16] with virtual agents, linking social cues and self-reported human personality collected judged the rapport of human-agent interaction, studying what kind of social cues infer on those judgments.

The Human Interaction (HINT) methodology, proposed by Sanchez-Cortes et al. [17], considered HCI using 1-min interaction videos that judged rapport. Rapport was collected to help extract social cues from audio-visual data. The social cues extraction was derived from HCI. The researchers used agents, Sensitive Artificial Listeners (SALs). Schroder et al. [18], stated that the SALs, these agents are the key to the social cue extraction because they evaluate by measuring facial expressions and social cue extraction, composed by verbal cues (language style) and no verbal cues (auditory cues, visual cues).

The Multi-Agent System Handling User and Device Adaptivity of Web Sites (MASHA), presented by Rosaci and Sarne [19], can be used as a handling user between a website and user, helped by different agents: Client Agent, Server Agent and Adapter Agent. The Client Agent creates a user profile, considering its interests, behaviors and desires. The MASHA uses the User Agent for monitoring user navigation; this profile increases with interactions. It is then helped by the Server Agent, which gathers information to have more relevant information about the Client Agent and also these agents can collaborate to improve the knowledge about navigation and user profile autonomously. The Agent Adapter analyzes gathered information and generates recommendations, based on user preferences. The MASHA delivers effective recommendation results of content-based filtering. This analysis is performed by the Adapter Agent. If the user increases its navigation, it consequently increases its profile and gets a better HCI based on its own predilections. The MASHA support constructs a community of agents; these communities are composed of two categories: C1, which links the human user and C2, which links with the website. The MASHA can deliver a novel tool for Web Site visitors and provide useful suggestions, considering its devices increase, and increase the satisfaction of user on web navigation.

When the related research is analyzed, we can see that they propose novel solutions to improve the HCI experience and provide different answers for greater adaptability supported by intelligent agents. The research combines different techniques, technologies, Argumentative Theory, Relevance Heuristics, Machine Learning and Transfer Learning; all these options can allow us to create powerful agents with intelligence, with abilities and functions to be intermediaries to interact with humans inclusive with hardware such as tracking systems. The creation of intelligent agents lets be ready to work and respond in different contexts from emergency scenario and predictions of human behavior. The advantage of agent-based approaches is that it allows the representation of an endless number of users from 1 user to thousands of users. Also, grants us to have a greater perspective on the creation of models and applications, based on agents. This allows for our proposed research adequate creation of HCI within the context of the user exhibition in an interactive museum. Likewise, it provides representation of our research, through agents and users involved in this context. We can create several simulations of different and possible scenarios of the real world. On the other hand, it also allows for the creation of agents with personalized features based on levels of user interaction. The creation of these agents will allow them to have greater autonomy, reactivity and adaptability, based on emerging changes in the context of user exhibition interaction. However, we must bear in mind that the proposed research only focuses on the context within an interactive museum and is limited to being tested in emergency contexts.

3. Methodology

This research proposes a model that allows for the representation of levels of interaction using a FIS to evaluate, in a qualitative and subjective manner, the values of the levels of interaction between a user and a museum exhibition module, bearing in mind uncertain results of exchanges of messages [1]. This approach helps avoid imperfect information when trying to provide services, information and content that the user requires, based on its interaction level, while seeking to supply a satisfactory level of interactive experience. To model the interaction levels in HCI using an intelligent hybrid system with interactive agents, we developed the following strategies.

Firstly, we approached the user-exhibition interaction module following some of the recommendations of Gaia, a methodology for agent-oriented analysis and design, to identify the roles and interactions in the referent/target system. In Figure 1, the process and models proposed by the agent-oriented modelling method is illustrated. We designed agents, relationship and services to build a prototype for experimentation. You can find further information about the full description of Gaia in [20].

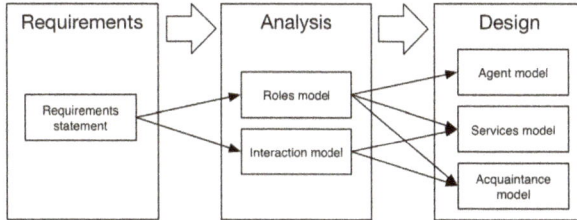

Figure 1. Process and models of Gaia, a methodology for agent-oriented analysis and design, applied to the interactive museum exhibition module case study.

Secondly, we approached the interaction levels in HCI using an intelligent hybrid system following a general methodology of computational modelling. It is an iterative process that begins with a referent system in the real world. Then, abstraction, formalisation, programming and appropriate data are used to develop a viable computational model. In Figure 2, we show the proposed steps by the modelling method applied to the case study. You can find further information about the full description of computational modelling in [21,22].

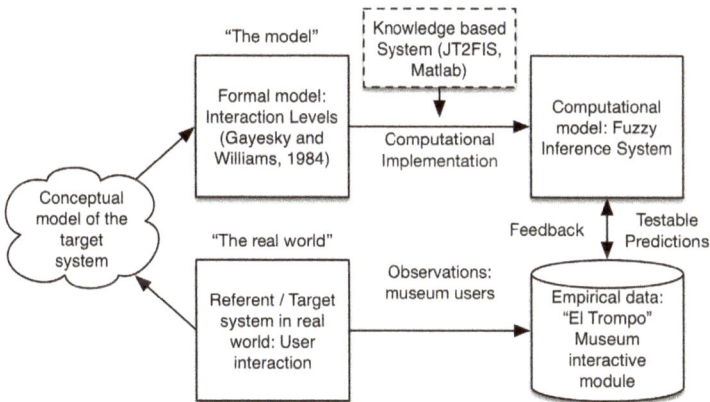

Figure 2. General computational modelling methodology applied to the interactive museum exhibition module case study.

Finally, we used an interactive the museum exhibition module case study to validate the proposed model. The interactions reported on in this study were simulated, performed, observed and analysed in an interactive museum in Tijuana, Mexico. We consider it appropriate to base our study on that addressed in [23]; the types of interaction that occur in this kind of environment are suitable for the proposed research. It is believed that by measuring levels of interaction, it allows for the strengthening of knowledge to determine the services or information that users require, based on predilections in conjunction with the level of interaction. After producing several computational models of the interaction levels applying different methods, we used confusion matrices to show the computational model predictions tests results. In machine learning, a confusion matrix is an error table that allows examining the performance of the data mining (or training process) algorithm and the produced inference system response. Each row of the array represents the instances in a predicted class while each column represents the cases in an actual class (or vice versa).

3.1. Case Study

3.1.1. Room and Exhibition Module Selection

To analyse the user-exhibition interaction, we studied both physical and theoretical aspects of the museum rooms, including their themes, objectives, methods of interaction and their location in the museum; additionally, we observed the methods of interaction found in each room in order to select a suitable room which allowed us to analyze the behaviour, actions, performance, interruption factors and interaction levels of users, as well as the interactive content type, information and/or services that the exhibitions provided. We also considered whether the content was adequate for users, suitable in relation to the kind of interactions of users and adequate in maintaining the attention of the user. We further examined whether the content was evidencing interaction and analysed the objective of the exhibition to determine whether it was adequate in encouraging a good interaction for the user and the media interface of the exhibition modules to determine whether they were adequate to have a suitable interaction.

After analysis of the different exhibition modules, an interesting interactive module was chosen with features that allowed us to obtain the majority of parameters that we wished to analyze in our research. The name of the exhibition module was "Move Domain". This educational exhibition involved users interacting and playing with one of four objects (car, plane, bike or balloon) which were displayed simultaneously on four separate screens, demonstrating the 4 different methods of moving in the simulated virtual world. Users were able to get the experience of using all 4 transportation means; they were able to interact in the virtual world and see how other users travelled and interacted around the virtual world. The exhibition's objective was to allow users to develop hand-eye coordination skills and spatial orientation using its technology. The content was based on eye coordination and interaction with electronic games, with the exhibition's message being "I can learn about virtual reality through playing". The suggested numbers of simultaneous users were 4.

3.1.2. Exhibition Module Interface

The module interface consisted of four sub-modules attached by connectors. Each module included a cover stand for the 32-inch screen, software that simulated the virtual world and a cabinet to protect the computer. The exhibition module was supported by a joystick to handle the plane, a steering wheel and pedals to drive the car, handlebars to ride the bike and a rope to fly the balloon. This interactive exhibition module, which is one of the most visited in the museum, allowed us to obtain important data for analysis, processing and validation of the proposed model. Figure 3 depicts the analyzed exhibition module.

Figure 3. Analyzed interactive museum exhibition module.

3.2. Study Subjects

As subjects for the study, users were randomly selected from those children and adults who participated in supervised tours as part of a permanent program of collaboration between local schools and the museum. The institution and schools involved have the necessary agreements in place to conduct non-invasive interactive module evaluations to improve their design. We evaluated user interaction behaviour by performing ethnographic research (notes style) to observe in a non-invasive manner. Personal data was not required in our data collection; therefore, information was produced directly in the museum room through real-time observation, in line with institution committee recommendations to guarantee the anonymity of users.

Evaluation Interaction Parameters

We analyzed and studied parameters such as Presence (do users have a constant presence?, do the users have intermittent presence?), Interactivity (do the users have interaction directly or indirectly with the museum exhibition?, do users have shared interactivity with the exhibition?), Control (do the users have full control over the exhibition?), Feedback (do the users receive some sort of feedback about the content viewed?), Creativity (do the users change the way they interact with the exhibition, according to their creativity?), Productivity (do the users propose something that changes their interaction?), Communication (do the users have communication directly from the exhibition?) and Adaptation (do the users adapt their actions according to the interactive content type delivered by the exhibition?).

All data collected was analyzed from user interaction behavior, which was compiled through ethnographic research that observed, in a non-invasive manner, the user's interaction. We obtained parameter values based on human expert evaluations with implicit uncertainty and calculated every user with arbitrary values, based on expert judgment.

Figure 4 depicts the average results of parameters for the 500 users analyzed. It shows the interaction parameters necessary to develop the adequate FIS to obtain the interaction level.

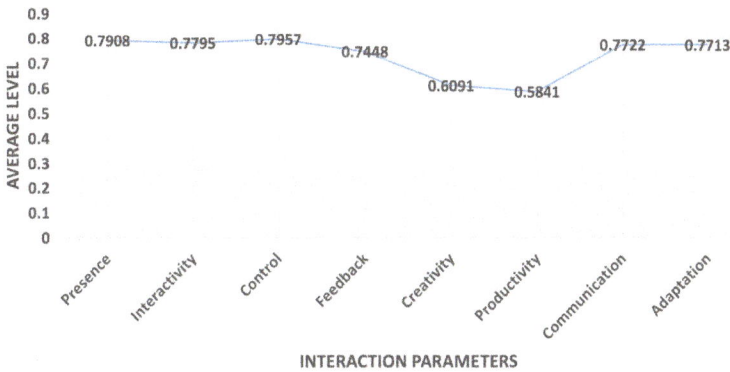

Figure 4. Average of Interaction Parameters.

4. The Model

To support user interactions, HCI is operating as a background process, using invisible sensing computational entities to interact with users. These entities are simulated by the User-Agent and Exhibition-Agent. The collaboration of these entities, permitted during HCI, deliver a customised interactive content type to users in a non-invasive manner and are context-aware. The relationships between Users (museum users) "User-Agent" and Computer (museum exhibition) "Exhibition-Agent" must be systematically modelled and represented to be ready for the emergent context; for this reason, we represent using user-exhibition relationships.

In our research, we represent HCI simulated on museum modelling with embedded agents (User-Agent, Exhibition-Agent, InteractionEvaluator-Agent) that allow the user-exhibition interaction to be supported. Our proposed modelling provides dynamic support for interactions and is aware not only of the user's physical context, but also of its social context i.e., when a user interacts with another user. Our model consists of contextual attributes, such as the location of the user and what they are actively doing during interaction.

The handling of uncertainty in information and service exchange environments presents numerous challenges in terms of imperfect information between the user-exhibition interaction; these interactions should not be predicted nor restricted to real-world applications, such as simulation processes in the exchanges occurred during interaction between user and exhibition. This research seeks to advance the following: To propose a model for representing interaction levels using a fuzzy inference system that helps to measure the level of interaction in order to identify the performance, actions and behaviour of users to offer information or services that are adequate, based on the theory of Gayesky and Williams [8] .

Established models exist which have been developed to process information based on classical logic where the propositions are either true or false. However, no model currently exists that addresses uncertainty generated in environments of information exchange and in imprecise services involved in user-exhibition interactions. We experiment with this proposed model using the FIS to address uncertainties involved in the process of information and service exchange to learn levels of interaction between the user-exhibition in an interactive museum context. We require a diffused input variable mechanism suitable to the environment; this is required to diffuse perceptions to define a fuzzy evaluation module to evaluate values generated among user–exhibition interactions. This evaluation module or diffused perception mechanism must be adapted to consider the method of Mamdani Fuzzy Inference [24] which will enable diffusion to the level of interaction.

4.1. Modelling User-Exhibition Elements

In this research, all interactions occurred on independent exhibitions providing different content which allowed for interactivity. We represented this using agent modelling. A user represented by User-Agent (UA) has complete freedom and infinite time to encounter different interactions (individual, group, accompanied etc.). In this sense, we can map more efficiently the proposed model using agents. First, we analysed a native user (User-Agent) in the environment in order to analyse the performance of the user and obtain inputs to the FIS. We then analysed the identified exhibition represented by Exhibition-Agent (GUI) and InteractionEvaluator-Agent (Interaction) with activities and content offered to measure the level of interaction, identifying available user activities, such as when the user–exhibition interaction arises. The InteractionEvaluator-Agent mediates between the stakeholders (user–exhibition), offering a status of the current state so that both can interact without problem. The InteractionEvaluator-Agent plays a "consultant" role, linking between the user and exhibition in order to provide enhanced integration. In this context, our reactive environment is ready at all times to obtain information. Figure 5 summarises the agent system prototype of the interactive museum exhibition module in a software agent platform [25].

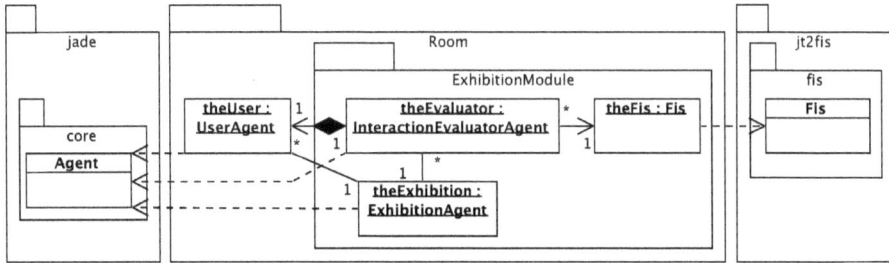

Figure 5. The agent system of the interactive museum exhibition module.

4.1.1. Representing Interaction Levels Using a Fuzzy Inference System

The Interaction Levels Scale, proposed by Gayesky and Williams [8], has been used as a basis for our proposal of Interaction Levels Scale, but it is still unknown how we represent this interaction level using a FIS? First, we defined our own interaction levels scale. The scale was defined against six levels: (1) Extremely Low Interaction (ELI), (2) Very Low Interaction (VLI), (3) Low Interaction (LI), (4) Medium Interaction (MI), (5) High Interaction (HI) and (6) Extremely High Interaction (EHI). Against the scale, we defined the key features of each level and assigned a linguistic value with the finality to represent these in a FIS as output variables. The following is a summary of the proposed scale.

Level 0. The user is present in the exhibition module area and is shown a welcome message and related content. The user does not answer, only presence is confirmed. No interaction exists, only the action of being present. **Key Features**. Interactivity Null. No significant movements, only presence. **Linguistic Value**. Extremely Low Interaction (ELI).

Level 1. The user hears or sees the content, but no meaningful action is perceived. The exhibition module only provides general content (welcome message or content and basic exhibition information). The user receives information, but does not control the interaction. **Key Features**. Very Low Interactivity. Few movements. **Linguistic Value**. Very Low Interaction (VLI).

Level 2. The user has mental reasoning of the content provided by the exhibition, while the exhibition can analyse interactions, raise questions, encourage feedback and summarise fundamental ideas or relevant passages. Approaches arise from responses to user questions. **Key Features**. Low Interactivity. Few movements, comment stimulation, mental analysis. **Linguistic Value**. Low Interaction (LI).

Level 3. The User reasons with the content offered by the exhibition, while the exhibition indicates pauses in which the user develops different types of activities, including oral queries, complementing support material etc., allowing them to control the sequence of the activity, its flow and its continuity. **Key Features**. Medium Interactivity. Pauses indicated, oral activities, queries. **Linguistic Value**. Medium Interaction (MI).

Level 4. At this level, there is greater control between user-exhibition. The user can alter the message they receive by means of feedback i.e., they can select the desired information to receive. The user has the option to decide how, when and what part of the activity they want to develop. **Key Features**. High Interactivity. Control, feedback, desired data selection. **Linguistic Value**. High Interaction (HI).

Level 5. The user has the ability to feedback, control, create, communicate, adapt and produce the information provided by the exhibition. This level represents all the qualities of interactivity. During parts of the interaction, "talk" can occur between the user and exhibition (a "talk" using different means of interaction). **Key Features.** Extremely High Interactivity. Control, feedback, creativity, adaptation, productivity and desired data selection. **Linguistic Value.** Extremely High Interaction (EHI).

The proposed scale was used as a reference to create the suitable model to measure the level of interaction. The ability to measure levels of interaction is essential to provide services or information that the user really needs; this is developed only by understanding the user. Nevertheless, how can we measure and represent this level of interaction using a FIS? It is desirable to have propositions of the reached level of interaction by the user for specific interactive activities.

Computational solutions involved in the development of real-time interaction can be implemented to help in this process. In our propositions, we represent the level of interaction that is assumed, where the user has a specified evaluated level. Within this context, we consider it relevant to integrate fuzzy logic modelling to formalise levels of representation. In this case, the level of interaction is not a result of interaction or non-interaction type; instead, it is a result of all elements that complement the interaction (user profile, preferences, actions, behaviour, performance etc.). The fuzzy logic maintains its knowledge base using rules, making the implementation process more appropriate for exhibition reasoning in order to measure the level of user interaction. This proposed format makes the rules easier to maintain and easier to update the knowledge base.

In this sense, this research analyses the data obtained directly from the user, in the context of interaction between user and exhibition using fuzzy logic to infer relevant information on the level of user interaction in relation to activities conducted. This information is obtained through fuzzy inputs that are used as inputs to the FIS; these inputs are: Presence (Pre), Interactivity (Int), Control (Ctl), Feedback (Fbk), Creativity (Cty), Productivity (Pdt), Communication (Com) and Adaptation (Ada). Each input value was collected using a scale from 0 (minimal value) to 1 (maximum value) derived from the user interaction behaviour. By performing ethnographic research, in a non-invasive way, we get the values based on human expert evaluations with inherent uncertainty.

This research is developed in such a way that the model can be applied to different environment scenarios. The integration of the proposed scale and the FIS helps measure the level of interactions performed by users during user-exhibition interaction. To recognise the level of interaction, the variables are defined as resources generated by the user performance, considering data analysis simulations and real-time monitoring. The obtained variables are evaluated to identify its level of interaction, analysing the proposed scale through applying the FIS. We analysed information available about the exhibition, including its features, media communication, content, etc. and studied the performance data of users, considering their individual actions. The changes that occurred are of great importance to the interaction as they are used to feedback information to the model. The user has a set level of interactions during a given period. This measurement can handle uncertainty determined by a set of membership functions. The states caused by the user are the states that can be induced. We define plans, with each being specified by different membership functions, by linguistic variables that receive the level of user interaction in the process of interaction. Through these, the model can determine the most appropriate content, service or information to be shown to the user.

The measurement of the level of interaction is composed of the input variables that are defined. Each variable has different membership values in the actions of users. When each user interaction begins, it results in different membership values being created that can vary the result of the actions. The interaction may have different levels (ELI, VLI, LI, MI, HI and EHI) to make a decision in order to provide services or information that are really needed by the user. The values of these levels can be interpreted in the calculations when monitoring and analysing the interaction between the user and exhibition; this is also used to determine subsequent interaction.

4.1.2. Implementing the Fuzzy Inference System

Unlike other models that use different paradigms through heuristics, our proposed model adopts a fuzzy set theory to build knowledge provided by users. We present uncertainty of information in a better and more appropriate manner. We have represented, in accordance with environmental inputs, variables (Pre, Int, Ctl, Fbk, Cty, Pdt, Com and Ada) with their respective membership functions which define the output (interaction level). The result of this implementation is written by a fuzzy value and, in this case, is given a linguistic value. The update process is dynamic and is altered according to a user's performance during interaction.

Membership functions were modelled by considering an initial user profile, ensuring a more accurate result for assessing the level of user interaction. Different activities were created with the intention to vary from low to high uncertainty to assess the level of interaction by each user. The implementation of a FIS for the purpose of effective utilization requires the use of programs that directly apply fuzzy logic functions. Some utility programs have specific modules to facilitate the accomplishment of this task, as is the case of the Fuzzy Logic Toolbox of MATLAB (MATLAB vR2017B. The MathWorks Inc., Natick, MA, USA, 2017) [26] that contains a library based on the C language, that contains a library based on the C language, providing the necessary tools to conduct effective fuzzification. JT2FIS (JT2FIS v1.0. Universidad Autónoma del Estado de Baja California, Mexicali, BC, Mexico, 2016) [27], a tool-kit for interval Type-2 fuzzy inference system, can be used to build intelligent object-oriented applications and provide an effective fuzzification method and tools; this utility was used to implement the proposed FIS (see Figure 5).

The model inputs are the variables that can be perceived by the exhibition; these are the performance data of the user interaction. As we consider the input variables (Pre, Int, Ctl, Fbk, Cty, Pdt, Com and Ada) and the output variable (interaction level) to the FIS, these are associated with a set of membership functions. The output function comprises six linguistic variables: (ELI, VLI, LI, MI, HI and EHI). Gaussian functions were used as this type of membership function has a soft non-abrupt decay. The FIS was implemented building inference rules covering all linguistic variables, composed by the operator associated with the minimum method. Aggregation rules are made by the maximum method. Table 1 depicts these base rules; thus, this is identified as the knowledge base representation.

At this stage, to enable the FIS, the fuzzy toolboxes of MATLAB [26] and JT2FIS [27] Tool-kits were used, simulating and entering the 500 users to be analysed; modifying user inputs generates a set of inputs with each one with set values, exemplifying the performance of the user's interaction. These values are subjected to the FIS that return an output variable (Interaction level).

The proposed fuzzy model provides a universe of six levels of interaction. These levels were defined with different values for the parameters of membership levels. This makes it possible to develop a knowledge base that allows a set of applications of membership functions that vary according to interaction and user performance; this is because the membership functions are altered to represent states with different degrees of uncertainty. One example is to build an initial function that is more flexible and categorizes users within sets (ELI, VLI LI MI, HI and EHI). Figure 6 depicts the variations of uncertainty from the first to last level of our fuzzy universe.

To verify the user's corresponding level of interaction, according to their inputs, we evaluate the defuzzification output of the resulting level. Thus, a user moves from one level to another when its membership function value is more inclined to the nearest integer e.g., if the level is at 0.2, it would remain at level 0 of interaction but, if the level is at 0.9, the level of interaction is moved to level 1; the system then updates the knowledge base for the next interaction. Another example is to represent the interaction at level 5, which is the highest level of interaction and presents all input variables near or at the maximum level. This value represents less uncertainty in measuring user performance. Therefore, the value forms the basis of the user behaviour in the environment; also, the value can change dynamically and functions of the membership values can be modified to characterize from a greater uncertainty to less uncertainty about the user's performance. Once the level of interaction is identified, information or services are sent, according to the interaction level result.

Table 1. Inference Fuzzy Rules of the Empirical FIS.

No	Inference Fuzzy Rules
1	If (Presence is Very Bad) and (Interactivity is Very Bad) and (Control is Very Bad) and (FeedBack is Very Bad) and (Creativity is Very Bad) and (Productivity is Very Bad) and (Communication is Very BAD) and (Adaptation is Very Bad) then (Level 0 is High) (Level 1 is Low) (Level 2 is Low) (Level 3 is Low) (Level 4 is Low) (Level 5 is Low).
2	If (Presence is Bad) and (Interactivity is Bad) and (Control is Bad) and (FeedBack is Bad) and (Creativity is Bad) and (Productivity is Bad) and (Communication is Bad) and (Adaptation is Bad) then (Level 0 is Low) (Level 1 is High) (Level 2 is Low) (Level 3 is Low) (Level 4 is Low) (Level 5 is Low).
3	If (Presence is Regular) and (Interactivity is Regular) and (Control is Regular) and (FeedBack is Regular) and (Creativity is Regular) and (Productivity is Regular) and (Communication is Regular) and (Adaptation is Regular) then (Level 0 is Low) (Level 1 is Low) (Level 2 is High) (Level 3 is Low) (Level 4 is Low) (Level 5 is Low).
4	If (Presence is Good) and (Interactivity is Good) and (Control is Good) and (FeedBack is Good) and (Creativity is Good) and (Productivity is Good) and (Communication is Good) and (Adaptation is Good) then (Level 0 is Low) (Level 1 is Low) (Level 2 is Low) (Level 3 is High) (Level 4 is Low) (Level 5 is Low).
5	If (Presence is Very Good) and (Interactivity is Very Good) and (Control is Very Good) and (FeedBack is Very Good) and (Creativity is Very Good) and (Productivity is Very Good) and (Communication is Very Good) and (Adaptation is Very Good) then (Level 0 is Low) (Level 1 is Low) (Level 2 is Low) (Level 3 is Low) (Level 4 is High) (Level 5 is Low).
6	If (Presence is Excellent) and (Interactivity is Excellent) and (Control is Excellent) and (FeedBack is Excellent) and (Creativity is Excellent) and (Productivity is Excellent) and (Communication is Excellent) and (Adaptation is Excellent) then (Level 0 is Low) (Level 1 is Low) (Level 2 is Low) (Level 3 is Low) (Level 4 is Low)(Level 5 is High).

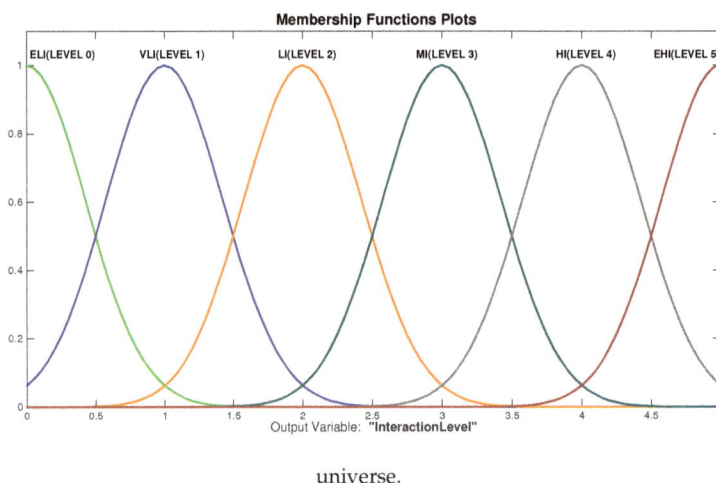

universe.

Figure 6. First and last level of our fuzzy

4.2. Validating the Fuzzy Inference System

The research results obtained can be considered closer to human intelligence as we consider linguistic variables from the users. We evaluated the input fuzzy set according to our knowledge base founded in the if-then rules of the FIS. As a result, the optimum outputs were obtained much closer to the target outputs. The building of the optimum results for the system depends on the experience of experts. If results are obtained that are similar to user performance, data or services can then be delivered according to user preference. The same data obtained and analysed under the same conditions of the 500 users was applied using the proposed Empirical FIS.

For validating each approach proposed in this article, we used a confusion matrix. A confusion matrix (error matrix) is a tool that objectively measures performance of a classification algorithm. Each column represents the number of predictions of each class, while each row shows the instances of the true classes. The diagonal elements represent the number of points for which the predicted label is equal to the true label, while off-diagonal elements are those that are mislabeled by the classifier. The higher the diagonal values of the confusion matrix, the better the result, indicating many correct predictions. To measure the performance of Empirical FIS, we compare the results obtained by our proposed FIS with the results of the expert. The Figure 7 shows the results of this classification. The bottom right cell indicates the overall accuracy, while the column on the far right of the plot illustrates the efficiency for each predicted class. The row at the bottom of the plot shows the accuracy for each true class.

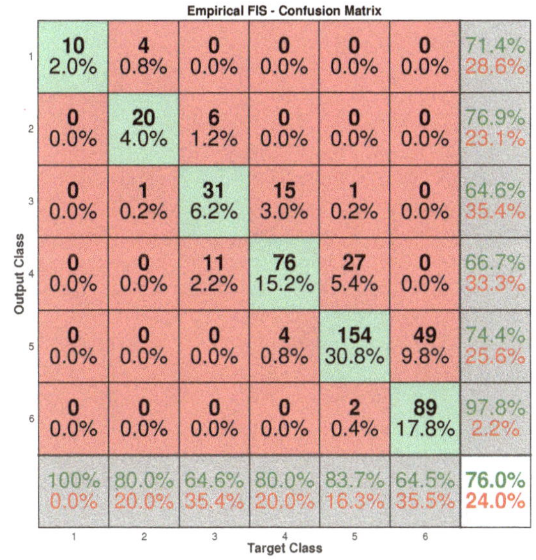

Figure 7. Confusion Matrix of the Empirical Fuzzy Inference System (FIS) Approach.

In Section 5, we evaluate the interaction of users using alternative approaches. For each, we use the confusion matrix to validate. To appropriately evaluate the interactions, we first separate the data into two sets, one set for the training model and the second set to test it. In all cases, 70% of the data is used for training and the rest to test. Each set was made by random selection of data.

We described the empirical FIS configuration in the supplementary material, where Table S1 shows the inputs configuration, Table S2 shows the outputs configuration and Table S3 shows the fuzzy inference rules of the empirical FIS.

5. Results

5.1. The Intelligent Hybrid System Approach

In this section, the results obtained from the sample of 500 users that visited the 'El Trompo' interactive museum in Tijuana, Mexico, are presented and analyzed. Users were evaluated and processed using an empirical FIS, Decision Tree, a fuzzy c-means method of data mining [4] named Data Mined Type-1 [23] and Neuro-Fuzzy System [7].

5.2. The Decision Tree Approach

First, the data collected was processed using a decision tree. We used the fitctree function of MatLab; this function returns a fitted binary classification decision tree, based on the input variables (also known as predictors, features or attributes) and output (response or labels). For this case study, the inputs selected were the presence, interactivity, control, feedback, creativity, productivity, communication and adaptation, while output is the level of interaction (Levels 0–5). We selected 70% of the data to fit the decision tree, while the remaining data was used to predict the level of interaction by users at the museum. Figure 8 show the results of this classification. The bottom right cell shows the overall accuracy, while the column on the far right of the plot shows the accuracy for each predicted class. The row at the bottom of the plot shows the accuracy for each true class.

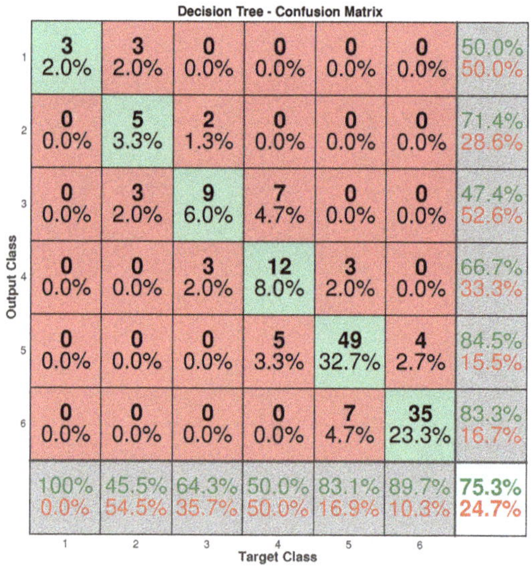

Figure 8. Confusion Matrix of the Decision Tree Approach.

5.3. The Data Mined Type-1 FIS Approach

A key aim of this research was to obtain more detailed and specific values, according to the performance and behavior of the user, taking into account uncertainty. For this reason, we used the Data Mined Type-1 approach, aided by the JT2FIS Tool-kit [27]. We selected 70% of the data for user sampling using a Fuzzy C-Means clustering algorithm for data mining [4]; once all data was mined, we obtained the configuration parameters of the FIS. In this case, FIS inputs were the Presence, Interactivity, Control, Feedback, Creativity, Productivity, Communication and Adaptation, while output was the level of interaction (Levels 0–5). Following this, we added six rules which enabled us to obtain a

FIS with a higher level of accuracy in the realized configuration. Consistent and accurate interaction levels were obtained to adhere, as much as possible, to the performance and behavior of the user to offer services, data, and content that is ultimately required by the user. Once the data mined Type–1 FIS was configured, we evaluated the 30% of remaining users, with their information used as input, to determine the level of interaction. Figure 9 shows the results of this approach; as seen, we can observe an improvement in the classification concerning the previous methods. The bottom right cell shows the overall accuracy, while the column in the far right of the plot shows the accuracy for each predicted class. The row at the bottom of the plot shows the accuracy for each true class.

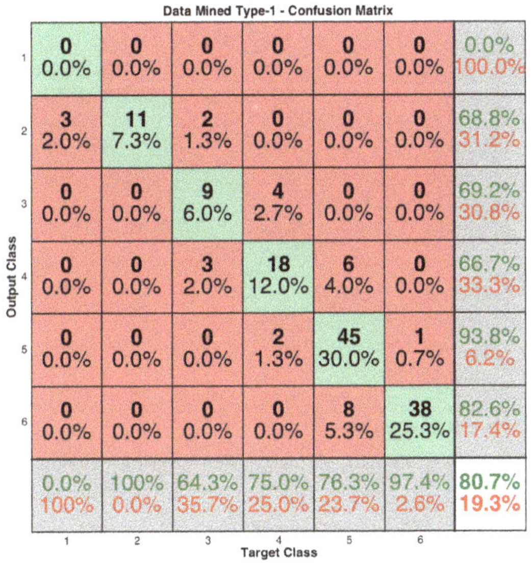

Figure 9. Confusion Matrix of the Data Mined Type-1 (DMT1F) Approach.

We described the Data Mined Type-1 configuration in the supplementary material, where Table S4 shows the inputs configuration, Table S5 shows the outputs configuration and Table S6 shows the fuzzy inference rules of the Data Mined Type-1 FIS.

5.4. Neuro-Fuzzy System Approach

To improve the accuracy in the classification of the interaction levels, we decided to generate a FIS using a Neuro-Fuzzy method. Neuro-Fuzzy systems encompass a set of techniques that share the robustness in handling of imprecise and uncertain information that exist in problems related to the real world e.g., recognition of forms, classification, decision making, etc. The main advantage of Neuro-Fuzzy systems is that they combine the learning capacity of neural networks with the power of linguistic interpretation of FIS, allowing the extraction of knowledge for a base of fuzzy rules from a set of data. In this case study, we generated a FIS combining the fuzzy-C means clustering and Least-Squares Estimate (LSE) algorithm. This method was proposed by Castro et al. [7].

In this approach, we selected 70% of the data based on fit. The rest of data was used to predict the level of interaction users had at the museum. Figure 10 shows the results of this classification. The bottom right cell shows the overall accuracy, while the column on the far right of the plot shows the accuracy for each predicted class. The row at the bottom of the plot shows the accuracy for each true class.

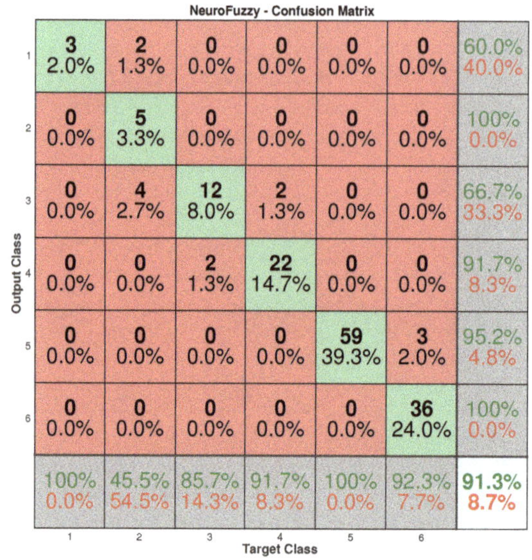

Figure 10. Confusion Matrix of the Neuro-Fuzzy System Approach.

We described the Neuro-Fuzzy FIS configuration in the supplementary material, where Table S7 shows the inputs configuration, Table S8 shows the outputs configuration and Table S9 shows the fuzzy inference rules of the Neuro-Fuzzy FIS.

5.5. Empirical FIS Approach Versus Hybrid FIS Approach

The use of artificial intelligence has been widely applied in most computational fields. The main feature of this concept is its ability to self–learn and self–predict desired outputs. This autonomous learning may be achieved in a supervised or unsupervised manner. The interaction level prediction of user data has been applied and processed using different approaches, including Empirical FIS, Desicion Tree, Data Mined Type–1 FIS and Neuro-Fuzzy System (NFS).

Table 2 shows the accuracy of each of these approaches used. We can see the precision of each approach for each level of interaction. The neuro-fuzzy system is identified as the one with best results.

Table 2. Accuracy Percent/Error Percent for Each Predicted Class for Each Method.

Predicted Class	Empirical FIS	Decision Tree	Data Mined Type-1	Neuro-Fuzzy System
Level 0	71.4/28.6	50/50	0/100	60/40
Level 1	76.9/23.1	71.4/28.6	68.8/31.2	100/0
Level 2	64.6/35.4	47.4/52.6	69.2/30.8	66.7/33.3
Level 3	66.7/33.3	66.7/33.3	66.7/33.3	91.7/8.3
Level 4	74.4/25.6	84.5/15.5	93.8/6.2	95.2/4.8
Level 5	97.8/2.2	83.3/16.7	82.6/17.4	100/0
Overall Accuracy	76/24	75.3/24.7	80.7/19.3	91.3/8.7

6. Discussion

In recent times, museum halls have experienced overcrowding with most users (students) only having a few opportunities to interact with the museum's exhibitions. The museum used to compensate this limited experience by providing guided tours by instructors, that accompanied groups of kids and maintained interest in demonstrations. In some way, the museum achieved its goals using this strategy, but the infrastructure could be considered under-utilised. A solution to this overcrowding is to make museums smart spaces with multi-user adaptive interaction exhibitions. Some museums in Mexico have based their interactive experience design on instructional activities and underlying technology (in general including touch screens with information and choices where the user plays selecting options or answers questions about a subject). However, the exhibitions often only offer the experience on an individual basis, and do not allow interaction by multiple users at the same time. If museums could expand interactive modules to multi-user experiences in Mexico, then guided tours by instructors may not be necessary.

6.1. 'El Trompo' as a Complex Sociotechnical System

The 'El Trompo' Interactive Museum has introduced interactive exhibitions using newly-available technologies. As an organisation, the museum recognizes that the interaction between its users and technology is crucial. The development of this educational institution has gone beyond its technological structure with the aim of extending its systems to end users and expanding the scope of its core business. In this sense, the interaction between users' complex infrastructures and human behaviour becomes paramount. We, therefore, consider it and most of its substructures as complex socio-technical systems. We further acknowledge social behaviour, spontaneous collaboration, feedback and adaptation, among users and technology as a complex system. The museum should be considered as a set of many interacting elements where the modelling of user behaviour is challenging due to its dependencies, relationships or interactions between users or between the user and their environment. As a result, we provide discussion on the different impressions encountered during the museum case study, with the aim of exploring the current state of the agent and multi-agent systems technology and its application to the complex socio-technical system domain.

6.1.1. Human-Agent Interaction

Firstly, in terms of the user and their environment, we considered that the HCI examined the intention and usage of computer technologies centered on the interfaces between users and devices. Thus, the behavioral sciences, media studies, sensors networks and other fields of study could help us to observe how humans interact with computers and to enable us to design technologies that let users interact with exhibitions in innovative ways. Interaction is the central subject of HCI, so we consider this to be crucial. Distinguishing levels of interaction is the first step to be taken in managing the interaction of a user and its technological environment and turning it into a socio-technical system. This could be a challenging task due to the subjective conceptual meaning.

The interaction-levels, proposed by Gayesky and Williams [8] , offer the advantage of being easy to interpret and implement in a software system, but the inputs could be difficult due to them requiring a set of qualitative measurements that could be difficult to observe through sensors, for example. As the level rises in the assessment, observations on user behaviour are often harder to obtain. For instance, at Level 0, we only need to sense if the user is present or not in the exhibition but, at level 5, we not only need to identify if the feedback, control, creativity, communication, adaptability and productivity is taking place, but we also need to identify the quality of these assessments. Of course, this is a challenge worth facing in order to create a more human validation of the user's competitiveness. By measuring interaction-levels, it can provide a simple communication among the involved elements. User preferences can be predicted to offer adequate information or services to complete their goals; this can increase the knowledge and productivity of the user, satisfying their needs. The understanding and knowledge of the human interaction allows for the development of an interactive system, which should provide the ability to choose and act, anticipating the possible actions of the user and coding them in the program, allowing for continuing interaction by the user.

The interaction-levels measuring can allow us a simple communication among the involved elements. The user's preferences can be predicted to offer adequate information or services to complete his goals; this can increase the knowledge and productivity, satisfying user's necessities. The understanding and knowledge of the human interaction step by step allowing development of an interactive system should take into account the ability to choose and act, anticipating the possible actions of the user and coding them in the program, continuing the interaction by the time.

6.1.2. The Intelligent Interactive-Exhibit System

Secondly, by using software systems that generate inferences from knowledge, we can use them to develop interactive displays in a museum to predict user performance. Reasoning systems play a significant role in the implementation of intelligent and knowledge-based interactive-exhibit systems. Thus, machine learning methods unfold user behaviour over time, based on activity in the exhibit room, particularly with the interactive displays. A learning process that searches for generalised rules or functions that users produce, in line with observations of actions, can be incorporated into the environment and used to manage predicted user behaviour. In our case, we built a FIS to represent the interaction-levels based on the observation rules proposed by Gayesky and Williams [8].

Further, considering that a Hybrid Intelligent System is a knowledge-based inference system that can combine data mining and knowledge discovery methods to produce an Inference System, we used it to build a FIS from real evaluator outcome data. We applied a neuro-fuzzy technique to produce an inference system in state of the art fashion. The advantage of a neuro-fuzzy system is that it convenes the neural network training process that researchers widely use in machine learning, but is hard to open to understand. Using a fuzzy inference system makes it easy to see what has happened inside the box. A fuzzy inference system could be built by the designer, if necessary, or by using a machine learning process. For this study, we used a dataset from an in situ observer with real museum visitors interacting with an actual exhibition-display to validate the hand-crafted FIS [8]. We then compared it to a FIS discovered form the data set through a neuro-fuzzy method. One vulnerability with this approach is that we assume that other sensor systems provide the inputs, as the fuzzy inference system expects it. A gap of this first approach is that we assume that input data is correct and the context awareness systems are capable of providing it, in the case of no human expert evaluations.

6.1.3. Knowledge-based Agent and Agent Architecture

Thirdly, incorporating the above into software agents, an intelligent or knowledge-based agent could perceive through sensors the motions and actions taken in an environment. In the museum case study, the strategy is to direct the user activity towards achieving instructional goals. The intelligent agents may further learn from the user and use the discovered knowledge to meet their aims. Further, multiple interacting intelligent agents can be used to address problems that are difficult or impossible for an individual agent to solve within a museum environment. In our case, we approached the museum as a complex socio-technical system by knowledge-based agents. With this strategy, we started with agent-based modelling of some components, but then went through a knowledge-based agent and agent-based architecture design to finally build an agent-based computational system. For this approach, all the museum components are considered agents that interact complexly and where the user is another agent and part of the community.

At this point of our study, we consider the inputs of user-module interaction in the agent architecture (presence, interactivity, control, feedback, creativity, productivity, communication and adaptation) as a simple behaviour evaluation performed by an observer. These attributes could be more intricate than first appears and we could conduct further in-depth study. For example, "Communication" could involve not only a user-module interaction, but also the talk between users. "Adaptation" could imply the negotiation of results in user collaboration processes to achieve common goals or "presence" could be determined by ubiquity in an infrastructure system and vicinity in a social network. In other words, we may evolve the exercise of a multi-agent system and consider further analysis techniques to approach complexity.

6.1.4. Agent-Oriented Software Engineering

Finally, as Agent-Oriented Software Engineering (AOSE) starts to support best practices in the development of complex Multi-Agent Systems (MAS), we can focus now on the use of agents and the combination of agents as the intermediate generalization of socio-technical systems at a museum, in an agent-oriented analysis, design and programming fashion. In the architecture, the modeler represented the intercommunication observer as an intelligent (knowledge-based) agent that evaluated the interaction behaviour of the user that performed in the exhibition module. This agent is an agent-oriented software that infers the interaction-level from environmental observations to send feedback to the user to support experience. In our case study, the interaction-evaluator agent is a Java software agent capable of qualifying the interaction level in real-time, and its prediction performance was tested and validated. Based on this practice, we believe that this type of knowledge-based engineering, hybridized with agent architectures, could form part of the AOSE.

7. Conclusions and Future Work

This paper has explored the evaluation of interaction on HCI using Gayesky and Williams' Interaction Levels Theory [8] to improve the user experience when interacting with museum exhibition modules. It has taken into account user behaviour based on presence, interactivity, control, feedback, creativity, productivity, communication and adaptation. In our experience, the Gayesky and Williams' interaction levels [8] were simple to understand and use.

Firstly, we modelled the interaction levels using an Intelligent Hybrid System to provide a classifier that evaluated user performance into interactive modules in HCI. We applied machine-learning techniques to set-up or automatically discover knowledge from a real observation data-set. The generated model was a FIS that described their interaction levels according to the Gayesky and Williams' user behaviour attributes. The Gayesky and Williams' interaction levels were simple to model by an inference system and we provided the obtained FIS configuration of all cases. We then used an empiric design from expert experience and an automatized method form on-site observed data-mining to generate the corresponding FIS. The prediction accuracy then validated and compared against the evaluators to recommend the best approach. We provided a confusion analysis and a comparative summary to highlight the advantages and disadvantages of each approach. We recommended that the method is the Neuro-Fuzzy System.

To show the applicability of the proposed model, we built software agents that represented a high-level abstraction of a gallery, specifically an interactive exhibition module at the 'El Trompo' museum in Tijuana, Mexico. In the agent architecture, the FIS performed as a decision-making system that helped the InteractionEvaluator-Agent to identify the interaction level from sensors in the environment and feedback the Exhibition-Agent to improve the user experience. We discussed different impressions when approaching the museum case study with the aim of showing the current state of the agent and multi-agent system technology and its application to the complex socio-technical system domain. We found that Agent-Based Models, with Intelligent Hybrid Systems (as an agent decision-making system), to approach complex socio-technical systems was beneficial.

Finally, we can see that the benefits of the proposed model help HCI agent-based systems to evaluate the user interaction in a high-level abstraction. Accurate feedback enhanced the user experience.

For future work, we must consider that Gayesky and Williams' user interaction attributes (presence, interactivity, control, feedback, creativity, productivity, communication and adaptation) should be further developed to add a new level of description. Each feature means new challenges to characterize and implement. The proposed architecture allowed us to add new evaluation fuzzy inference systems on cascade in each performed input to escalate the model and consequently improve the Interaction Evaluator-Agent. Considering approaching complexity, we will also evolve the model to a multi-agent system. From this perspective, the interaction between user-agents to coordinate and collaborate to achieve common goals and describe the relationship between them, is an essential improvement to enhance HCI. Social and network theory will contribute to new epistemological approaches to user interaction modelling as users' social nature involves them in complex social systems.

Supplementary Materials: The following are available online at www.mdpi.com/2076-3417/8/3/446/s1, Table S1: Inputs configuration of the empirical FIS. s = standard deviation, m = average; Table S2. Outputs configuration of the empirical FIS. s = standard deviation, m = average; Table S3. Inference Fuzzy Rules of the Empirical FIS; Table S4. Inputs configuration of the Data Mined Type-1 FIS. s = standard deviation, m = average; Table S5. Outputs configuration of the Data Mined Type-1 FIS. s = standard deviation, m = average; Table S6. Rules configuration of the Data Mined Type-1 FIS; Table S7. Inputs configuration of the Neuro-Fuzzy FIS. s = standard deviation, m = average; Table S8. Outputs configuration of the Neuro-Fuzzy FIS. s = standard deviation, m = average; Table S9. Rules configuration of the Neuro-Fuzzy FIS.

Acknowledgments: We would like to thank the Mexican National Council for Science and Technology, The Autonomous University of Baja California and "El Trompo" Museo Interactivo Tijuana, A.C., in Mexico, for their support throughout this research project.

Author Contributions: All authors conceived and designed the study; Manuel Castañón-Puga and Felipe Lara-Rosano provided methodological advice. Ricardo Rosales, Nora Osuna-Millan and Miguel Josue Flores-Parra completed the field survey and analyzed the collected data. Ricardo Rosales, Richard Evans and Carelia Gaxiola-Pacheco wrote and proof-read the paper. All authors read and approved the manuscript before submission.

Conflicts of Interest: The authors declare no conflict of interest.

References

1. Mendel, J.; Wu, D. *Perceptual Computing, Aiding People in Making Subjetive Judgments*; IEEE Press: Hoboken, NJ, USA, 2010.

2. Zadeh, L. Fuzzy sets. In *Information and Control*; Prentice Hall: Upper Saddle River, NJ, USA, 1965; Volume 8, pp. 338–353.

3. Zadeh, L. The concept of a linguistic variable and its application to approximate reasoning. *Inf. Sci.* **1975**, *8*, 199–249.

4. Bezdek, J. FCM: The fuzzy c-means clustering algorithm. *Comput. Geosci.* **1984**, *10*, 191–203,

5. Yin, X.; Khoo, L.; Chong, Y. A fuzzy c-means based hybrid evolutionary approach to the clustering of supply chain. *Comput. Ind. Eng.* **1984**, *66*, 768–780.

6. Rantala, J.; Koivisto, H.A.N.N.U. Optimised Subtractive Clustering for Neuro-Fuzzy Models. In Proceedings of the 3rd WSEAS International Conference on Fuzzy Sets and Fuzzy Systems, Interlaken, Switzerland, 11–14 February 2002, 3971-3976.

7. Castro, J.R.; Castillo, O.; Melin, P.; Rodríguez-Díaz, A. A hybrid learning algorithm for a class of interval type–2 fuzzy neural networks. *J. Inf. Sci.* **2008**, *179*, 2175–2193.

8. Gayesky, D.; Williams, D. Interactive Video in Higher Education. In *Video in Higher Education*; Zubber-Skerrit, O., Ed.; Kogan Page: London, UK, 1984.

9. Rosenfeld, A.; Agmon, N.; Maksimov, O.; Kraus, S. Intelligent agent supporting human-multi-robot team collaboration. *Artif. Intell.* **2017**, *252*, 211–231.

10. Rosenfeld, A.; Sarit, K. Providing arguments in discussions on the basis of the prediction of human argumentative behavior. *ACM Trans. Interact. Intell. Syst.* **2016**, *6*, 30.

11. Casell, J. *Embodied Conversational Agents*; MIT Press: Cambridge, MA, USA, 2000.

12. Berg, M.M. Modelling of Natural Dialogues in the Context of Speech-based Information and Control Systems. Ph.D. Thesis, Christian-Albrechts University of Kiel, Kiel, Germany, 2015.

13. Garruzo, S.; Rosaci, D. Agent Clustering based on Semantic Negotiation. *ACM Trans. Auton. Adapt. Syst.* **2008**, *3*, 7.

14. Rosaci, D. CILIOS: Connectionist Inductive Learning and Inter-Ontology Similarities for Recommending Information Agents. *Inf. Syst.* **2007**, *32*, 793–825.

15. Cerekovic, A.; Aran, O.; Gatica-Perez, D.D. Rapport with virtual agents: What do human social cues and personality explain? *IEEE Trans. Affect. Comput.* **2017**, *8*, 382–395.

16. Tickle-Degnen, D.L.; Rosenthal, R. The nature of rapport and its nonverbal correlates. *Psychol. Inq.* **1990**, *1*, 285–293.

17. Sanchez-Cortes, D.; Aran O.; Jayagopi, D.; Mast, M.S.; Gatica-Perez, D. Emergent leaders through looking and speaking: From audio-visual data to multimodal recognition. *IEEE J. Multimodal User Interfaces* **2017**, *7*, 39–53.

18. Schroder M.; Bevacqua E.; Cowie R.; Eyben, F.; Gunes, H.; Heylen, D.; Ter Maat, M.; McKeown, G.; Pammi, S.; Pantic, M.; et al. Building autonomous sensitive artificial listeners. *IEEE Trans. Affect. Comput.* **2012**, *3*, 165–183.

19. Rosaci D.; Sarne, G.M.L. MASHA: A Multi Agent System Handling User and Device Adaptivity of Web Sites. In *User Modeling and User-Adapted Interaction (UMUAI)*; Springer: Berlin, Germany, 2006; Volume 16, pp. 435–462.

20. Wooldridge, M.; Jennings, N.R.; Kinny, D. The Gaia methodology for agent-oriented analysis and design. *Auton. Agents Multi-Agent Syst.* **2000**, *3*, 3, 285-312.

21. Gilbert, N. *Agent-Based Models*; SAGE Publications: Newcastle upon Tyne, UK, 2008.

22. Cioffi-Revilla, C. Simulations I: Methodology. In *Introduction to Computational Social Science. Texts in Computer Science*; Springer: London, UK, 2014.

23. Rosales, R.; Castañón-Puga, M.; Lara-Rosano, F.; Evans, R.D.; Osuna-Millan, N.; Flores-Ortiz, M.V. Modelling the Interruption on HCI Using BDI Agents with the Fuzzy Perceptions Approach: An Interactive Museum Case Study in Mexico. *Appl. Sci.* **2017**, *7*, 832.

24. Barros, L.; Bassanezi, R. Topicos de Logica Fuzzy e Biomatematica. In *Universidad de Estadual de Campinas (Unicamp): IMECC*; Universidade Estadual de Campinas: Campinas, Brazil, 2006.

25. Bellifemine, F.; Caire, G.; Greenwood, D. *Developing Multi-Agent Systems with JADE*; John Wiley and Sons: Hoboken, NJ, USA, 2007.

26. Sivanandam, S.; Sumathi, S.; Deepa, S. *Introduction to Fuzzy Logic Using Matlab*; Springer: Berlin/Heidelberg, Germany, 2007.

27. Castanon-Puga, M.; Castro, J.; Flores-Parra, J.; Gaxiola-Pacheco, C.G.; Martínez-Méndez, L.G.; Palafox-Maestre, L.E. JT2FIS A Java Type-2 Fuzzy Inference Systems Class Library for Building Object-Oriented Intelligent Applications. In Proceedings of the 12th Mexican International Conference on Artificial Intelligence, Mexico City, Mexico, 24–30 November 2013; Volume 8266, pp. 204–215.

applied
sciences

MDPI

Article

MOVICLOUD: Agent-Based 3D Platform for the Labor Integration of Disabled People

Alberto L. Barriuso *, Fernando De la Prieta, Gabriel Villarrubia González, Daniel H. De La Iglesia and Álvaro Lozano

BISITE Digital Innovation Hub., University of Salamanca, Edificio Multiusos I+D+i, C/Espejo SN, 37007 Salamanca, Spain; fer@usal.es (F.D.l.P.); gvg@usal.es (G.V.G.); danihiglesias@usal.es (D.H.D.L.I.); loza@usal.es (Á.L.)
* Correspondence: albarriuso@usal.es; Tel.: +34-923-29-44-00

Received: 18 January 2018; Accepted: 23 February 2018; Published: 27 February 2018

Featured Application: The main application of this work is to perform 3D simulations of different work environments, which can help to establish the accessibility problems that may occur in the corresponding real-world environments.

Abstract: Agent-Based Social Simulation (ABSS), used in combination with three-dimensional representation, makes it possible to do near-reality modeling and visualizations of changing and complex environments. In this paper, we describe the design and implementation of a tool that integrates these two techniques. The purpose of this tool is to assist in creating a work environment that is adapted to the needs of people with disabilities. The tool measures the degree of accessibility in the place of work and identifies the architectural barriers of the environment by considering the activities carried out by workers. Thus, thanks to the use of novel mechanisms and simulation techniques more people with disabilities will have the opportunity to work and feel comfortable in the environment. To validate the developed tool, a case study was performed in a real environment.

Keywords: multi-agent system; agent-based simulation; 3D representation

1. Introduction

According to the World Health Organization [1], around 15% of the total world population suffers from some kind of physical disability. From the total number of disabled people, 2% to 4% experience serious difficulties in performing daily activities. At present, disability is a factor that often impedes people from enjoying equal work opportunities, despite having all the necessary qualifications. This is due to the employers' negative attitudes, the lack of information and poorly adapted working environments. Statistics show that this is a serious problem, in developed countries around 50–70% of disabled people in working age are unemployed (reaching 80–90% in developing countries) [2].

Our society is becoming increasingly aware of the challenges faced by people with disabilities, including difficulties in finding employment. As a result of this collective awareness, large institutions and governments have created proposals related to the employment of people with disabilities. Many of these proposals have been put into practice, especially since the declaration of the *Standard Rules on the Equalization of Opportunities for Persons with Disabilities* issued by the UN [3]. We can find various concrete regulatory frameworks, which aim to promote social justice and facilitate access to work for people with disabilities. Two important frameworks can be distinguished, the *European Disability Strategy 2010–2020* [4] and the *Americans with Disabilities Act* [5]. Not only governments have taken initiatives, many companies are also committed to including and promoting people with disabilities in the private sector. An example of this is the *ILO Global Business & Disability Network Charter* [6], signed by 11 large companies worldwide.

Despite these efforts, there is still a long way to go before people with disabilities are given equal opportunities in employment. Thus, it is necessary to continue providing advanced technological tools that will enable company human resource (HR) managers to integrate employees with disabilities in the workforce. For successful management, it is of course necessary for HR managers to have the necessary experience and knowledge. However, advanced tools and software models would facilitate their work and make it even more effective. Social machines are an adequate computational model, since they allow for the development of computer mechanisms which facilitate social studies [7] and social interaction between human beings and computer systems [8]. In short, social machines are capable of balancing the human input (experience) and computational systems (computational capacity) [9].The MOVICLOUD platform is part of the framework of this proposal. framework of this research, the MOVICLOUD platform is proposed: a multi-agent system (MAS) based on virtual organizations (VO) with the capacity to perform social simulations over a 3D environment. The use of organizational MAS allows the implementation of social machines, facilitating the work of humans (HR departments) and machines for the accomplishment of common objectives (integration of disabled people). Likewise, it also allows the inclusion of advanced techniques and algorithms such as social simulation techniques and the modeling of human behavior.

In the framework of the labor integration of disabled people, simulations allow to discover in advance the problems that workers with disabilities will face before these problems happen. The two key elements to face this kind of difficulties are, on the one hand, the lack of accessibility of the buildings, and on the other hand, the need to adapt the different processes that are carried out in the company, so that they are adapted to people with disabilities but maintaining or increasing their level of efficiency or effectiveness.

To face these challenges, the MOVICLOUD platform proposes the use of a specifically developed 3D tool that allows to create three dimensional representations of the information extracted from an office environment. New mechanisms and techniques are employed to make realistic simulation models. The results of these simulations, which are interactive, facilitate the work of those responsible for HR, making decisions about the improvement of work processes that occur in the environment. For this purpose, a collaborative work framework between machines and humans is proposed, where experts can modify the simulations interactively, while the machines have the capacity to provide adequate and relevant information to the problem to be simulated, such as the identification of the different architectural barriers, the improvement of production processes, location of jobs, etc.

This paper it is structured as follows: in Section 2 a review of the state of the art in social simulation based on agents and agent development platforms is made, Section 3 details the design of the platform presented in this article, Section 4 describes a case of study with which the platform has been validated and finally, in Section 5 the obtained conclusions are exposed, as well as the possible lines of future work.

2. Background

Performing simulations of heterogeneous systems where a large number of elements intervenes is a complex task, but has found in the agent-based modeling a technique that facilitates its design and implementation. For this reason, agent-based modeling has been widely used to model different active objects, such as vehicles [10,11], markets [12,13], people [14], industrial units [15], animals [16], health [17] etc. The phenomenon to be simulated is composed by a set of sequences, events or tasks that may exist in the environment, which will be configurable through the simulation model. These tasks are interrelated, since in the end, they depend on the order or time in which the events occur. The complexity lies in the fact that these models allow the simulation of changing and complex environments, so the use of open MAS is ideal, thus emerging the concept of Agent-Based Social Simulation (ABSS) [18].

ABSS is a discipline that deals with the simulation of social phenomena by using multi-agent computational models. Therefore, ABSS is a combination of social sciences, agent-based computing and computational simulation. In recent times, other areas of study whose interests interact with each other have also emerged, as shown in Figure 1 [19].

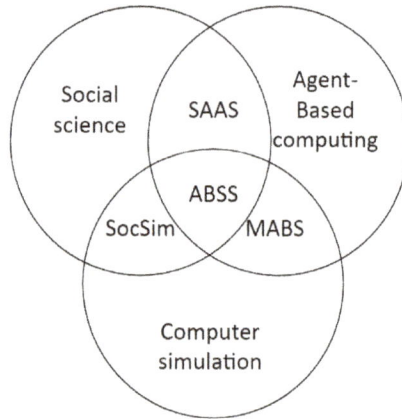

Figure 1. Intersection of different areas within the Agent-Based Social Simulation (ABSS).

The use of the theory of agents in the framework of simulations is especially indicated when it is necessary to capture the different tasks, elements, objects or people of complex and dynamic environments. This is because these can be implemented without needing to have a great knowledge about the global interdependencies. Another reason is the ease they offer when building them when facing changes in the model, since generally it is only necessary to make local changes, not global [20]. This paradigm is used in many disciplines; some examples can be found in the field of health [21], urbanism [22], economics [23], architecture [24], criminology [25] or biology [26] among others.

The Foundation of Intelligent Physical Agents (FIPA) is an international non-profit organization whose main objective is the development of different standards to promote the use of intelligent agents. The specifications made by FIPA have been grouped into three sets: FIPA 97, FIPA 98 and FIPA 2000. In these groups, the characteristics of agents and multi-agent systems are defined in relation to their management, communication languages, interactions between agents and human, mobility or security among others. The architecture proposed by FIPA for the development of multi-agent systems is defined in terms of abstract designs, as an interface. In this way, multiple implementations of this architecture can be possible, attending to the interests of concrete developments. The pursued idea is to obtain different implementations of the same architecture, which defines common elements of the different systems, so that they operate under the same specification, in an interoperable way. There are several agent development platforms based on FIPA specifications such as Lightweight and Extensible Agent Platform (LEAP) [27], FIPA-OS [28], Reusable Task Structure-based Intelligent Network Agents (RETSINA) [29] or MOLE [30]. However, the most widespread platform is JADE (Java Agent Development Environment) [31]. It is a framework for the development of multi-agent systems whose objective is to simplify the development of this type of systems through a set of system services and auxiliary agents, making use of the FIPA Interaction protocols, which facilitates the communication between agents. JADE is implemented in Java and is distributed as free software with Lesser General Public License (LPGL). The communication architecture of JADE offers an efficient and flexible message communication, where the platform is responsible for creating and managing a queue of arrival Agent Communication Language (ACL) messages, which are private for each agent.

Currently, the idea of modeling systems through the iterations that occur between their agents is opening the way to a new trend: modeling the behavior of the system from an organizational point of view. Following this trend, it is necessary to define concepts of rules [32], norms and institutions [33] and social structures [34], which arise from the need to have a higher level of abstraction-independent of the agent-, which explicitly defines the organization in which agents coexist. In this context, the concept of agent society emerges [35], as the set of interrelated and interacting artificial entities, which are governed by certain rules and conditions. The architectures used to model and build multi-agent systems based on organizations must support the coordination between agents, so that they can dynamically adapt to changes in their structure, objectives or interactions [36]. By doing this, it is possible to allow the coexistence of agents in shared environments, in such a way that they can carry out their objectives. When an agent is part of a society of agents, it takes a collectivist vision, so it must take into account the global functioning of the system, how some agents influence others, so that he does not focus on itself. The current trend towards computer developments with collaborative work environments, distributed knowledge management, has sparked interest in virtual organizations: a set of people and institutions that need to coordinate resources and services across different institutional boundaries [37]. There is a clear analogy between human societies and agents; we can define a human society as a formation or social entity with a number of members that can be specified and an internal differentiation of the functions that are performed by these members [38] and as an agent organization a collection of roles, which maintain certain relationships among themselves, and that take part in patterns of interaction with other roles in an institutionalized and systematic way [39].

On the one hand, ABSS has been applied with good results for the design of work methods [40] or training of people with disabilities [41] and on the other hand, VO have also been previously applied to similar scenarios such as the monitoring of elderly and disabled people in geriatric residences [42]. However, the use of this techniques (together or separated) has not yet been applied to the integration of people with disabilities in their working place. Though, considering the current state of the art, ABSS together with the social aspects that VO provide (analogous to the human societies), are suitable techniques for modeling work environments where there are people with disabilities, since they allow modeling large and distributed problems where environments are open and dynamic and a set of actors with different roles interact. By following this kind of approach to perform simulations, it will be possible experiment on a model, which if is good enough, will respond in a similar way to the studied system. This would allow to carry out different models of experiments on multiple occasions, permitting to test and evaluate the different configurations, besides of being able to include random components. Thanks to this, it is possible to analyze the simulation context from different perspectives, obtaining much more realistic results from the analysis. In the case of the labor integration of disabled people, these techniques can be applied to the simulation of different work scenarios in order to detect, from these simulations, which are the obstacles that these people may find in their work, thus being able to take preventive actions to eliminate these limitations.

Therefore, in view of the limitations observed in the existing platforms, within the framework of this work, the MOVICLOUD platform will incorporate a novel simulation environment, based on 3D techniques that allow ABSS to be made from advanced models of physical spaces, that would allow to perform simulations in a closer way to reality. Next, the design of this platform is detailed.

3. MOVICLOUD Platform

Zambonelli et al. [39] introduce the idea that MAS must accept the participation of agents of all types, regardless of architectures and languages. This statement goes to the extreme in the framework of social computing, where humans and machines work collaboratively [9]. However, it is necessary to determine rules or social rules for agents, since a priori their behavior is not trustful. Thus, the idea of modeling systems through iterations has evolved to the point of being necessary to perform an organizational model, where artificial societies exist [43] in which humans and machines collaborate to achieve the common objectives of the society. Following this trend, it is necessary to define concepts of rules [32], norms and institutions [33] and social structures [34], which arise from the need to have a level of abstraction, which explicitly defines the organization in which the agents they coexist. The current trend towards computer developments in collaborative work environments and with distributed knowledge management, has generated interest in VO: a group of people and institutions that need to coordinate resources and services across different institutional boundaries [37]. There is a clear analogy between human and agents' societies; we can define a human society as a foundation or social entity with a certain number of members that can be specified and where an internal differentiation of the functions that are performed by these members is done [38], on the other hand, an agent organization is a collection of roles, which maintain certain relationships among themselves, and that take part in patterns of interaction with other roles in an institutionalized and systematic way [39].Thanks to these capabilities provided by the VO of agents, it is possible to model social machines, creating artificial societies of humans and machines that share tasks, objectives and norms. Human and artificial agents in a MAS try to cooperatively carry out tasks in the environment of the system to which they belong. The tasks that can be performed within the system will be established by the knowledge and skills of the agents and the restrictions imposed by the environment.

Laying on this theoretical base, the development of a tool for modeling simulations of human processes that take place in different environments in a flexible and dynamic way has been carried out. The proposed system is based on a novel approach, consisting of the use of MAS based on VO and a three-dimensional representation environment. The combination of both strategies is key to the purpose of the simulations to be carried out since on the one hand, agents can implement different tasks, objectives, purposes, etc.; while the three-dimensional environment allows the physical modeling of environments. In this work, a simulation tool oriented to model productive processes for workers in office environments is presented. Thus, this approach allows simulations to be carried out on the work environment itself, as well as the visualization of the results of the simulations. The possibility of integrating the tools for the generic construction of 3D environments into a platform is a differentiating factor with respect to platforms such as MASON [44], Repast [45], Pangea [46] or Swarm [47] that allow the visualization of the interactions between the agents themselves, as well as with the environment. These models and tools allow to model a wide variety of environments and tasks. In fact, they have even been applied in previous studies, obtaining good results [48], however, although they can be used as a basis for the construction of simulations with 3D elements, they are not specifically designed for it, so they do not offer the proper utilities that facilitate the deployment of these simulations in an agile and simple way.

The main novelties included in the developed platform are (a) the possibility to easily model 3D environments using a tool based on *cloud computing*; (b) the use of virtual organizations that model the behavior of people in a work environment (making use of the FIPA ACL communication mechanisms provided by JADE); (c) the design of a visualization platform in the field of office environments, which offers a vision closer to reality than other generic agent platforms. In addition, through a case study is intended to validate the use of this tool to obtain useful information to reduce accessibility barriers found by disabled people based on simulations carried out in the three-dimensional environment.

In the following Sections, the presented architecture is described in detail: in Section 3.1 an overview of the components architecture is done, Section 3.2 describes the implementation of the organizational model, Section 3.3 shows what the structure of the agents that intervene in the simulations is like and finally, Section 3.4 describes how the communication between the agent platform and the three-dimensional visualization tool takes place.

3.1. Components of the Architecture

The general architecture of the system is made up of two large blocks: a MAS and a tool for designing and visualizing three-dimensional environments. On the one hand, the MAS is composed of a set of intelligent agents that allow to perform the simulation of the desired model. For the implementation of this component, the JADE platform [31] has been used, as it will be detailed later. The developed platform is oriented towards the modeling of the MAS as a VO of agents. From this perspective, it is able to define a structure, roles, and a set of rules that regulate the interactions between agents. To do this, a higher-level layer to JADE, responsible for managing the MAS, is included in the system. Through this, it has been possible to provide the system with a set of additional functionalities that maximize the capabilities of MOVICLOUD, allowing a better interactivity and configuration on the part of human experts. These functionalities are detailed below:

- Ability to use **virtual organizations of agents**, which allow agents to adopt different topologies, such as hierarchies, oligarchies, groups, congregations, etc. [49]. VO allow the artificial representation of human societies in work environments. Thus, the structure of the multi-agent system in each simulation may vary, generating specific components depending on the environment to be modeled. Attending to the simulation that it is intended to be performed, is necessary that the multi-agent system would be able to adopt one of these topologies, so different types of agents must co-exist. JADE does not support the management of VO, so it has been necessary to introduce additional functionality with which to perform this task. To support this functionality, a database has been incorporated into the system in which the topology of the agents present in the system will be stored, making it possible to reuse it in the future.
- The tasks that the agents can carry out within the artificial societies that are generated within the framework of the simulations can vary, so that the objectives of the agents can be the same, but the decision-making process to carry them out may differ depending on the kind of disability. In this way, agents with advanced capabilities have been developed for **planning the distribution of tasks** according to the needs and configuration of the organizational structure. The user of the platform is responsible (through a series of functionalities implemented in the visualization tool) to model the different agents that act in representation of the elements that they want to model, thus allowing human-artificial collaborative work. For example, when defining an organization with a hierarchical structure, two kinds of agents will exist. The agents in charge of the decision-making are located at the higher level of the hierarchy–determining which tasks need to be carried out and who should carry them out–and the agents that have a more basic functionality–those in charge of carrying out the designated tasks–will be placed in the lower layers.
- **Communication with the 3D environment**, so that all the changes occurred can be reported between the agent platform and the three-dimensional environment in a bidirectional way.

Figure 2 shows in detail the different components present in the multi-agent system, as well as the relationships between them.

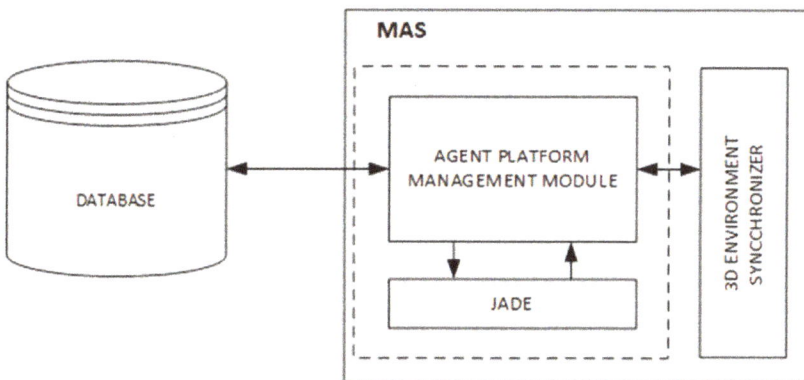

Figure 2. Multi-agent system (MAS) Components.

Finally, the **3D environment** that allows real information representation collected from the simulated environment is composed of an editor and a viewer. An organization is surrounded by an environment that is constantly changing. In this case the model of the environment is determined by the design phase of the building in which the simulation is developed. This process is carried out through a tool based on Unity 3D [50].

The **3D editing environment** (Figure 3) allows the modeling of virtual environments for urban areas (buildings, flats, apartments, gardens, parks, etc.) in three dimensions with a high level of realism and quality. This tool is able to generate 3D virtual environments in scale and define elements such as floors or walls. It also includes a set of default furniture models that can help to reduce the virtual environment modeling time. This editor uses different services from a *cloud computing* platform [51], being the main ones (i) a repository where the different 3D objects are stored, (ii) a repository of the scenarios created by the users of the platform and a (iii) user management service. Thanks to the possibility of remotely storing of the different Unity 3D packages that contain the 3D objects and textures used in the editing tool, the editor is able to obtain these models on demand. This allows the tool to be lighter, since it is not necessary that all the models offered by the platform are previously stored in the user's computer. There is a total amount of 14 different categories of 3D resources in the platform, but it is possible to include new categories if it is required. Besides, in order to facilitate the handling of different types of objects, a classification has been established according to their placement and behavior within the 3D editor:

- TYPE 0: Objects whose pivot is in the lower part of its mesh, that is, correspond to the objects that are placed at ground level.
- TYPE 1: Objects whose pivot is at the top of it, such as a lamp or any other object that hangs from the ceiling.
- TYPE 2: The pivot is on the side of the mesh to facilitate its placement on a wall.
- TYPE 3: The pivot is in its lower part, as in the case of type 0, including the possibility of placing these objects over other objects.
- TYPE 4: Objects of type 0 including an accessibility property of to them. That is, these objects will be used to perform the different interactions between avatars and furniture.
- TYPE 5: Objects of type 2 including the accessibility property.
- TYPE 6: Objects of type 3 including the accessibility property.

On the other hand, after completing the process of generating the virtual environments through the 3D editing tool, a json file is generated, in which all the environment characteristics are stored. By using these files, it is possible to generate the same scene in Unity, but without the need of store the unity project, (which is usually heavy in terms of storage). This lightweight file is sent to the *cloud computing* platform, so that, if the same scene has to be generated again, it is only necessary to download this file through the 3D editing tool or the 3D viewer, and 3D model is generated again using the 3D models that are provided by the platform, thus favoring the possible reuse of all resources by different users. The platform users can manage if the generated scenarios are public, private, or shared with certain users. To do this, thanks to the user management system, users of the platform can log into the system, accessing the list of projects for which they have access. Thanks to the use of the functionality offered by the *cloud computing* platform, several benefits are obtained such as a reduction in the amount of data to be stored by users, an improvement of the elasticity and scalability of the use of resources, or the dynamic adjustment to the load of the system and the needs on the part of the users of the platform. In addition, more effective use of computer resources can be made, since they can be shared among several entities.

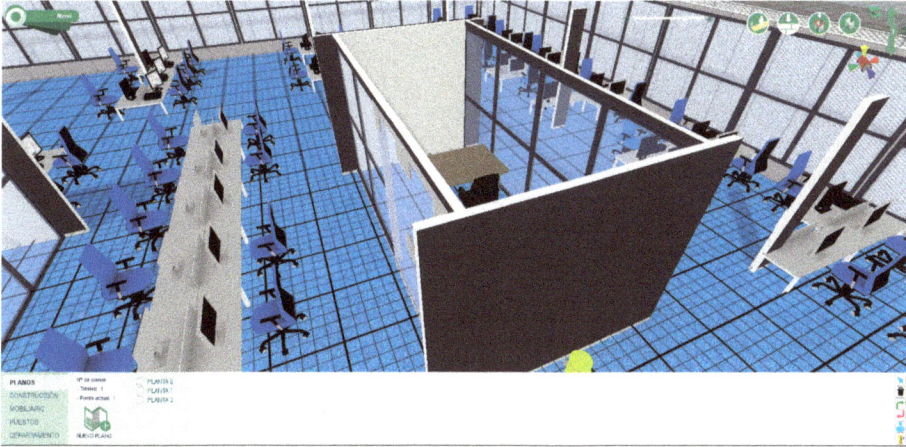

Figure 3. The 3D edition environment.

The **3D viewer**, which supports the representation and rendering of the environment and the features that were previously built with the 3D editor. It is also responsible for showing the different agents that are part of the system, making it possible to visualize the performance of the different activities of each agent. One of the main problems that must be faced in 3D scenes is to move one model from one point to another, so it is necessary to use an algorithm that calculates the route. In this case CritterAI [52] was used. It is a navigation system for Unity, which offers functionality related to the generation of navigation meshes, road search and movement management. This tool provides different options of the route calculation algorithm to be used: Dijsktra and A*. These algorithms are used to calculate the lowest cost path between an origin and a destination node. The nodes on which these algorithms will search for the minimum cost path will be the ones generated in the navigation mesh.

It has been necessary to add synchronization mechanisms between the 3D viewer and the agent platform, so that there is a correspondence between what happens on the agent platform and what is displayed in this tool. In addition, this platform is the main way of interaction between humans and machines, allowing experts in HR to modify the simulations based on their objectives. The operations that can be carried out are the creation and elimination of agents and assignment of tasks. As discussed in the previous chapter, all the operations requested from the 3D environment are communicated to the agent platform, which will attend these requests and confirm or not its completion in order to update the visualization status accordingly. Figure 4 shows the set of modules that make up the 3D editing environment and 3D viewer, previously explained.

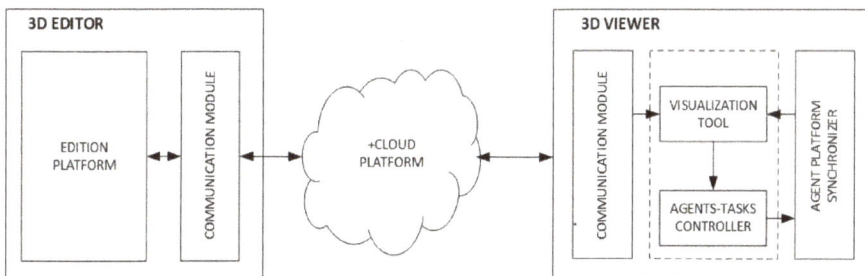

Figure 4. The 3D environment modules.

3.2. Implementation of the Organizational Model

After the analysis of the current state of the art in agent specification languages, platforms and specifications for the development of agents, we consider that the standard promoted by FIPA is flexible enough to define with guarantees any type of interaction between agents, so it adapts perfectly to the requirements that were initially marked to develop the platform. Since JADE [31] is a free open source system, it is well documented, and it is the most widely used solution within the development of multi-agent systems based on the FIPA standard, it has been taken as the basis for the implementation of the MAS, adapting it to support the management of VO. Thus, for the management of the existing agents in the VO, a specially designed database is used, which allows to store the agents' characteristics: the roles which have been assigned to them, the services they offer, as well as additional features that may be presented by the agents.

The control of the information stored in the database, and therefore of the organizational models used in each simulation, is encapsulated in a JADE agent called the Database Agent (DBA). This agent offers a set of services that facilitate all forms of access to the database, preventing inappropriate access by agents who should not be able to consult or modify the content of the database. Among the services offered by this agent, is to add agents, roles, services, etc., as well as query, modify or delete information. To access the different services, a module to check permissions is implemented, making the services accessible only to the agents that are specified.

Next, the structure of the database that will be used to store the structure of the VO is shown in Figure 5. Through a relational model, it is possible to define the different roles associated with the agents, as well as the tasks that can be carried out by an agent with a specific role. Additionally, three specific tables have been incorporated for the case study in which the proposed model will be evaluated, thus being able to define which agents intervene in each project, if they present some type of disability and associate roles with the different departments of a company.

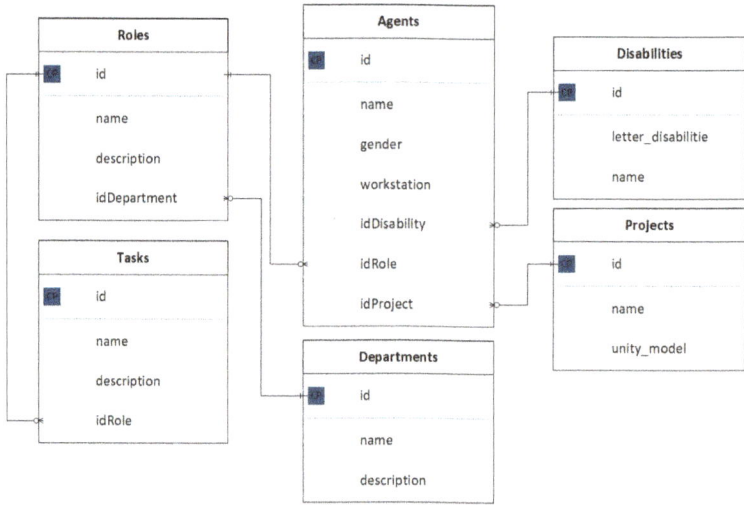

Figure 5. Virtual Organization database schema.

3.3. Structure of the Agents

Figure 6 shows in detail the structure of each of the agents, which consists of three large blocks: (i) a communication module, which will allow the agent to communicate with the other agents of the platform; (ii) a reasoning module, which is detailed below; and (iii) an agent communication module with the 3D environment, formed by a sensor and an effector.

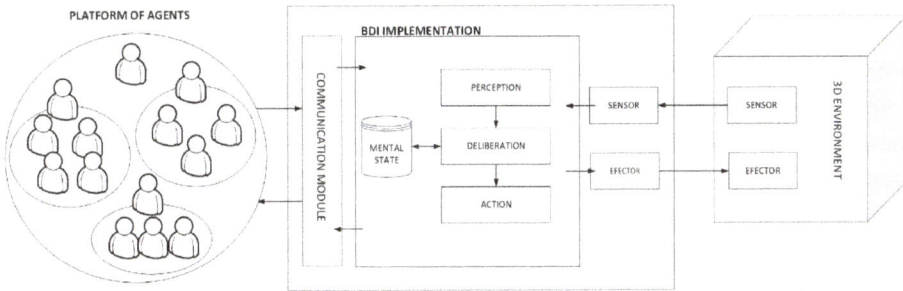

Figure 6. Agent structure.

For the definition and implementation of the reasoning module of agents within the platform, the belief–desire–intention (BDI) model [26] has been selected. The behavior of the system is therefore determined by the mental attitudes of the agents. For this reason, it is fundamental to achieve a correct operation of the system when the deliberation processes are limited by the resources.

In each iteration, the interpreter first reads an event queue and generates a list of options (possible goals to be achieved). Then it selects the best of these actions and adds them to the structure of intentions. If there is any intention to perform an atomic action, the agent executes it. Events that may occur during the interpreter's execution cycle are added to the queue. Finally, the agent modifies the structures of intentions and desires by eliminating both the intentions and desires that have been successfully carried out and those that are not possible. With this model, deliberative systems are more appropriate for dynamic environments and are also more suitable for applications that work in real time. In this way, this model perfectly adapts to the requirements of the platform that is presented.

In addition to the reasoning module, it is necessary to introduce two additional modules that allow the agent to obtain information about the environment (**sensor**), as well as to be able to modify it (**effector**). It is necessary that these modules are distributed, having one component present in the agent's own implementation within the agent platform, and another component within the 3D environment. In this way, the agent is able to be aware of the state of the three-dimensional environment that has been rendered for the simulation in progress, monitoring the changes that are produced in the virtual world.

The agent model presented above is designed to represent different actors involved in the simulation in representation of elements such as human beings, or similar. However, it is also necessary to introduce another type of agents: **environment agents**. These agents represent elements of the environment with which the other agents can interact directly. Examples of these agents can be telephones, elevators, photocopiers, etc. These agents are not aware of the environment around them, they will only be aware of their internal state, so they will lack of sensors. Any interaction with other agents cannot be initiated by an environment agent. When an agent requests an interaction with them, it will consult their internal state, and will communicate the pertinent information to the agent who finally, will determine according to this information if they can carry out the action that has previously been assigned to them.

The communication between agents is important in order to achieve the objectives of the system, where the objectives of the individual agents coexist. There is a set of stages in which the need for the existence of communication between agents is obvious. For example, when defining the problem to be solved, as well as its decomposition and distribution among the agents that make up the system. To carry out the communication between the agents, they are equipped with a FIPA ACL **communication module**. Thanks to the use of JADE it has not been necessary to redefine additional functionality, since this platform facilitates a complete implementation of these interaction protocols between agents.

3.4. Comunication of the Agent Platform with the Visualization Tool

One of the essential tasks within the platform is to establish a communication between JADE and the 3D environment, so that there is a real correspondence between what happens in the agent platform and the visualization that will take place on the 3D platform. Within the agent platform, there is an agent in charge of managing communications in the JADE side.

For this purpose, the communication between the agent platform and the 3D environment is done through Transmission Control Protocol (TCP) sockets. In the Unity side, a module dedicated to the management of the requests made by the agent platform and vice versa is included. The tasks of creation, elimination and interaction between agents executed on the platform must be updated in the 3D environment, and the tasks of creation, elimination and interaction between agents executed from the 3D interface, must be carried out first in the agent platform for being later updated in Unity. For the implementation of the communication module hosted in Unity, IKVM.NET is used. It allows the execution of compiled Java code in MICROSOFT.NET. The communication will be divided into 2 blocks:

1. Sockets dedicated to carrying out the tasks in JADE and their corresponding update in Unity.
2. Sockets dedicated to the performance of the tasks initialized from the Unity interface, which must be done first in JADE and then updated in Unity.

For the simulation of tasks in JADE that will be updated later in Unity, three elements are used: two sockets, one for sending tasks and another for the confirmation of the performed tasks, and a table with which will keep a record of these tasks (Figure 7).

*   **UnityTasks**, stores tasks that have not yet been updated.
*   **JadeProducer**, it allows to send tasks to Unity through a TCP socket. In Unity, there will be a resource (**UnityConsumer**) which is listening to the requests that are being produced.
*   **JadeConsumer**, opens a TCP socket to receive information regarding the performed tasks. In Unity, **UnityProducer** is in charge of sending the confirmation of the performed tasks.

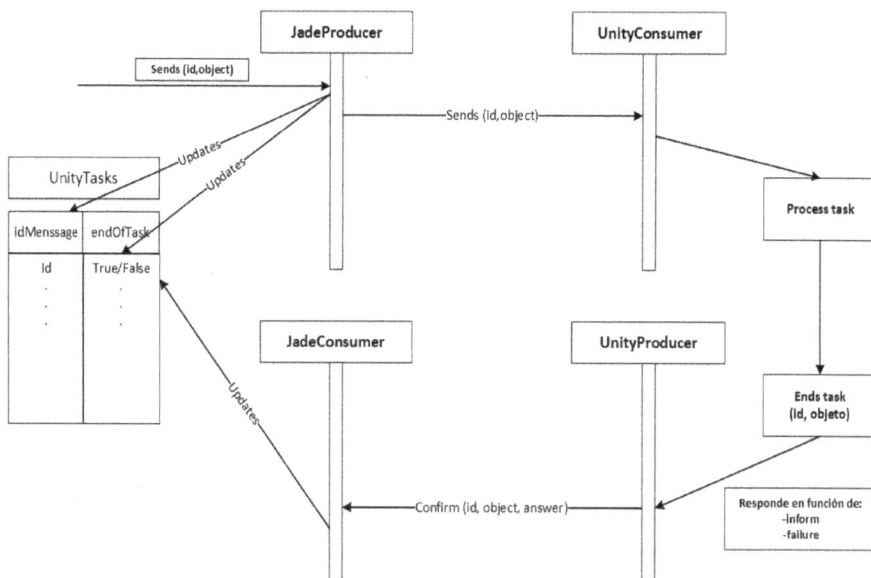

Figure 7. Java Agent Development Environment (JADE)-Unity communication.

Regarding the tasks that are performed in Unity and later updated in JADE, UnityTasks resides on the Unity side, since it is in charge of making requests and must verify that everything that is done in JADE is updated later in Unity. Figure 8 shows the scheme that follows this communication, where a TCP socket will be used. JADE is the server which will listen to the requests performed from Unity. Each request in Unity is a new client that requests a connection to the server, so throughout the execution several clients can coexist. The requests made from Unity can be of three different types: creation of agents, elimination of agents or assignment of tasks to agents.

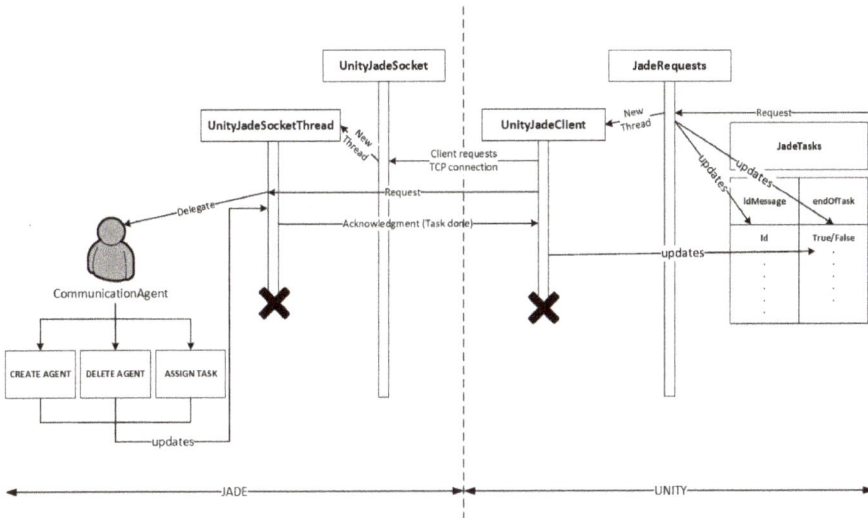

Figure 8. Unity-JADE communication.

The **Communication Agent (CA)** is in charge of attending the requests that arrive from Unity, carrying out the tasks that correspond according to the frame exchanged. The frame that is sent in each request contains the components shown in Figure 9.

Frame type	Id task	Agent name	Extra 1	Extra 2

Frame type:
-"C" for creation frames
-"D" for deletion frames
-"T" for task assignment frames

Id task: identification number

Agent name: name of the agent which performs the request

Extra 1: this field meaning may vary depending on the frame type:
-"C" frames: kind of agent
-"D" frames: null value
-"T" frames: kind of task to be implemented

Extra 2: this field meaning may vary depending on the frame type:
-"C" frames: extra information attending to the case of study
-"D" fames: null value
-"T" frames: true/false, indicates if the task performance has been finished in Unity

Figure 9. Communication frame.

4. Case Study

This section presents the case study which was conducted in a real environment to assess the performance of the developed system. The scope of the case study is the analysis of the accessibility of the different jobs in a real company: the offices of Indra Sistemas S.A. in Salamanca. By this way, through the performance of a set of simulations and the supervision of experts in HR and accessibility, knowledge about the different problems that disabled workers of the company may encounter can be extracted.

For the performance of the simulations, it was necessary to define the characteristics of the VO, in such a way that it would be able to modeling the processes carried out in the company as closely as possible to reality [53]. To do this, the different roles that agents can acquire, the services necessary for the proper functioning of the organization, the rules that will govern the society and the agents' own messages and iterations must be defined.

In addition, the environment in which the simulation will be carried out must be defined, in this case the building of the company under study. For this purpose, the 3D editing tool mentioned above was used. The 3D editing process begins from the plans of the building, so the environment in which the simulation takes place is as similar as possible to the real world. Furthermore, the disposition of the furniture inside the building must also be defined in order to evaluate, for each job, if each of the working places are accessible for people with different types of disabilities.

Afterwards, a series of simulations that show the behavior of the organization in different situations can be carried out, finally getting into the validation of the proposed model. First, we define the interaction model, analyzing the needs of the users who use the system and the way in which the exchange of information takes place. We define the following roles:

- **User:** represents the user or customer of the system. In this context, the user may be responsible for defining the distribution of the different elements in the building. Users can be the human resources responsible for assigning jobs to employees or the architect in charge of designing the building in which the professional activity will be carried out, in order to obtain information about the accessibility of the environment. User:

  - Is responsible for initiating the simulation process.
  - Will model the different agents that act as actors representing the workers of the company, as well as the tasks they will perform during the simulation.
  - Can access tob the information that is generated after the processes that are simulated.

- **Manager:** this agent is responsible for carrying out the task planning that will be carried out by the whole organization. It may be present or not in the system, depending on the organizational structure of the company to be represented. Manager:

  - Generates the tasks that represent the objectives to be fulfilled by the whole organization.
  - Delegates the task distribution to the different area managers according to the nature of each task.

- **Department/Area managers:** these agents are in charge of planning the distribution of tasks assigned to the department or area that manages. Department managers:

  - Receive the tasks from the manager.
  - Plan the task distribution among the agents that make up the department based on their availability (agents with a lower number of tasks will have a higher priority) and capacities (not all agents offer the same services).

- **Workers:** agents who represent each of the workers involved in the simulation.

- ○ The agents will have a stack where they store all the tasks they must perform, which have been previously assigned by the area manager.
- ○ They will send information regarding the degree of success with which they have carried out each of them.
- ○ All these agents have a set of elements that define their common behaviors and cognitive abilities. In addition to these common characteristics, the agents will have their own characteristics, defined according to their role within the organization and their disability.
- ○ The assigned roles will determine the tasks they can perform and the specific behaviors for each one of them, so different executions can be modeled for the same task according to the type of disability of the worker.

- **Environment agents**: these agents represent those elements of the building with which the worker agents can interact directly. Some examples of this type of agents can be found in different elements of the environment such as telephones, elevators or photocopiers.

  - ○ These agents are not aware of the environment around them, they are only aware of their internal state.
  - ○ Any interaction with other agents will be initiated in any case by a worker agent.
  - ○ When an agent requests its use, it will consult its internal status, checking if it is free and if it has the necessary resources so that the requested task can be carried out. It will communicate this information to the working agent, determining if it can complete the requested task.

- **Database Agent**: this agent offers a set of services that facilitate all forms of access that are held on the data base that stores the information regarding the structure of the VO. Database Agent:

  - ○ Will take care of inadequate accesses by agents that should not be able to query or modify the contents of the database. These services will allow adding agents, roles, services, etc. to the database, as well as query, modify or delete information.
  - ○ Includes a small module which is in charge of checking will require to check permissions, making the services accessible only to the specified agents.

- **Human resources agent**: The proposed VO can be classified within the so-called semi-open organization, since it will have a mechanism to control the admission of agents. In this way, it is necessary to make a request to evaluate the inclusion or not of an agent to society. The human resources agent is in charge of carrying out this task. The human resources agent:

  - ○ Evaluates the candidatures of those agents which request access to the organization, approving the different proposals based on the vacancies that exist in the department to which the agent intends to access.
  - ○ The number of spaces in each department will be established by the working places that are defined in the work environment. Each of the working place will be associated with a single department and an agent if they are occupied.

- **Communication agent**: it is responsible for communicating the multi-agent system with the visualization tool, in order to facilitate the message-passing between the two modules of the architecture.

  - ○ It guarantees that there is a real correspondence between what happens in the agent platform and the visualization that takes place in the 3D viewer.

- **Supervising agent**: analyzes the different behaviors and evolutions of the agents involved in the simulation to carry out an analysis of the processes that take place in the organization.

○   It will communicate with the other agents after the completion of a task, collecting whether they could carry it out or not.

○   If a task has been carried out satisfactorily, it will be communicated which task was performed, which agent intervened, and the time and resources that have been used to carry it out. Otherwise, the agent involved shall indicate to the supervising agent the reasons why it could not be carried out.

Within the processes that are carried out in the company under study, three departments have been identified, according to the tasks that workers can perform: reception, administration and maintenance. The worker-type agents will have associated a series of tasks that they can carry out depending on the department to which they belong. We define the following tasks as well as well as possible additional environmental agents that may be necessary for each of them (see in Table 1. Relations between tasks, departments and environment agents).

Table 1. Relations between tasks, departments and environment agents.

Task Type	Department	Environment Agents
Make photocopies	Reception, Administration, Maintenance	Photocopier
Go to the bathroom	ALL	Toilet
Fire alarm	ALL	Siren
Answer Telephone	Reception, Administration	Telephone
Carry letters	Reception	-
Drink coffee	ALL	Coffee maker
Eat	ALL	Microwave
Throw Trash	Maintenance	Trash can
Collect documentation	Reception, Administration	-
PC	Reception, Administration	-
Check elevator	Maintenance	Elevator
Watering plants	Maintenance	-
Sweep	Maintenance	-
Collect garbage	Maintenance	Trash can, container

Regarding the relationship of a person with the physical environment, the activities that are carried out during its daily life are widely varied, but the accessibility difficulties that arise when doing them are repeated. Therefore, the activities are analyzed from the perspective of accessibility to detect what difficulties they generate in order to find an alternative. To that end, two components have been established for all the activities that workers can carry out: displacement and use. **Displacement** is the transfer from a point to the ideal place where to perform an action, which implies being able to move around the environment without the presence of limitations or obstacles. Two types of displacement can be differentiated: (i) *horizontal* (displacements made in rooms, corridors, etc.) and (ii) *vertical* (displacements that imply a change in height, such as steps, stairs or ramps). The **use** is the performance of the action itself. Two types of use are differentiated: (i) *preparation*: situation process or approach to the object to be used and (ii) *execution*: the performance of the desired activity; it is the final objective of the process. If the person has any limitation that hinders the performance of their activity with respect to an individual without disabilities, we can use these as a reference to define the specific needs of these individuals in order to project the buildings. Difficulties of different types may appear:

Attending the displacement:

- **Maneuver:** limitation of the ability to access to spaces and move within them.
- **Change of level:** difficulties that arise when overcoming unevenness.

Considering the use of spaces, the following difficulties may appear:

- **Reach:** when there are limitations in the possibilities of reaching objects and perceiving sensations.

- **Control**: difficulties that arise as a direct consequence of the loss of ability to perform actions or precise movements with the extremities.

In order to guarantee the movement within a building, it is necessary to define accessible routes that link the different spaces. These itineraries must meet three basic functional criteria:

- That they are flat or with a gentle slope.
- That they have a passage area free of obstacles: they must be wide enough to fit wheelchairs, and also be free of obstacles of medium height, which are dangerous for the blind people.
- That they have safe support and guidance elements: that is non-slip pavements, handrails that serve as support, differentiated textures for the blind, etc.

For each action, four functional objectives have been distinguished:

- That it can be carried out, also, by a seated person. Thus, it can be performed by wheelchair users, elderly people with difficulties, etc.
- Without leaving the site while it lasts. Most people with limitations use their upper limbs to help themselves with movement; therefore, they can hardly move and act at the same time.
- Disregarding fine joints. The faucets, switches, door handles, control elements, etc., must have a design that allows their manipulation by those who have lost strength and dexterity in the hands.

Once the building model and the interaction model have been defined, we can proceed to perform the simulation. To do this, the visualization tool must be deployed, which at the beginning of the process, has perform different operations: (i) loading the building model; (ii) generating of the navigation mesh; (iii) synchronizing with the agent platform. Figure 10 shows a screenshot of the simulation tool.

Figure 10. Simulation tool.

The execution of the simulation will depend on the set of tasks that the MAS has to perform, which may be specified by the user, or generated by the management agent, as well as the process of the tasks assignment. Overall, the simulation will be defined by the definition of:

- The set of agents responsible for each department: $R = \{R_0, \ldots, R_i\}$
- The set of worker agents: $A = \{A_0, \ldots, A_j\}$
- The set of tasks to be performed by the worker agents: $T = \{T_0, \ldots, T_k\}$
- The type of resources: $l \ni \{1, \ldots, L\}$
- The set of resources (environment agents): $E = \{E_0, \ldots, E_m\}$
- The amount of available resources of n type at a certain t time: $L_{n,t}$

The result space after the performance of a certain task by a worker agent has only two possible output states (success and failure): $O \in \{0,1\}$, where 1 represents the success of the performance of a task, and 0 represents its failure.

First, a simulation of 15 min is planned. For the initial set-up of the different agents that represent actual workers within the organization, a Human Resources manager established a simulation of a total of 8 agents, whose department was assigned proportionally to the amount of workers of each department that were working at the company −1 receptionist, 2 maintenances and 5 administrative. Three of them have reduced mobility problems, three are visually impaired and one is deaf. The assignment of the work places for each of the agents was made just according to the productive processes that take place in the organization, without taking the possible architectural barriers that they might find into account. Once the environment and the agents' distribution and characteristics were set up, the simulation can take place in the 3D viewer. After the simulation, the tool generates a detailed report with the accessibility problems that the avatars have encountered, which have impeded the achievement of the tasks assigned to them. 118 were successful and 79 failed (Figure 11A). Some of the deficiencies found were: (i) The lack of adapted bathrooms in the building; (ii) working places not adapted for people with visual or hearing disabilities: impossibility to perform tasks such as answering the telephone, or being aware that the fire alarm is active; (iii) lack of ramps that enable the access to higher floors as an alternative to the elevator, which in case of breakdown would obstruct the access or exit to the building; (iv) limited access to certain work stations-spaces not large enough for access with wheelchairs or; (v) not accessible furniture elements–file cabinets, photocopiers.

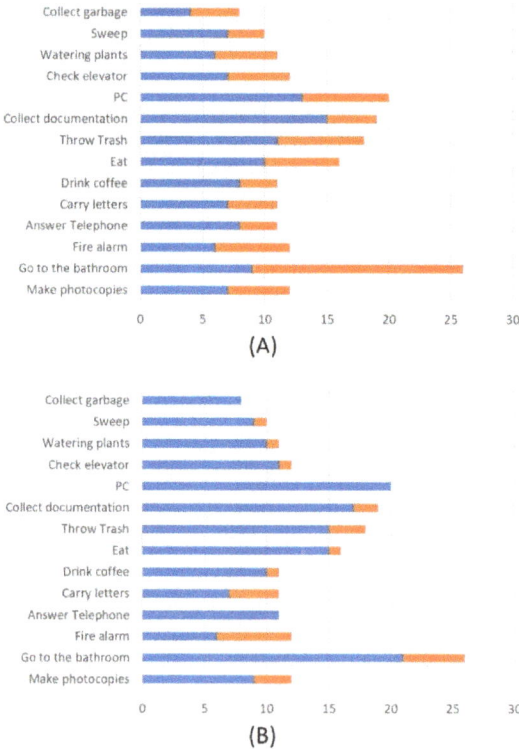

Figure 11. Successful and failed tasks after simulations. (**A**) results after first simulation; (**B**) results after second simulation.

After the detection of these accessibility problems, all the information obtained from the tool was used by the Human Resources manager in order to perform several changes (both in the distribution of environment elements and workers) that could emerge into an improvement of the accessibility of the working place. First, the 3D editor was used to perform a redistribution of the elements of the furniture susceptible to improvement and the adaptation of the work places for users with auditory and visual deficiencies. Secondly, a relocation of the work places of those users of wheelchairs that could not perform their designed activities due to several displacement criteria such as maneuver, change of level, reach or control was made using the 3D viewer. After doing this, a new simulation with the same agents and task-assignment was made, obtaining as a result a total of 169 tasks performed satisfactorily and 28 failed (Figure 11B), going through a reduction of 40.1% of failed tasks in the initial simulation to 14.2% in the second simulation. Obviously, there are elements of the environment that will continue to limit their accessibility, such as the lack of access ramps or adapted toilets, but the possibility of improving the accessibility of the environment through the restructuring of employees' jobs or furniture in the critical points detected through the simulation. Since there are infinite possibilities of furniture distribution, it is difficult to establish an optimal solution regarding the accessibility levels of work environments. However, it would be possible, given a fixed distribution of the elements of the environment, to make all possible assignments of types of employees, jobs and tasks, thus being able to know what the optimal distribution of employees under a certain structure of the elements of the environment is.

5. Conclusions and Future Lines of Research

Once the simulation platform has been presented and evaluated, it can be affirmed that it allows the performance of simulations oriented to the labor integration of the disabled, allowing not only to analyze the tasks that are carried out in a work environment, but also the physical spaces where these tasks take place.

From an academic point of view, it has been possible to model a real organization through an VO of agents with self-adaptive capacities. This system has been implemented using JADE, but adapting this platform to specific needs. In the same way, a 3D environment has been created, and it is capable of modeling and enabling simulations in work environments. In the same way, a system that allows communication between both platforms, which allows sharing relevant information about communications, has been developed. In short, the implementation of MOVICLOUD allows to simulate different processes in a work environment, and it has been proven to be appropriated in order to establish the accessibility problems that may occur in it. In this sense, the representation of the employees working activity is performed, such and as presented in the case study. Thanks to the flexibility provided by the editing tool, which allows to quickly model virtually any office environment using specifically designed models for rapid deployment, MOVICLOUD is valid for simulations carried out in office environments. More specifically, thanks to the case study, the tool has been able to define accessibility problems and architectural constraints following specific criteria of displacement. This tool has been evaluated in a work office, proving to be a useful tool when it comes to identifying accessibility problems. The definition of the models to be used and the problems to be detected in each task based on defined criteria facilitates the inclusion of the study of different new tasks. That is why this tool allows to model and include new actions to be developed by the actors with ease. However, it should be noted that MOVICLOUD is aimed at simulating processes within office environments, so that its use is currently limited to this area. In the event that the tool is intended for simulation in an open environment and with non-human actors, it would be necessary to include additional functionality.

Based on the obtained results, we can conclude that the objectives set at the beginning of the project have been met. With the successful achievement of this work, it is possible to obtain useful information to facilitate the accessibility of work environments to disabled people. This has been possible through the use of agent-based social simulation techniques, using organizational concepts to model the interactions between the actors and a three-dimensional environment. Although this tool initially was conceived for the detection of accessibility problems in office environments, thanks to the generic modeling of the platform and the use VO of agents, it is considered a valid tool for the simulation of any agent production process representing human beings in the field of office environments. Once the platform has been validated, it can be considered that within the scope of accessibility problems, it can be very useful at the time of: (i) early detection of failures in the design time of the buildings, facilitating new designs of the distribution of building elements, such as rooms, corridors, ramps or elevators to avoid in advance future problems related to the displacement of people with disabilities; (ii) prevent people with disabilities from facing accessibility problems before starting work at their place of work, being able to adapt their position and environment in a more appropriate way according to their capabilities (iii) improve the accessibility of people who are already working in a certain place.

Below, the different lines of research and future work that have been raised in the context of this work are shown:

- Include the possibility of making an abstraction of the building model through ontologies, so that the information of the environment does not only reside in the three-dimensional environment, but the agent platform also has knowledge of it.
- Allow the development of different three-dimensional environments, not only limited to representation in buildings.
- Improve the extraction of knowledge by improving the algorithms that intervene in the accessibility recommendations, offering real-time statistics and a report generation services.
- Include additional functionality to the visualization, so that there are alternatives when extracting information about the processes, such as graphs that show how the relationships between those agents interacting with each other or the generation of heat maps that show which are the routes taken by each worker in the three-dimensional environment.
- Perform stress tests to check the limitations of the platform in relation to communication processes and the size of the 3D environments. In this way, it will be possible to estimate the maximum number of agents and the characteristics of the 3D model supported by the platform over different hardware infrastructures.

Acknowledgments: This work has been supported by Cátedras de Tecnologías Accesibles of INDRA and Fundación Adecco. The research of Alberto L. Barriuso and Daniel H. de la Iglesia has been co-financed by the European Social Fund (Operational Programme 2014–2020 for Castilla y León, EDU/128/2015 BOCYL). Álvaro Lozano is supported by the pre-doctoral fellowship from the University of Salamanca and Banco Santander.

Author Contributions: Alberto L. Barriuso and Fernando De la Prieta conceived and designed the multi agent system, while Daniel H. De La Iglesia and Álvaro Lozano have contributed with the development and implementation of the system; Gabriel Villarrubia has designed and coordinated the case study. All authors have participated in the writing process of the article.

Conflicts of Interest: The authors declare no conflict of interest.

References

1. World Health Organization. *World Report on Disability*; WHO: Geneva, Switzerland, 2011.
2. Parmenter, T. *Promoting Training and Employment Opportunities for People with Intellectual Disabilities: International Experience*; International Labour Organization: Geneva, Switzerland, 2011.
3. *Standard Rules on the Equalization of Opportunities for Persons with Disabilities*; Technical Report; United Nations: New York, NY, USA, 1994.
4. European Comission. *Communication from the Commission to the European Parliament, the Council, the European Economic and Social Committee and the Committee of the Regions*; European Disability Strategy 2010–2020: A Renewed Commitment to a Barrier-Free Europe (COM/2010/0636); European Comission: Brussels, Belgium, 2010.

5.	*Americans with Disabilities Act of 1990*; S.933—101st Congress; The United States Congress: Washington, DC, USA, 1990.

6.	International Labour Organization. *The ILO Global Business & Disability Network Charter*; International Labour Organization: Geneva, Switzerland, 2015.

7.	Wang, F.-Y.; Carley, K.M.; Zeng, D.; Mao, W. Social computing: From social informatics to social intelligence. *IEEE Intell. Syst.* **2007**, *22*, 2. [CrossRef]

8.	Erickson, T.; Kellogg, W.A. Social translucence: An approach to designing systems that support social processes. *ACM Trans. Comput. Interact.* **2000**, *7*, 59–83. [CrossRef]

9.	Robertson, D.; Giunchiglia, F. Programming the social computer. *Philos. Trans. R. Soc. A* **2013**, *371*, 20120379. [CrossRef] [PubMed]

10.	Barthélemy, J.; Carletti, T. An adaptive agent-based approach to traffic simulation. *Transp. Res. Procedia* **2017**, *25*, 1238–1248. [CrossRef]

11.	HÖrl, S. Agent-based simulation of autonomous taxi services with dynamic demand responses. *Procedia Comput. Sci.* **2017**, *109*, 899–904. [CrossRef]

12.	Aliabadi, D.E.; Kaya, M.; Sahin, G. Competition, risk and learning in electricity markets: An agent-based simulation study. *Appl. Energy* **2017**, *195*, 1000–1011. [CrossRef]

13.	Immonen, E. Simple agent-based dynamical system models for efficient financial markets: Theory and examples. *J. Math. Econ.* **2017**, *69*, 38–53. [CrossRef]

14.	Kamara-Esteban, O.; Azkune, G.; Pijoan, A.; Borges, C.E.; Alonso-Vicario, A.; López-de-Ipiña, D. MASSHA: An agent-based approach for human activity simulation in intelligent environments. *Pervasive Mob. Comput.* **2017**, *40*, 279–300. [CrossRef]

15.	Terán, J.; Aguilar, J.; Cerrada, M. Integration in industrial automation based on multi-agent systems using cultural algorithms for optimizing the coordination mechanisms. *Comput. Ind.* **2017**, *91*, 11–23. [CrossRef]

16.	Anderson, J.H.; Downs, J.A.; Loraamm, R.; Reader, S. Agent-based simulation of Muscovy duck movements using observed habitat transition and distance frequencies. *Comput. Environ. Urban Syst.* **2017**, *61*, 49–55. [CrossRef]

17.	Montagna, S.; Omicini, A. Agent-based modeling for the self-management of chronic diseases: An exploratory study. *Simulation* **2017**, *93*, 781–793. [CrossRef]

18.	Davidsson, P. *Multi Agent Based Simulation: Beyond Social Simulation*; Springer: Berlin/Heidelberg, Germany, 2000; pp. 97–107.

19.	Davidsson, P. Agent Based Social Simulation: A Computer Science View. *J. Artif. Soc. Soc. Simul.* **2002**, *5*, 1–7.

20.	Borshchev, A.; Filippov, A. From system dynamics and discrete event to practical agent based modeling: Reasons, techniques, Tools. In Proceedings of the 22nd International Conference of the System Dynamics Society, Oxford, UK, 25–29 July 2004; p. 45.

21.	Venkatramanan, S.; Lewis, B.; Chen, J.; Higdon, D.; Vullikanti, A.; Marathe, M. Using data-driven agent-based models for forecasting emerging infectious diseases. *Epidemics* **2017**. [CrossRef] [PubMed]

22.	Haman, I.T.; Kamla, V.C.; Galland, S.; Kamgang, J.C. Towards an Multilevel Agent-based Model for Traffic Simulation. *Procedia Comput. Sci.* **2017**, *109*, 887–892. [CrossRef]

23.	Han, T.; Zhang, C.; Sun, Y.; Hu, X. Study on environment-economy-society relationship model of Liaohe River Basin based on multi-agent simulation. *Ecol. Model.* **2017**, *359*, 135–145. [CrossRef]

24.	Wagner, N.; Agrawal, V. An agent-based simulation system for concert venue crowd evacuation modeling in the presence of a fire disaster. *Expert Syst. Appl.* **2014**, *41*, 2807–2815. [CrossRef]

25.	Devia, N.; Weber, R. Generating crime data using agent-based simulation. *Comput. Environ. Urban Syst.* **2013**, *42*, 26–41. [CrossRef]

26.	De Almeida, S.J.; Ferreira, R.P.M.; Eiras, Á.E.; Obermayr, R.P.; Geier, M. Multi-agent modeling and simulation of an Aedes aegypti mosquito population. *Environ. Model. Softw.* **2010**, *25*, 1490–1507. [CrossRef]

27.	Bergenti, F.; Poggi, A. *LEAP: A FIPA Platform for Handheld and Mobile Devices*; Springer: Berlin/Heidelberg, Germany, 2002; pp. 436–446.

28.	Poslad, S.; Buckle, P.; Hadingham, R. The FIPA-OS agent platform: Open source for open standards. In Proceedings of the 5th International Conference and Exhibition on the Practical Application of Intelligent Agents and Multi-Agents, Salamanca, Spain, 4–7 July 2000; Volume 355, p. 368.

29. Sycara, K.; Pannu, A.; Willamson, M.; Zeng, D.; Decker, K. Distributed intelligent agents. *IEEE Expert* **1996**, *11*, 36–46. [CrossRef]

30. Strasser, M.; Baumann, J.; Hohl, F. Mole: A Java based mobile agent system. In Proceedings of the 2nd ECOOP Workshop on Mobile Object Systems, Linz, Austria, 8−9 July 1997; pp. 28–35.

31. Bellifemine, F.; Poggi, A.; Rimassa, G. JADE—A FIPA-compliant agent framework. Proceedings of PAAM, London, UK, 19–21 April 1999; pp. 97–108.

32. Zambonelli, F. Abstractions and infrastructures for the design and development of mobile agent organizations. In Proceedings of the International Workshop on Agent-Oriented Software Engineering, Limerick, Ireland, 10 June 2001; pp. 245–262.

33. Esteva, M.; Rodríguez-Aguilar, J.A.; Sierra, C.; Garcia, P.; Arcos, J.L. On the formal specifications of electronic institutions. *AgentLink* **2001**, *1991*, 126–147.

34. Van Dyke Parunak, H.; Odell, J. Representing social structures in UML. In Proceedings of the Fifth International Conference on Autonomous Agents, Montreal, QC, Canada, 29 May 2001; pp. 100–101.

35. Annunziato, M.; Pierucci, P. The emergence of social learning in artificial societies. Proceedings of Workshops on Applications of Evolutionary Computation, Essex, UK, 14−16 April 2003; pp. 293–294.

36. Dignum, M.V. A Model for Organizational Interaction: Based on Agents, Founded in Logic. Ph.D. Thesis, Utrecht University, Utrecht, The Netherlands, 2004.

37. Foster, I.; Kesselman, C.; Tuecke, S. The Anatomy of the Grid: Enabling Scalable Virtual Organizations. *Int. J. High Perform. Comput. Appl.* **2001**, *15*, 200–222. [CrossRef]

38. Peiro, J.M.; Prieto, F.P.; Zornoza, A.M. Nuevas tecnologías telemáticas y trabajo grupal. Una perspectiva psicosocial. *Psicothema* **1993**, *5*, 287–305.

39. Zambonelli, F.; Jennings, N.R.; Wooldridge, M. Developing multiagent systems: The Gaia methodology. *ACM Trans. Softw. Eng. Methodol.* **2003**, *12*, 317–370. [CrossRef]

40. Rodriguez, S.; Julián, V.; Bajo, J.; Carrascosa, C.; Botti, V.; Corchado, J.M. Agent-based virtual organization architecture. *Eng. Appl. Artif. Intell.* **2011**, *24*, 895–910. [CrossRef]

41. Bajo, J.; Fraile, J.A.; Pérez-Lancho, B.; Corchado, J.M. The THOMAS architecture in Home Care scenarios: A case study. *Expert Syst. Appl.* **2010**, *37*, 3986–3999. [CrossRef]

42. Zato, C.; Rodriguez, S.; Tapia, D.I.; Corchado, J.M.; Bajo, J. Virtual Organizations of agents for monitoring elderly and disabled people in geriatric residences. In Proceedings of the 2013 16th Internation Conference on Information Fusion (FUSION), Istanbul, Turkey, 9–12 July 2013; pp. 327–333.

43. Davidsson, P.; Johansson, S. On the potential of norm-governed behavior in different categories of artificial societies. *Comput. Math. Organ. Theory* **2006**, *12*, 169–180. [CrossRef]

44. Luke, S.; Cioffi-Revilla, C.; Panait, L.; Sullivan, K. Mason: A new multi-agent simulation toolkit. In *Proceedings of the 2004 Swarmfest Workshop*; George Mason University: Fairfax, VA, USA, 2004; Volume 8, pp. 316–327.

45. North, M.J.; Howe, T.R.; Collier, N.T.; Vos, J.R. The repast simphony runtime system. In Proceedings of the Agent 2005 Conference On Generative Social Processes, Models, and Mechanisms, Argonne, IL, USA, 13−15 October 2005; Volume 10, pp. 13–15.

46. Zato, C.; Villarrubia, G.; Sánchez, A.; Barri, I.; Rubión, E.; Fernández, A.; Rebate, C.; Cabo, J.A.; Álamos, T.; Sanz, J.; et al. *PANGEA—Platform for Automatic coNstruction of orGanizations of intElligent Agents*; Springer: Berlin/Heidelberg, Germany, 2012; pp. 229–239.

47. Gilbert, N.; Terna, P. How to build and use agent-based models in social science. *Mind Soc.* **2000**, *1*, 57–72. [CrossRef]

48. García, E.; Rodríguez, S.; Martín, B.; Zato, C.; Pérez, B. *MISIA: Middleware Infrastructure to Simulate Intelligent Agents*; Springer: Berlin/Heidelberg, Germany, 2011; pp. 107–116.

49. Dignum, V. *Handbook of Research on Multi-Agent Systems: Semantics and Dynamics of Organizational Models*; Information Science Reference; IGI Global: Hershey, PA, USA, 2009.

50. Unity Official Web Page. Available online: https://unity3d.com/es/ (accessed on 22 September 2017).

51. La Prieta, F.; Rodríguez, S.; Corchado, J.M.; Bajo, J. Infrastructure to simulate intelligent agents in cloud environments. *J. Intell. Fuzzy Syst.* **2015**, *28*, 29–41.

52.   CritterAI Documentation. Available online: http://www.critterai.org/projects/cainav/doc/ (accessed on 22 September 2017).

53.   Corchado, J.M.; Corchado, E.S.; Aiken, J.; Fyfe, C.; Fernandez, F.; Gonzalez, M. Maximum Likelihood Hebbian Learning Based Retrieval Method for CBR Systems. In *Case-Based Reasoning Research and Development*; Springer: Berlin/Heidelberg, Germany, 2003; pp. 107–121.

applied
sciences

MDPI

Article

Linked Data Aware Agent Development Framework for Mobile Devices

İlker Semih Boztepe and Rıza Cenk Erdur *

Department of Computer Engineering, Ege University, 35100 Bornova, İzmir, Turkey;
ilker.semih.boztepe@ege.edu.tr
* Correspondence: cenk.erdur@ege.edu.tr; Tel.: +90-232-311-2597

Received: 28 August 2018; Accepted: 1 October 2018; Published: 6 October 2018

check for
updates

Abstract: Due to advances in mobile device and wireless networking technologies, it has already been possible to transfer agent technology into mobile computing environments. In this paper, we introduce the Linked Data Aware Agent Development Framework for Mobile Devices (LDAF-M), which is an agent development framework that supports the development of linked data aware agents that run on mobile devices. Linked data, which is the realization of the semantic web vision, refers to a set of best practices for publishing, interconnecting and consuming structured data on the web. An agent developed using LDAF-M has the ability to obtain data from the linked data environment and internalize the gathered data as its beliefs in its belief base. Besides linked data support, LDAF-M has also other prominent features which are its peer-to-peer based communication infrastructure, compliancy with Foundation for Intelligent Physical Agents (FIPA) standards and support for the Belief Desire Intention (BDI) model of agency in mobile device agents. To demonstrate use of LDAF-M, an agent based auction application has been developed as a case study. On the other hand, LDAF-M can be used in any scenario where systems consisting of agents in mobile devices are to be developed. There is a close relationship between agents and linked data, since agents are considered as the autonomous computing entities that will process data in the linked data environment. However, not much work has been conducted on connecting these two related technologies. LDAF-M aims to contribute to the establishment of the connections between agents and the linked data environment by introducing a framework for developing linked data aware agents.

Keywords: agent development framework; mobile device agent; linked data; semantic web

1. Introduction

An agent is a computational entity that is capable of autonomous action on behalf of its users in order to satisfy its delegated objectives [1]. Autonomy is generally considered as an essential property of an agent. On the other hand, Wooldridge and Jennings suggest three other additional properties that an intelligent agent is expected to have [2]. These properties are reactivity, proactivity, and social ability. While reactivity addresses the ability of an agent to perceive its environment and respond to changes that occur in its environment, proactivity is related with exhibiting goal-oriented behavior. Finally, social ability is the capability of an agent for interacting with other agents and possibly human users by exchanging messages represented in a specific agent communication language. Agents have found many application areas in various domains such as human computer interaction (e.g., digital personal assistants), information retrieval, electronic commerce, and manufacturing since the mid-1990s [3,4]. Recent studies show that use of agent technology is still in progress in various domains such as cyber physical systems and internet of things that are popular research areas in the context of industry 4.0 [5,6].

Architecture of an agent typically consists of two layers. The first one is the communication layer, which provides an agent with the necessary means for communicating with other agents by exchanging agent communication language messages over a network infrastructure. The second one is the agency layer, where deliberation and planning mechanisms of an agent are implemented. In other words, properties of reactivity and proactivity are provided in the agency layer. It would of course be impractical to develop an agent by implementing these layers from scratch each time a new agent based application is being developed. Thus, agent development frameworks have been introduced to ease and expedite the implementation of agent based solutions. JADE [7], JADEX [8], and Jason [9] are the most widely known open source agent development frameworks. Interested readers can refer to Reference [10] which is a comprehensive survey of agent development frameworks.

Due to advances in wireless network and hardware technologies, mobile devices such as mobile phones and tablets, through which people can access information from anywhere and anytime, have already become an indispensable part of our lives. This situation has inevitably affected the agent research community and studies have been conducted on deploying agent development frameworks into mobile devices. As a result of these studies, several agent development frameworks that address mobile devices have been released. Jade LEAP [11], 3APL-M [12], and Agent Factory Micro Edition (AFME) [13] are examples of such frameworks. As will be discussed in the related work section, each of these frameworks has its own prominent features. For example, while Jade LEAP is distinguished by the agent communication and execution infrastructure that it provides, 3APL-M and AFME employ deliberation mechanisms which are based on the Belief-Desire-Intention (BDI) model of agency [14] for developing intelligent agents. Recently, JADE and JADEX released new versions that support the Android Operating system [15,16].

Another technology that has close relationship with the agent technology is the semantic web. Since the beginning of the semantic web movement, agents have been considered as the main type of software that would consume the data on the semantic web. In their original article about the semantic web, Berners-Lee, Hendler and Lassila [17] emphasized the connection between agents and semantic web by defining agents as the autonomous computing entities that would be used in the semantic web environment. Linked data, which presents a set of techniques and standards to publish, link and query data on the web, is considered the realization of the semantic web vision, since it had a great impact on the development of real-life semantic web applications [18,19]. Thus, the number of real-life semantic web applications has begun to increase with the introduction of linked data.

On the other hand, although the close relationship has been emphasized from the very beginning, in practice there has been little collaboration between agent and semantic web/linked data communities.

In this paper, we introduce the agent development framework that we have developed for mobile devices. Using this framework, developers can build intelligent agents that have the attributes of social ability, reactivity, proactivity, and autonomy. The main contribution of the presented framework is its linked data support. Linked data support in an agent simply corresponds to the ability to supply its beliefs from the linked data environment, to keep the obtained beliefs internally, and use those beliefs during planning process.

By providing linked data support, we also aim to contribute to the establishment of the links between agents and the linked data environment. Since the ability to supply its beliefs from the linked data environment is the developed agent framework's distinguishing feature, we called it Linked Data Aware Agent Development Framework for Mobile Devices (LDAF-M). The belief component of LDAF-M uses the Resource Description Framework (RDF) model to represent the beliefs of an agent and provides a graph data structure to hold the beliefs internally. By this way, semantic data stored in ontologies as well as in open data stores, such as Freebase or DBpedia in the linked data environment, can be used to represent an agent's beliefs. Linked data support, which is the distinguishing property of LDAF-M, makes it the first linked data aware agent framework for mobile devices to the best of our knowledge.

LDAF-M agents can be qualified as intelligent agents according to the definition of Wooldridge [2], since they support the attributes of autonomy, reactivity, proactivity, and social ability. Social ability is related to the ability of an agent to exchange agent communication language messages with others in the system. Social ability is provided through the LDAF-M communication infrastructure. Reactivity is the ability to respond to the changes in the environment, and proactivity is related to exhibiting goal directed behavior. Reactivity and proactivity are satisfied through the use of the BDI model of agency [14]. Finally, autonomy is satisfied through the use of a planner within the context of the BDI architecture. Communication infrastructure and the agency layer are discussed in detail in Section 2, where the architecture of LDAF-M is given. On the other hand, some highlights related to the communication infrastructure, standards followed during the development, and BDI model of agency are given below.

Support for peer-to-peer (P2P) paradigm: In the communication layer, LDAF-M provides a P2P based communication and resource sharing infrastructure where each mobile device behaves as a peer. The P2P infrastructure provided by LDAF-M also includes a mechanism that can detect a peer that becomes offline due to a problem and can handle this situation so that the system continues to operate seamlessly.

FIPA compliant execution platform: FIPA (Foundation for Intelligent Physical Agents) is an IEEE Computer Society standards organization that issues a set of specifications which intend to promote interoperability among heterogeneous agent systems [20]. LDAF-M provides a FIPA compliant platform for mobile device agents. Jade LEAP and AFME also support FIPA standards. On the other hand, during the design of LDAF-M, we have made slight modifications which are discussed in Section 2.2.

Support for BDI model of agency: Belief-Desire-Intention (BDI) model of agency [14,21,22] is a framework for describing the behavior of agents in terms of their beliefs, desires (options) and intentions (adopted options). This model of agency draws its inspiration from practical reasoning which is a model of decision making proposed by Bratman [23]. The agency layer of LDAF-M provides the necessary infrastructure to develop BDI agents for mobile devices. 3APL-M and AFME also support BDI model of agency. However, different from those frameworks, LDAF-M has the ability of representing beliefs of an agent using semantic web standards (i.e., RDF) as mentioned before.

The rest of the paper is organized as follows. The layered architecture of LDAF-M is explained in Section 2. P2P communication 1. Section 2.2, first overviews FIPA standards and then discusses compliancy of LDAF-M to the FIPA standards. In Section 2.3; the agency layer, where the BDI model is implemented, is explained. Linked data support is discussed in Section 3. Section 4, which is the case study and discussion section, consists of three parts. The first part illustrates how LDAF-M is used in developing agent systems. In the second part, an agent based auction application that has been developed using LDAF-M is introduced by giving a detailed interaction diagram. In the third part, the case study given is discussed. Related work and conclusion are given in Sections 5 and 6 respectively. Finally, references are listed.

2. Architecture of LDAF-M

LDAF-M has a layered architecture which mainly consists of two layers, as shown in Figure 1. The bottom layer provides the communication infrastructure. LDAF-M also follows the FIPA specifications which mostly affect the communication and deployment infrastructure. FIPA compliancy is also explained in this section following the communication infrastructure. On top of the communication layer, the agency layer is located. The agency layer contains the implementation for the BDI model of agency.

The communication layer and the agency layer basically interact by means of agent communication layer messages. The agency layer extracts the objectives that it will try to satisfy by examining the incoming messages stored in the incoming message queue of the agent. On the other hand, as a result of executing a plan, some messages need to be sent to other agents. In this case, the agency layer places the messages to be sent in the outgoing message queue, which then are sent to the related agents by the communication layer.

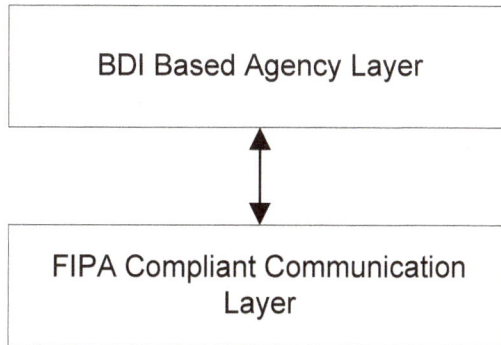

Figure 1. Linked Data Aware Agent Development Framework for Mobile Devices (LDAF-M) layers.

2.1. Communication Infrastructure

In this subsection, LDAF-M's communication mechanism, which has been developed following the P2P paradigm, is introduced. In addition, it is also discussed how the case of a peer that becomes inaccessible due to a problem arising from a resource constraint in a mobile device is handled in this P2P infrastructure.

P2P is an approach for network communication where shared data is processed by two or more client computers, each of which have equal responsibilities. In contrast to the client-server architecture where there are separate dedicated clients and servers, in a P2P infrastructure each of the peers can play the role of both a server and a client.

There are three approaches regarding the architecture of a P2P communication infrastructure [24–26]. These are central, decentralized, and hybrid approaches.

A dedicated central server and a group of peers are used in the implementation of a central P2P architecture. The central server maintains a database where it stores the indexes of all peers and the resources that belong to each peer. In the decentralized P2P architecture, there is not any central server that controls the network. Instead, each peer has equal functionality and holds information about its own resources. In other words, each of the peers acts as the index server. The hybrid P2P approach, which is the third-generation architecture, provides a hierarchical structure in which there are super peers having the properties of a central server.

Considering the constraints of mobile devices and the scale of the systems that would be constructed using LDAF-M, the communication infrastructure has been based on the centralized P2P approach. In the alternative (i.e., decentralized) P2P approach, since each peer keeps track of its own resources and needs to search resources of other peers, an extra load is introduced for peers. Additionally, a certain level of message traffic occurs on the network because of the circulating queries. On the other hand, using a centralized approach brings in several advantages such as easy implementation and timely responses to queries among peers. In addition, in our case, the connectivity status of a mobile device may dynamically change, as a result of running out of battery or going out of the range of the network. Using the centralized approach, we can easily keep track of and handle the connectivity problems, as it is explained in the following paragraphs.

As it can be seen from Figure 2, a group of peers and one central dedicated server have been used for the realization of the central P2P architecture. The dedicated server, which is called the "index server", is responsible for controlling and sustaining the P2P network. When any one of the peers becomes online, information associated with resources of this peer are recorded in the database within the index server. In the same way when any of the peers becomes offline, information associated with that peer is removed from the database within the index server. In addition, each of the online peers sends notifications to the index server periodically in order to indicate that they still actively use the system. If the index server cannot receive notifications from a peer for a certain amount of time, the index server updates its database to indicate that this peer becomes offline.

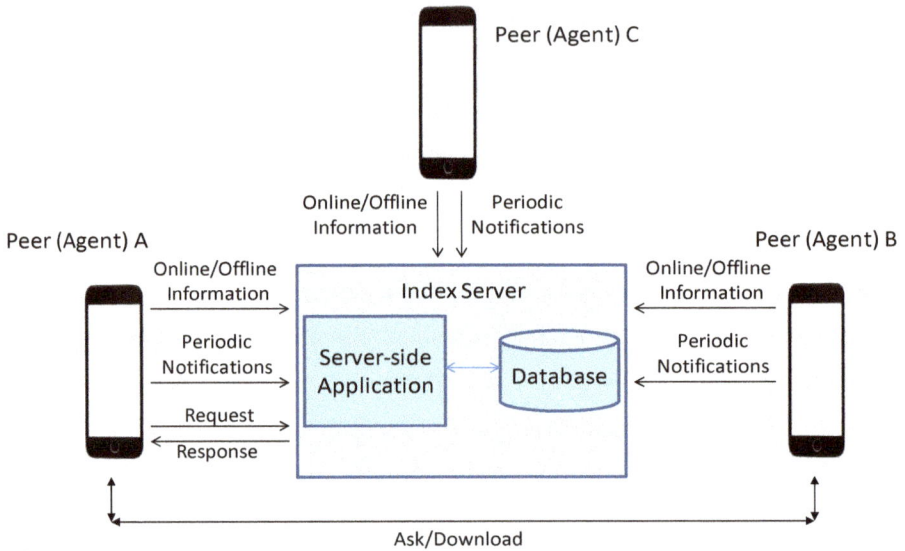

Figure 2. Peer-to-peer (P2P) communication infrastructure in LDAF-M.

While determining the length of the time interval for periodic notifications of the mobile device agents, we need to consider the timeliness of the notification, but at the same time we need to avoid introducing too much load for the network and the index server. A peer being offline may not be detected for a long time if periodic time intervals are too long. On the other hand, an extra load comes on the index server and the network when the periodic time intervals are too short. Therefore, the length of periodic time intervals was determined to minimize the negative effects of these two cases, which are opposite of each other. In particular, we have defined a specific notification time interval of 60 s based on our experiences in an agent based auction application that has been developed using LDAF-M.

The server-side application (e.g., Java Servlet based) on the index server may receive three main types of notification from a peer. The first type of notification is sent by a peer when it first becomes online. The second type of notification is sent by a peer when it decides to leave the system and will become offline. In both cases, the index server updates records in its local database according to the notification received. The third type of notification is sent as an acknowledgement that the peer is still actively participating in the system. This type of notification is sent periodically as explained above. When the index server realizes that a peer has not sent a notification for a long time, it understands that there is a problem with it and updates its local database to record the status of this peer as being offline.

Figure 2 also shows the working mechanism of the central P2P approach. The peer labeled as "Peer A" sends a request to the index server. Upon receiving this request, the index server queries its local database and sends the obtained result to "Peer A". This result includes information about the peers that can provide the information in response to the request of "Peer A". In Figure 2, it is assumed that the requested data is provided by the "Peer B". Consequently, the "Peer A" communicates directly with "Peer B" to obtain the requested data.

Both handling the notifications coming from the peers and behaving as the resource finder, the index server plays an important role in the communication of peers and provision of information and services for the whole platform.

2.2. FIPA Compliant Architecture

FIPA (Foundation for Intelligent Physical Agents) is an IEEE Computer Society standards organization that issues a set of specifications about agents and multi-agent systems. These specifications, which mainly aim to promote interoperability among agent systems, can be classified in terms of different categories. Agent communication, agent management, agent message transport and abstract architecture are the main categories of FIPA specifications [20].

During the design and implementation of LDAF-M, one of the specifications that we have followed to a great extend was the FIPA agent management specification [27]. FIPA agent management specification introduces the "agent management reference model", using which we could have mapped the central P2P communication infrastructure into a FIPA compliant architecture.

The agent management reference model introduced by the FIPA agent management specification [24] consists of the following components:

Directory Facilitator (DF), provides yellow page services to other agents. Agents register their services with the DF using a specific registration template or ontology and may query the DF to find out the required services offered by other agents.

Agent Management System (AMS), manages the life cycle of agents. Tasks such as creation, termination and suspension of agents are under the responsibility of the AMS. The AMS also keeps the Agent Identifiers (AIDs) which contain transport addresses of agents that registered with the platform. Initially, each agent registers with the AMS to get an AID for itself.

Message Transport Service (MTS), is responsible for delivering messages among agents residing on the same or remote agent platforms. Agents exchange messages expressed in an agent communication language and use transport addresses in delivering messages.

Agent Platform (AP), provides the physical environment on which agents can run properly. It consists of DF, AMS, MTS and agents.

The mapping made to a FIPA compliant architecture following the FIPA agent management reference model is shown in Figure 3. The index server and the peers (i.e., mobile devices), which are the main components of the central P2P communication infrastructure, have been implemented as different FIPA agent platforms, as shown in Figure 3.

Each of the agent platforms that reside on the mobile devices (peers) contains its own Agent Management System (AMS) as the managing authority in the agent platform. An AMS neither has any knowledge about other agent platforms, nor communicates with an AMS located in a different agent platform. Information about agents on remote platforms can be accessed through DF (Directory Facilitator) located in the agent platform that contains the index server. The tasks such as the creation of a new agent, handling the changes in the life cycle of an agent and the killing of an agent are under the responsibility of the AMS. When a new agent is created, Agent Identifier (AID) information that includes transport address(es) for that agent are recorded by the AMS. These records are kept within the AMS during the lifetime of the agent. More than one agent can run on an AP, and AIDs of the agents located within an AMS, can be queried by other agents which run on the same AMS.

The DF is in charge of providing yellow page services to all of the agents. Agents from different platforms can publish through the DF the services that they provide and these services are searchable from all of the mobile device agent platforms. Unlike similar frameworks such as Jade LEAP [11], LDAF-M contains only one DF and it is located in the agent platform that contains the index server. There does not exist an AMS in the index server's agent platform, rather there is an AMS in each mobile device agent platform. Since AMS manages life cycles and naming of agents in the system, it is obvious that placing a single AMS in the index server's agent platform may cause overloading of this single AMS. In LDAF-M, an agent looking for a specific agent first makes a search using the AMS in its local platform. If it is understood that the specific agent being looked for is not in the same platform, then the search is extended to the DF residing in the index server's agent platform. With such a design, we aim to reduce the load on the index server's agent platform.

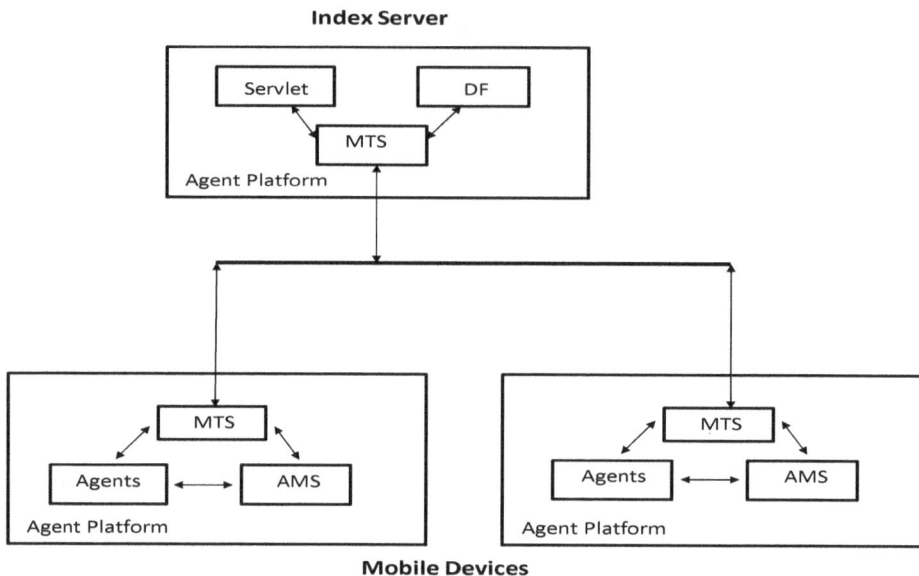

Figure 3. Mapping of the P2P communication infrastructure into Foundation for Intelligent Physical Agents (FIPA) compliant architecture.

As discussed in detail in the previous section, there is a notification mechanism between peers and the index server. By this way, peers that go offline because of battery or out of network range problems can be tracked. In such cases, the index server updates the entries in the DF to provide the most accurate information to agents located on mobile device agent platforms.

Message Transport Service (MTS), is the component that handles the message delivery between agents on the same or different platforms. The type of message transport protocol that is used in the communication between agents residing in a mobile device is the socket technology, which is actually on top of TCP/IP transmission protocol. On the other hand, the type of message transport protocol that is used in the communication between the index server and any mobile device is Hypertext Transfer Protocol (HTTP), since the agent platform running on the index server receives messages sent to it from the mobile devices using Java Servlet technology.

Agents exchange messages expressed in an agent communication language. FIPA also issued standards related with agent communication language and message structures. FIPA Agent Communication Language (FIPA-ACL) is the proposed agent communication language, and hence used in LDAF-M for the communication of agents. The structure of a standard FIPA message consists of two parts which are the message envelope and the message content. To overcome the memory limitations on the mobile devices, a slight modification has been performed on the structure of messages exchanged. This slight modification includes combining the envelope and the content parts of the messages to form a single deliverable unit and using only one of the optional message parameters that are similar. For example, the "to" and "from" parameters that can be used within the message envelope, and the "receiver" and "sender" parameters that are used in the content of a FIPA-ACL message have been combined into two single parameters.

Finally, agents running on the mobile device agent platforms are implemented as independent Java threads. Internally, each agent supports the agent life cycle standard proposed by FIPA. According to this specification, an agent can be in one of five states, which are "initiated", "active", "suspended", "waiting", and "transit". Transitions among these states are also defined. When an agent is constructed, the finite state machine representing the life cycle begins at the initial state. The active state is the one where an agent continues to behave in order to meet its delegated objectives. In the active state, transport messages are transmitted to the agent in a normal fashion. Within the scope of suspended and waiting states, all activities of an agent are terminated temporarily. In suspended and waiting states, the message transport service stores the messages locally on behalf of the agent. After the agent passes to the active state again, the stored messages are delivered to it. In the transient stage, an agent is able to move to different platforms. In the current version of LDAF-M, agent mobility is not supported, since agents that travel between the platforms is a research topic in its entirety.

2.3. Agency Layer

The agency layer in LDAF-M framework is based on the Belief-Desire-Intention (BDI) architecture. BDI is a model of agency which has been widely used in the development of agents and multi-agent systems, including commercial agent software. Hence, the design and implementation of the basic modules in this layer are inevitably similar with the existing BDI supported agent frameworks. On the other hand, the main difference of LDAF-M is that it can obtain knowledge from linked data resources and handle the obtained knowledge. In this regard, new functionalities and internal constructs has been added to a standard BDI implementation. These variations are explained in Section 3, where linked data support is discussed.

The BDI architecture's origins lie in the theory of practical reasoning [23] developed by Michael Bratman. As stated in Reference [22], this kind of reasoning is always directed towards doing an action and consists of two activities. The first activity is deliberation which is about deciding what state of affairs the agent wants to achieve. The second activity is means-ends reasoning (planning) and it decides how to achieve those selected state of affairs using an available set of actions.

Practical reasoning is realized in agent systems based on the concepts of beliefs, desires and intentions. This is why these architectures are called BDI architectures.

Everything that an agent knows and believes about the world in which it is located, including itself and also other agents around it, is represented by the belief component within the BDI architecture. Agents obtain their beliefs by continuously perceiving the environment in which they operate.

The motivational stance of an agent is represented by the desire component located within the BDI agent architecture. The desire concept characterizes the goals of an agent which define what the agent wants to achieve. In other words, desires are the possible options of an agent [14,21,22,28].

At the end of the deliberation mechanism, desires which are adopted by an agent become intentions. This means that the adopted desire would lead to an action. In this regard, during the process of identifying the actions that an agent will perform, the intention component within the BDI agent architecture has a much stronger role than the desire component. An agent may host many desires which can possibly be transformed into intentions and then to actions in the near future. The agent evaluates desires in terms of agent's beliefs, other desires and intentions, and then suitable desires are adopted as intentions that would lead to actions.

In LDAF-M, beliefs may come from the environment in which the agent runs, or from the linked data environment. The messages delivered from other agents may also cause the beliefs of an agent to be updated. The ability to obtain its beliefs from the linked data environment (e.g., from interrelated RDF documents) is the distinguishing feature of LDAF-M. The beliefs are used in option generation together with the current intentions to produce desires. The filter function uses beliefs, desires and intentions to produce new intentions. The filter component has a typical rule base which is used in determining which goals to adopt. Primarily, the beliefs in the active state at the belief base are processed using this rule base. Then the filter component determines what goals to adopt by matching the beliefs in the active state and the rules defined in its own context. Since an intention is directed towards an action, the related actions corresponding to an intention are executed next. Basic components of the BDI architecture that are implemented in LDAF-M are shown in Figure 4.

In fact, desire and intention which is a strong form of desire are similar notions, and in the implementation of BDI architectures they are represented by the "goal" concept. An agent may have many goals (i.e., desires) and the adopted goals are its intentions. The planning mechanism is the second stage of practical reasoning as mentioned above, and in order to transform an adopted goal into an action, the planning mechanism constructs a plan to fulfill that goal and executes the plan step by step.

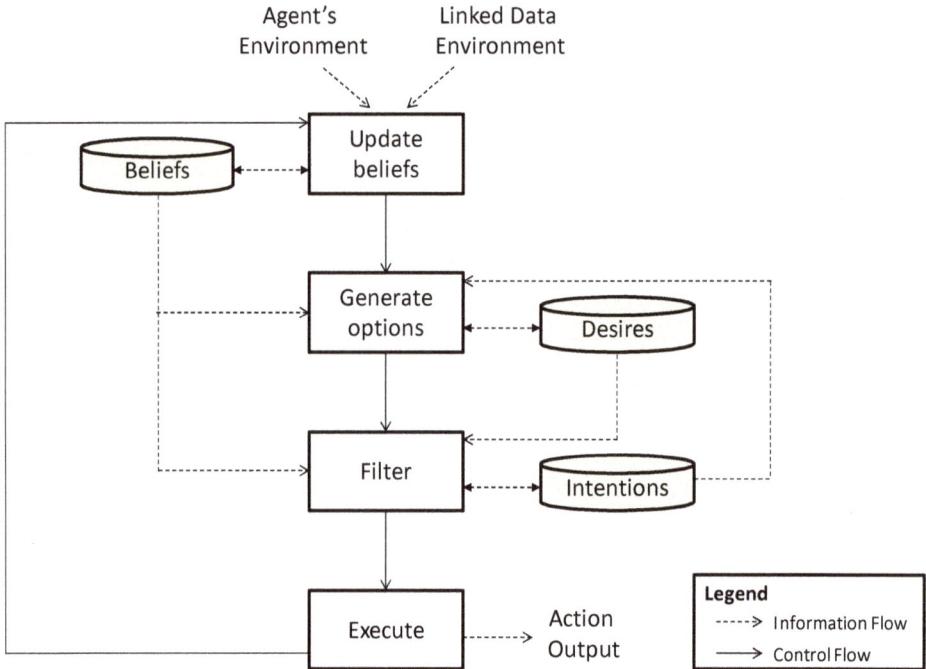

Figure 4. Control and information flow in the Belief-Desire-Intention (BDI) model based agency layer in LDAF-M.

Each plan consists of two parts. The header part includes information about which goal the plan is associated with and under what conditions it is applicable. The body part defines the actions to be performed or the sub-goals to be obtained. In this respect, there is a tight relationship between goals and plans, since a plan is executed after an agent's adoption of the goal to which that plan is related.

JADEX [8] is a well-known open source agent development framework that supports the development of BDI agents. The agency layer that we have implemented in LDAF-M was inspired by JADEX. On the other hand, since we target mobile devices, the agency layer that we have implemented works on a mobile device supporting the Java platform. As mentioned before, the other property that makes LDAF-M different from other BDI agent development frameworks, is the ability for collecting its beliefs from the linked data environment. Linked data support is discussed in the following section.

Both the agency layer and the communication layer of LDAF-M consist of several software packages that contain many classes. In the communication layer; platform, core, gui, AMS, DF, MTS, Message, MessagePattern, and MessageBase packages are included. These packages support agent platform operations on a mobile device, provide several graphical user interfaces that can be used in managing the platform, define the agent communication language messaging infrastructure, and provide the FIPA compliancy. The agency layer consists of several software packages such as BeliefBase, GoalBase, PlanBase, Deliberation, LinkedData, DataModel, Conditions (drop, suspend, triggering conditions) that includes the classes necessary for implementing the BDI model of agency and linked data support. These packages have open source access. Please refer to GitHub page for the project [29].

3. Linked Data Support in LDAF-M

The linked data concept, which has been originally introduced by Tim Berners-Lee in his technical note [30], refers to best practices for publishing, interlinking and consuming machine processable data on the web. In practice, Resource Description Framework (RDF) is the data model used for publishing structured and machine processable data on the web. A dataset represented using RDF can also include links to other datasets. These links can then be followed to aggregate data from different datasets [31,32]. To perform queries on linked data, SPARQL query language [33] is used.

RDF uses triples to represent resources and their relationships. Each triple consists of a subject, predicate and an object. The subject is the resource being defined and hence is represented by a Uniform Resource Identifier (URI). On the other hand, the object in a triple may also have a Uniform Resource Identifier (URI) value, since it is possible for an object to identify other resources. Otherwise, the object part of the triple contains a literal value. The predicate plays a key role in determining the relationship of the subject and the object in the triple. Using URIs to identify resources makes it possible to represent RDF statements as a graph where nodes represent subjects and objects, and arcs represent predicates. These relationships are illustrated in Figure 5. In the figure, the upper left node is a subject representing a resource. This subject is connected via predicates to the two objects labeled as object 1 and object 2. These two objects have literal values. Object 3 that the subject is connected to is also a resource. Thus, it is connected to another literal object labeled as object 4. For example, the subject that we are defining information about may be a URI representing a faculty in a university department. Object 1 and object 2 may be strings representing name and surname for that faculty. The faculty resource may be connected to an object representing a resource about a PhD student via the "advisorOf" predicate. Finally, the resource representing the PhD student may be connected to an object having a string value and representing "UniversityId" of the student.

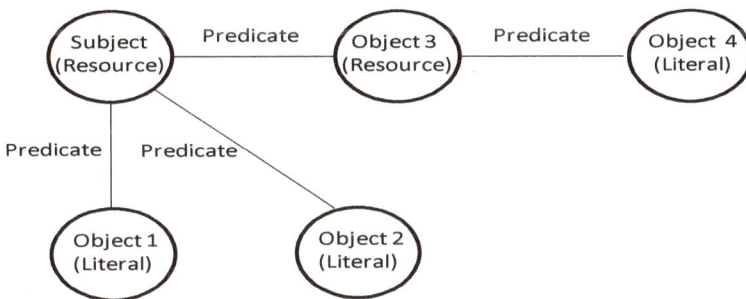

Figure 5. Resource Description Framework (RDF) statements on a graph data structure.

Linked data aims to transform a current web of documents into a global information space called "web of data". As a result of using RDF as the common standard for publishing and linking data, the "web of data" emerges as a big knowledge graph. Agents are the autonomous computing entities that will process the data in the linked data environment.

We think that gathering data from the linked data environment and internalizing that data as an agent's beliefs is one of the most important requirements that should be fulfilled in order to establish a connection between agents and linked data. Since data is gathered from the linked data environment by parsing the RDF documents, we need a graph based data model that will be used for representing the RDF triples internally in an agent.

Within the BDI architecture of LDAF-M, everything that an agent believes about its environment, including knowledge about other agents, is represented by the belief component. Any LDAF-M agent firstly obtains data from its environment. Subsequently, it associates a significant part of the obtained data with the belief component it contains. On the other hand, a LDAF-M agent can also obtain data from resources which have been developed in accordance with the principles of the linked data. Initially, a LDAF-M agent has been designed and implemented so that it can take its beliefs from Freebase [34]. Freebase provides documents and their relations expressed following the linked data principles. However, the belief component and architecture of LDAF-M has been designed and developed in such a flexible structure that data from other linked data sources such as DBpedia can also be obtained by easily integrating the necessary implementation to the system. LDAF-M includes an RDF parser for parsing these RDF documents and their relations. Since RDF documents form a knowledge graph, in LDAF-M a graph data structure that can represent data obtained in the form of RDF has been designed and implemented. The implementation model for the internal graph data structure is shown in Figure 6.

There are four components of the graph based data structure. These components are "GraphContainer", "GraphNode", "GraphPredicate" and "GraphProperty". A "GraphContainer" contains a certain number of "GraphNodes" (i.e., subjects or objects) and "GraphPredicates". The "GrapPredicate" concept establishes links between two nodes of a graph, hence represents the edges. Literals has the task of representing non-URI data related with a resource and is modeled using the "GraphProperty" class in Figure 6.

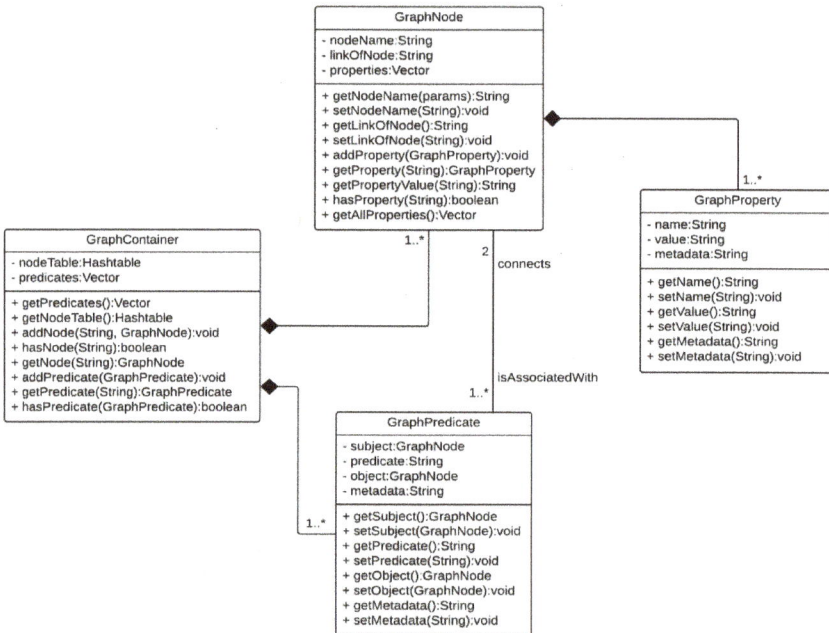

Figure 6. Implementation model for the internal graph data structure.

In the process of obtaining the data, first, the URI based links between RDF documents are discovered using a crawler-style mechanism. Then, each of the RDF documents discovered is parsed using the parser within the agent framework. Finally, a graph container is instantiated using the data obtained after parsing. Each "GraphContainer" represents a specific part of the beliefs stored in the belief base of an agent. Thus, an agent may contain more than one "GraphContainer" instance.

4. Case Study and Discussion

In the previous sections, we have given detailed information about the layers and structure of LDAF-M. In this section, we mainly introduce the agent based auction application that we have implemented as a case study.

Auctions are mechanisms by which buyers and sellers are in a continuous interaction and negotiate on scarce resources such as goods or utilities that they are going to sell or buy. As being autonomous computing entities, agents can carry out the bidding process on behalf of their users, can help in finding an optimal bidding strategy, and can collect information for their users during auctions. As a result, agents can facilitate many aspects in an online auction, and hence auction design and implementation is one of the well-known application areas of agents. We have chosen auction application, since there is a large number of interactions with many messages passing between agents in an auction scenario and this would be a good test environment for the communication infrastructure and planning mechanism of LDAF-M agents. The implemented scenario is an English auction.

We have structured this section in four parts. The first part introduces the main scenario of the agent based auction application that we have developed. The second part illustrates how to use the LDAF-M framework while building such an application. In particular, construction of a new agents, and definition of beliefs, goals and plans are explained. The third part gives a collaboration diagram that shows the interactions between agents and administrative components of the FIPA compliant LDAF-M agent platform as well as the interactions between agents themselves. Finally, a discussion constitutes the last part of this section.

4.1. Main Scenario

The auction application that we have implemented mainly allows users to exchange products among themselves. In this application, each one of the users may behave in two ways: They can either request to start an auction to sell a product or participate in an auction to buy a product.

A mobile device can play the role of a server which is responsible for managing an auction. In particular, it evaluates the requests coming from the clients for starting a new auction or participating in any of the ongoing auctions, and tracks the information flow in the auction. On the other hand, agents that run as clients on the mobile devices also play an important role for completing the deal. Information about the products and buyer and seller agents is represented using the belief component provided by the BDI model based agency layer of LDAF-M. Those beliefs that agents have about the products that are desired to be purchased or to be sold are obtained after processing Freebase data which is represented using the linked data principles.

4.2. Using the LDAF-M Framework

4.2.1. Constructing a New Agent

Using LDAF-M, developers are able to build agent based software which is mainly composed of agents running on mobile devices in various application domains.

When a developer needs to construct a new agent, he/she should extend the "Agent" class which serves as the main component for defining agents in LDAF-M. For example, the code fragment that is going to be used for building a client agent of the auction application is shown in Listing 1.

```
01    public class clientAgent extends Agent {
02        public void start() { ... }
03        public void finish() { ... }
04    }
```

Listing 1. Template for building a client agent.

The BDI related components of an agent, such as belief base, plan base, and goal base together with the other components, such as message base and agent life cycle related data, are held in the Agent class. The definitions of these components are made in the start() method. On the other hand, operations related with the termination of an agent are defined in the finish() method.

Agents constructed using LDAF-M framework communicate with each other using FIPA Agent Communication Language messages. An agent may receive many messages and to give the agent the ability to select the meaningful ones for itself among the incoming messages, a message filter class (e.g., FactoryMessageFilter) is defined in the framework. Using the methods in that class, developers can define patterns for indicating the significant messages that an agent expects. These pattern definitions are also made in the start() method of the agent class.

4.2.2. Defining Beliefs

As mentioned before, beliefs of an agent can be obtained dynamically from linked data resources as well as the other data sources defined by the developers. In this context, two different classes are used in defining the belief component of an agent. LinkedDataMetaBelief class which is extended from the MetaBelief class is responsible for representing beliefs coming from the linked data environment. It has a three-parameter constructor, using which a belief coming from a linked data resource is defined. The first parameter is the name given to the belief, while the second parameter is the URI of one of the related RDF documents that includes data that constitutes the content of the belief component. The third parameter is the name of the belief category that the defined belief belongs. The code fragment illustrating this process on an example belief in the auction application is given in Listing 2.

```
01   getBeliefBase().addGroupName("sell_product");
02   getBeliefBase().addGroupName("receive_product");
03   MetaBelief metaBelief = new LinkedDataMetaBelief("sell_product1",
04   "http://rdf.freebase.com/rdf/m.0pd3pb_", "sell_product");
05   getBeliefBase().addMetaBelief(metaBelief);
06   metaBelief = new LinkedDataMetaBelief("receive_product1",
07   "http://rdf.freebase.com/rdf/m.0pd3tfx", "receive_product");
08   getBeliefBase().addMetaBelief(metaBelief);
```

Listing 2. Defining linked data beliefs.

As shown in Listing 2, initially two belief categories, one for selling and the other for buying, are defined. Then, information about a product that is going to be sold is obtained from an RDF document in the Freebase environment and added as a belief to the agent's belief base. Similarly, another belief about buying a product is also added to the belief base of the agent.

The other class used in defining beliefs is the UserDefinedMetaBelief class which is extended from the MetaBelief class. This class is responsible for obtaining beliefs from the user defined data resources. It has a two-parameter constructor used in defining beliefs. The first parameter is the name given to the belief and the second parameter is the type of the data that constitutes the content of the belief component. The code fragment illustrating this process on an example belief in the auction application is given in Listing 3. The definition of two beliefs which are about an action and the participant list are given in Listing 3 respectively.

```
01   MetaBelief metaBelief = new UserDefinedMetaBelief("auction", Auction.class);
02   getBeliefBase().addMetaBelief(metaBelief);
03   metaBelief = new UserDefinedMetaBelief("participants", Vector.class);
04   getBeliefBase().addMetaBelief(metaBelief);
```

Listing 3. Defining other beliefs.

Both kinds of beliefs defined above are linked to the belief base of the agent by calling the addMetaBelief method that adds a belief passed to it as a parameter to the belief base of the agent.

4.2.3. Defining Goals

MetaGoal class plays the key role in defining an agent's goals. As you can see from the code fragment given in Listing 4, the MetaGoal class has a two-parameter constructor. The first parameter is the name of the goal, and the second parameter specifies the type of the goal. Some of the goals of an agent can be adopted either from the very beginning when the agent is activated or just before the agent terminates. These goals can be added to the goalbase of the agent using the methods addInitialGoalName and addFinishingGoalName as shown in Listing 4.

```
01  MetaGoal metagoal = new MetaGoal("finishing", MetaGoal.PERFORMGOAL);
02  getGoalbase().addMetaGoals(metagoal);
03  metagoal = new MetaGoal("initialization", MetaGoal.PERFORMGOAL);
04  getGoalbase().addMetaGoals(metagoal);
05  getGoalbase().addInitialGoalName("initialization");
06  getGoalbase().addFinishingGoalNames("finishing");
```

Listing 4. Adding goals to the goalbase of the agent.

As mentioned in Section 2.3, the life cycle of a goal can be controlled by the developer. The necessary conditions that are required for the transitions between the phases of the goal life cycle are defined in a class extended from the TriggeredCondition class. Such a class defined within the context of the developed auction application is shown in Listing 5. The AuctioneerAgentSuspendCondition class defines the conditions for an auctioneer agent to make a transition to suspended state and these conditions are written in the body of the method triggeredCondition().

```
01  public class AuctioneerAgentSuspendCondition extends TriggeredCondition{
02      public Boolean triggeredCondition(){
03          .....
04      }
05  }
```

Listing 5. Defining a condition for making a transition between two states.

The metaGoal class provides the methods for relating the defined conditions with the goal at hand. For example, setSuspendConditionClassName relates a suspend condition definition to a goal. On the other hand, the conditions that define when an agent has to drop a goal, is related with the goal via the method setDropConditionClassName. Examples for defining conditions for a goal that is related with starting an auction are shown in Listing 6.

```
01  metagoal = new MetaGoal("auction_process", MetaGoal.PERFORMGOAL);
02  metagoal.setSuspendConditionClassName
03  ("tr.edu.ege.bilmuh.application.sample.serverside.AuctioneerAgentSuspendCondition");
04  metagoal.setDropConditionClassName
05  ("tr.edu.ege.bilmuh.application.sample.serverside.AuctioneerAgentDropCondition");
06  getGoalbase().addMetaGoals(metagoal);
```

Listing 6. Defining conditions for a goal.

Another property in LDAF-M, which is a result of its BDI architecture, is that a goal can block some or all of the other goals during deliberation. For example, some of the goals of the agent can be transferred to the option phase from the active phase, based on the blocking list of another goal that is being adopted by the agent. These blocked goals can never pass to the active state unless the blocking goal continues its execution. This information is provided through the addInhibitGoalName method located in the MetaGoal class. In addition, the number of maximum goals the agent can handle while executing a specific goal is defined using the setCardinality method as shown in Listing 7. In Listing 7, one can see that searching for an auction while sending an auction generation request is inhibited.

```
01   MetaGoal metagoal = new MetaGoal("send_request_generate_auction",
02   MetaGoal.PERFORMGOAL);
03   metagoal.addInhibitGoalName("search_auction");
04   metagoal.setCardinality(5);
05   getGoalbase().addMetaGoals(metagoal);
```

Listing 7. Setting properties for a goal.

4.2.4. Defining Plans

The MetaPlan class is the main class that supports the definition of plans which play an active role in transforming the goals into a series of actions. As illustrated in the code fragment given in Listing 8, MetaPlan class has a constructor with two parameters. The first parameter is the name of the plan being defined, and the second parameter is the name of the class containing the body of the plan. The class containing the body of a plan is an extension of the PlanContent class. In addition, it is also important to define the relationships between plans and goals, and the MetaPlan class includes a method using which such relationships can be defined. In Listing 8, the goal called "auction_process" is associated with the auction processing of the auctioneer agent which is responsible for managing the auction. Finally, the plan defined is added to the plan base of the agent.

```
01   MetaPlan metaplan = new MetaPlan("auction_process_plan",
02   "tr.edu.ege.bilmuh.application.sample.serverside.AuctioneerAuctionProcessPlan");
03   metaplan.addProbablyGoalName("auction_process");
04   getPlanbase().addMetaPlan(metaplan);
```

Listing 8. Defining plans.

In LDAF-M, execution of a plan not only starts with the adoption of a goal which associated with that plan, but also starts based on the occurrence of an event within the agent platform. For example, when an agent receives an agent communication language message from another agent, an event is created. Then, the plan associated with that event begins to execute provided that the relationship between that plan and the message generated the event have already been defined. This relationship is defined using the method addMessageEventTrigger which is located in the Planbase class. The use of that method is shown in Listing 9. The first parameter of the method is the reference name of the message trigger, and the second parameter is the name of the plan.

```
01   getPlanbase().addMessageEventTrigger("subscribe_auction_msg", "subscribe_auction");
02   getPlanbase().addMessageEventTrigger("get_auction_name_msg", "get_auction_name");
```

Listing 9. Adding event trigger to a plan.

The PlanContent class has also an important role in the execution of the plans, since the detailed operation of a plan is defined within that class. A plan content example is given in Listing 10.

The actions to be done within the scope of the plan are defined in the method called "body". Execution of a plan can end with success or failure. The actions to be done in case of success, just before the plan is terminated, are defined in the method called "passed", while the actions to be done in case of a failure are defined in the method called "failed". On the other hand, some plans cannot be completed due to some reasons. When this is the case, the actions that should be done are defined in the method called "aborted".

```
01   public class EvaluateAuctionOfferPlan extends PlanContent {
02        public void body() { ...... }
03        public void passed() { ...... }
04        public void failed() { ...... }
05        public void aborted() { ...... }
06   }
```

Listing 10. Plan content.

From the agent developer's perspective, the plan content plays an important role, since a developer is able to define the following tasks and access the main components of an agent within a plan body:

- Interacting with the Agent Management System (AMS) to create an agent, to kill an agent, or to query the AMS about an agent.
- Accessing belief base, goal base, and plan base of the agent that is executing the plan body.
- Registering the services of an agent with the Directory Facilitator (DF), searching the DF for discovering the agents giving a specific service.
- Sending agent communication language messages to other agents.

4.3. Collaboration Diagram for the Case Study

The auction application which has been developed as a case study using the LDAF-M platform, is in fact a multi-agent system consisting of a number of agents. Each of these agents interacts with each other as well as with the LDAF-M platform's main components. We illustrate these interactions in a collaboration diagram which is given in Figure 7.

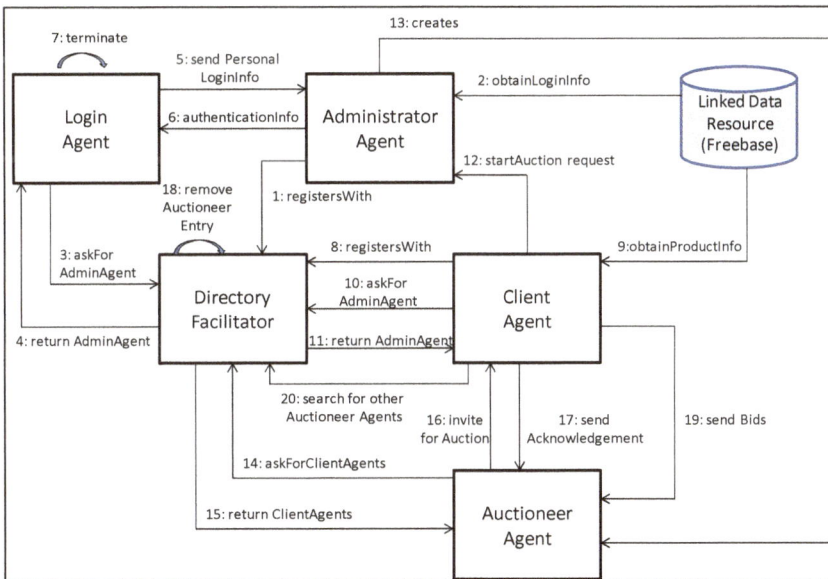

Figure 7. Collaboration diagram for the agent based auction application.

Each numbered interaction in the collaboration diagram is explained below. Please note that initially the agent platform is initialized, the AMS which will keep the agent identifiers and manage agent life cycle and the DF which is like a yellow page service are created. These steps are not shown for the sake of simplicity.

1: After the auction application has begun to execute on the mobile device that would serve as the server within the context of the developed application, an administrator agent is created. Upon creation, the administrator agent registers with the Directory Facilitator (DF) so that agents that are going to be created in future within the context of the auction application can discover it. The DF information console is shown in Figure 8.

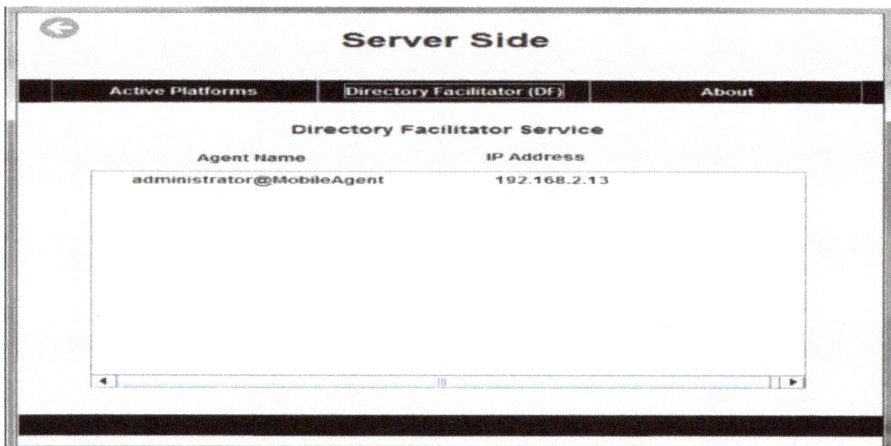

Figure 8. The Directory Facilitator (DF) information console.

2: The administrator agent obtains the login information belonging to all of the system's users from a data resource. In our case, the administrator agent obtains this knowledge from the RDF documents that we have already recorded to Freebase and then internalizes that data in its belief base using the graph data structure introduced in Section 3.

3. On the client side, users that want to use the auction system need to create a login agent. Upon creation, the login agent asks the DF for the administrator agent.

4: The DF sends the information about the administrator agent to the login agent.

5: The login agent transmits its user's login information to the administrator agent.

6: If the authentication process ends with success, the user receives a login success message.

7: The login agent terminates after successful login.

8: After successful login, client agents which represent participants in an auction are created. Then they register with the DF.

9: A client agent obtains the data about the products that it is going to sell or buy from the RDF documents residing in Freebase. After the client agent parses that data, it maps the parsed data into the LDAF-M's graph based data structure to record them as its beliefs.

10: The client agent asks the DF for the administrator agent of the application.

11: The DF sends the information about the administrator agent to the client agent.

12: The client agent communicates directly with the administrator agent to send it a request for starting an auction for a product that it decides to sell. The client agent repeats sending this request for each of the products that it wants to sell.

13: After processing this request, the administrator agent creates an auctioneer agent which is responsible for managing the auction for that specific product.

14: The auctioneer agent asks the DF for the client agents registered.

15: The DF sends information about all the client agents which registered with it.

16: The auctioneer agent sends invitation messages to all of the client agents (other than the client agent which is the seller) asking them whether they prefer to participate in the auction for that specific product or not.

17: A client agent evaluates this invitation considering the beliefs in its belief base and makes a decision as a result. If it decides to participate in that auction, it notifies the auctioneer agent about this decision.

18: Each auctioneer agent registers itself with the DF for only a pre-specified time period. When the time period for an auctioneer agent completes, its registration record is removed from the DF. Upon this removal operation, the auction starts and participant client agents get a notification as shown in the screen given in the left part of Figure 9. The alert includes information about the product that is subject to the auction and the name and address of the client agent who plays the seller role.

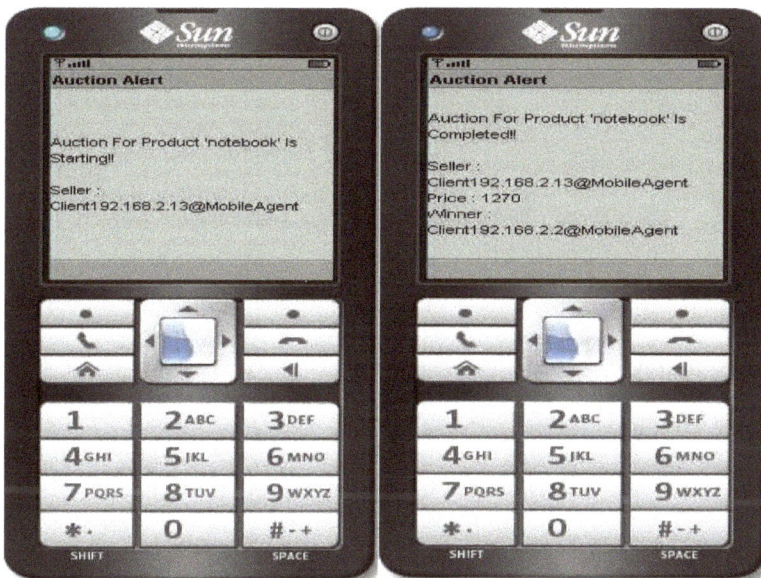

Figure 9. Alerts indicating the start and end of an auction for a specific product.

19: Each participating client agent submits a bid to the auctioneer agent in sequence until the time period for the auction completes. The client agent who submitted the highest price bid wins the auction. Each participating client agent receives an auction completed alert including the seller, winner, and the price as shown in the screen given at the right part of Figure 9.

20: A client agent can also want to know about auction proposals that are suggested before that client agent has been created. When this is the case, the client agent communicates with the DF to discover auctioneer agents that are related to the products that the client agent wants to buy. After discovering the related auctioneer agents, the client agent sends a request to the auctioneer agent to indicate its decision in participating with the ongoing auction.

4.4. Discussion

There is no standard benchmark for evaluating the components and services provided by the LDAF-M, which is a FIPA-compliant and linked data aware agent development framework targeting mobile devices. Thus, we need to define a basic evaluation framework which defines different perspectives for the implemented framework. Those perspectives are listed below:

Perspective 1: We need to verify that the system fulfills the basic requirements given in the sub list below:

Requirement 1.1: The case when an agent(s) located in a mobile device becomes inaccessible should be detected and handled.

Requirement 1.2: Compliancy with the FIPA Agent Management Reference Model should be ensured.

Requirement 1.3: Agents supporting the BDI model can be developed using LDAF-M.

Requirement 1.4: Belief component within the BDI architecture should be able to obtain its beliefs also from the linked data environment and use them in the deliberation process.

Perspective 2: We need to observe the physical limits of the system in terms of the number of agents created together with the agent platforms.

Agent based auction domain has been selected, on purpose, as the domain of the application implemented as the case study, since this application includes several types of agents located on different platforms and constitutes a multiagent system where member agents intensively interact with each other as well as with the managerial components of the platform.

Several client agents playing the role of buyers in an auction and auctioneer agents playing the role of sellers of a specific product have been created. Some of the auctioneer agents have been intentionally made offline. Then, a client agent that has been created before those auctioneer agents is selected. When this client agent queries the DF to discover the auctioneer agents created before itself so that it can send a request to each of them for participating in ongoing auctions, it has been verified that the offline auctioneer agents are removed from the system and their related auctions are cancelled. This observation fulfills requirement 1.1.

The developed application also tests the FIPA-compliancy related components of the framework. Agent platforms have been constructed on several mobile devices and the services such as AMS and DF has been used. For example, life cycles of agents in the auction application are managed through AMS services. On the other hand, DF has been intensively used in the auction application. FIPA-ACL agent communication language messages have been sent during the interactions of agents. As a result, requirement 1.2 is fulfilled.

Within the scope of the developed auction application; beliefs, goals, plans and actions are defined for different kinds of agents such as buyer, seller, administrator and log-in agents. Those agents are executed successfully on their platforms and it has been observed that the deliberation mechanism inside each agent works. This fulfills the requirement 1.3.

To fulfill requirement 1.4, Freebase has been used as the linked data resource. Login information for the system users, as well as the data about products that are going to be exchanged, is supplied from the RDF documents located in Freebase. It has been observed that this data can be internalized as beliefs of the agents using the graph data structure defined inside the agents for this purpose.

The auction application has been implemented by installing the LDAF-M to several computers. One of them functioning as the server is the platform where the index server and the DF of the system reside. The index server plays a major role in managing the P2P communication infrastructure. The other ones, on which the LDAF-M has been installed to, correspond to mobile device agent platforms that include their own AMS services. Two auctioneer agents and four client agents are created and distributed over those platforms. Java multi-threading, which is used especially in the agency layer to support the BDI based deliberation process, is the primary factor affecting the physical limits of the system. The source code for agent based auction application is provided at the GitHub page of the project [29].

5. Related Work

In the literature, there are widely-known open source mobile-device agent frameworks which are JADE-LEAP, AFME, and 3APL-M.

JADE-LEAP (Java Agent Development Environment-Lightweight Extensible Agent Framework) [11] allows developing agent systems on mobile devices. In fact, LEAP is an add-on for the JADE framework. FIPA-compliant agent communication and execution infrastructure provided is its prominent feature. The main container in the JADE-LEAP framework includes both the DF and the AMS. On the other hand, in LDAF-M the index server agent platform, which is the equivalent component, includes only the DF. Instead, each mobile device platform includes its own AMS to prevent a possible overloading in the index server agent platform. In addition, the BDI model of agency is not supported in JADE-LEAP; it provides its own task execution infrastructure based on the different behavior types defined

AFME (Agent Factory Micro Edition) [13] is an agent framework which provides software developers the ability to develop agent-based applications in mobile platforms. AFME is built on top of the Agent Factory framework developed by the same research group. AFME supports the BDI model of agency and represents beliefs as logical sentences in its belief base.

3APL-M [12], is another mobile device agent development framework that supports the BDI model of agency. It is built upon the Artificial Autonomous Agents Programming Language (3APL) language from which it takes the deliberation mechanism. The beliefs of agents in 3APL-M are represented using Prolog language.

However, data that represent the beliefs of the agents running within LDAF-M platform can be obtained from the data sources in the linked data environment. This gathered data is internalized using the graph data structure defined in LDAF-M for this purpose. This support for linked data is the main difference of LDAF-M.

On the other hand, after the introduction of Android operating system, several studies have been conducted to develop agents for Android based mobile devices. While some of these studies aimed to extend previous agent development frameworks for supporting Android based devices, the other studies aimed to develop agent platforms for Android based devices from scratch. A new JADE configuration that addresses Android running mobile devices [15] is an example for the first category. Additionally, a new JADEX version that supports implementing agents on Android based mobile devices has been released [16].

Andromeda (ANDROid eMbeddED Agent) platform [35] is an example for the second category. The aim of the Andromeda project is to provide a development platform to support users in developing embedded agents for mobile devices using the Android operating system. It provides the basic building blocks such as Agent, Behavior, Task, and Capability. These components are integrated with the Android system's building blocks. On the other hand, Andromeda supports neither FIPA standards nor BDI model of agency.

JaCa-Android [36] is another platform for developing mobile applications. As its developers also emphasize, JaCa-Android is not a new agent development platform and does not aim to port existing agent technologies into mobile devices. It rather presents a programming model for developing Android based smart mobile applications using the Jason agent development framework [9] and the CArtAgO environment programming framework [37].

When this study was initiated, we decided to begin the development of LDAF-M using Java technology for mobile devices (e.g., Java Platform, Micro Edition–Java ME) [38]. Java ME is supported by many different kinds of mobile devices and mobile device operating systems by means of the different Java ME profiles. On the other hand, undoubtedly the Android operating system has found widespread use in recent years. Based on this fact, there is a tendency towards developing an Android configuration for mobile device agent frameworks that have originally been developed using Java technology, as seen in the case of Android configurations for JADE and JADEX. There are several bridges to convert Java ME applications into Android applications, but we foresee that we might spent serious manual effort for porting LDAF-M which is rather a comprehensive Java ME application. Being focused on the linked data support in the current version of LDAF-M, but accepting the fact that Android operating system occupies an important place in the mobile applications ecosystem, porting the presented framework into the Android environment is left as the primary future work.

A comparison table that compares LDAF-M and other frameworks from four perspectives is given in Table 1 below. Please note that JaCa-Android is not included in the comparison table, since it is not a new agent framework intended to be developed specifically for mobile devices.

Table 1. Comparison table for LDAF-M and other mobile device agent frameworks.

Framework Name	P2P Communication Infrastructure	FIPA Compliancy	BDI Model of Agency	Linked Data Support
JADE LEAP	YES	YES	NO	NO
3 APL-M	NO	NO	YES	NO
AFME	YES	YES	YES	NO
JADEX-Android	NO	YES	YES	NO
ANDROMEDA	NO	NO	NO	NO
LDAF-M	YES	YES	YES	YES

An important research area in the multi-agent systems field is agent based simulation. There are many techniques proposed and tools developed in this area [39]. One such technique proposed for developing agent based simulation applications is TABSAOND [40]. TABSAOND has a software framework that allows the deployment of agent based simulation models as mobile applications. This is an important characteristic for a simulation tool, since users' access to desktop computers can be limited in domains such as crisis management. The software framework of TABSAOND has been developed using the JavaScript programming language. This is also an important property, since it is possible to generate cross-platform native mobile applications from JavaScript codes. LDAF-M does not support the development of agent based simulations, it is rather a general agent development framework that can either be used in developing multi-agent systems in any domain that requires cooperative or competitive behavior, such as information retrieval, and e-commerce as long as the appropriate plans are defined, or be used in developing intelligent personal digital assistants that are in continuous interaction with the other components and data resources in the system. On the other hand, we will be inspired by TABSAOND, when we think of porting the presented framework to JavaScript environment to take the advantages of cross-platform development as a future work.

Another perspective that we want to cover in the related work section is the coordination of agents. Coordination in multi-agent systems is a closely related topic with communication, since a coordination mechanism is built on top of a communication infrastructure. In a nutshell, in LDAF-M, agents use the communication infrastructure for exchanging agent communication language messages with other agents. In addition, life cycle management of an agent (i.e., detecting whether an agent is alive or off-line) is handled through this infrastructure. Since an agent can behave autonomously on behalf of its users, and can interchangeably play the roles of a client and a server, it is possible to describe an agent as a "peer". The communication infrastructure employs the appropriate network programming and server-side programming technologies to give agents (peers) the ability to communicate with each other, and to support agent life cycle management. In LDAF-M, coordination can be realized through the agent plans that are manually defined by the developers for a specific application.

In the literature, there are several coordination patterns defined for multi-agent systems. One such coordination pattern is the one described by Magariño and Gutiérrez, considering a crisis management scenario where there are collaborative agents interacting with both each other and the environment [41]. The implementation of this pattern includes coordination, network and information agents. Coordination agents live in the mobile devices and are responsible for coordinating people in the area in case of a crisis. Network agents behave as an intermediary in communication of the agents. Finally, information agents keep knowledge of locations related with the crisis and the city map. In LDAF-M, the peers can communicate directly with each other, there is no intermediary service. The index server in LDAF-M platform is responsible for maintaining the life cycle of agents, it does not behave as an intermediary during communications.

On the other hand, coordination patterns can be included in LDAF-M as pre-built interaction protocols. In this way, developers can choose the appropriate protocol during design. Support for built-in interaction protocols has already been provided by the JADE framework. Contract net and brokering are examples for JADE framework's pre-built interaction protocols. The same approach can also be employed in LDAF-M.

Studies that only aim to use semantic web/linked data technologies from mobile applications form the final perspective regarding the related work, though these studies do not consider agent technology. For example, DBpedia mobile is a location aware application that uses location data obtained from DBpedia [42]. A case study in the tourism domain has been developed to demonstrate the DBpedia mobile system. Although this study is a valuable effort that aims to use semantic web technologies in mobile devices, it does not address agent technology.

6. Conclusions

In this study, the FIPA-compliant agent framework which has been developed for mobile devices is introduced. Providing P2P communication infrastructure, following FIPA Agent Management Reference Model specification, and having an agency layer that implements the BDI model of agency are prominent features of this framework. On the other hand, an agent which is developed using the provided framework is able to construct its belief base by internalizing the data obtained from the resources in the linked data environment. Linked data support is the distinguishing feature of the developed framework that makes it different from other mobile device agent frameworks.

An agent based auction application has been developed as the case study using the framework. This application having a number of different agents, requiring many interactions between agents themselves and the administrative components of the platform, and making agents to obtain their beliefs from Freebase as the linked data resource, has sufficiently provided the context to test the framework. On the other hand, the implementation is flexible enough to add new linked data sources such as DBpedia.

As mentioned in the introduction section, when recent studies are inspected, one can see that use of agent technology is still in progress in popular domains such as cyber physical systems and internet of things among others. We think that agent development frameworks, which specifically address mobile devices, will be of great importance for being used in developing agent based applications in such domains.

Author Contributions: Conceptualization, İ.S.B. and R.C.E.; Software, İ.S.B.; Supervision, R.C.E.; Validation, İ.S.B.; Writing—original draft, İ.S.B. and R.C.E.; Writing—review and editing, R.C.E.

Funding: This research received no external funding.

Acknowledgments: This work has been conducted during the Master degree thesis study of İlker Semih Boztepe under the supervision of Rıza Cenk Erdur at Ege University Computer Engineering Department. During the Master degree education period, İlker Semih Boztepe has been awarded a scholarship by the Scientific and Technological Research Council of Turkey (TUBITAK).

Conflicts of Interest: The authors declare no conflict of interest.

References

1. Wooldridge, M. *An Introduction to Multiagent Systems*, 2nd ed.; John Wiley & Sons Ltd.: Chichester, UK, 2009; pp. 22–47, ISBN 9780470519462.
2. Wooldridge, M.; Jennings, N.R. Intelligent agents: Theory and practice. *Knowl. Eng. Rev.* **1995**, *10*, 115–152. [CrossRef]
3. Wooldridge, M. *An Introduction to Multiagent Systems*, 2nd ed.; John Wiley & Sons Ltd.: Chichester, UK, 2009; pp. 201–219, ISBN 9780470519462.
4. Parunak, H.V.D. A practitioners' review of industrial agent applications. *Autonom. Agents Multi-Agent Syst.* **2000**, *3*, 389–407. [CrossRef]

5. Leitao, P.; Karnouskos, S.; Ribeiro, L.; Lee, J.; Strasser, T.; Colombo, A.W. Smart agents in industrial cyber–physical systems. *Proc. IEEE* **2016**, *104*, 1086–1101. [CrossRef]

6. Carlier, F.; Renault, V. IoT-a, Embedded Agents for Smart Internet of Things. Application on a Display Wall. In Proceedings of the IEEE/WIC/ACM International Conference on Web Intelligence Workshops (WIW), Omaha, NE, USA, 13–16 October 2016; IEEE Computer Society: Los Alamitos, CA, USA, 2016; pp. 80–83. [CrossRef]

7. Bellifemine, F.; Caire, G.; Greenwood, D. *Developing Multi-Agent Systems with JADE*; John Wiley & Sons Ltd.: Chichester, UK, 2007; pp. 1–286, ISBN 978-0470057476.

8. Pokahr, A.; Braubach, L.; Lamersdorf, W. Jadex: A BDI reasoning engine. In *Multi-Agent Programming. Multiagent Systems, Artificial Societies, and Simulated Organizations*; Bordini, R.H., Dastani, M., Dix, J., El Fallah Seghrouchni, A., Eds.; International Book Series; Springer: Boston, MA, USA, 2005; Volume 15, pp. 149–174, ISBN 978-0387245683. [CrossRef]

9. Bordini, R.H.; Hübner, J.F.; Wooldridge, M. *Programming Multi-Agent Systems in AgentSpeak Using Jason*; John Wiley & Sons Ltd.: Chichester, UK, 2007; pp. 1–273, ISBN 978-0470029008.

10. Kravari, K.; Bassiliades, N. A survey of agent platforms. *J. Artif. Soc. Soc. Simul.* **2015**, *18*, 1–18. [CrossRef]

11. Leap User Guide. Available online: http://jade.tilab.com/doc/tutorials/LEAPUserGuide.pdf (accessed on 19 August 2018).

12. Koch, F.; Meyer, J.J.C.; Dignum, F.; Rahwan, I. Programming Deliberative Agents for Mobile Services: The 3APL-M Platform. In Proceedings of the Programming Multi-Agent Systems (ProMAS 2005), Utrecht, The Netherlands, 26 July 2005; Bordini, R.H., Dastani, M.M., Dix, J., El Fallah Seghrouchni, A., Eds.; Springer: Berlin/Heidelberg, Germany, 2006; pp. 222–235. [CrossRef]

13. Muldoon, C.; O'Hare, G.M.P.; Collier, R.; O'Grady, M.J. Agent Factory Micro Edition: A Framework for Ambient Applications. In Proceedings of the International Conference on Computational Science (ICCS2006), Reading, UK, 28–31 May 2006; Alexandrov, V.N., van Albada, G.D., Sloot, P.M.A., Dongarra, J., Eds.; Springer: Berlin/Heidelberg, Germany, 2006; pp. 727–734. [CrossRef]

14. Rao, A.S.; Georgeff, M.P. BDI Agents: From Theory to Practice. In Proceedings of the First International Conference on Multi-Agent Systems (ICMAS-95), San Francisco, CA, USA, 12–14 June 1995; Lesser, V., Gasser, L., Eds.; MIT Press: Cambridge, MA, USA, 1995.

15. Bergenti, F.; Caire, G.; Gotta, D. Agents on the Move: JADE for Android Devices. In Proceedings of the XV Workshop From Objects to Agents ("Dagli Oggetti agli Agenti"), Catania, Italy, 26 September 2014; Volume 1260.

16. Jadex Android. Available online: https://download.actoron.com/docs/releases/jadex-3.0.69/jadex-mkdocs/android/android/ (accessed on 25 September 2018).

17. Berners-Lee, T.; Hendler, J.; Lassila, O. The Semantic Web. *Sci. Am.* **2001**, *284*, 34–43. [CrossRef]

18. Wood, D.; Zaidman, M.; Ruth, L.; Hausenblas, M. *Linked Data: Structured Data on the Web*, 1st ed.; Manning Publications Co.: Shelter Island, NY, USA, 2014; pp. 1–276, ISBN 978-1617290398.

19. Hausenblas, M. Exploiting linked data to build web applications. *IEEE Internet Comput.* **2009**, *13*, 68–73. [CrossRef]

20. FIPA—The Foundation for Intelligent Physical Agents. Available online: http://www.fipa.org/ (accessed on 19 August 2018).

21. Rao, A.S.; Georgeff, M.P.; Kinny, D. A Methodology and Modelling Technique for Systems of BDI Agents. In Proceedings of the Seventh European Workshop on Modelling Autonomous Agents in a Multi-Agent World (MAAMAW' 96), Eindhoven, The Netherlands, 22–25 January 1996; Van de Velde, W., Perram, J.W., Eds.; Springer: Berlin/Heidelberg, Germany, 1996; pp. 56–71. [CrossRef]

22. Wooldridge, M. *An Introduction to Multiagent Systems*, 2nd ed.; John Wiley & Sons Ltd.: Chichester, UK, 2009; pp. 65–84, ISBN 978-0470519462.

23. Bratman, M.E. *Intention, Plans, and Practical Reason*, 1st ed.; Harvard University Press: Cambridge, MA, USA, 1987; pp. 1–200, ISBN 9780674458185.

24. Abiona, O.O.; Oluwaranti, A.I.; Anjali, T.; Onime, C.E.; Popoola, E.O.; Aderounmu, G.A.; Oluwatope, A.O.; Kehinde, L.O. Architectural Model for Wireless Peer-to-Peer (WP2P) File Sharing for Ubiquitous Mobile Devices. In Proceedings of the IEEE International Conference on Electro/Information Technology (EIT 2009), Windsor, ON, Canada, 7–9 June 2009; IEEE: Piscataway, NJ, USA, 2009; pp. 35–39. [CrossRef]

25. Vu, Q.H.; Lupu, M.; Ooi, B.C. *Peer-to-Peer Computing—Principles and Applications*, 1st ed.; Springer: New York, NY, USA, 2010; pp. 1–317, ISBN 9783642035142. [CrossRef]

26. Steinzment, R.; Wehrle, K. *Peer-to-Peer Systems and Applications*, 1st ed.; Springer-Verlag: Berlin/Heidelberg, Germany, 2005; pp. 1–615, ISBN 9783540320470. [CrossRef]

27. FIPA Agent Management Specification. Available online: http://www.fipa.org/specs/fipa00023/ (accessed on 19 August 2018).

28. Wooldridge, M. Intelligent agents. In *A Modern Approach to Distributed Artificial Intelligence*, 1st ed.; Weiss, G., Ed.; MIT Press: Cambridge, MA, USA, 2000; p. 58, ISBN 9780262731317.

29. GitHub Page for LDAF-M. Available online: https://github.com/ilkersemih/SemanticWebAgent (accessed on 1 October 2018).

30. Berners-Lee, T.; Linked Data. World Wide Web Design Issues. Available online: https://www.w3.org/DesignIssues/LinkedData.html (accessed on 19 August 2018).

31. Bizer, C.; Heath, T.; Berners-Lee, T. Linked data–the story so far. *Int. J. Semant. Web Inf. Syst.* **2009**, *5*, 1–22. [CrossRef]

32. Liyang, Y.A. *Developer's Guide to the Semantic Web*, 2nd ed.; Springer-Verlag: Berlin/Heidelberg, Germany, 2014; pp. 1–829, ISBN 9783662437964.

33. SPARQL Query Language for RDF. Available online: http://www.w3.org/TR/rdf-sparql-query/ (accessed on 19 August 2018).

34. Freebase. Available online: https://developers.google.com/freebase/ (accessed on 19 August 2018).

35. Agüero, J.; Rebollo, M.; Carrascosa, C.; Julian, V. Developing intelligent agents in the Android platform. In Proceedings of the 6th European Workshop on Multi-Agent Systems, Bath, UK, 18–19 December 2008.

36. Santi, A.; Guidi, M.; Ricci, A. Jaca-Android: An Agent-based Platform for Building Smart Mobile Applications. In Proceedings of the Languages, Methodologies, and Development Tools for Multi-Agent Systems (LADS 2010), Lyon, France, 30 August–1 September 2010; Dastani, M., El Fallah Seghrouchni, A., Hübner, J., Leite, J., Eds.; Springer: Berlin/Heidelberg, Germany, 2011; pp. 95–114. [CrossRef]

37. Ricci, A.; Piunti, M.; Viroli, M. Environment programming in multi-agent systems: An artifact-based perspective. *Autonom. Agents Multi-Agent Syst.* **2011**, *23*, 158–192. [CrossRef]

38. Java Platform, Micro Edition (Java ME). Available online: https://www.oracle.com/technetwork/java/embedded/javame/index.html (accessed on 27 September 2018).

39. Abar, S.; Theodoropoulos, G.K.; Lemarinier, P.; O'Hare, G.M.P. Agent based modeling and simulation tools: A review of the state-of-art software. *Comput. Sci. Rev.* **2017**, *24*, 13–33. [CrossRef]

40. Magariño, I.G.; Palacios-Navarro, G.; Lacuesta, R. TABSAOND: A technique for developing agent-based simulation apps and online tools with nondeterministic decisions. *Simul. Model. Pract. Theory* **2017**, *77*, 84–107. [CrossRef]

41. García-Magariño, I.; Gutiérrez, C. Agent-oriented modeling and development of a system for crisis management. *Expert Syst. Appl.* **2013**, *40*, 6580–6592. [CrossRef]

42. Becker, C.; Bizer, C. DBpedia Mobile: A Location-enabled Linked Data Browser. In Proceedings of the WWW2008 Workshop on Linked Data on the Web (LDOW 2008), Beijing, China, 22 April 2008; Bizer, C., Heath, T., Idehen, K., Berners-Lee, T., Eds.; Volume 369.

applied
sciences

MDPI

Article

Development of Semantic Web-Enabled BDI Multi-Agent Systems Using SEA_ML: An Electronic Bartering Case Study

Moharram Challenger [1,*], Baris Tekin Tezel [1,2], Omer Faruk Alaca [1], Bedir Tekinerdogan [3] and Geylani Kardas [1]

[1] International Computer Institute, Ege University, Bornova, Izmir 35100, Turkey;
baris.tezel@deu.edu.tr (B.T.T.); omerfarukalaca@gmail.com (O.F.A.), geylani.kardas@ege.edu.tr (G.K.)
[2] Department of Computer Science, Dokuz Eylul University, Buca, Izmir 35390, Turkey
[3] Information Technology Group, Wageningen University & Research, 6706 KN Wageningen,
The Netherlands; bedir.tekinerdogan@wur.nl
* Correspondence: m.challenger@gmail.com or moharram.challenger@ege.edu.tr; Tel.: +90-541-918-8836

Received: 4 March 2018; Accepted: 25 April 2018; Published: 28 April 2018

check for
updates

Abstract: In agent-oriented software engineering (AOSE), the application of model-driven development (MDD) and the use of domain-specific modeling languages (DSMLs) for Multi-Agent System (MAS) development are quite popular since the implementation of MAS is naturally complex, error-prone, and costly due to the autonomous and proactive properties of the agents. The internal agent behavior and the interaction within the agent organizations become even more complex and hard to implement when the requirements and interactions for the other agent environments such as the Semantic Web are considered. Hence, in this study, we propose a model-driven MAS development methodology which is based on a domain-specific modeling language (called SEA_ML) and covers the whole process of analysis, modeling, code generation and implementation of a MAS working in the Semantic Web according to the well-known Belief-Desire-Intention (BDI) agent principles. The use of new SEA_ML-based MAS development methodology is exemplified with the development of a semantic web-enabled MAS for electronic bartering (E-barter). Achieved results validated the generation and the development-time performance of applying this new MAS development methodology. More than half of the all agents and artifacts needed for fully implementing the E-barter MAS were automatically obtained by just using the generation features of the proposed methodology.

Keywords: multi-agent system; BDI agents; model-driven development; agent development methodology; semantic web service; ontology; SEA_ML; electronic bartering system

1. Introduction

Autonomous, reactive, and proactive agents have social ability and can interact with other agents and humans to solve their problems. To perform their tasks and interact with each other, intelligent agents constitute systems called Multi-Agent Systems (MASs) [1]. In addition, autonomous agents can evaluate semantic data and collaborate with semantically defined entities of the Semantic Web, such as semantic web services (SWS), by using content languages [2]. The implementation of agent systems is naturally a complex task when considering their characteristics. The internal agent behavior model and any interaction within the agent organizations become even more complex and hard to implement when the requirements and the interactions for other agent environments such as the Semantic Web [3,4] are considered.

Therefore, it is natural that methodologies are being applied to master the problem of defining such complex systems. One of the possible alternatives is represented by domain-specific modeling languages (DSMLs) [5,6] that have notations and constructs tailored towards an application domain (e.g., MAS). DSMLs raise the abstraction level, expressiveness, and ease of use.

The application of model-driven development (MDD) and use of DSMLs for MAS development emerged in agent-oriented software engineering (AOSE) research field especially for the last decade [7]. Researchers developed various metamodels (e.g., [8–10]) and DSMLs (e.g., [11–15]) to cope with the challenges encountered on design and implementation of MASs. Moreover, some fully fledged DSMLs (e.g., [16,17]) exist for developing software agents especially working in semantic web environments where agents can handle the Semantic Web content on behalf of their human users and interact with other semantic web environment entities, such as SWS. One of these MAS DSMLs is Semantic Web Enabled Agent Modeling Language (SEA_ML) [17] which has a built-in support for the modeling interactions of agent and semantic web services by including several specialized viewpoints. SEA_ML aims to enable domain experts to model their own MASs on the Semantic Web without considering the limitations of using existing MAS development frameworks (e.g., JADE [18], JADEX [19] or JACK [20]). The evaluations [21], conducted for the assessment of SEA_ML, show promising results considering the generation performance and the development time reduction during MAS design and implementation. According to the experiences gained from the multi-case study [21] conducted by using SEA_ML, the developers can benefit more from this DSML when they use different viewpoints of SEA_ML in a proper way in the development of MAS. Therefore, in this study, we propose a model-driven MAS development methodology which is based on an extended version of SEA_ML and covers the whole process of analysis, modeling, code generation and fully implementation of a MAS working in the Semantic Web according to the well-known Belief-Desire-Intention (BDI) [22] agent principles. The use of the new SEA_ML-based MAS development methodology is exemplified with the development of a semantic web-enabled MAS for electronic bartering (E-barter).

An agent-based E-barter system consists of agents that exchange goods or services of owners according to their preferences without using any currency. Although there are some studies developing agent-based E-barter systems such as [23–28], none of them use BDI agents and their internal reasoning mechanism which can bring extra intelligence in the procedure of matchmaking for the E-bartering. Also, these studies do not use SWSs as the automatic selection mechanism for the categories. Finally, while the methodologies applied in above studies are mostly code-centric and do not consider MDD, our study benefits from MDD and uses a DSML and its tool for the rapid implementation of the MAS. As discussed in this paper, this new model-driven MAS development methodology based on SEA_ML makes the design and development of the MAS system easier and less-costly since the agent developers work with the domain concepts and utilize generative capability of the tool.

The rest of the paper is organized as follows: The next section presents the proposed MAS development methodology based on SEA_ML modeling language. The analysis, design, and implementation of the E-barter system, using the proposed methodology are discussed in Section 3. Section 4 gives a demonstration of the implemented system. In Section 5, the related work is reported and compared with our study. Finally, the paper is concluded in Section 6.

2. SEA_ML-Based MAS Development Methodology

In this study, a model driven approach is adopted, and a model-based methodology is proposed for design and implementation of semantic web-enabled MASs. To this end, the proposed methodology covers the use of a DSML. In this way, the complex systems including SWSs and MAS components are modeled at a higher level of abstraction. In addition, these languages can model the interaction between SWSs and Agents. As a result, the system can be analyzed, and the required elements can be designed using the terms and notations very close to the domain. These domain-specific elements and their relations to each other creates the domain-specific instance models which pave the way to implement the system. As these models are persisted in a structural and formal way, they can be transformed to other proper paradigms, such as mathematical logics. In this way, they can be formally analyzed and validated based on formal methods. This can decrease the number of semantic errors later in the developed system. Furthermore, these models can be used to automatically generate the architectural codes for agents and artifacts of the complex systems which can end up with less syntactical errors and speed up the development procedure. According to the definition of artifacts in Agents & Artifacts (A&A) metamodel [8], artifacts in our study are environmental components and entities providing services, such as OWL-S documents (including process, grounding, and interface documents) for SWS. Faster development requires less efforts and it brings cost reduction in the projects. Moreover, fewer syntactical and semantical errors mean less iterations in the development phase and less testing phase which also reduce the cost and effort. Therefore, the system can be checked, and the errors can be partially found in the early phases of development, namely analysis and design phases, instead of finding them in the implementation and testing phases.

In the scope of this study, SEA_ML [17] is used as a DSML for the construction of semantic web-enabled MASs. SEA_ML enables the developers to model the agent systems in a platform independent level and then automatically achieve codes and related documents required for the execution of the modeled MAS on target MAS implementation platforms. To support MAS experts when programming their own systems, and to be able to fine-tune them visually, SEA_ML covers all aspects of an agent system from the internal view of a single agent to the complex MAS organization. In addition to these capabilities, SEA_ML also supports the model-driven design and implementation of autonomous agents who can evaluate semantic data and collaborate with semantically defined entities of the Semantic Web, such as SWSs. Within this context, it includes new viewpoints which specifically pave the way for the development of software agents working on the Semantic Web environment. Modeling agents, agent knowledge-bases, platform ontologies, SWS and interactions between agents and SWS are all possible in SEA_ML.

SEA_ML's metamodel is divided into eight viewpoints, each of which represents a different aspect for developing Semantic Web-enabled MASs. Agent's Internal Viewpoint is related to the internal structures of semantic web agents (SWAs) and defines entities and their relations required for the construction of agents. Interaction Viewpoint expresses the interactions and the communications in a MAS by taking messages and message sequences into account. MAS Viewpoint solely deals with the construction of a MAS as a whole. It includes the main blocks which compose the complex system as an organization. Role Viewpoint delves into the complex controlling structure of the agents and addresses role types. Environmental Viewpoint describes the use of resources and interaction between agents with their surroundings. Plan Viewpoint deals with an agent Plan's internal structure, which is composed of Tasks and atomic elements such as Actions. Ontology Viewpoint addresses the ontological concepts which constitute agent's knowledgebase (such as belief and fact). Agent—SWS Interaction Viewpoint defines the interaction of agents with SWS including the definition of entities and relations for service discovery, agreement, and execution. A SWA executes the semantic service finder Plan (SS_FinderPlan) to discover the appropriate services with the help of a special type of agent called SSMatchMakerAgent who executes the service registration plan (SS_RegisterPlan) for registering the new SWS for the agents. After finding the necessary service, one SWA executes an agreement plan (SS_AgreementPlan) to negotiate with the service. After negotiation, a plan for service execution (SS_ExecutorPlan) is applied for invoking the service. Table A1 lists the important SEA_ML concepts (meta-entities) and their brief descriptions for the comprehension of the corresponding visual notations used in the diagrams throughout this paper.

Based on SEA_ML, the analysis and the design of the software system can be realized using the application domain's terms and notations. This helps the end users to work in a higher level of abstraction (independent of target platform) and close to expert domain. Also, generative feature of SEA_ML paves the way to produce the configured templates from the designed models for the software system in the underlying languages and technologies. Currently, SEA_ML can generate architectural code for JADE [18], JADEX [19], and JACK [20] agent programming languages and OWL-S [29] and WSMO [30] SWS documents. This is realized by model to model transformation of the designed platform independent instance models to the instance models of the target MAS languages and SWS technologies. Then, these platform specific models are transformed to the platform specific codes by model to code transformations. This generation capability of SEA_ML can increase the development performance of the software system considerably. Finally, by constraints checking provided in SEA_ML, the instance models are controlled considering domain-specific syntactic and semantic rules. These rules are applied in the abstract and the concrete syntaxes of the language. This feature helps to reduce the number of errors during the analysis and design of the software system and avoid postponing them to the development and the testing phases.

In this section, the SEA_ML-based MAS development approach is discussed. Although this new development methodology also considers the adoption of SEA_ML, it differentiates from the previous development approach [17] as being a complete development methodology covering the analysis, design, and implementation of the MAS. Analysis phase, which does not exist previously, is now included in the methodology and both analysis and design phases are improved with two types of iterations. Such an iterative development process is not considered in the previous methodology. In addition to the modification of models, new methodology also supports the changes in auto-generated codes if required. The proposed SEA_ML-based MAS development methodology includes several steps following each other (see Figure 1): MAS Analysis, MAS Modeling, Model-to-Model (M2M) and Model-to-Code (M2C) Transformations, and finally code generation for exact MAS implementation. Following subsections discuss the methodology's phases covering those steps.

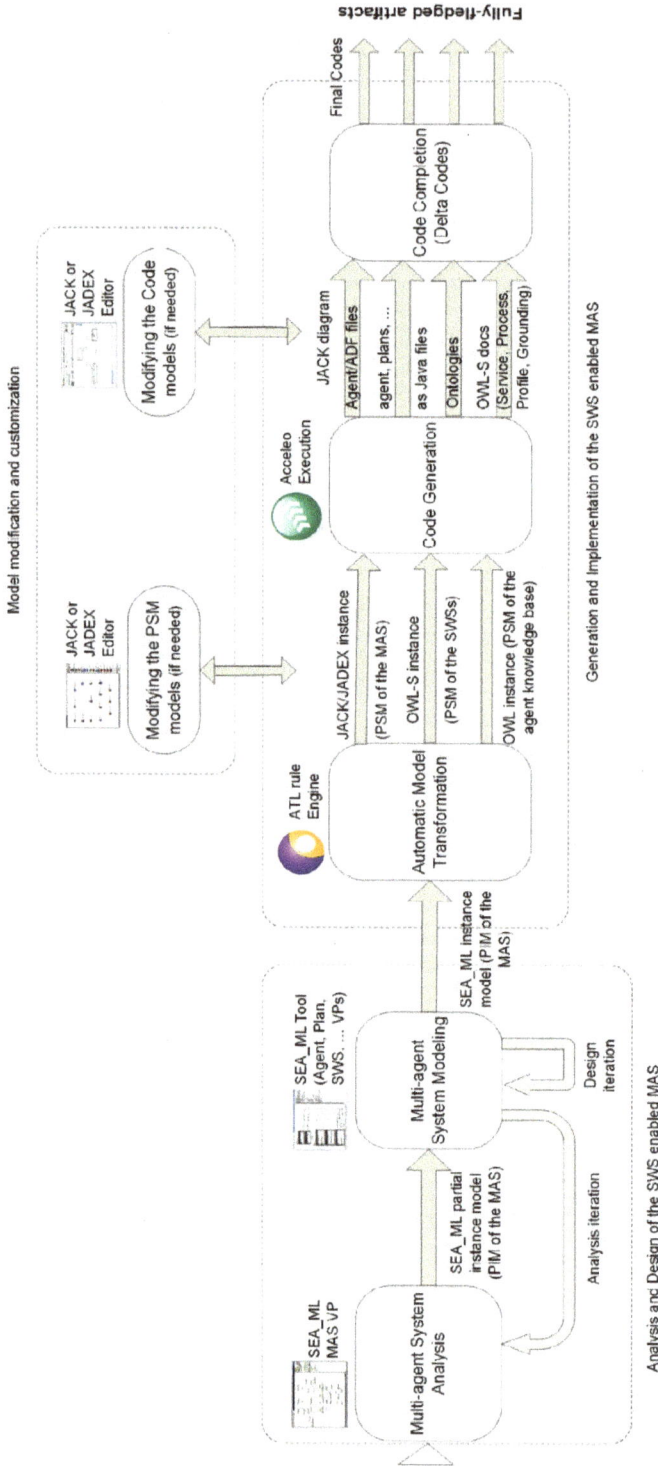

Figure 1. SEA_ML-based MAS development methodology. SEA_ML: semantic web enabled agent modeling language; MAS: Multi-Agent System; VP: View Point; SWS: semantic web services; PIM: platform independent model; PSM: platform-specific model; OWL-S: Web Ontology Language for Services; ADF: agent definition file.

2.1. MAS Analysis and Design

Based on the proposed methodology, the development of a semantic web-enabled MAS starts with the analysis of the system by considering the MAS viewpoint of SEA_ML (see Figure 1). This viewpoint includes MAS elements such as organizations, environments, agents, and their roles. The viewpoint provides the eagle-view of the system and shapes the high-level structure of the system. The result is a partial platform independent instance model of the system covering the analysis phase of the system development and providing a preliminary sketch of the system.

In the system modeling step the agent developer can use the fully functional graphical editors of SEA_ML to elaborate the design of the system, which includes 7 viewpoints of the SEA_ML's syntax, in addition to the MAS viewpoint used in the analysis phase. These viewpoints cover both multi-agent part of the system (using Agent Internal, Plan, Role, Interaction, and Environment viewpoints) and semantic web aspect of the system (using Agent-SWS Interaction and Ontology viewpoints). Each viewpoint has its own palette which provides various controls leading the designers to provide more accurate models. By designing each of these models for viewpoints, additional details are added to the initial system model provided in the analysis phase. These modifications immediately are updated in the diagrams of all other viewpoints. As the other viewpoints may have some constraint checks to control some properties related to the newly added element, the developer will be directed to complete those other viewpoints to cover the errors and warnings (coming from the constraint checks). This can lead to several iterations in the design phase. The result of this phase is the development of a complete and accurate platform independent model for the designed MAS.

2.2. Transformation and Implementation

The next step in the MAS development methodology based on SEA_ML is the automatic model transformations. The models created in the previous step need to be transformed from platform independent level into the platform-specific level, e.g., to the JACK and OWL-S models as in the case of this study. These transformations are called M2M transformations.

According to OMG's well-known Model-driven Architecture (MDA) [31], SEA_ML metamodel can be considered as a Platform Independent Metamodel (PIMM) and JACK and OWL-S metamodels can be considered as Platform-specific Metamodels (PSMM). The model transformations between these PIMMs and PSMMs pave the way for the MDD of the Semantic Web-enabled MASs. These transformations are implemented using ATL Language [32] to produce the intermediate models which enable the generation of architecture code for the agents and SWS documents. An agent developer does not need to know both the details of these transformations written in ATL and the underlying model transformation mechanism. Following the creation of models in the previous modeling steps, the only thing requested from a developer is to initiate the execution of these transformations via the interface provided by SEA_ML's Graphical User Interface (GUI).

Upon completion of model transformations, the developers have two options at this stage: (1) They may directly continue the development process with code generation for the achieved platform independent MAS models or (2) if they need, they can visually modify the achieved target models to elaborate or customize them, which can lead to gain more accurate software codes in the next step, code generation. In either case, the achieved result of this step are platform-specific system models for JACK platform, OWL-S and OWL instances.

The next step in the proposed methodology is the software code generation for the MAS. To this end, the developers' platform-specific models (conforming to PSMMs) are transformed into the code in the target languages. The M2T transformation rules are automatically executed on the target models and the codes are obtained for the implementation of the MAS. In SEA_ML, it is possible to generate code for BDI agent languages such as JACK from SEA_ML models. In addition, semantic web components of the system can be obtained through other transformations to generate OWL-S documents. Based on the initial models of the developer, the generated files and codes are also interlinked during the transformations where it is required. To support the interpretation of SEA_ML models, the M2T transformation rules are written in Acceleo [33]. Acceleo is a language to convert models into text files and uses metamodel definitions (Ecore Files) and instance files (in XMI format) as its inputs. More details on how mappings and model transformation rules between SEA_ML and the target PSMMs are realized as well as how codes are generated from PSMs can be found in [17].

As the last step, the developer needs to add his/her complementary codes, aka delta codes, to the generated architectural code to have fully functional system. However, some agent development languages, such as JACK, have their graphical editor in which the developer can edit the structure of MAS code. The generated codes achieved from the previous step can be edited and customized to add more platform specific details which helps to reach more detailed agents and artifacts. Then the delta code can be added to gain the final code.

It is important to note that, although all above mentioned steps are supported by SEA_ML to be done automatically, at any stage the developer may intervene in this development process if he/she wishes to elaborate or customize the achieved agents and artifacts.

3. E-barter Case Study

In this paper, the design and the implementation of SWS-enabled agent-based E-barter system were realized using JACK agent language [20] and OWL-S SWS technology [29].

JACK is a BDI oriented MAS development language providing a framework in Java. It is a third-generation agent platform building on the experiences of the Procedural Reasoning System (PRS) [34] and Distributed Multi-Agent Reasoning System (dMARS) [35]. JACK is one of the MAS platforms that uses the BDI software model and provides its own Java-based plan language and graphical planning tools.

OWL-S (Semantic Markup for Web Services) enables the discovery, invocation, interoperation, composition, and verification of services. It builds on the formerly developed DAML-S [36] and was the first submission for describing SWS submitted to the W3C. Each SWS in OWL-S consists of a service profile, a service model, and a grounding. The service profile describes what the service does and is used to advertise the service. The service model answers the question "how it is used?" and describes how the service works internally. Finally, the service grounding specifies how to access the service. OWL-S is based on the Web Ontology Language OWL [37] and supplies web service providers with a core set of markup language constructs for describing the properties and capabilities of their web services in an unambiguous, computer-interpretable form.

The following subsections discuss the details of analysis, design, generation, and implementation of E-barter case study using the detailed methodology proposed in this study and benefiting from SEA_ML platform.

3.1. System Analysis and Design with SEA_ML

In this subsection, we discuss the analysis and the design of the agent-based E-barter system. System analysis is realized by specifying bartering elements using agents and their components in SEA_ML, while system design is realized by providing diagrams of different viewpoints of the system using SEA_ML.

3.1.1. System Analysis with MAS and Organization Viewpoint

An agent-based E-barter system consists of agents that exchange goods or services of owners according to their preferences without using any currency. The system analysis phase is performed by considering the MAS viewpoint of SEA_ML language. In fact, this viewpoint provides an overview of the system which is shown in Figure 2. When considering the structure of the system, the EbarterSystem constitutes of two semantic web organizations called Customers organization which include Customer agents, and Management organization where the Matchmaker agent and ServiceAgent agents reside. It is worth indicating that entities given in this overview can be considered as stereotypes and in the real system implementation, there can be many instances of these entities. For example, there can be many agents of type ServiceAgent working in this system.

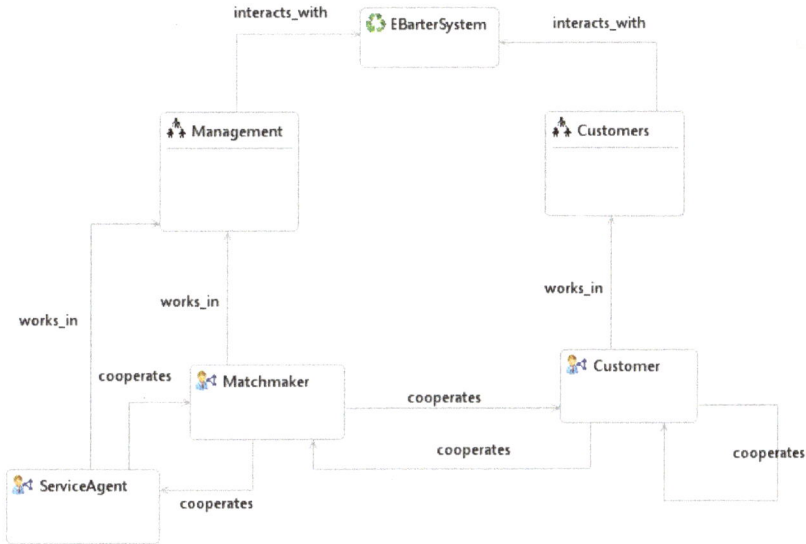

Figure 2. Overview of E-Barter MAS.

In this MAS, a Matchmaker agent, which is defined as a SWA, handles the interaction between Customer agents and ServiceAgents. This agent is responsible for registering SWS provided by each ServiceAgent in the system and matching proper services with customers. To infer about semantic closeness between offered and purchased items based on some defined ontologies, Matchmaker may use SWS. Conforming to its matchmaking definition, Matchmaker needs to discover the proper SWS, interact with the candidate service and realize the execution of SWS after an agreement.

Customer agents represent the end users in the E-barter system. This agent receives the user's offer and purchase items and interacts with Matchmaker and other Customers to realize bartering. At the first stage, this agent interacts with Matchmaker to find out if there is a proper service containing candidate customers. In case of success, it receives the service addresses and interacts with those services to get the list of suitable customers. These services contain ontologically close customers with our customer needs. In case of failure, the Matchmaker simply registers the customer into the proper service. A Customer agent, having the list of candidate customers for bartering in hand, starts to negotiate with them one by one to make an agreement and realize bartering.

The ServiceAgent agents represent the E-bartering SWSs in the system. They interact with the Matchmaker agent to register, update, and un-register the SWSs used in the system.

3.1.2. MAS Design by Modeling in SEA_ML

In accordance with the SEA_ML-based MDD methodology, we start by creating system models based on different viewpoints. The information needed for designing these models is gathered with the appropriate requirements engineering within the domain of the E-barter System.

In this study, we present three viewpoints of the E-barter system in SEA_ML namely MAS, Agent Internal and AgentSWS interaction viewpoints which represent both the MAS and SWS aspect of the system. The diagrams representing the models in these viewpoints are shown in Figures 2–4 respectively.

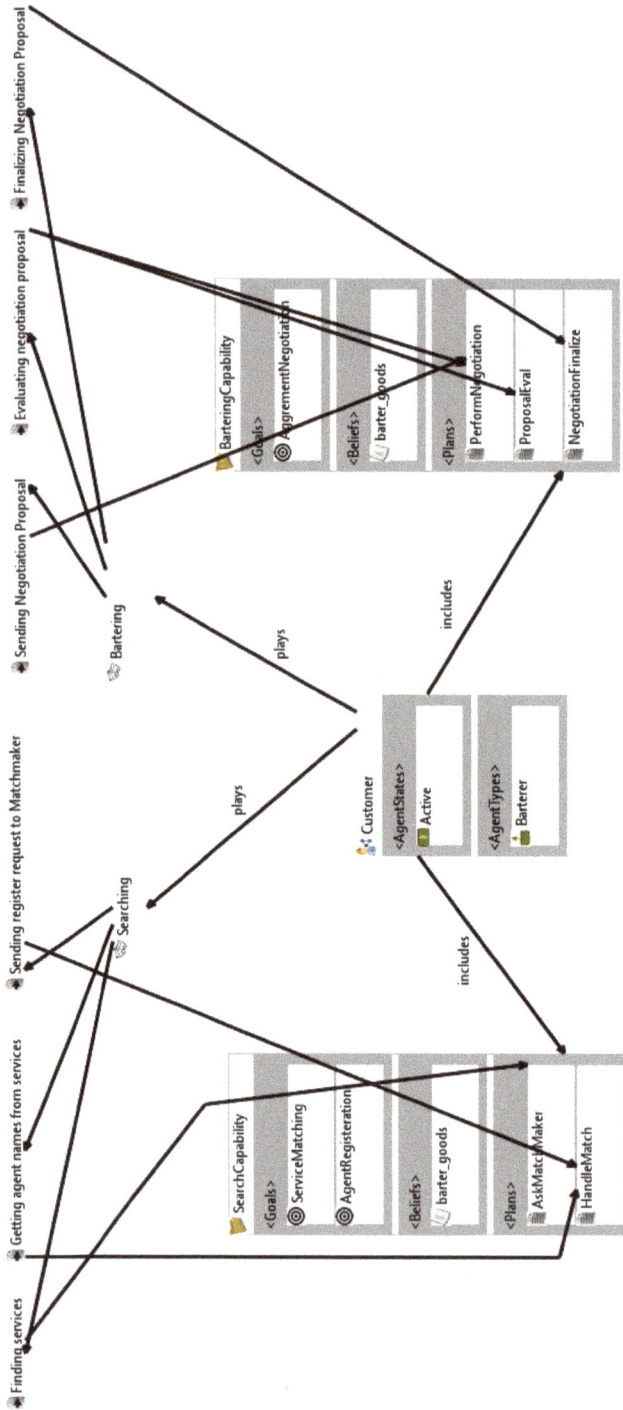

Figure 3. Agent Internal Diagram.

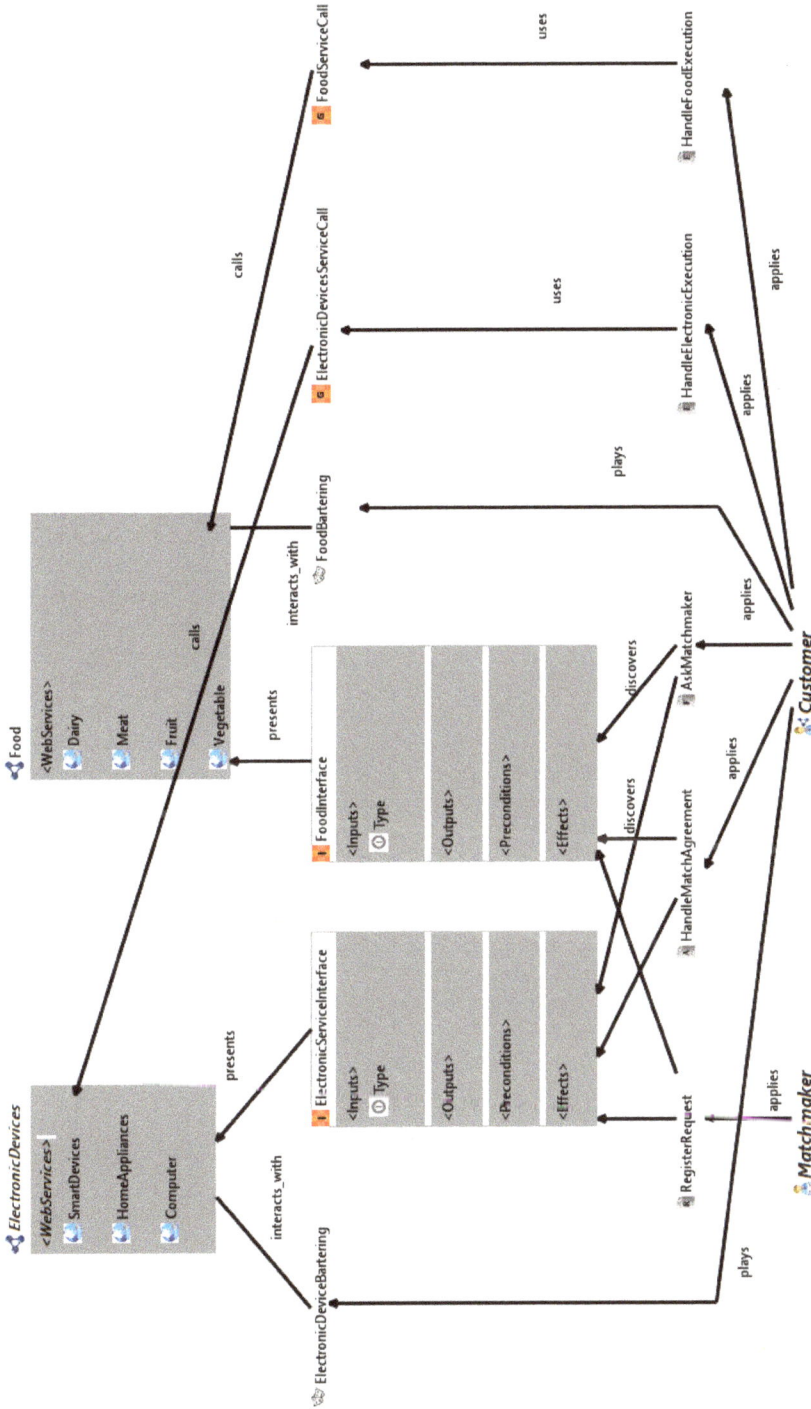

Figure 4. Agent—Semantic Web Service Interaction Diagram.

An agent's general interaction and bartering processes are modeled with using SEA_ML based on the performed analysis discussed above. In this step, we evaluate the SWA agent instances in the MAS and then model the internal structure of each agent using Agent Internal Viewpoint. It is worth noting that, the instance models and the instances of specific elements (such as SWA) are related to example models conforming to the SEA_ML meta-model within the model-driven approach. After that, we model the interactions of these agents with services, using the internal components of the semantic web services of Agent-SWS interaction viewpoint.

According to the system analysis realized in the previous phases (discussed in Section 3.1.1), the system constitutes of two semantic web organizations: Management and Customers. Each semantic web organization is a composition of SWAs having similar goals or duties. These organizations need access to some resources in the EBarterSystem environment. For this reason, there are interactions with the EBarterSystem environment to get access permission. In addition, the interactions of agents with each other are modeled. In this case, study, the MAS-to-be-implemented consists of 3 types of semantic web agents: Matchmaker, Customer and ServiceAgent. All Customers and ServiceAgents cooperate with Matchmaker to access the E-Barter system. In addition, customer agents interact with each other to negotiate for bartering. For instance, a Customer agent cooperates with a Matchmaker agent to get the list of Customer agents who have the requested product(s).

When the system's agents are determined in MAS viewpoint, the internal structure of each semantic web agent is modelled. The instance model should cover all the required roles, behaviors, plans, beliefs, and goals of an agent. Figure 3, illustrates an instance model of the agent internal viewpoint for a Customer agent.

Customer agent has two Capabilities called SearchCapability and BarteringCapability. The SearchCapability includes its Goals ("ServiceMatching" and "AgentRegisteration"), Belief ("barter_goods"), and Plans ("AskMatchmaker" and "HandleMatch"). The BarteringCapability includes its own Goal ("AggrementNegotiation"), Belief ("barter_goods"), and Plans ("PerformNegotiation", "ProposalEval" and "NegatiationFinalize"). When considering the Beliefs, the agent uses them to know which goods it has, and which goods it needs. Therefore, the agent decides what to offer and what to require for in the bartering process. Also, the agent could play Searching and Bartering roles. The Searching role could realize its task over "Finding Services" behavior by calling the AskMatchmaker plan. If this plan is executed successfully, "Getting Agent Names from Services" behavior is performed with the "HandleMatch" plan. Otherwise, the agent performs "Sending Register Request to Matchmaker" behavior. The Bartering role realizes all behaviors associated with the bartering transaction. The bartering transaction is carried out among the Customer agents and the "Bartering Role" covers all these process behaviors which are realized by relevant plans.

Figure 4 shows the instance model which includes semantic services and the required plan instances of the Agent-SWS interaction viewpoint. The instance model contains all the plans for discovering, negotiating with and executing the candidate services. Customer and Matchmaker agents are modeled with relevant plan instances to find, make the agreement with, and execute the services which are the instances of the SS_FinderPlan, SS_AgreementPlan, and SS_ExecutorPlan, respectively. The services could also be modeled for interaction between the SWS's internal components (such as Process, Grounding, and Interface), and the SWA's plans.

Therefore, when considering Customer agent request for bartering Foods, the agent should play the FoodBartering role. While playing this role, the agent applies AskMatchmaker plan for finding a suitable service interface of a Food SWS. This plan is realized by interacting with the Matchmaker agent which applies RegisterRequest plan to register services. Therefore, Customer agent cooperates with Matchmaker to receive some services for getting name of Customer agents who are candidates to barter. Finally, the Customer agent applies its FoodServiceCall plan to collect candidate customer agents with whom it can negotiate.

3.2. System Implementation with Model Transformations and Delta Code Development

In this study, the proposed multi-agent E-barter system is implemented using the JACK platform [20,38]. JACK is selected as it is one of the widely accepted Java-based BDI MAS development platforms. Also, it is as a mature and robust commercial product and meets the appropriate needs for industry adoption, such as scalability and integration.

OWL, the standard language of the W3C for the definition and development of ontologies, is employed in the realization of ontological concepts of SEA_ML. OWL is built on RDF and RDF Schema [39] and adds more vocabulary for describing the properties and classes such as relationship between classes, cardinality, equality, richer typing of properties, and the characteristics of properties and enumerated classes. SEA_ML adopts the Ontology Definition Metamodel (ODM) of OMG [40] as the metamodel of OWL and that metamodel is used as target PSMM during the transformation.

```
01   <?xml version="1.0" encoding="ISO-8859-1"?>
02   <owls:OWLSplatform
03   xmi:version="2.0"
04   xmlns:xmi="http://www.omg.org/XMI"
05   xmlns:xsi="http://www.w3.org/2001/XMLSchema-instance"
06   xmlns:owls="http://owls.com">
07   <containsService name="Food">
08   <presentedBy name="FoodServiceCall">
09   <containsInput name="Type"/>
10   <containsOutput name=" "/>
11   <containsCondition name=" "/>
12   <containsEffect name=" "/>
13   </presentedBy>
14   ...
15   <supportedBy name="FoodServiceCall"/>
16   </containsService>
17   <containsService name="ElectronicDevices">
18   <presentedBy name="ElectronicServiceInterface"/>
19   <containsInput name="Type"/>
20   <containsOutput name=" "/>
21   <containsCondition name=" "/>
22   <containsEffect name=" "/>
23   </presentedBy>
24   ...
25   <supportedBy name="ElelctronicDevicesServiceCall"/>
26   </containsService>
27   ...
28   </owls:OWLSplatform>
```

Listing 1. Part of the generated OWL-S model for the E-barter system.

The semantic web services modeled in SEA_ML are transformed into OWL-S services to enable the implementation of these services. OWL-S offers a high-level service ontology that can store three basic information about a service. The Service Profile tells you what the service is doing and provides information to discover a service. The Service Model describes how the service can be used and the composition of the service. Finally, Service Grounding provides information on how to interact with the service. Therefore, in our study, each SWS modeled in SEA_ML is transformed into an OWL-S

Service element and the appropriate Service Profile, Service Model and Service Grounding documents are created for the related SWS.

3.2.1. Model Transformations

Based on the proposed methodology, M2M transformations are applied for transforming the models designed in SEA_ML as platform independent models and JACK BDI agent and OWL-S models as platform-specific models. A part of generated OWL-S instance model for E-barter system is depicted in Listing 1. In this model the Food service is defined in Lines 7–16, and Electronic Devices service in Lines 17–26 with their interfaces and groundings.

```
01    <?xml version="1.0" encoding="ISO-8859-1" ?>
02    <!DOCTYPE uridef [
03    <!ENTITY rdf "http://www.w3.org/1999/02/22-rdf-syntax-ns">
04    <!ENTITY rdfs "http://www.w3.org/2000/01/rdf-schema">
05    <!ENTITY owl "http://www.w3.org/2002/07/owl">
06    <!ENTITY service "http://www.daml.org/services/OWL-S/1.0/Food.owl">
07    <!ENTITY congo_profile "http://www.daml.org/services/OWL-S/1.0/FoodProfile.owl">
08    <!ENTITY congo_grounding "http://www.daml.org/services/OWL-S/1.0/FoodGrounding.owl">
09    <!ENTITY DEFAULT "http://www.daml.org/services/OWL-S/1.0/FoodService.owl"> ]>
10    <rdf:RDF
11    xmlns:rdf = "&rdf;#"
12    xmlns:rdfs ="&rdfs;#"
13    xmlns:owl = "&owl;#"
14    xmlns:service= "&service;#"
15    xmlns:profile= "&profile;#"
16    xmlns:process= "&process;#"
17    xmlns:grounding= "&grounding;#"
18    xmlns:tradingServiceProfile=&profile;#
19    xmlns:tradingServiceModel=&process;#
20    xmlns:tradingServiceGrounding=&grounding;#
21    xmlns ="&DEFAULT;#"
22    xml:base="&DEFAULT;">
23    <owl:Ontology rdf:about="">
24    <owl:versionInfo>  $Id:Service.owl generated at: 20/02/2018 10:12:13   </owl:versionInfo>
25    <rdfs:comment>
26    This ontology represents the OWL-S service description for the FoodService web service.
27    </rdfs:comment>
28    <owl:imports rdf:resource= "&FoodService_service;" />
29    <owl:imports rdf:resource= "&FoodService_profile;" />
30    <owl:imports rdf:resource= "&FoodService_process;" />
31    <owl:imports rdf:resource= "&FoodService_grounding;" />
32    </owl:Ontology>
33    <service:Service rdf:ID= "FoodService">
34    <service:presents rdf:resource="&Food_profile; #ServiceProfile /> <!-- Reference to the Profile -->
35    <!-- Reference to the Process Model -->
36    <service:describedBy rdf:resource=&food_process; #ServiceModel/>
37    <!-- Reference to the Grounding -->
38    <service:supports rdf:resource= &FoodServiceCall_grounding; #ServiceGrounding'/>
39    </service:Service>
40    …
41    </rdf:RDF>
```

Listing 2. An excerpt of the generated OWL-S service file.

The M2T rules are applied on platform-specific models (JACK and OWL-S models) for the generation of JACK BDI agent codes and OWL-S documents (including Service, Profile, Process and Grounding documents) corresponding to semantic web agents and semantic web services designed in the system. As an example of the generated code, an excerpt of OWL-S Service file ("Service.owl") is shown in Listing 2 This file consists of the definition of the other documents for the service.

Although the codes generated for the MAS can be executed directly in the JACK environment, additional codes should be added into these generated codes, called delta code, by the developer to have the fully functional system.

The Generated Codes in the Target Language Environment

According to the proposed MAS development methodology, the generated code can be modified in the target language environment, JACK editor in the case of our study. JACK environment has a built-in graphical user interface that represents classes and their relations. After model-to-text transformations, JACK Java classes are produced for the customer agent from intermediate models. Apart from creating a Java class for the customer agent, separate Java classes are generated for this agent's capabilities, plans, events, and beliefs. Part of the generated codes demonstrated in the JACK editor is depicted in Figures 5 and 6. In the generated codes, the customer agent has two capabilities, namely Bartering and Searching. There are separate Java classes produced for these capabilities which are interlinked to the generated architecture code. Also, the Java classes that are generated for each capability of the agent, are linked to the plan, event, and belief classes that this capability requires.

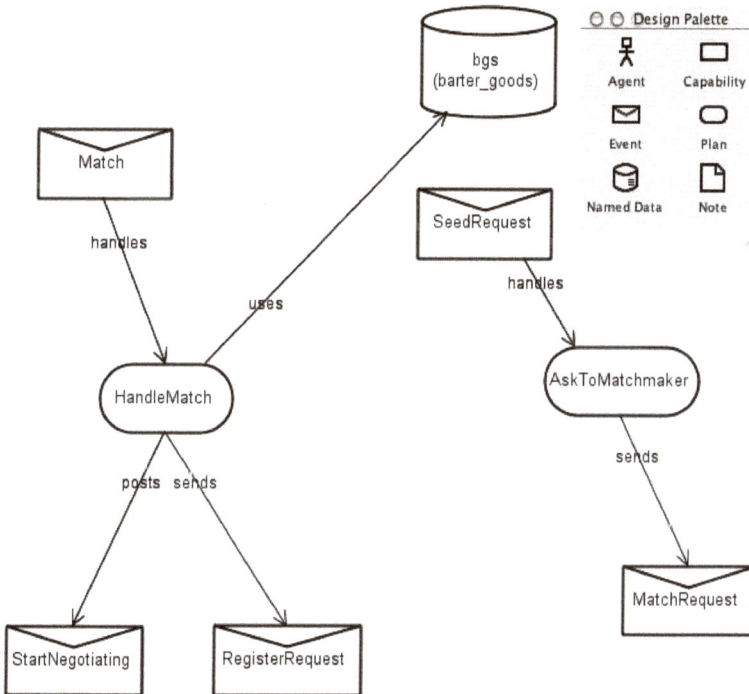

Figure 5. Searching capability in JACK editor.

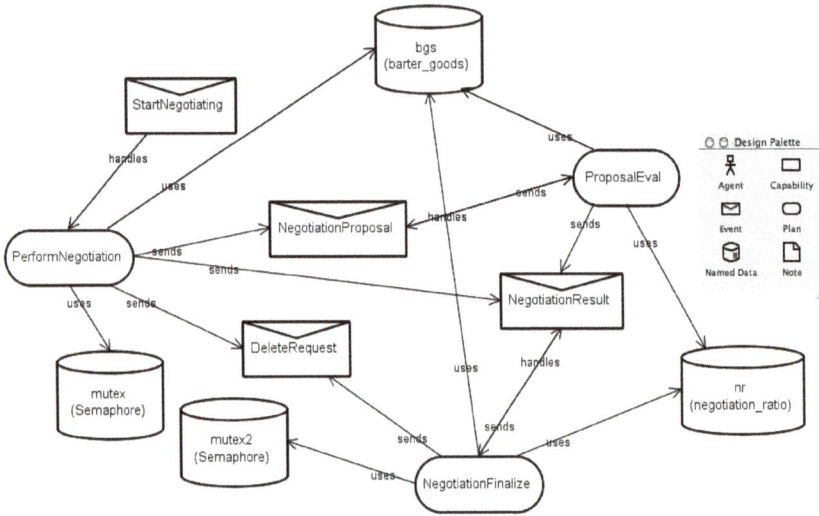

Figure 6. Bartering capability in JACK editor.

As it is shown in Figure 5, the searching capability has two plans. One of them is AskToMatchmaker plan. This plan handles SeedRequest event and sends MatchRequest event to the Matchmaker agent for getting suitable services list. The HandleMatch plan handles Match event which is sent from the Matchmaker agent. The Match event encapsulates a service list. If the service list is null, it means that there is no suitable service for MatchRequest sent from Customer agent. Otherwise, the HandleMatch plan try to get appropriate agents from the services for bartering. If the Customer agent finds suitable agents using the HandleMatch plan, it will start negotiation with these agents, if not, it will send a request to Matchmaker agent to register itself in a suitable service.

Bartering capability (see Figure 6) is responsible for the negotiations between the agents. When StartNegotiating event is posted, the PerformNegotiation plan is executed to handle it. NegotiationProposal event is created and sent to the relevant Customer agent. ProposalEval plan is responsible for evaluating incoming proposals and responding them. If the answer of the proposal is positive, the NegotiationFinalize plan is used to finalize the negotiation between two agents.

3.2.2. Delta Code Development

The codes generated by SEA_ML are architectural codes and the relations are established by the language considering the model which are controlled by the language at the semantic control stage that prevents most of the semantic errors in the code. Also, the codes do not have any syntactical errors at this level. The delta codes should be added manually to establish behavioral logic. Such code completion is need for both MAS and SWS parts of the system.

3.2.2.1. Delta Code for MAS Part of the E-barter System

The codes containing the negotiation logic of the Customer agents are critical to the system. These codes are located mostly in the plans of the agents, such as ProposalEval, NegotiationProposal, and PerformNegotiation, where most of the delta codes are added. In Listing 3 the delta code for the reasoning method of ProposalEval plan that allows an agent to evaluate proposals, is shown.

```
01    body(){
02    double actR=nr.getActRatio(ev1.from,$seed.getValue(),$offer.getValue());
03    double l=$lratio.getValue();
04    double u=$uratio.getValue();
05    double r=1.0/ev1.ratio;
06    self.guiMessage("incoming r:"+r+" from "+ ev1.from+" actR:"+actR);
07    if(r<=l || Math.abs(r-actR)<0.0001){
08    nr.add(ev1.from,$seed.getValue(),$offer.getValue(),r);
09    @reply(ev1,ev2.result(0,$seed.getValue(),$offer.getValue(),r));
10    self.guiMessage("The proposal came from "+ev1.from+" was accepted");
11    }
12    else if(r>u) {
13    @reply(ev1,ev2.result(1,$seed.getValue(),$offer.getValue(),r));
14    self.guiMessage("The proposal came from "+ev1.from+" was refused");
15    } else if(actR>0) r=(r+actR)/2.0;
16    else if(r<=(l+u)/2) r=(r+(l+u)/2)/2;
17    self.guiMessage("Sended r: "+r+ " to "+ev1.from );
18    nr.add(ev1.from,$seed.getValue(),$offer.getValue(),r);
19    @reply(ev1,ev2.result(2,$seed.getValue(),$offer.getValue(),r));
20    }
```

Listing 3. The generated and delta code for ProposalEval plan of Customer agent.

In this Listing, the templates of communicating messages in Lines 1, 9, 13, and 18–20 are generated and the other lines are added by the developer as delta code. In ProposalEval plan, there is a lower limit and an upper limit for the ratio between offered and needed products of each agent desiring a deal. In Lines 7 and 11, if the incoming bid is below the lower limit or there is an epsilon (0.0001) difference with the previous incoming bid, then the proposal is accepted by the agent. On the other hand, if the offer is above the upper limit, the agent will refuse the offer, as shown in Lines 12–14. If the incoming bid remains between the lower and upper limits, the Customer agent sends a new proposal to the agent which the initial proposal came from (Lines 15 to 20).

3.2.2.2. Delta Code for SWS Part of the E-barter System

In this case, study, each service has an ontology to help matchmaking. As there are 2 semantic web services developed for the E-barter system, we have two ontologies called electronic devices and food. These ontologies are used to demonstrate affinity relations in directing appropriate customer agents to appropriate services for barter processing. In the process of matching between the customer needs with services, each semantic web service uses its own ontology.

In the generation procedure, a structure is generated from the system model for each ontology. These structures are extended to develop the complete ontologies for the services. Part of the ontology developed for the food service is depicted in Figure 7. In this ontology, there are 4 product categories under the basic food node. These are the fruit, diary, vegetable and meat categories. These categories are divided into subcategories within themselves to obtain a tree structure in which the closeness relation can be established semantically.

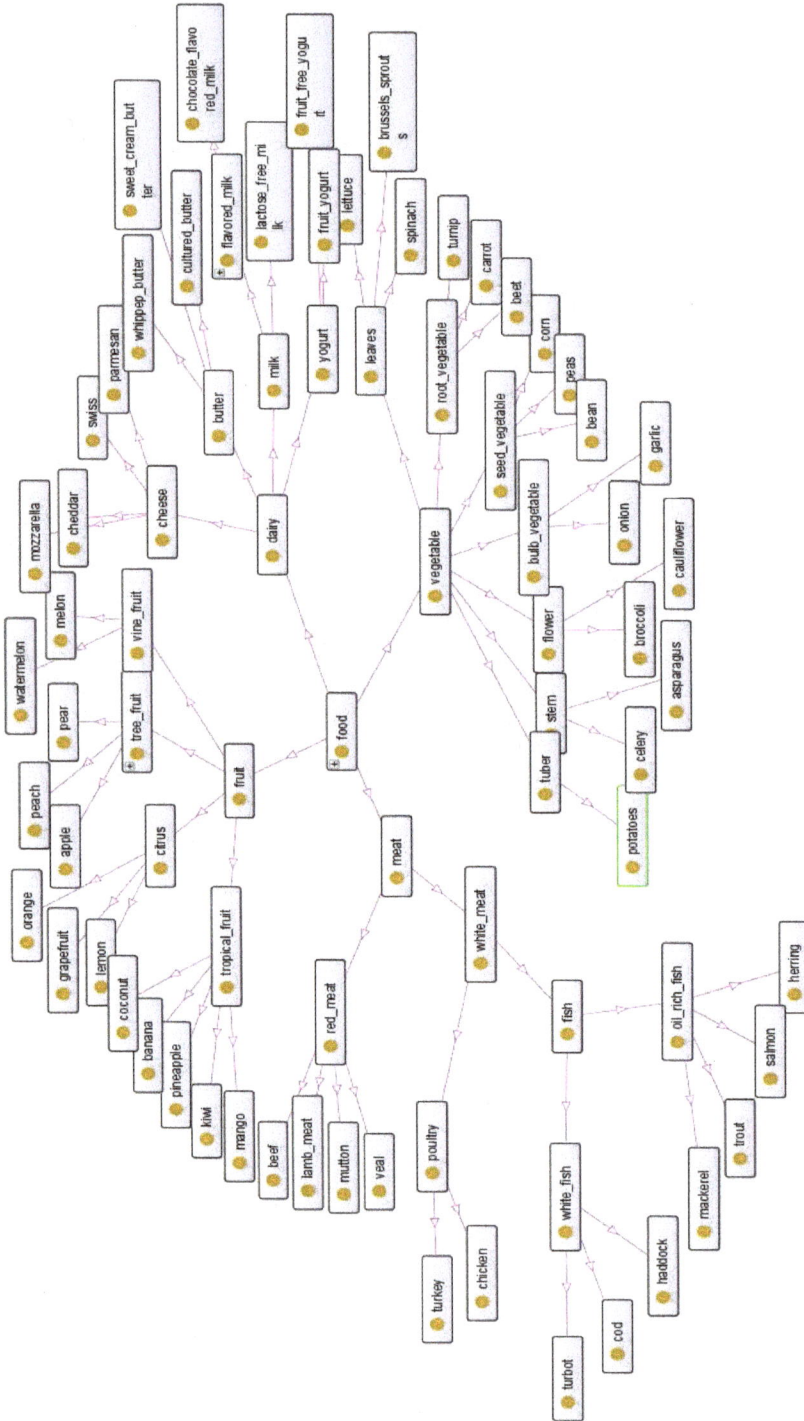

Figure 7. The ontology for food semantics web service.

The services used in E-barter system also need to have semantic web service documents to provide semantic web service functionality. For this purpose, the draft documents produced through SEA_ML have been used. Listing 4 shows the draft Profile.owl document produced by SEA_ML for the food service. The document generated by SEA_ML basically provides a draft of the resources imported with the owl: ontology tag in the rdf:RDF tag (Listing 4—Lines 16–25).

```
1     <?xml version="1.0"?>
2
3     <rdf:RDF xmlns:rdf= "&rdf;#"
4         xmlns:rdfs= "&rdfs;#"
5         xmlns:owl = "&owl;#"
6         xmlns:actor= "&actor;#"
7         xmlns:service= "&service;#"
8         xmlns:process= "&tradeFlyer;#"
9         xmlns:profile= "&tradeFlyer;#"
10        xmlns:profileHierarchy= "&profileHierarchy;#"
11        xmlns:xsd= "&xsd;#"
12        xmlns= "&DEFAULT;#"
13        xml:base= "&DEFAULT;">
14
15        <owl:Ontology rdf:about="">
16            <owl:imports rdf:resource="&service;" />
17            <owl:imports rdf:resource="&process;" />
18            <owl:imports rdf:resource="&profile;" />
19            <owl:imports rdf:resource="&foodService_process;" />
20            <owl:imports rdf:resource="&foodService_profile;" />
21
22            <owl:imports rdf:resource= "&time; />
23            <owl:imports rdf:resource= "&profileHierarchy; />
24        </owl:Ontology>
25    </rdf:RDF>
26
27    <!-- ################################################################ -->
28    <!-- #        Instance Definitions of the Serivce goes here        # -->
29    <!-- ################################################################ -->
30
```

Listing 4. Generated Profile for the food service.

This draft document was modified by the system developer in accordance with the ontology and hence the complete Profile.owl document (see Listing 5) was obtained. In this document, the profile tag, which is the label we used to determine the semantic proximity, was added to the draft to obtain the document in Listing 5 (Lines 18–39). Within the profile:Profile tag, the methods provided by that service are addressed which are based on the top-level concepts of the food ontology.

For the semantic proximity detection in the E barter system, the methods that provide the main categories (the top-level concepts in the food ontology) are mapped to the profile documents. Thus, the Matchmaker agent can determine the appropriate method for the Customer agent through the profile document to find the appropriate service. Within this case study, only the Profile.owl document was used for SWS operations. The Matchmaker agent can propose the appropriate service to the Customer agent with the help of the profile:hasOutput tag in this document. The profile document contains categories at the method level, and these services correspond to the top-level concepts on the ontology. If the product category searched by the Matchmaker cannot be found in the profile, the service's ontology is traversed to find out if there is an upper category containing this product and the search in the profile document is repeated.

```
1   <?xml version="1.0" encoding="UTF-8"?>
2   <rdf:RDF
3       xml:base="http://localhost:8080/FoodService/FoodWebService/FoodWebService_Profile.owl#"
4       xmlns:owl="http://jamsci.servehttp.com/owlsedit/owl.rdf#"
5       xmlns:process="http://staff.um.edu.mt/cabe2/supervising/undergraduate/owlseditFYP/owls11/Process.owl#"
6       xmlns:profile="http://staff.um.edu.mt/cabe2/supervising/undergraduate/owlseditFYP/owls11/Profile.owl#"
7       xmlns:rdf="http://www.w3.org/1999/02/22-rdf-syntax-ns#"
8       xmlns:rdfs="http://jamsci.servehttp.com/owlsedit/rdf-schema.rdf#"
9       xmlns:service="http://staff.um.edu.mt/cabe2/supervising/undergraduate/owlseditFYP/owls11/Service.owl#">
10  <owl:Ontology rdf:about="">
11      <owl:versionInfo>Version 1.0</owl:versionInfo>
12      <rdfs:comment>Add Ontology Comment</rdfs:comment>
13      <owl:imports rdf:resource="http://www.w3.org/1999/02/22-rdf-syntax-ns"/>
14      <owl:imports rdf:resource="http://jamsci.servehttp.com/owlsedit/owl.rdf"/>
15      <owl:imports rdf:resource="http://jamsci.servehttp.com/owlsedit/rdf-schema.rdf"/>
16      <owl:imports rdf:resource="http://staff.um.edu.mt/cabe2/supervising/undergraduate/owlseditFYP/owls11/Profile.owl"/>
17      <owl:imports rdf:resource="http://staff.um.edu.mt/cabe2/supervising/undergraduate/owlseditFYP/owls11/Service.owl"/>
18  </owl:Ontology>
19  <profile:Profile rdf:ID="FoodWebService_Profile">
20      <service:presentedBy rdf:resource="FoodWebService_Service"/>
21      <profile:serviceName>FoodWebService</profile:serviceName>
22      <profile:textDescription/>
23      <profile:hasInput rdf:resource="/FoodWebService/FoodWebService_ProcessModel#FoodWebService_register_agent_IN"/>
24      <profile:hasInput rdf:resource="/FoodWebService/FoodWebService_ProcessModel#FoodWebService_register_need_IN"/>
25      <profile:hasInput rdf:resource="/FoodWebService/FoodWebService_ProcessModel#FoodWebService_register_offer_IN"/>
26      <profile:hasOutput rdf:resource="/FoodWebService/FoodWebService_ProcessModel#FoodWebService_register_return_OUT"/>
27      <profile:hasInput rdf:resource="/FoodWebService/FoodWebService_ProcessModel#FoodWebService_delete_agent_IN"/>
28      <profile:hasInput rdf:resource="/FoodWebService/FoodWebService_ProcessModel#FoodWebService_delete_need_IN"/>
29      <profile:hasInput rdf:resource="/FoodWebService/FoodWebService_ProcessModel#FoodWebService_delete_offer_IN"/>
30      <profile:hasOutput rdf:resource="/FoodWebService/FoodWebService_ProcessModel#FoodWebService_delete_return_OUT"/>
31      <profile:hasInput rdf:resource="/FoodWebService/FoodWebService_ProcessModel#FoodWebService_meat_need_IN"/>
32      <profile:hasOutput rdf:resource="/FoodWebService/FoodWebService_ProcessModel#FoodWebService_meat_return_OUT"/>
33      <profile:hasInput rdf:resource="/FoodWebService/FoodWebService_ProcessModel#FoodWebService_dairy_need_IN"/>
34      <profile:hasOutput rdf:resource="/FoodWebService/FoodWebService_ProcessModel#FoodWebService_dairy_return_OUT"/>
35      <profile:hasInput rdf:resource="/FoodWebService/FoodWebService_ProcessModel#FoodWebService_fruit_need_IN"/>
36      <profile:hasOutput rdf:resource="/FoodWebService/FoodWebService_ProcessModel#FoodWebService_fruit_return_OUT"/>
37      <profile:hasInput rdf:resource="/FoodWebService/FoodWebService_ProcessModel#FoodWebService_vegetable_need_IN"/>
38      <profile:hasOutput rdf:resource="/FoodWebService/FoodWebService_ProcessModel#FoodWebService_vegetable_return_OUT"/>
39      <profile:qualityRating/>
40  </profile:Profile>
41  </rdf:RDF>
```

Listing 5. Completed Profile for the food service.

The Matchmaker returns the WSDL addresses of the appropriate services to the Customer agent after determining the appropriate services. When the Customer agent sends a request to a service which is just found, its repository is used which is an XML file containing the names, requests, and offers of agents eligible for bartering. The service searches for those agents which can match with the Customer agent on this XML file. If some agents are found, the service will return the names of these agents to the Customer agent.

4. Demonstration Scenario

For a detailed demonstration of the implemented system, this section illustrates a system execution consisting of three Customer agents and two semantic web services.

When the system is started to run on the client side, the initial interface is shown. On this interface, the user can add a service agent that represents a semantic web service or a Customer agent which represents himself/herself to propose a bargaining.

First, service agents, that represent semantic web services, should be added to the system. When the "Add Service Agent" in the initial user interface is selected, the interface for adding a service agent appears (see Figure 8) to get the information necessary for registering the semantic web service. Then an agent is run to represent the relevant semantic web service and sends a message to Matchmaker agent to register the semantic web service.

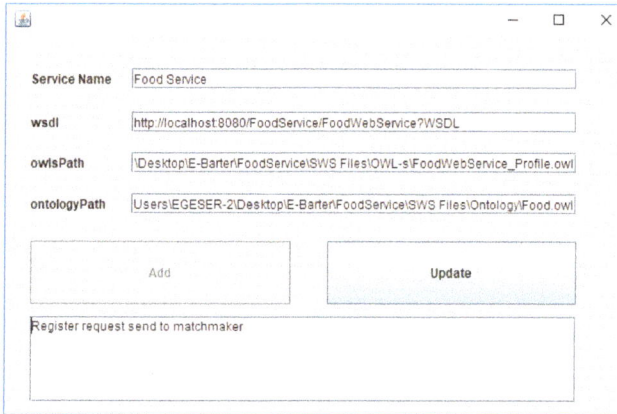

Figure 8. The user interface for adding a service agent to the system.

In our demonstration scenario, there are two semantic web services called "Food Service" and "Electronic Device". For each one, a separate service agent must be established.

After the services are registered in the system, Customer agents can be included in the system. For this purpose, "Adding Customer Agent" interface (an example shown in Figure 9) needs to be launched from the initial user interface. Using this interface, the agent is created after the information for the Customer agent is entered. To instantiate a Customer agent, the product which is needed and offered as well as the lower and upper limits must be prescribed by the user for the bartering. In this scenario, three agents named CUSTOMER A, CUSTOMER B, and CUSTOMER C, are included to the system with the details (such as Name, Need, Offer, Lower and Upper limits) provided in Table 1. Figure 9 shows, as an example, the instantiation of "CUSTOMER C" agent in the system.

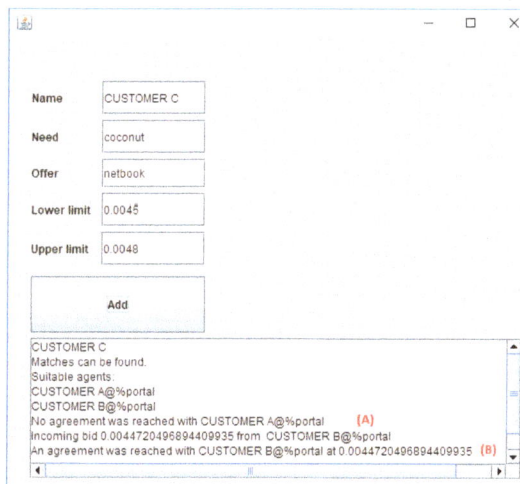

Figure 9. Instantiation of "CUSTOMER C" agent.

Table 1. Details of the Customer Agents in the E-barter System.

Name	Need	Offer	Lower Limit	Upper Limit
CUSTOMER A	netbook	Coconut	150	200
CUSTOMER B	netbook	Coconut	200	250
CUSTOMER C	coconut	Netbook	0.0045	0.0048

CUSTOMER A agent would like to barter with some other agents who offer netbooks. Meanwhile, this agent offers coconuts for netbooks. To achieve this, CUSTOMER A creates a barter request. This barter request is sent to the Matchmaker agent which uses each service's own ontology to send the customer a list of the most appropriate semantic web services. In this way, the agent gets semantically close services that accommodate the agents providing the needed product. However, till this point of the scenario, no agent has been introduced to the system before, so if CUSTOMER A agent finds appropriate services, these services will not return any agents that CUSTOMER A can barter. Therefore, CUSTOMER A will be registered to an appropriate service and wait for the new agents to be added to the system.

In this scenario, after CUSTOMER A, CUSTOMER B and CUSTOMER C agents are involved in the system. The requirements of the CUSTOMER B are the same as the CUSTOMER A, except for the lower and upper limits that the CUSTOMER A determines for bartering. In this case, CUSTOMER B would register to a suitable service because it will not find a suitable agent for bartering. The CUSTOMER C agent offers "netbook" and requests "coconut", unlike CUSTOMER A and CUSTOMER B. Therefore, a semantic web service, which is found by the Matchmaker agent, sends suitable agents for bartering to CUSTOMER C. In this case, CUSTOMER A and CUSTOMER B are the suitable agents for CUSTOMER C.

At this point, the negotiation between agents starts. First, CUSTOMER C sends a proposal to CUSTOMER A. However, as seen in Line (A) of Figure 9, the CUSTOMER A rejects this offer because the offer sent by the CUSTOMER C is higher than the upper limit of the CUSTOMER A. Then, CUSTOMER C sends the same proposal to the CUSTOMER B. Since the proposal is in the acceptable range of CUSTOMER B, the negotiation between them begins and eventually, an agreement was reached by these two agents as shown in Line (B) of Figure 9.

5. Related Work

The work conducted in this study is mainly related with two research fields: MAS development methodologies and e-barter systems. Hence, in the following, we first discuss the efforts given on deriving MAS development methodologies as similar to our proposal and then give some noteworthy studies on developing e-barter systems which especially consider employing agents.

There are various AOSE methodologies which can be used for the development of MAS. Methodologies such as ADELFE [41], Gaia [42], INGENIAS [43], PASSI [44], Prometheus [45] and Tropos [46] provide systematic processes including analysis and design of agent systems. It is also possible to integrate the outcomes of these methodologies with various agent execution platforms/frameworks such as JADE [18], JACK [20], Jason [47], CArtAgO [48], MOISE [49] and JaCaMo [50] to implement designed agents. However, neither of these methodologies nor platforms directly support model-driven MAS development. In fact, re-engineered and improved versions of some of these methodologies (e.g., [51] for ADELFE, [52] for INGENIAS, [53] for Prometheus, [54] for Tropos) enable MAS development according to MDD paradigm as indicated in [55]. Although MAS modeling from different viewpoints and code generation for various agent platforms are also covered in these updated methodologies, model-driven development of semantic web services and interactions between agents and these semantic web services are not included. Moreover, an iterative process supporting modeling and implementation are not considered and it is too difficult to modify intermediate models and update auto-generated codes by using most of these methodologies. The only exception is INGENIAS, which deems supporting the iterative development and modification of models and codes as similar to the MDD methodology proposed in this study.

In addition to abovementioned AOSE methodologies, the researchers have significant efforts on using model-driven approaches for agent development and the derivation of DSMLs for MAS. For instance, Agent-DSL [56] is used to specify the agency properties that an agent needs to accomplish its tasks. However, the proposed language is presented only with its metamodel and provided just a visual modeling of the agent systems according to agent features, such as knowledge, interaction, adaptation, autonomy and collaboration. The A&A metamodel introduced in [8] considers the notion of artifacts for agents. In the A&A metamodel, agents are modeled as proactive entities for the systems' goals and tasks while the artifacts represent the reactive entities providing the services and the functions and, hence, constitute the environment for MAS. The FAML metamodel, introduced in [9], is a synthesis of various existing metamodels for agent systems. Design time and runtime concepts for MASs are given and validation of these concepts is provided by their use at various MAS development methodologies.

Hahn [11] introduces a DSML for MAS called DSML4MAS. The abstract syntax of the DSML is derived from a platform independent metamodel structured into several aspects, each focusing on a specific viewpoint of a MAS. To provide a concrete syntax, the appropriate graphical notations for the concepts and relations are defined. Furthermore, DSML4MAS supports the deployment of modeled MASs both in JACK and JADE agent platforms by providing an operational semantics over model transformations [57]. DSML4MAS also guides MDD of different agent applications. For instance, Ayala et al. [58] use DSML4MAS for the development of agent-based ambient intelligence systems. The metamodel of DSML4MAS is employed as a source metamodel to support the modeling of context aware systems and conforming models are transformed into target models which are instances of an aspect-oriented agent metamodel called Malaca. Code generation enables the implementation of Malaca models to run in the ambient intelligence devices.

Another model driven MAS development approach is provided in [12] with introducing a new DSML. The abstract syntax of the DSML is presented using the Meta-object Facility (MOF), the concrete syntax and its tool is provided with Eclipse Graphical Modeling Framework (GMF), and finally the code generation for the JACK agent platform is realized with model transformations using Eclipse JET. The language supports modeling of agents according to Prometheus methodology [45]. A similar study is performed in [59] which proposes a technique for the definition of agent-oriented engineering process models and can be used to define processes for creating both hardware and software agents. This study also offers a related MDD tool based on INGENIAS methodology [52].

The work conducted in [13] aims at creating a UML-based agent modeling language, called MAS-ML, which can model the well-known types of agent internal architectures, namely simple reflex agent, model-based agent, reflex agent, goal-based agent and utility-based agent. Representation and exemplification of all supported agent architectures in the concrete syntax of the introduced language are given. MAS-ML is also accompanied with a graphical tool which enables agent modeling. However, the current version of MAS-ML does not support any code generation for MAS frameworks which prevents the execution of the modeled agent systems.

Wautelet and Kolp [60] investigate how a model-driven framework can be constructed to develop agent-oriented software by proposing strategic, tactical and operational views. Within this context, they introduced a Strategic Services Model in which strategic agent services can be modeled and then transformed into the dependencies modeled according to the well-known i* early phase system modeling language [61] for a problem domain. In addition, generated i* dependencies can be converted to BDI agents to be executable on appropriate agent platforms such as JACK. However, implementation of the required transformations and code generation are not included in the study.

Bergenti et al. [15] propose a language, called JADEL, for the MDD of agents on JADE platform. Instead of covering all features of JADE, JADEL only provides high-level agent-oriented abstractions, namely agents, behaviors, communication ontologies, and interaction protocols. JADEL is supported with a compiler which enables source code generation for implementing agents on JADE platform. However, the related code generation feature of JADEL is not currently functional enough to fully implement JADE agents as also indicated in [15].

The new MAS development methodology, introduced in this paper, differentiates from many of the above MAS development approaches with presenting a complete development process including analysis, design and implementation phases according to MDD principles. In most of these studies (e.g., [8,9,13,15,56]) only the derivation of metamodels and/or DSMLs is considered without a guide for how those metamodels/DSMLs can be utilized within a structural development process. Only the remaining works in [11,12,59] and can be said to describe some sort of MDD processes along with the proposed DSMLs. Benefiting from the features of SEA_ML, the development process, discussed in this paper, enables both modeling and automatic generation of semantic constructs required for the discovery and execution of semantic web services by the agents. Such a development opportunity for agent-semantic web services interactions is not considered in those MDD processes.

On the other hand, there are some studies in literature addressing the development of E-barter systems with different approaches. Generally, these studies aim at formalizing the domain and increasing its effectiveness.

For example, Lopez et al., performed two consecutive studies to create a formal framework for E-barter systems [62,63]. In their first study [62], they propose a formal framework in which customers are grouped in local markets according to their location, so that a global market takes a tree-like shape. While all these processes are identified, algebraic notation and some microeconomic theory concepts have been used. In addition, a utility function has been defined to indicate the valuation of customers in the exchange of goods. The use of algebra and micro-economy help to eliminate ambiguity and get the scheme of the system. The second study [63] focuses on transactions and shipping costs which are not considered in the first study. The early framework has been extended to include these concepts.

Another study to formalize E-barter systems is the study of Nunez et al. [64]. This study presents a classical algebra-based language to identify and analyze E-barter systems. This framework also suggests a hierarchical market structure. Product exchanges are made using the agents representing the customers. It is shown that the barter balance of the goods provides a Pareto optimum.

These studies focus on the formal representation of E-barter systems while our study mostly focuses on the efficient development of these systems. The market structure in our system is adopted from [64]. However, any other structure and formalism can be integrated into our study.

Cavalli and Maag [65] have developed an approach and its supporting tool that generates test scenarios suitable for the E-Barter system. These scenarios are intended to test the compatibility of the system with the intended functions. System specs were implemented using the Specification and Description Language (SDL) [66]. With this method, design mistakes are prevented in the early stages of development. However, test case generation is not the aim of our study.

In the study by Bravetti et al. [23], an E-barter system have been designed using multi-agent architecture with web services. In this design, BPEL4WS [67] web services are used. The focus of this study is filling the gap between formal representation and design of the system. This study is based on the formal representations provided in [62,63]. However, in [23], the authors focus on the problems in the use of these formal representations in design time. Bravetti et al. also designed the E-barter system using WS-BPEL [67] web services [68]. The studies of [23,68] uses web services as the base element for the development of the E-barter system, however, our study benefits from semantic web services and provides semantically matching capability for bartering.

Ragone et al. [69] focused on E-barter systems with a new knowledge-based approach in their study. The goal is to ensure that multiple barter situations are performed with optimal matching. In this study, a logical language was introduced that provides more complex specifications of agents' requests. It is also intended to simulate the semantic similarity between proposals that will be presented in a logic-based utility function [69]. On the other hand, in [70], a game concept was defined to describe the interactions in the barter system. According to this concept, there are several agents, and these agents have vectors with parameters specifying their requests and bids. Bartering is performed according to the matching of these vectors.

The studies presented in [69,70] focuses on improving the matching mechanism for bartering using more specific parameters in the definition of requirements and logic-based utility functions. However, our study tackles semantic matching in two levels: one in finding the closest semantic web service and another in the level of items to be bartered using ontologies.

Abdalla et al. [25] have designed an agent-based application called Bartecell. In this application, software agents can work on wireless networks and reach mutually beneficial barter agreements. New negotiation algorithms have also been introduced for transactions between agents. An E-barter architecture compatible with mobile devices has been introduced that provides location-based services [25]. This study focusses on the use of agents and benefiting from wireless networks in the negotiation of those agents for E-bartering propose. However, unlike our study, Bartecell does not utilize semantic web services and neither considers the semantic discovery of these services nor semantic matching of the bartering items.

Dhaouadi et al. [26] have designed and developed a MAS for supply chain automation. The system automatically recommends suitable suppliers for handicraft women (HDWs). The recommendation procedure is based on two supplier selection levels and then a negotiation phase. The first level is the process of selecting vendors that sell the necessary products. On the second level, it only specifies vendor profiles that can successfully match HDW. During the negotiation phase, the relevant actors will conduct discussions on the required quantity, quality, cost and delivery processes. Ontologies have also been utilized in the operation of these processes. Although this study addresses a different domain than ours, the general approach is close when considering the two selection levels. However, they use the ontologies only in the second level where they specify vendor profiles which match HDWs. In our study, selection of the categories in the first level is also done with the help of semantic matching of services. Moreover, the selection mechanism in the first level is automatized by using semantic web services.

In [24], a MAS was developed for the E-barter systems. Unlike other studies, an architecture has been designed and implemented that uses ontology-based comparisons in bid mapping. This architecture introduces a type of agent named Barter Manager Agent in addition to the E-barter agents. This agent determines the barter partners according to semantic proximity. With this approach, it is aimed to find the best match, not just based on the price and quantity of goods but also considering the relation between supply and demand. There are two groups of agents in this design. The Service Management Agent group including the Barter manager agent, SWS agent, and Cargo agent, which are responsible for managing the barter operations. The barter manager Agent is responsible for the management of the agent barter operations and matching. The SWS agent is responsible for the mapping based on the ontological proximity. The cargo agent plays a role in the exchange of products in the next stage after the barter operation is completed. The User Agents group is the group that contains the Customer agents. In this group, there are agents that request barter. In this study, the system was implemented using the JADE language [18].

In [71], an E-barter system was designed using MAS-CommonKADS methodology [72]. The focus of this system is in the negotiation phase. A bargaining protocol between two matched agents was presented.

In [27], an E-barter system was designed and developed by using Prometheus methodology [44]. In this system, BDI agents are used which is not the case in the previous studies in the literature. System development is realized using JACK intelligent agent platform [20,38]. Ontologies have been used for the matching purpose. There are two basic agents in the system. The Matchmaker agent is responsible for performing appropriate matching between customers. It benefits from ontologies for these mappings. The customer agents request bartering and negotiate with each other to realize the barter transaction. This study proposes a new rationale for the negotiations which takes the ratio of offer and need into consideration.

We believe that our work on the development of E-barter systems contributes to the abovementioned efforts by first utilizing the semantic web services and capabilities of BDI agents for E-bartering instead of reactive agents which is preferred in most of the previous work. In addition, the remaining studies (e.g., [27]) which utilizing BDI model for E-bartering do not benefit from internal reasoning mechanism of BDI agents which can bring extra intelligence in the procedure of matchmaking for the E-bartering. Also, these studies do not use SWSs as the automatic selection mechanism for the categories. Finally, while the methodologies of the other studies are generic for agent development, our study uses SEA_ML and its tool inside a domain-specific methodology. This makes the design and development of the MAS system easier and less-costly as the developer works with the domain concepts and benefits from generative capability of the provided tool.

6. Discussion and Conclusions

In this paper, a development methodology is proposed for development of MASs working in semantic web environments. This methodology is based on a DSML, called SEA_ML. The study is demonstrated using a case study for E-barter. To this end, the BDI agents and the SWSs for the E-barter system are analyzed, designed and developed using different viewpoints and features of SEA_ML. Also, a demonstration scenario is provided for the implemented system.

In the traditional E-barter systems [62–64], customers and their products' information are stored in databases in a monolithic way. This approach has two major disadvantages. First, the system is not scalable. By adding different product categories, the maintenance effort and cost of the system will be increased. The second disadvantage is that a semantic approach cannot be achieved with the traditional methods. In this study, these two disadvantages have been overcome by using SWS. The use of SWS primarily ensures that the customer and the product information are stored categorically in the external services. This leads to a more scalable system. In addition, these services with the semantic structure allow a semantic logic to be implemented in the matching and gives the opportunity to the customers to communicate only with the appropriate services. This increases the likelihood that the barter process has successful result in a limited time.

In this study, we also experienced that the proactive behavior of the BDI agents may help the fruitful application of E-barter systems by especially preventing bartering the goods ineffectively with undesired exchange ratios. Implemented BDI agents in here aim at choosing the most appropriate plan to achieve the maximum gain out of the bargaining on behalf of their users. It is possible to develop a similar MAS for the same purpose with agent models other than BDI which probably leads to provide desired efficiency in bargaining. However, in addition to the achieved fruitfulness, we also found modeling and implementation of the MAS for e-bartering convenient by utilizing BDI constructs and their relations.

On the other hand, using the proposed methodology, an efficient implementation of the system is possible through an accurate design with few errors in the analysis and the design phases [21]. In our work, semantic errors can be detected during the analysis and design phase using a SEA_ML-based methodology and the implementation process is completed in a shorter time with fewer problems. Achieved results validated the generation and the development-time performance of SEA_ML discussed in [21]. For the E-barter case study discussed in this paper, application of the SEA_ML-based MAS development methodology enabled the generation of approximately more than half of the whole agents and artifacts.

Despite the abovementioned advantages, the proposed approach in this work can be improved in several ways. Our future work consists of the followings: The semantic web services in the system are currently added to the system in a static way to facilitate the development These services can be expected to be added to the system automatically by communicating with the E-barter system. Moreover, the negotiation logic between the customers is presented at a basic level. Successful barter transactions may require a more complex negotiation rationale [73]. In addition, the business logic in the agent plans are mostly developed as the delta codes and hence our aim is to leverage the comprehensiveness of the models. In this way, it is possible to reduce the delta code and gain more generation performance using the proposed methodology. We also plan to empower the current methodology by using the formal methods discussed in [74,75] for SEA_ML to validate the models and check some domain properties in the design level for the MAS.

As another future study, our aim is to provide the execution of SEA_ML agents on Jason platform [47]. Similar to operational semantics of SEA_ML currently provided for JACK, MAS models prepared in SEA_ML can be transformed into Jason model instances and code generation can also be possible for this platform. Recently, we derived a metamodel [76] for Jason agents and we plan to use this metamodel as another PSMM and prepare a series of M2M transformations between SEA_ML metamodel and this metamodel which will lead to generation of Jason agents inside the MAS development methodology proposed in this paper.

Author Contributions: Moharram Challenger and Geylani Kardas conceived the proposed MAS development methodology. Moharram Challenger also led the writing of the paper. Baris Tekin Tezel and Omer Faruk Alaca implemented the E-barter system and performed the experiments. Bedir Tekinerdogan contributed to the analysis and assessment of the results.

Acknowledgments: This work is partially funded by the Scientific and Technological Research Council of Turkey (TUBITAK) under grant 115E591.

Conflicts of Interest: The authors declare no conflict of interest.

Appendix Descriptions of Selected SEA_ML Concepts

Table A1. SEA_ML concepts, their notations and descriptions.

Icon.	Concept	Description
	Semantic Web Agent (SWA)	Semantic web agent in the SEA_ML stands for each agent which is a member of semantic web-enabled MAS. It is an autonomous entity which can interact with both the other agents and the semantic web services, within the environment.
	Semantic service matchmaker agent (SSMatchmakerAgent)	It is a SWA extension. This meta-element represents matchmaker agents which store the SWS' capabilities list in a MAS and compare it with the service capabilities required by the other agents, in order to match them.
	Belief	Beliefs represent the informational state of the agent, in other words its knowledge about the world (including itself and other agents).
	Goal	A goal is a desire that has been adopted for active pursuit by the agent.
	Role	An agent plays different roles to realize different behaviors in various situations, such as organizations, or domains.
	Capability	Taking BDI agents into consideration, there is an entity called Capability which includes each agent's Goals, Plans and Beliefs about the surroundings.
	Fact	The statement about the agent's environment which can be true. Agents can decide based on these facts.
	Plan	Plans are sequences of actions that an agent can perform to achieve one or more of its intentions.

Table A1. *Cont.*

Icon.	Concept	Description
R	Semantic service register plan (SS_RegisterPlan)	The Semantic Service Register Plan (SS_RegisterPlan) is the plan used to register a new SWS by SSMatchmakerAgent.
F	Semantic service finder plan (SS_FinderPlan)	Semantic Service Finder Plan (SS_FinderPlan) is a Plan in which automatic discovery of the candidate semantic web services take place with the help of the SSMatchmakerAgent.
A	Semantic service agreement plan (SS_AgreementPlan)	Semantic Service Agreement Plan (SS_AgreementPlan) is a concept that deals with negotiations on quality of service (QoS) metrics (e.g., service execution cost, duration and position) and contract negotiation.
E	Semantic service executor plan (SS_ExecutorPlan)	After service discovery and negotiation, the agent applies the Semantic Service Executor Plan (SS_ExecutorPlan) to invoke appropriate semantic web services.
	Send	An action to transmit a message from an agent to another. This can be based on some standard such as FIPA_Contract_Net
	Receive	An action to collect a message from an agent. This can be based on some standard such as FIPA_Contract_Net
	Task	Tasks are groups of actions which are constructing a plan in an agent.
	Action	An action is an atomic instruction which constitutes a task.
	Message	A package of information to be send from an agent to another; possibly to deliver some information or instructions. Two special types of actions, namely Send and Receive, are used to handle these messages.
	Agent state	This concept refers to certain conditions in which agents are present at certain times. An agent can only have one state (Agent State) at a time, e.g., waiting state in which the agent is passive and waiting for another agent or resource.
	Resource	It refers to the system resources that the MAS is interacting with. For example, the database.
	Service	Any computer-based service presented to the users.
	Web Service	Type of service which is presented via web.
	Semantic Web Service	Semantically defined web services which can be interpreted by machines.
P	Process	It describes how the SWS is used by defining a process model. Instances of the SWS use the process via described_by to refer to the service's ServiceModel.
I	Interface	This document describes what the service provides for prospective clients. This is used to advertise the service, and to capture this perspective, each instance of the class Service presents a Service Interface.
G	Grounding	In this document, it is described how an agent interact with the SWS. A grounding provides the needed details about transport protocols. Instances of the class Service have a supports property referring to a Service Grounding.
	Input	Defines the inputs for processes and interfaces of a SWS.

Table A1. *Cont.*

Icon.	Concept	Description
	Output	Defines the output for processes and interfaces of a SWS.
	Precondition	Defines the pre-conditions for processes and interfaces of a SWS.
	Effect	Defines the post-conditions or effects for processes and interfaces of a SWS.
	Semantic web organization	Refers to an organized group of semantic web agents (SWAs).
	Interaction	For communication and collaboration of agents, they can use series of messages via a message sequence which results to an agent interaction.
	Environment	The agent's surroundings including digitized resources, fact, and services.
	Registration Role	A specialized type of architectural role which is used to register SWSs in the multi agent systems.
	Behavior	In re-active agents, a behavior is a re-action of an agent towards an external or internal stimulus.
	Agent type	The agents in a multi-agent system can have different types taking various responsivities and representing various stakeholders.

References

1. Wooldridge, M.; Jennings, N.R. Intelligent agents: Theory and practice. *Knowl. Eng. Rev.* **1995**, *10*, 115–152. [CrossRef]
2. Kardas, G.; Goknil, A.; Dikenelli, O.; Topaloglu, N.Y. Model Driven Development of Semantic Web Enabled Multi-agent Systems. *Int. J. Coop. Inf. Syst.* **2009**, *18*, 261–308. [CrossRef]
3. Berners-Lee, T.; Hendler, J.; Lassila, O. The Semantic Web. *Sci. Am.* **2001**, *284*, 34–43. [CrossRef]
4. Shadbolt, N.; Hall, W.; Berners-Lee, T. The Semantic Web Revisited. *IEEE Intell. Syst.* **2006**, *21*, 96–101. [CrossRef]
5. Mernik, M.; Heering, J.; Sloane, A. When and how to develop domain-specific languages. *ACM Comput. Surv.* **2005**, *37*, 316–344. [CrossRef]
6. Fowler, M. *Domain-Specific Languages*; Addison-Wesley Professional: Boston, MA, USA, 2001; pp. 1–640.
7. Kardas, G.; Gomez-Sanz, J.J. Special issue on model-driven engineering of multi-agent systems in theory and practice. *Comput. Lang. Syst. Struct.* **2017**, *50*, 140–141. [CrossRef]
8. Omicini, A.; Ricci, A.; Viroli, M. Artifacts in the A&A meta-model for multi-agent systems. *Autonom. Agents Multi-Agent Syst.* **2008**, *17*, 432–456.
9. Beydoun, G.; Low, G.C.; Henderson-Sellers, B.; Mouratidis, H.; Gomez-Sanz, J.J.; Pavon, J.; Gonzalez-Perez, C. FAML: A Generic Metamodel for MAS Development. *IEEE Trans. Softw. Eng.* **2009**, *35*, 841–863. [CrossRef]
10. Garcia-Magarino, I. Towards the integration of the agent-oriented modeling diversity with a powertype-based language. *Comput. Stand. Interfaces* **2014**, *36*, 941–952. [CrossRef]
11. Hahn, C. A Domain Specific Modeling Language for Multiagent Systems. In Proceedings of the 7th International Conference on Autonomous Agents and Multi-Agent Systems, Estoril, Portugal, 12–16 May 2008; pp. 233–240.
12. Gascuena, J.M.; Navarro, E.; Fernandez-Caballero, A. Model-Driven Engineering Techniques for the Development of Multi-agent Systems. *Eng. Appl. Artif. Intell.* **2012**, *25*, 159–173. [CrossRef]
13. Goncalves, E.J.T.; Cortes, M.I.; Campos, G.A.L.; Lopes, Y.S.; Freire, E.S.S.; da Silva, V.T.; de Oliveira, K.S.F.; de Oliveira, M.A. MAS-ML2.0: Supporting the modelling of multi-agent systems with different agent architectures. *J. Syst. Softw.* **2015**, *108*, 77–109. [CrossRef]

14. Faccin, J.; Nunes, I. A Tool-Supported Development Method for Improved BDI Plan Selection. *Eng. Appl. Artif. Intell.* **2017**, *62*, 195–213. [CrossRef]
15. Bergenti, F.; Iotti, E.; Monica, S.; Poggi, A. Agent-oriented model-driven development for JADE with the JADEL programming language. *Comput. Lang. Syst. Struct.* **2017**, *50*, 142–158. [CrossRef]
16. Hahn, C.; Nesbigall, S.; Warwas, S.; Zinnikus, I.; Fischer, K.; Klusch, M. Integration of Multiagent Systems and Semantic Web Services on a Platform Independent Level. In Proceedings of the 2008 IEEE/WIC/ACM International Conference on Web Intelligence and Intelligent Agent Technology (WI-IAT 2008), Sydney, Australia, 9–12 December 2008; pp. 200–206.
17. Challenger, M.; Demirkol, S.; Getir, S.; Mernik, M.; Kardas, G.; Kosar, T. On the use of a domain-specific modeling language in the development of multiagent systems. *Eng. Appl. Artif. Intell.* **2014**, *28*, 111–141. [CrossRef]
18. Bellifemine, F.L.; Caire, G.; Greenwood, D. *Developing Multi-Agent Systems with JADE*; John Wiley & Sons: Hoboken, NJ, USA, 2007; Volume 7.
19. Pokahr, A.; Braubach, L.; Lamersdorf, W. Jadex: A BDI Reasoning Engine. In *Multi-Agent Programming Languages, Platforms and Applications*; Bordini, R.H., Dastani, M., Dix, J., El Fallah Seghrouchni, A., Eds.; Springer: Berlin, Germany, 2005; pp. 149–174.
20. Howden, N.; Ronnquist, R.; Hodgson, A.; Lucas, A. Jack intelligent agents- summary of an agent infrastructure. In Proceedings of the 5th International Conference on Autonomous Agents (AGENTS'01), Montreal, QC, Canada, 28 May–1 June 2001.
21. Challenger, M.; Kardas, G.; Tekinerdogan, B. A systematic approach to evaluating domain-specific modeling language environments for multi-agent systems. *Softw. Qual. J.* **2016**, *24*, 755–795. [CrossRef]
22. Rao, A.S.; Georgeff, M.P. Decision procedures for BDI logics. *J. Log. Comput.* **1998**, *8*, 293–343. [CrossRef]
23. Bravetti, M.; Casalboni, A.; Nunez, M.; Rodriguez, I. From Theoretical e-barter Models to an Implementation Based on Web Services. *Electr. Notes Theor. Comput. Sci.* **2006**, *159*, 241–264. [CrossRef]
24. Demirkol, S.; Getir, S.; Challenger, M.; Kardas, G. Development of an agent-based E-barter system. In Proceedings of the International Symposium on INnovations in Intelligent SysTems and Applications, Istanbul, Turkey, 15–18 June 2011; pp. 193–198. [CrossRef]
25. Abdalla, S.; Swords, D.; Sandygulova, A.; O'Hare, G.M.P.; Giorgini, P. BarterCell: An agent-based bartering service for users of pocket computing devices. In Proceedings of the International Conference on Industrial Applications of Holonic and Multi-Agent Systems, Prague, Czech Republic, 26–28 August 2013.
26. Dhaouadi, R.; Salah, K.B.; Miled, A.B.; Ghédira, K. Ontology based multi-agent system for the handicraft domain e-bartering. In Proceedings of the 28th Bled eConference: Wellbeing, Bled, Slovenia, 7–10 July 2015; pp. 353–367.
27. Cakmaz, Y.E.; Alaca, O.F.; Durmaz, C.; Akdal, B.; Tezel, B.; Challenger, M.; Kardas, G. Engineering a BDI Agent-based Semantic e-Barter System. In Proceedings of the International Conference on Computer Science and Engineering (UBMK'17), Antalya, Turkey, 5–8 October 2017.
28. Miyashita, K. Incremental Design of Perishable Goods Markets through Multi-Agent Simulations. *Appl. Sci.* **2017**, *7*, 1300. [CrossRef]
29. Martin, D.; Burstein, M.; Hobbs, J.; Lassila, O.; McDermott, D.; McIlraith, S.; Narayanan, S.; Paolucci, M.; Parsia, B.; Payne, T.; et al. OWL-S: Semantic Markup for Web Services. W3C Member Submission. Available online: http://www.w3.org/Submission/OWL-S/ (accessed on 7 April 2018).
30. WSMO: Web Service Modeling Ontology. Available online: http://www.w3.org/Submission/WSMO/ (accessed on 27 April 2018).
31. Frankel, D.S. *Model Driven Architecture: Applying MDA to Enterprise Computing*; Wiley Publishing: Hoboken, NJ, USA, 2003.
32. Jouault, F.; Allilaire, F.; Bezivin, J.; Kurtev, I. ATL: A model transformation tool. *Sci. Comput. Programm.* **2008**, *72*, 31–39. [CrossRef]
33. Acceleo Code Generator. Available online: https://www.eclipse.org/acceleo/ (accessed on 7 April 2018).
34. Myers, K.L. *User Guide for the Procedural Reasoning System*; SRI International AI Center Technical Report; SRI International: Menlo Park, CA, USA, 1997.
35. d'Inverno, M.; Luck, M.; Georgeff, M.; Kinny, D.; Wooldridge, M. The dMARS architecture: A specification of the distributed multi-agent reasoning system. *Autonom. Agents Multi-Agent Syst.* **2004**, *9*, 5–53. [CrossRef]

36. Anupriya, A.; Mark, B.; Jerry, R.H.; Ora, L.; David, M.; Drew, M.; Sheila, A.M. DAML-S: Web service description for the semantic web. In *International Semantic Web Conference*; Springer: Berlin/Heidelberg, Germany, 2002; pp. 248–363.

37. World Wide Web Consortium, OWL 2 Web Ontology Language—Document Overview (Second Edition), W3C Recommendation, 2012. Available online: https://www.w3.org/TR/owl2-overview/ (accessed on 7 April 2018).

38. Agent Oriented Software Inc. JACK Intelligent Agents. Available online: http://www.aosgrp.com/products/jack/ (accessed on 7 April 2018).

39. Broekstra, J.; Kampman, A.; Van Harmelen, F. Sesame: A generic architecture for storing and querying rdf and rdf schema. In Proceedings of the International Semantic Web Conference, Sardinia, Italy, 9–12 June 2002; Springer: Berlin/Heidelberg, Germany, 2002; pp. 54–68.

40. Object Management Group, Ontology Definition Metamodel (ODM) Version 1.1. 2014. Available online: https://www.omg.org/spec/ODM/ (accessed on 7 April 2018).

41. Bernon, C.; Gleizes, M.-P.; Peyruqueou, S.; Picard, G. ADELFE: A methodology for adaptive multi-agent systems engineering. *Lect. Notes Artif. Intell.* **2003**, *2577*, 70–81.

42. Zambonelli, F.; Jennings, N.R.; Wooldridge, M. Developing multiagent systems: The Gaia methodology. *ACM Trans. Softw. Eng. Methodol.* **2003**, *12*, 317–370. [CrossRef]

43. Pavón, J.; Gómez-Sanz, J.J. Agent oriented software engineering with INGENIAS. In *CEEMAS*; Mark, V., Müller, J.P., Pechoucek, M., Eds.; Springer: Berlin/Heidelberg, Germany, 2003; pp. 394–403.

44. Cossentino, M. From requirements to code with the PASSI methodology. In *Agent-Oriented Methodologies*; Henderson-Sellers, B., Giorgini, P., Eds.; Idea Group Publishing: Hershey, PA, USA, 2005; pp. 79–106.

45. Padgham, L.; Winikoff, M. Prometheus: A methodology for developing intelligent agents. In Proceedings of the 1st International Joint Conference on Autonomous Agents and Multiagent Systems: Part 1, Bologna, Italy, 15–19 July 2002; pp. 37–38.

46. Bresciani, P.; Perini, A.; Giorgini, P.; Giunchiglia, F.; Mylopoulos, J. Tropos: An agent-oriented software development methodology. *Autonom. Agents Multi-Agent Syst.* **2004**, *8*, 203–236. [CrossRef]

47. Bordini, R.H.; Hübner, J.F.; Wooldridge, M. *Programming Multi-Agent Systems in AgentSpeak Using Jason*; John Wiley & Sons: Hoboken, NJ, USA, 2007; Volume 8.

48. Ricci, A.; Viroli, M.; Omicini, A. CArtAgO: A framework for prototyping artifact-based environments in MAS. In Proceedings of the International Workshop on Environments for Multi-Agent Systems, Hakodate, Japan, 8 May 2006; Springer: Berlin/Heidelberg, Germany, 2006.

49. Hannoun, M.; Boissier, O.; Sichman, J.S.; Sayettat, C. MOISE: An organizational model for multi-agent systems. In *Advances in Artificial Intelligence*; Springer: Berlin/Heidelberg, Germany, 2000; pp. 156–165.

50. Boissier, O.; Bordini, R.H.; Hübner, J.F.; Ricci, A.; Santi, A. Multi-agent oriented programming with JaCaMo. *Sci. Comput. Programm.* **2013**, *78*, 747–761. [CrossRef]

51. Rougemaille, S.; Migeon, F.; Maurel, C.; Gleizes, M.-P. Model Driven Engineering for Designing Adaptive Multi-Agent Systems. In Proceedings of the 8th Annual International Workshop on Engineering Societies in the Agents World (ESAW 2007), Athens, Greece, 22–24 October 2007.

52. Pavon, J.; Gomez, J.; Fuentes, R. Model Driven Development of Multi-Agent Systems. *Lect. Notes Comput. Sci.* **2006**, *4066*, 284–298.

53. Thangarajah, J.; Padgham, L.; Winikoff, M. Prometheus design tool. In Proceedings of the 4th International Joint Conference on Autonomous Agents and Multiagent Systems, Utrecht, The Netherlands, 25–29 July 2005; pp. 127–128.

54. Penserini, L.; Perini, A.; Susi, A.; Mylopoulos, J. From Stakeholder Intentions to Software Agent Implementations. *Lect. Notes Comput. Sci.* **2006**, *4001*, 465–479.

55. Kardas, G. Model-driven development of multi-agent systems: A survey and evaluation. *Knowl. Eng. Rev.* **2013**, *28*, 479–503. [CrossRef]

56. Kulesza, U.; Garcia, A.; Lucena, C.; Alencar, P. A generative approach for multi-agent system development. *Lect. Notes Comput. Sci.* **2005**, *3390*, 52–69.

57. Hahn, C.; Madrigal-Mora, C.; Fischer, K. A Platform-Independent Metamodel for Multiagent Systems. *Autonom. Agents Multi-Agent Syst.* **2009**, *18*, 239–266. [CrossRef]

58. Ayala, I.; Amor, M.; Fuentes, L. A model driven engineering process of platform neutral agents for ambient intelligence devices. *Autonom. Agents Multi-Agent Syst.* **2014**, *28*, 214–255. [CrossRef]

59. Fuentes-Fernandez, R.; Garcia-Magarino, L.; Gomez-Rodriguez, A.M.; Gonzalez-Moreno, J.C. A technique for defining agent-oriented engineering processes with tool support. *Eng. Appl. Artif. Intell.* **2010**, *23*, 432–444. [CrossRef]

60. Wautelet, Y.; Kolp, M. Business and model-driven development of BDI multi-agent system. *Neurocomputing* **2016**, *182*, 304–321. [CrossRef]

61. Yu, E.; Giorgini, P.; Maiden, N.; Mylopoulos, J. *Social Modeling for Requirements Engineering: The Complete Book*; MIT Press: Cambridge, MA, USA, 2011.

62. López, N.; Nunez, M.; Rodriguez, I.; Rubio, F. A Formal Framework for E-Barter Based on Microeconomic Theory and Process Algebras. In *International Workshop on Innovative Internet Community Systems*; Springer: Berlin/Heidelberg, Germany, 2002; pp. 217–228.

63. López, N.; Núñez, M.; Rodríguez, I.; Rubio, F. A Multi-Agent System for e-barter including Transaction and Shipping Costs. In Proceedings of the 2003 ACM Symposium on Applied Computing, Melbourne, FL, USA, 9–12 March 2003; pp. 587–594.

64. Núñez, M.; Rodríguez, I.; Rubio, F. Formal specification of multi-agent e-barter systems. *Sci. Comput. Programm.* **2005**, *57*, 187–216. [CrossRef]

65. Cavalli, A.; Maag, S. Automated test scenarios generation for an e-barter system. In Proceedings of the 2004 ACM Symposium on Applied Computing—SAC '04, New York, NY, USA, 14–17 March 2004; p. 795.

66. International Telecommunication Union (ITU). *Telecommunication Standardization, Specification and Description Language (SDL)*; Technical Report Z-100; ITU: Geneva, Switzerland, 2011; p. 206.

67. Andrews, T.; Curbera, F. Web Service Business Process Execution Language, Working Draft, Version 2.0, 1. 2004. Available online: https://www.uni-ulm.de/fileadmin/website_uni_ulm/iui.emisa/Downloads/emisaforum06.pdf (accessed on 7 April 2018).

68. Bravetti, M.; Casalboni, A.; Nunez, M.; Rodriguez, I. From Theoretical e-Barter Models to Two Alternative Implementations Based on Web Sevices. *J. UCS* **2007**, *13*, 2035–2075. [CrossRef]

69. Ragone, A.; Di Noia, T.; Di Sciascio, E.; Donini, F.M. Increasing bid expressiveness for effective and balanced e-barter trading. *Lect. Notes Comput.* **2009**, *5397 LNAI*, 128–142. [CrossRef]

70. Tagiew, R. Barter Double Auction as Model for Bilateral Social Cooperations. In Proceedings of the 1st Computer Science and Electronic Engineering Conference (CEEC'09), Colchester, UK, 19–21 September 2009; pp. 1–7.

71. Dagdeviren, Z.A.; Kardas, G. A Case Study on the Development of Electronic Barter Systems using Software Agents. In Proceedings of the 5th Turkish National Software Engineering Symposium (UYMS 2011), Ankara, Turkey, 26–28 September 2011; pp. 123–126.

72. Arenas, A.E.; Barrera-Sanabria, G. Applying the MAS-CommonKADS Methodology to the Flights Reservation Problem: Integrating Coordination and Expertise. In Proceedings of the 5th Joint Conference on Knowledge-Based Software Engineering, Maribor, Slovenia, 11–13 September 2002.

73. Ye, D.; Zhang, M.; Vasilakos, A.V. A survey of self-organization mechanisms in multiagent systems. *IEEE Trans. Syst. Man Cybern. Syst.* **2017**, *47*, 441–461. [CrossRef]

74. Getir, S.; Challenger, M.; Kardas, G. The Formal Semantics of a Domain-specific Modeling Language for Semantic Web enabled Multi-agent Systems. *Int. J. Coop. Inf. Syst.* **2014**, *23*, 1–53. [CrossRef]

75. Challenger, M.; Mernik, M.; Kardas, G.; Kosar, T. Declarative specifications for the Development of Multi-agent Systems. *Comput. Stand. Interfaces* **2016**, *43*, 91–115. [CrossRef]

76. Kardas, G.; Tezel, B.T.; Challenger, M. A domain-specific modeling language for belief-desire-intention software agents. *IET Softw.* **2018**. [CrossRef]

MDPI
St. Alban-Anlage 66
4052 Basel
Switzerland
Tel. +41 61 683 77 34
Fax +41 61 302 89 18
www.mdpi.com

Applied Sciences Editorial Office
E-mail: applsci@mdpi.com
www.mdpi.com/journal/applsci

www.ingramcontent.com/pod-product-compliance
Lightning Source LLC
Chambersburg PA
CBHW041214220326
41597CB00032BA/5406